Push your Career Publish your Thesis

Science should be accessible to everybody. Share the knowledge, the ideas, and the passion about your research. Give your part of the infinite amount of scientific research possibilities a finite frame.

Publish your examination paper, diploma thesis, bachelor thesis, master thesis, dissertation, or habilitation treatises in form of a book.

A finite frame by infinite science.

Infinite Science
Publishing

An Imprint of
Infinite Science GmbH
MFC 1 | Technikzentrum Lübeck
BioMedTec Wissenschaftscampus
Maria-Goeppert-Straße 1
23562 Lübeck
book@infinite-science.de
www.infinite-science.de

recruiticon

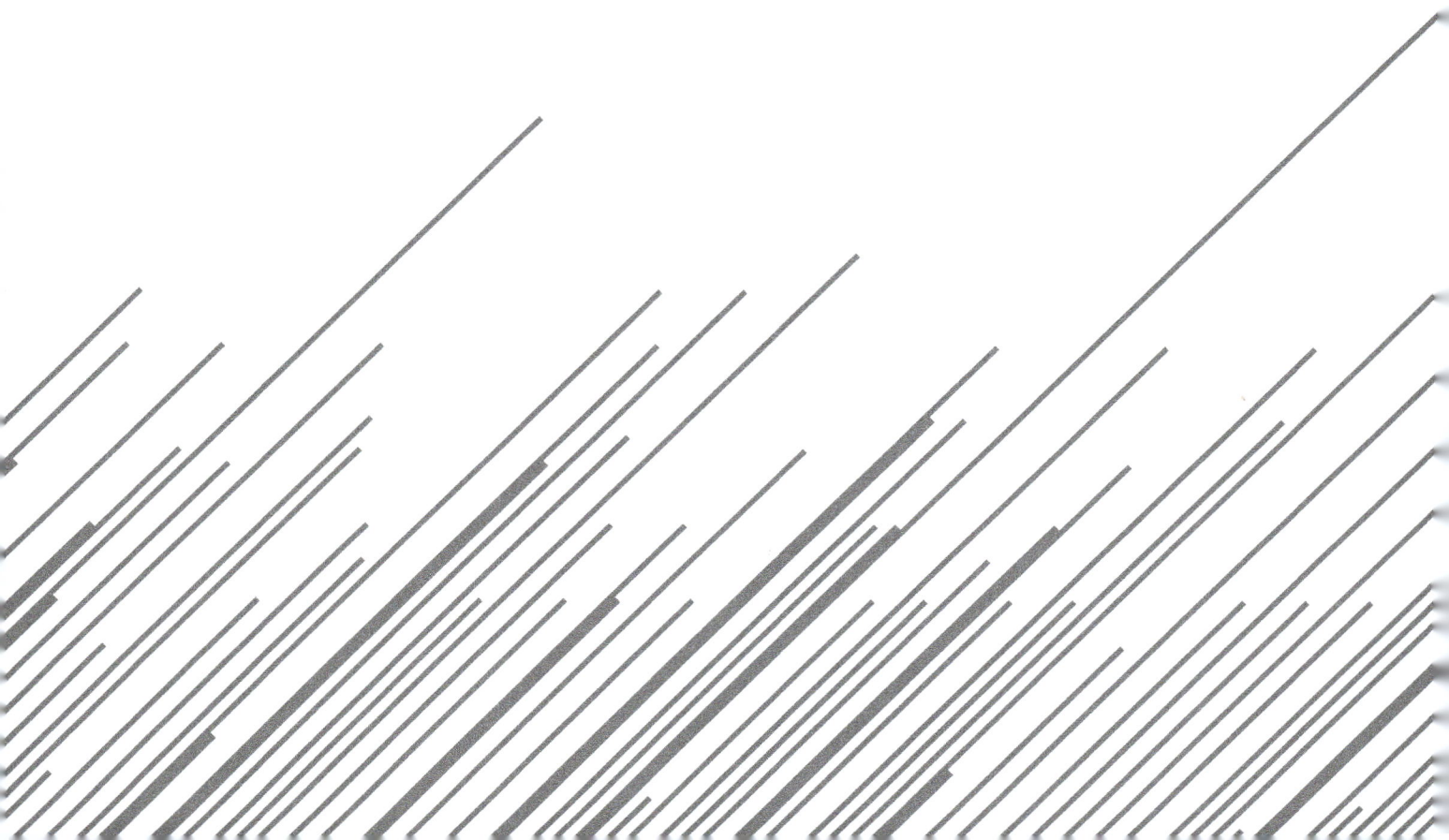

© 2017 Infinite Science Publishing
University Press and
Academic Printing

Imprint of Infinite Science GmbH,
Technikzentrum | MFC 1
Maria-Goeppert-Straße 1
23562 Lübeck, Germany

Cover Design and Illustration: Uli Schmidts, metonym
Editorial and Copy Editing: University of Lübeck

Publisher: Infinite Science GmbH, Lübeck, www.infinite-science.de
Printed in Germany, BoD, Norderstedt

ISBN: 978-3-945954-33-1

Bibliografische Information der Deutschen Nationalbibliothek:
Die Deutsche Nationalbibliothek verzeichnet diese Publikation in der Deutschen Nationalbibliografie; detaillierte bibliografische Daten sind im Internet über http://dnb.d-nb.de abrufbar.

Student Conference Proceedings 2017

6th Conference on
Medical Engineering Science

2nd Conference on
Medical Informatics

Lübeck, March 7-9, 2017

Editors in Chief

T. M. Buzug and H. Handels

Associate Editors

C. Debbeler, J. Degen, C. Kaethner, J.-H. Wrage

Editors

I. Angelova-Fischer, E. Barth, R. Birngruber, R. Brinkmann,
T. M. Buzug, R. Duden, J. Ehrhardt, F. Ernst, S. Fischer, T. Fischer, H. Gehring,
S. Groppe, T. Gutsmann, H. Handels, M. Heinrich, H. Hellbrück, R. Huber,
C. Hübner, G. Hüttmann, J. Ingenerf, M. Kleemann, S. Klein, M. A. Koch,
M. Leucker, K. Lüdtke-Buzug, A. Madany Mamlouk, A. Mertins, J. Modersitzki,
J. Obleser, H. Paulsen, R. Rahmanzadeh, F. Reinholz, P. Rostalski, Y.-H. Song

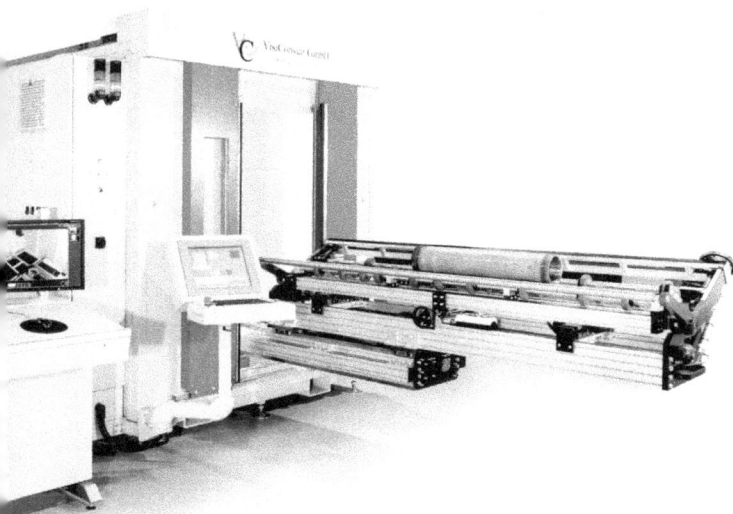

stryker

Entdecker
Querdenker
Spitzenkollege
Lebensretter

Ihr Talent – unsere Leidenschaft als Toparbeitgeber in der Medizintechnik

33.000+ Mitarbeiter

4.898 Patente

$ 11,3 Mrd. Umsatz

Vielfältige Perspektiven und Entwicklungsmöglichkeiten

Flexible Arbeitszeitmodelle

Einzigartige und engagierte Teamkollegen

Stryker ist ein breit aufgestellter Weltmarktführer in der Medizintechnik; unsere Artikel erleichtern in über 100 Ländern medizinischen Fachkräften ihre Arbeit und Patienten das Leben. Mit ihrer Motivation, ihrer Kompetenz und ihren Ideen haben unsere Mitarbeiter Stryker zu dem gemacht, was es heute ist: zu einem innovativen Marktführer auf dem Gebiet der Medizintechnik. Unsere Produkte für Orthopädie, Neurochirurgie, medizinische Technik und Ausrüstung unterstützen weltweit Mediziner dabei, ihren Patienten die Rückkehr in ein aktives und erfülltes Leben zu ermöglichen. Gemeinsam mit unseren Kunden streben wir nach Verbesserung der medizinischen Versorgung. Unsere Werte Integrität, Zuverlässigkeit, Menschen und Leistung sind bei uns keine leeren Worthülsen, sondern werden tagtäglich von unseren Mitarbeitern weltweit gelebt. Ihr Talent ist unsere Leidenschaft als Toparbeitgeber in der Medizintechnik!

Helfen Sie uns zu helfen!

Als Mitarbeiter bei Stryker profitieren Sie von den Vorteilen einer innovativen und mitarbeiterorientierten Unternehmenskultur sowie internationalen Karrierechancen in der Wachstumsbranche Medizintechnik. Unser Unternehmenserfolg basiert darauf, engagierte, talentierte und verantwortungsbewusste Mitarbeiter (w/m) zu gewinnen, die permanent danach streben, die Messlatte zu erhöhen, um zu den Besten zu gehören.

Wir freuen uns auf Ihre Onlinebewerbung über unser Karriereportal auf www.stryker.de!

Pia Jedamzik
04348 702 577
www.stryker.de

kununu TOP COMPANY

GALLUP GREAT WORKPLACE AWARD

top EMPLOYER DEUTSCHLAND 2017

Jetzt bewerben!

Dräger

Damit Lebensretter keine Sekunde verlieren.
Dafür lohnt es sich zu arbeiten.

Mit mobiler Beatmungstechnik unterstützen wir Notfallhelfer bei der Versorgung ihrer Patienten in lebensbedrohlichen Situationen. Als Praktikant (m/w) finden Sie bei uns viel Freiraum für eigene Ideen, übernehmen erste Verantwortung und werden optimal auf den späteren Berufseinstieg vorbereitet. Bei Dräger sind Sie Teil eines internationalen Teams. Leben schützen, unterstützen und retten sind die Ziele, die uns alle verbinden. **www.draeger.com/karriere**

Dräger. Technik für das Leben®

Wir fördern Studenten und Young Professionals

YXLON für Young Professionals
- **Traineeprogramm**
- **Direkteinstieg**

Bewerbung an:

Job@hbg.yxlon.com

YXLON International GmbH
Frau Simten Bascetincelik
Essener Bogen 15
22419 Hamburg
Tel.: 040 527 29-225

Sie haben Ihr Studium oder Ihre Ausbildung abgeschlossen?
Sie möchten Ihr Wissen nun in die Praxis umsetzen?
Sie möchten Ihre Leidenschaft für Technologie mit uns teilen?

Dann starten Sie bei uns als Trainee oder mit einem Direkteinstieg!

- Sie übernehmen von Anfang an interessante Aufgaben und erarbeiten Lösungen im Team
- Sie können Verantwortung übernehmen und direkt zum Unternehmenserfolg beisteuern
- Sie arbeiten in einem internationalen Umfeld
- Als YXLON Mitarbeiter/in werden Ihre Fähigkeiten gezielt gefördert

YXLON für Studenten
- **Praktika von 3 bis 6 Monaten**
- **Betreuung von Abschlussarbeiten (Bachelor, Master, Doktorarbeiten)**
- **Werkstudenten**

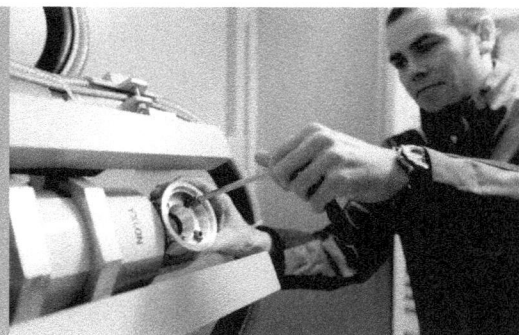

Sie suchen während Ihres Studiums einen interessanten Praktikumsplatz oder ein spannendes Thema für die Abschlussarbeit?

Dann sind Sie bei YXLON genau richtig!

Wir bieten Ihnen Einblicke in folgende Abteilungen:

Entwicklung	**Vertrieb**
Produktmarketing	**Produktion**
Supply Chain	**Personal**

YXLON
Technology with Passion

V⟨⟨

VKK® Patentanwälte

PATENTANWÄLTE · EUROPEAN PATENT ATTORNEYS
HAMBURG · KEMPTEN · MÜNCHEN · FRANKFURT

Vkk Patentanwälte sind eine traditionsreiche, international tätige Patentanwaltskanzlei, spezialisiert auf die Erlangung, Verteidigung und Durchsetzung gewerblicher Schutzrechte weltweit mit Kompetenz im nationalen und internationalen Patent- und Gebrauchsmusterrecht sowie im nationalen und internationalen Marken- und Designrecht.

Ansprechpartner Dr. Philipp Knoop, Absolvent des Institutes für Medizintechnik, Universität zu Lübeck 1999, Patentanwalt, European Patent Attorney.

An der Alster 84
20099 Hamburg
Tel.: +49 - (0) 40 - 28 08 13 0
Fax: +49 - (0) 40 - 28 08 13 31
hamburg@vkkpatent.com

Edisonstraße 2
87437 Kempten
Tel.: +49 - (0) 831 - 232 91
Fax: +49 - (0) 831 - 177 15
kempten@vkkpatent.com

Sckellstraße 6
81667 München
 +49 - (0) 89 - 23 08 93 0
muenchen@vkkpatent.com

www.vkkpatent.com

senTec

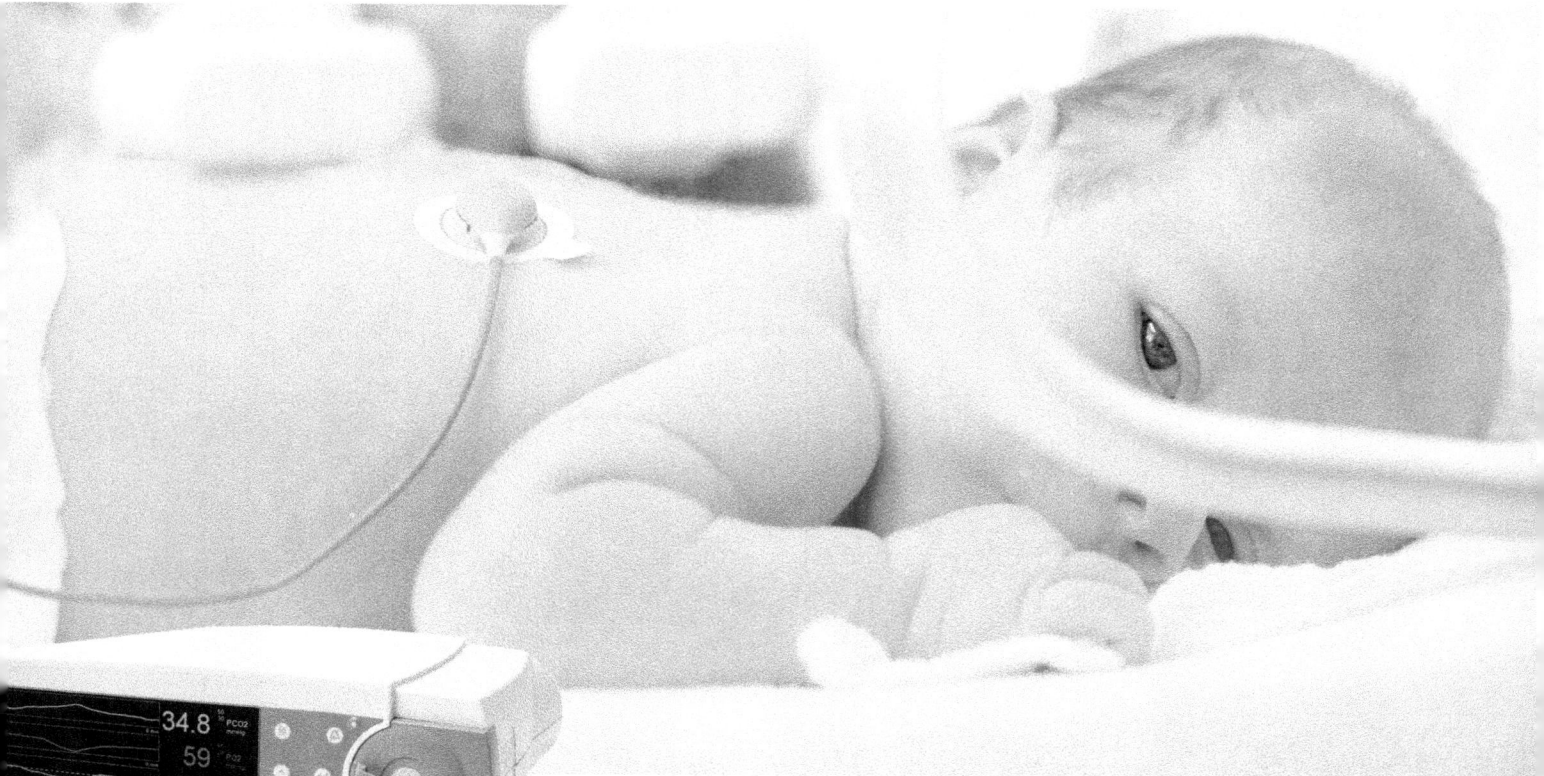

Advancing Noninvasive Patient Monitoring™

SenTec ist ein junges, innovatives und international tätiges Unternehmen im Bereich Medizintechnik.

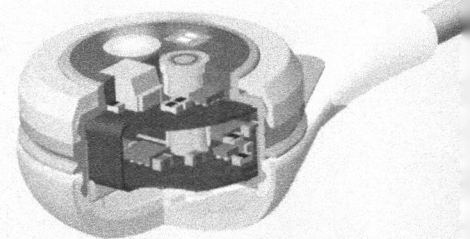

Mit unseren engagierten Mitarbeitern in Rostock und in der Schweiz, entwickeln und produzieren wir unsere high-tech Produkte weltweit. Unsere SenTec DigitalMonitoringSystems werden zur nicht-invasiven und kontinuierlichen Überwachung von Patienten eingesetzt – sei es zur Überwachung der Atmung von Patienten mit schweren Lungenerkrankungen oder der Beatmung von Frühgeborenen.

Getrieben von unserem hohen Anspruch an die Qualität und Zuverlässigkeit unserer Produkte, streben wir mit unseren Forschungs- und Entwicklungstätigkeiten stetig nach neuen innovativen Messtechnologien.

Aufgabengebiete
- Konstruktion/ Mikrotechnik
- Informatik & Software Entwicklung
- Betriebsmittelentwicklung & Industrialisierung
- Produktzulassungen
- Qualitätsmanagement

Gesuchte Fachrichtungen
- Elektrotechnik
- Mechanik
- Informatik
- Mikrotechnik

Einstiegsmöglichkeiten
- Praktika, Semester-/Masterarbeiten
- Berufseinsteiger
- Professionals

Kontakt
SenTec GmbH
Carl-Hopp-Strasse 19A
D-18069 Rostock | Deutschland
+49 (0) 381 3677 96120
jobs@sentec.com | www.sentec.com

Student Conference Proceedings 2017

Conference Organization

Conference Chairs

Thorsten M. Buzug (Chair), Institute of Medical Engineering, Universität zu Lübeck
Heinz Handels (Chair), Institute of Medical Informatics, Universität zu Lübeck
Hartmut Gehring (Co-Chair), Department of Anesthesiology, UKSH
Stephan Klein (Co-Chair), Center for Biomedical Technology, Lübeck University of Applied Sciences

Local Coordination

Christian Kaethner, Institute of Medical Engineering, Universität zu Lübeck
Christina Debbeler, Institute of Medical Engineering, Universität zu Lübeck
Johanna Degen, Institute of Medical Informatics, Universität zu Lübeck
Jan-Hinrich Wrage, Institute of Medical Informatics, Universität zu Lübeck
Gisela Thaler, Institute of Medical Engineering, Universität zu Lübeck
Susanne Petersen, Institute of Medical Informatics, Universität zu Lübeck

Scientific Program Committee

Angelova-Fischer, Dr. Irina	Klinik für Dermatologie, Allergologie und Venerologie
Barth, Prof. Dr. Erhardt	Institute für Neuro- und Bioinformatik
Birngruber, Prof. Dr. Reginald	Institut für Biomedizinische Optik
Brinkmann, Dr. Ralf	Institut für Medizinisches Laserzentrum
Buzug, Prof. Dr. Thorsten M.	Institut für Medizintechnik
Duden, Prof. Dr. Rainer	Institut für Biologie
Ehrhardt, Dr. Jan	Institut für Medizinische Informatik
Ernst, Dr. Floris	Institut für Robotik und Kognitive Systeme
Fischer, Prof. Dr. Stefan	Institut für Telematik
Fischer, Dr. Tobias	Klinik für Dermatologie, Allergologie und Venerologie
Gehring, Prof. Dr. Hartmut	Klinik für Anästhesiologie
Groppe, Dr. Sven	Institut für Informationssysteme
Gutsmann, Prof. Dr. Thomas	Forschungszentrum Borstel
Handels, Prof. Dr. Heinz	Institut für Medizinische Informatik
Heinrich, Prof. Dr. Mattias	Institut für Medizinische Informatik
Hellbrück, Prof. Dr. Horst	Institut für Telematik und Fachbereich Elektrotechnik und Informatik, FH Lübeck
Huber, Prof. Dr. Robert	Institut für Biomedizinische Optik
Hübner, Prof. Dr. Christian	Institut für Physik
Hüttmann, PD Dr. Gereon	Institut für Biomedizinische Optik
Ingenerf, Prof. Dr. Josef	Institut für Medizinische Informatik
Kleemann, Prof. Dr. Markus	Klinik für Allgemeine Chirurgie
Klein, Prof. Dr. Stephan	Labor für Medizinische Sensor- und Gerätetechnik, FH Lübeck
Koch, Prof. Dr. Martin A.	Institut für Medizintechnik
Leucker, Prof. Dr. Martin	Institut für Softwaretechnik und Programmiersprachen
Lüdtke-Buzug, Dr. Kerstin	Institut für Medizintechnik
Madany Mamlouk, PD Dr. Amir	Institut für Neuro- und Bioinformatik
Mertins, Prof. Dr. Alfred	Institut für Signalverarbeitung
Modersitzki, Prof. Dr. Jan	Institut of Mathematics and Image Computing
Obleser, Prof. Jonas	Institut für Psychologie
Paulsen, PD Dr. Hauke	Institut für Physik
Rahmanzadeh, Dr. Ramtin	Institut für Biomedizinische Optik
Reinholz, Dr. Fred	Institut für Biomedizinische Optik
Rostalski, Prof. Dr. Philipp	Institut für Medizinische Elektrotechnik
Song, Dr. Young-Hwa	Institut für Physik

Preface and Acknowledgements

After the great success of the previous meetings from 2012 to 2016, the Student Conference 2017 shows continuing growth both in quality and quantity of scientific contributions. In this year, the 6th Student Conference on Medical Engineering Science is hold together with the 2nd Student Conference on Medical Informatics.

The organization team of the Institute of Medical Engineering and the Institute of Medical Informatics in cooperation with the Life Science North Management GmbH, the North German Life Science Cluster Agency, has spared no effort to provide an excellent conference, where master students of the campus present their recent research results to a broad public of academics and industry.

The contributions show, how new approaches and methods in medical engineering and medical informatics can advance medicine, health, and health care. Moreover, this conference offers a good opportunity for both students and companies to get in touch at the Recruiticon, i.e. a satellite recruiting fair with industrial exhibition. Students from the Life Sciences programs present their results from projects carried out at the laboratories, clinics and institutes of Lübeck's Universities, in international research facilities, or research-oriented industrial companies. The conference focus has been placed on topics from medical engineering and medical informatics. The interdisciplinary field of medical engineering has also been established at the Lübeck University of Applied Sciences for decades and Medical Engineering Science (Medizinische Ingenieurwissenschaft – MIW) is an important bachelor and master program at the Universität zu Lübeck as well. Both universities jointly offer the international master degree course Biomedical Engineering (BME). Furthermore, in the young master program Medical Informatics (Medizinische Informatik – MI) the 2nd Student Conference on Medical Informatics is integrated as an important element where project results in the emerging field of digital medicine are presented by the students.

As Conference Chairs, we thank all the people who worked with enthusiasm and dedication to make the conference a successful event. We want to thank the companies who support the meeting. Moreover, our thanks go to Infinite Science for producing these proceedings and organizing the Recruiticon meeting supporting the student conference. Personally and on behalf of all colleagues of the Student Conference Committee, we especially want to thank Christian Kaethner and Christina Debbeler from the Institute of Medical Engineering, they have been the central contact points for all questions of students and the program committee members as well as Johanna Degen and Jan-Hinrich Wrage from the Institute of Medical Informatics editing the proceedings. Their in-depth overview of all details of this event is the key to the success of the Student Conference 2017.

Lübeck, March 7-9, 2017

Prof. Dr. Thorsten M. Buzug
Chair of the 6rd Student Conference
on Medical Engineering Science

Lübeck, March 7-9, 2017

Prof. Dr. Heinz Handels
Chair of the 2nd Student Conference
on Medical Informatics

Exhibitors and Sponsors

senTec

Dräger

stryker®

YXLON

Sponsors

VKK® Patentanwälte

VisiConsult
X-ray Systems & Solutions

LIFE
SCIENCE
NORD

Infinite Science

Contents

Biomedical Optics

Biochemical Physics

Safety and Quality

Biomedical Engineering

E-Health

Image Processing

Signal Processing

1

Biomedical Optics

Decreasing progressive myopia by Rose Bengal mediated crosslinking of the sclera

E. Beck[1,3], T. G. Seiler[2,3], M. Engler[1,3], I. E. Kochevar[3], and R. Birngruber[3,4]

[1] Medizinische Ingenieurwissenschaft, Universität zu Lübeck, eric.beck@student.uni-luebeck.de

[2] Institute for Refractive and Ophthalmic Surgery (IROC), 8002 Zurich, Switzerland

[3] Wellman Center for Photomedicine, Massachusetts General Hospital, Harvard Medical School, 50 Blossom Street, Boston, Massachusetts, USA, 02114

[4] Institut für Biomedizinische Optik, Universität zu Lübeck, Peter-Monnik-Weg 4, 23562 Lübeck, bgb@bmo.uni-luebeck.de

Abstract

Scleral crosslinking (SXL), a photochemical technique, is a potential strategy to mechanically reinforce the sclera and therefore prevent progressive axial elongation of the eye, which is responsible for high myopia. Previous approaches of SXL used Riboflavin as photosensitizer. We evaluated the potential of Rose Bengal, excited by green light, as another photosensitizing agent for SXL.

To quantify the effect of photochemical crosslinking we performed tensiometry. We compared rabbit sclera treated with 0.05%, 0.1%, 0.5% Rose Bengal and 532 nm light at 150 J/cm^2. Furthermore, differences between 10 and 20 minutes imbibition time were evaluated.

Scleral stiffness increased after SXL with Rose Bengal at concentrations of 0.05% or 0.1%. Longer imbibition time achieved additional stiffening. Rabbit sclera stained with 0.05% Rose Bengal solution for 20 minutes showed the highest increase in rigidity up to a factor of 2 compared to control samples. Therefore, Rose Bengal SXL is a promising procedure to prevent progressive eye elongation.

1 Introduction

Myopia is a rapidly growing visual disorder. Patients with severe myopia have higher risk for serious problems such as glaucoma, retinal detachment, and cataract [1]. Since the severity of myopia increases with earlier onset, strategies to control myopia progression in at-risk children and young patients with high myopia are of great interest. Although scleral deformities are thought to be in response to aberrant retinal signaling, pathologic changes in the sclera have been identified as an important feature in the early development of myopia and considered possible targets for preventive therapy. In particular, studies have suggested that matrix remodeling of the sclera plays an active role leading to thinning and biomechanical weakening of scleral tissue, and subsequent axial lengthening of the vitreous chamber [2],[3]. A promising strategy to decrease eye elongation and thereby inhibit myopia development is to mechanically reinforce the sclera.

Corneal collagen crosslinking is a low-invasive, successful treatment already adopted in the clinic to arrest the progression of keratoconus primarily by improving the mechanical stability of corneal tissue. Collagen crosslinking should also be a technique for strengthening the sclera to prevent the progression of myopia. Light-activated photosensitizers induce intermolecular crosslinks between neighboring collagen fibrils, strengthening the collagen network and enhancing tissue rigidity. Scleral crosslinking (SXL) has been demonstrated in several animal models [4],[5],[6], most commonly using riboflavin in conjunction with UVA (360-370 nm) or blue light (445 nm).

Rose Bengal is a photosensitizing agent activated by green light [7] and strongly binds to collagen in tissue [8]. Fig. 1 shows the absorption spectrum of Rose Bengal. The lowest excited singlet state is short-lived and relaxes primarily to the lowest excited triplet state [9]. The triplet state can generate singlet oxygen, which initiates protein-protein crosslinks [10]. It has been used as an agent for photochemical crosslinking which was shown to increase corneal stiffness [11] and was used for bonding amniotic membrane to cornea [8].

The aim of this study was to evaluate the potential of SXL treatment to change the mechanical properties of rabbit sclera, using Rose Bengal and a green light source. A protocol with optimized parameters, such as Rose Bengal concentration and imbibition time, was developed.

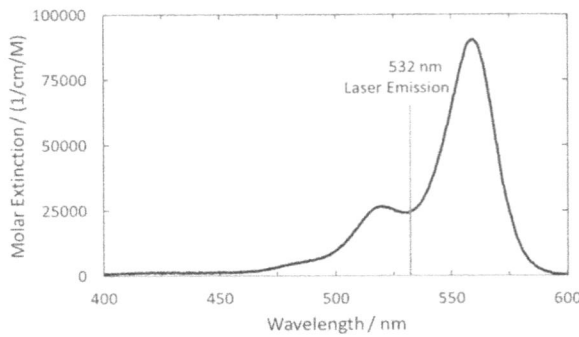

Figure 1: Absorption spectrum of Rose Bengal in Ethanol

2 Material and Methods

A treatment protocol with varying imbibition times and dye concentrations was used to perform Scleral Crosslinking (SXL) on rabbit eyes. To quantify the mechanical properties of the sclera tensiometry was performed.

2.1 Materials

Frozen albino rabbit eyes purchased from Pel-freeze Biologicals (Rogers, AR) were used. A continuous wave KTP-frequency doubled Nd:YAG laser (Oculight; OR;IRIDEX Corporation, Mountain View, CA) provided light with a wavelength of 532 nm. Light of this wavelength is highly absorbed by Rose Bengal (Sigma-Aldrich, St. Louis) [11], which was used as 0.05%, 0.1% and 0.5% wt/wt solutions in Phosphate Buffered Saline (Sigma-Aldrich) to perform the crosslinking procedure. An optical setup using a 400 μm multimode fiber, a 40X microscope objective (Edmund Optics, Barrington NJ), a system of Fresnel lenses and a 30 degree diffuser (Edmund Optics, Barrington NJ) was used to create a highly homogenous irradiation profile with a diameter of 12 mm. A scheme of the setup is shown in Fig.2. For

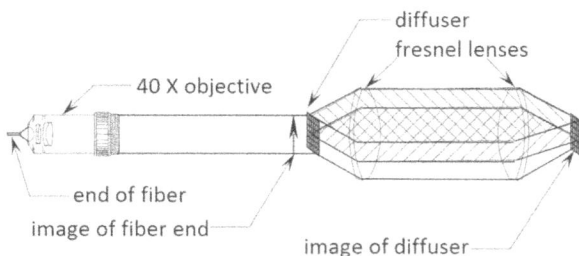

Figure 2: Scheme of optical setup, a homogenous beam profile of 12 mm diameter is created.

the tensiometry measurement, a 10 N load cell (Micro EP Miniature; Admet, Norwood, MA) with a MTest-Quattro controller (Admet, Norwood, MA) was used.

2.2 Crosslinking Procedure

After defrosting, the rabbit eyes were prepared by removing the conjunctiva and remaining muscles. A total of 10 to 15 prepared eyes per group were placed in 0.05%, 0.1% and 0.5% Rose Bengal solution for 10 and 20 minutes to stain the sclera. Serving as control, a group of 15 eyes were treated with PBS only for an imbibition time of 20 minutes. Subsequently, the eyes were rinsed with PBS. To perform photochemical crosslinking, a spot of a 12 mm diameter at the equator of the eye was irradiated with an irradiance of 250 mW/cm^2 and a radiant exposure of 150 J/cm^2 at a wavelength of 532 nm. To prevent the sclera from drying, the irradiated area was moistened with destilled water every minute.

2.3 Tensiometry

A scleral strip of 4.9 mm width was cut at the irradiated area from proximal to distal, using two parallel razor blades. The thickness of the sample was measured using a spring-loaded micrometer (No.1010; Starrett Co., Athol, MA). The scleral strip was mounted at the sample holder of the load cell, shown in Fig. 3. The distance between the clamps was

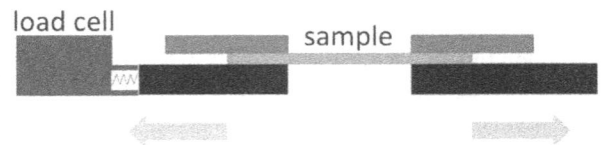

Figure 3: Scheme of tensiometer setup. The sample is mounted on the jaws, the distance is increased while the load cell detects the force.

2.4 mm. The distance was increased with 1 mm/min until the load reached 0.075 N, then decreased to a load of 0.01 N to determine the starting position of the measurement cycle and increased again until the load reached 6 N. Finally, it decreased to a load of 0 N. During the measurement cycle, the load was detected 30 times per second by the load cell. In Fig. 4 the typical viscoelastic stress / strain behaviour of sclera is depicted. The stress P was given by the quotient

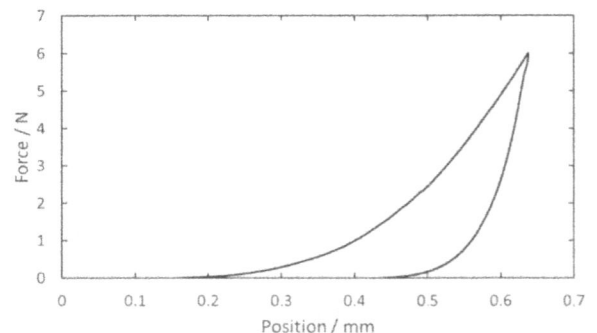

Figure 4: Raw data of a tensiometry measurement

of the force F at each position and the sectional area of the sample, which was defined by the product of the width x and the thickness z of the scleral strip (eq. 1).

$$P(F) = \frac{F}{x \times z} \qquad (1)$$

2.4 Statistics

Mann–Whitney U test was used for the comparison of the mean values of the strain between the different groups. In Mann–Whitney U test, a p-value less than 0.05 was indicated as significant.

3 Results

As shown in Fig. 5, crosslinking treatment with a 10 minutes imbibition time did not increase scleral stiffness significantly after SXL using 0.1% or 0.5% Rose Bengal. In contrast, for eyes treated with a 0.05% Rose Bengal solution (p=0,049) scleral stiffness increased by 145% to 175%, depending on the strain, compared to the control, shown in Fig. 6.

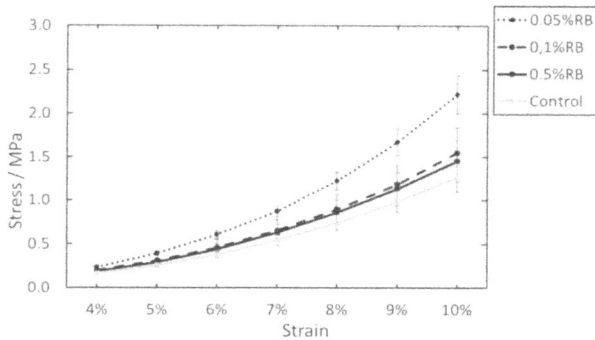

Figure 5: Stress-Strain diagram of rabbit sclera after SXL. 0.05, 0.1, 0.5% Rose Bengal for PBS only and 10 minutes inhibition time.

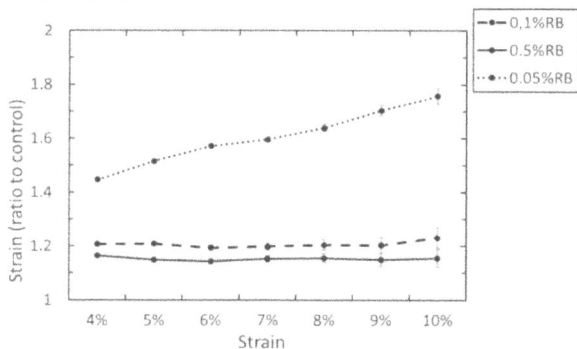

Figure 6: Stress-Strain diagram of rabbit sclera after SXL. 0.05, 0.1, 0.5% Rose Bengal for 10 minutes inhibition time, in relation to PBS only.

Fig. 7 shows average scleral rigidity after SXL using 20 minutes imbibition time. Using both imbibition times, scleral stiffness did not increase significantly after treatment with 0.1% or 0.5% Rose Bengal concentration. Again,

treatment with 0.05% Rose Bengal achieved the highest rigidity and a significant increase (p=0.027) of 170% to 195% depending on strain, shown in Fig. 8.

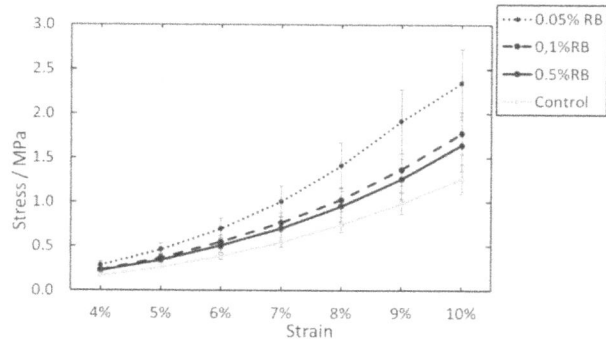

Figure 7: Stress-Strain diagram of rabbit sclera after SXL. 0.05, 0.1, 0.5% Rose Bengal or PBS only and 20 minutes inhibition time.

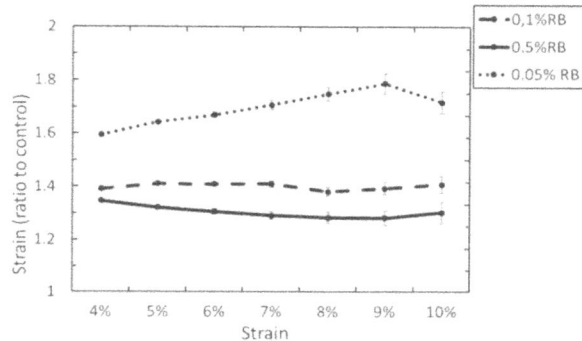

Figure 8: Stress-Strain diagram of rabbit sclera after SXL. 0.05, 0.1, 0.5% Rose Bengal for 20 minutes inhibition time, in relation to PBS only.

4 Discussion

The mean scleral stiffness increased up to 195% using 0.05% Rose Bengal. The lower the concentration was, the higher was the increase in stiffness. This may be caused by the limited penetration depth of Rose Bengal. Cherfan et al. and Yao et al. showed a penetration depth of Rose Bengal of 100 to 120 μm in corneal stroma [11] and 100 μm in dermis [12], respectively. Due to the average sample thickness of 330-360 μm and the very superficially localized crosslinking mechanism [8], we assume that only about 30% of the scleral tissue is affected by the crosslinking process. By developing a technique to increase the penetration depth of Rose Bengal into sclera a much higher amount of tissue could be affected by photocrosslinking and, therefore, result in a higher increase of rigidity. With higher imbibition times the penetration depth of dye increased which may be the reason for the slightly higher increase in stiffness after treatment with 0.05% and 20 minute imbibition time. However, the increase in penetration depth by longer imbibition time is limited to the strong binding of Rose Bengal to collagen [11].

At high dye concentration, light does not penetrate into the tissue but is absorbed near the surface because of the high absorbance of Rose Bengal at 532 nm (see Fig. 1). Therefore, only a small part of the stained tissue is irradiated by a sufficient amount of light. In summary, not all stained tissue is treated by SXL. At lower concentrations the absolute light absorption gradient is more homogenous and a higher amount of tissue is affected by the photochemical crosslinking. Using a wavelength with smaller light absorption (around 500 or 580 nm) should result in a more effective stiffening of the sclera.

5   Conclusion

Rose Bengal / green light SXL was shown to increase scleral rigidity and, therefore, be a possible treatment to prevent progressive myopia. The results are highly influenced by dye concentration. To maximize the stiffening effect, further investigations are recommended to investigate the penetration depth of the dye into scleral tissue and develop techniques to increase it. Additionally, irradiation parameters such as irradiance, radiant exposure and wavelength should be varied in further studies. For in vivo and clinical investigations, treatment parameters should be optimized and a light-delivery technique developed to treat the equator as homogenous as possible during surgery. After optimization, an in vivo study should be performed to determine the safety of procedure and to determine whether that the stiffening effect can prevent eye elongation.

Acknowledgement

The work has been carried out at Wellman Center for Photomedicine, Boston, Massachusetts and supervised by the Institute of Biomedical Optics, Universität zu Lübeck.
We thank William Farinelli and Joe Musacchia for the inspiring discussions and their support during the experiments.

6   References

[1] B. Holden, et al. *Myopia, an underrated global challenge to vision: where the current data takes us on myopia control*. In: Eye 28, London, pp. 142–6, 2014.

[2] W. Meng, J. Butterworth, F. Malecaze and P. Calvas, *Axial length of myopia: A review of current research*. In: Ophthalmologica 225, pp. 127–134, 2011.

[3] D. S. Schultz, J. C. Lotz, S. M. Lee, M. L. Trinidad and J. M. Stewart, *Structural factors that mediate scleral stiffness*. In: Investigative Ophthalmolology Visual Science 49, pp. 4232–4236, 2008.

[4] H. P. Iseli, et al., *Scleral cross-linking by riboflavin and blue light application in young rabbits: damage thresh-old and eye growth inhibition*. In: Graefe's Arch. Clin. Exp. Ophthalmol. 254, pp. 109–122, 2016.

[5] A. Dotan, et al. *Scleral cross-linking using riboflavin and ultraviolet-A radiation for prevention of progressive myopia in a rabbit model*. In: Exp. Eye Res. 127, pp. 190–195, 2014.

[6] S. Liu, et al., *Scleral Cross-Linking Using Riboflavin UVA Irradiation for the Prevention of Myopia Progression in a Guinea Pig Model: Blocked Axial Extension and Altered Scleral Microstructure*. In: PLoS One 11, e0165792, 2016.

[7] M. T. Allen, M. Lynch, A. Lagos, R. W. Redmond, I. E. Kochevar, *A wavelength dependent mechanism for rose bengal-sensitized photoinhibition of red cell acetylcholinesterase*. In: Biochim Biophys Acta 1075, pp. 42-49, 1991.

[8] E. E. Verter, T. E. Gisel, P. Yang, A. J. Johnson, R. W. Redmond and I. E. Kochevar, *Light-Initiated Bonding of Amniotic Membrane to Cornea*. In: Investigative Ophthalmology and Visual Science 52, pp. 9470-9477, 2011.

[9] M. A. J. Rodgers, *Picosecond fluorescence studies ofrose bengal in aqueous micellar dispersions*. In: Chem Phys Lett 78, pp. 509-514, 1983.

[10] H. R. Shen, J. D. Spikes, P. Kopeckova and J. Kopecek, *Photodynamic crosslinking of proteins*. In: J Photochem Photobiol B34, pp.203-219, 1996.

[11] D. Cherfan, E. E. Verter, S. Melki, T. E. Gisel, F. J. Doyle Jr., G. Scarcelli, S. H. Yun, R. W. Redmond, and I. E. Kochevar, *Collagen Cross-Linking Using Rose Bengal and Green Light to Increase Corneal Stiffness*. In: Investigative Ophthalmology and Visual Science. 54, pp. 3426–3433, 2012.

[12] M. Yao, A. Yaroslavsky, F. P. Henry, R. W. Redmond and I. E. Kochevar, *Phototoxicity is not associated with photochemical tissue bonding of skin*. In: Lasers Surg Med. 42, pp. 123-131, 2011.

Strengthening the cornea post-LASIK using Rose Bengal mediated crosslinking

M. Engler[1,2], T. G. Seiler[2], E. Beck[1,2], I. E. Kochevar[2], and R. Birngruber[2,3]

[1] Medizinische Ingenieurwissenschaft, Universität zu Lübeck, marleen.engler@student.uni-luebeck.de

[2] Wellman Center for Photomedicine, Massachusetts General Hospital, Harvard Medical School, Boston, MA, mengler@mgh.harvard.edu

[3] Institut für Biomedizinische Optik, Universität zu Lübeck, birngruber@bmo.uni-luebeck.de

Abstract

Laser assisted in situ keratomileusis (LASIK) disturbs the corneal integrity and in rare cases may induce severe complications like corneal ectasia. Bonding the LASIK-flap to the stroma may at least partially compensate for the loss in bio-mechanical stiffness. Therefore, the traditional LASIK protocol was extended by staining the stroma with Rose Bengal (RB) before repositioning the flap and subsequent irradiation. The aim of this study was to determine the most efficient RB concentration for photobonding ex vivo. To quantify the bonding effect, the force to separate LASIK-flap from stroma after the procedure was measured. The maximal adhesion achieved was up to 2-fold higher than in the standard control group. By means of optical transmission measurements, a RB concentration of 0.1% was identified as the best compromise between a sufficient total absorbance and a transmission of enough photons to the interface.

1　Introduction

Laser assisted in situ keratomileusis (LASIK) was introduced in the 1990's and is the most commonly performed refractive surgery today [1],[2]. The surgery consists of 2 steps: Cutting a circular flap using a microkeratome, or more routinely now a femtosecond laser, and removing a lenticule of the remaining cornea by laser ablation in order to correct the refractive error. Afterwards the flap is repositioned onto the stroma bed [2]. By creating the flap and removing tissue from the stromal bed the cornea is weakened. When the stability loss is too high the cornea can not sustain the intra-ocular pressure and a keratectasia develops. Keratectasia is a progressive thinning and bulging forward of the cornea that compromises the vision. The first report about an induced iatrogenic keratectasia after LASIK was published in 1998 [3]. In order to exclude borderline patients, risk scores were evaluated by Randleman and others to identify patients who are suspect for developing keratectasia and thus shouldn't be treated with LASIK [4].

Parallel to that evolution, corneal crosslinking (CXL) was introduced experimentally by the Dresden group in 1996 [5],[6] to treat primary keratectasia by stiffening the cornea. To prevent keratectasia post-LASIK both procedures have been combined using Riboflavin and UVA-light [7],[8]. A stiffening effect is achieved but more importantly a bonding effect between flap and stroma bed is induced by crosslinks between proteins of the two different layers [8].

In this experimental study rose bengal (RB) , a photosensitizing agent that is activated by green light, and used for many types of tissue photocrosslinking [10],[13], is used instead of riboflavin. The main advantage is its strong binding to collagen in tissue and, therefore, its limited penetration which enables localized photo-activation in targeted structures [9]. The association with cornea results in a red shift from an absorption maximum at a wavelength of 550 nm in solution to 562 nm [10]. By absorbing green light, RB is excited from the ground singlet state to the lowest very unstable excited singlet state. By intersystem crossing it relaxes primarily to the lowest long-lived excited triplet state [11]. Two possible reactions appear depending on the availability of oxygen. If the dye reaches molecular oxygen, which is naturally in the triplet state, energy is transferred and the dye as well as the oxygen change into their singlet state. The singlet oxygen initiates protein-protein crosslinking by oxidizing amino acid side chains [12],[13]. Without the presence of oxygen electrons are transferred from electron-rich amino acids to triplet RB. Radicals are produced that initiate crosslinking (unpublished results). In this study different RB concentrations were analyzed.

2　Material and Methods

2.1　Crosslinking procedure

Adult New Zealand White fresh frozen rabbit eyes (Pel-Freez Biologicals, Roger, AR, USA) were used after defrosting in air. Prior to all experiments the epithelium was removed using a blunt hockey knife and a corneoscleral disk was excised. To establish physiological hydration conditions samples were placed in a 15% dextran wt/wt aqueous solution (Sigma-Aldrich Corp., St. Louis, MO, USA) for 60 minutes. Corneal pachymetry was measured by an ultrasound pachymeter (SP-100; Tomey, Nagoya, Japan) to proof steady conditions, which were assumed after 3 equal consecutive measurements 5 minutes apart. After mounting the cornea onto an artificial anterior chamber (Ziemer Ophthalmic Systems, Port, Switzerland), an intraocular pressure of 20 mmHg was set. A 90 μm flap with a diameter of 9 mm was created by means of the clinical used femtosecond laser Femto LDV model Z6 (Ziemer Ophthalmic Systems, Port, Switzerland). Parameters were set to a velocity of 10 mm/s, output power of 650 mW for the stroma and 845 mW for the sidecut, and a hinge with a height of 0.3 mm. The flap was manually lifted under a microscope. RB (Sigma-Aldrich, St. Louis, MO, USA) was dissolved in phosphate-buffered saline (PBS; Sigma-Aldrich, St. Louis, MO, USA). Solutions at concentrations of 0.02%, 0.1% and 0.5% wt/wt and PBS were applied onto the stromal bed using a blunt cannula . After an imbibition of 2 minutes the exceed dye was dried off using an ophthalmologic sponge (K-Sponge II, Katena Products, Denville, NJ, USA) and the flap was repositioned.

The cornea was irradiated by a green high-power light emitting diode (LED) (M565D2, ThorLabs, Newton, NJ, USA) with a broad emission spectrum (Fig.4). The light was collimated, and focused by two fresnel lenses to a spot size of 10 mm generating a top-hat profile. An irradiance of 180 mW/cm^2 and a radiant exposure of 150 J/cm^2 were used and the cornea was misted with distilled water every 3 minutes to prevent drying.

2.2 Separation measurement

Seven corneas for each concentration group were treated. After the irradiation a 7 mm disc containing flap and underlying stroma was punched out of the cornea (Donor Cornea Punch, Katena, Denville, NJ, USA). The force needed to separate flap and stroma was then measured by a modified tensiometer (Admet, Norwood, MA, USA) with a 9 N load cell.The stromal side of the sample was mounted onto a self-built needle cushion. The needle cushion consisted of needle tips glued into perforated plastic. It was mounted on the left jaw of the tensiometer, which was connected to the load cell. Figure 1 shows the modified tensiometer. On top of the right jaw another self-built modification was mounted, in order to attach the flap-side of the corneal sample. On its rough surface, which consisted of sandpaper, a drop of acrylic glue (Mini Super Glue Gel, CVS, RI, USA) was applied to glue the flap to the jaw. On top of the mounted sample, a 100 g weight was applied to guarantee consistent measurement conditions for all samples. As soon as the sample was properly mounted, the measurement cycle was started so the distance between the both jaws was increased with a velocity of 10 mm/min. The load cell collected the corresponding force every 10 ms. In the end the data were exported into a MS excel file for further processing.

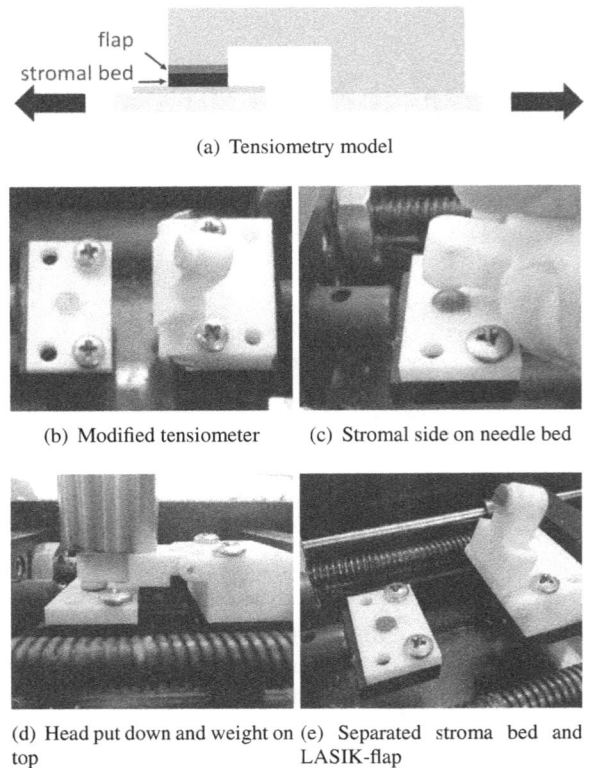

(a) Tensiometry model

(b) Modified tensiometer   (c) Stromal side on needle bed

(d) Head put down and weight on   (e) Separated stroma bed and
top   LASIK-flap

Figure 1: Measurement of bonding strength using a modified tensiometer.

2.3 Spectral transmission measurement

The transmission measurements were performed on 3 corneas per investigated group before and after irradiation with a radiant exposure of $150 J/cm^2$. Again a 7 mm disc was punched out of the cornea, put on quartz glass and the transmission spectrum was measured using the CARY 300 Scan spectrophotometer (Aligent Technologies, Santa Clara, CA, USA) with an integrating sphere. Subsequently, flap and stroma bed were separated and the transmission was measured again separately.

2.4 Statistical analysis

Mann-Whitney-U-Test (Winstat, R. Finch, Germany) was used to compare groups regarding the bonding strength of stroma bed and flap. A p-value less than 0.05 indicates a significant difference between two groups.

3 Results

3.1 Separation measurement

In Fig.2, two typical data sets obtained from separation measurements are depicted. The detected force is plotted versus the corresponding displacement at that time point. The control group measurement showed a non-linear elastic behavior of the sample followed by a smooth transition into sliding friction between the flap and the fixated stroma.

The crosslinked cornea was deformed until the induced crosslinked layer between stroma bed and flap started to rupture. The main disconnection at the peak force was followed by sliding friction between the two generated corneal layers. The average peak force is shown in Fig.3 for each RB concentration. In the control eyes, the maximal detected forces were used for the evaluation. Eyes treated with all 3 concentrations showed a significantly higher adhesion compared to untreated control eyes ($p<0.05$). The maximal factor achieved in the 0.1% group was 2-times higher compared to control eyes.

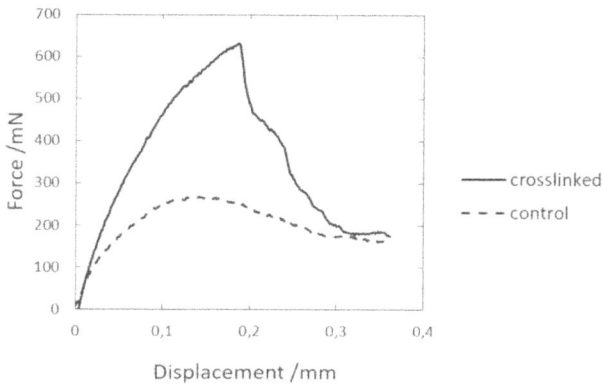

Figure 2: Bonding strength measurement.

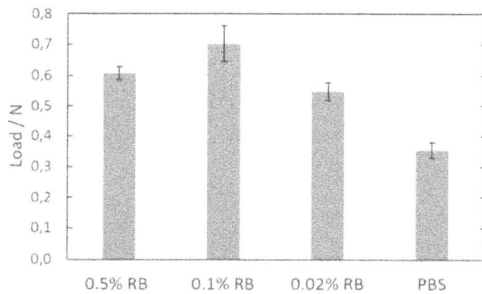

Figure 3: Bonding strength of LASIK-flap and stroma bed after crosslinking with different RB concentrations and green light with a radiant exposure of 150 J/cm^2.

3.2 Spectral transmission measurement

The transmission spectra of the differently stained corneas before and after irradiation are shown in Fig.4. For the lower concentrations of 0.02% and 0.1% an increase in transmission occurred during irradiation. After treatment using a dye concentration of 0.02% the transmission through the flap at the absorption maximum increased from 65% to 85% after irradiation and the transmission through the entire cornea from 32% to 70% (Fig.4a). Using a RB concentration of 0.1% the transmission through the flap increased from 23% to 39% and through stroma and flap from 2% to 6% (Fig.4b). In eyes treated with a RB concentration of 0.5% only 2% of the light was transmitted through the

flap and 0.2% through stroma and flap. After the irradiation no change in transmission was detected (Fig.4c).

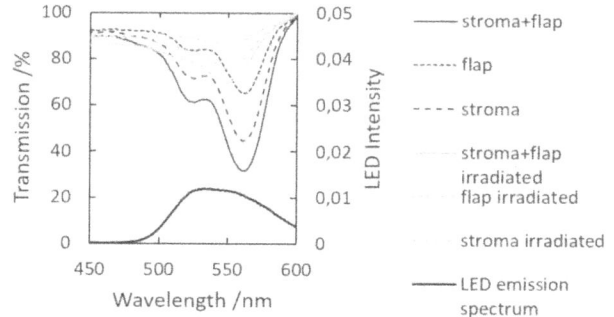

(a) RB concentration of 0.02%, LED emission spectrum

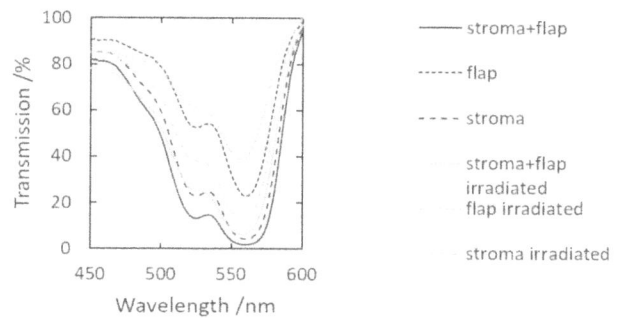

(b) RB concentration of 0.1%

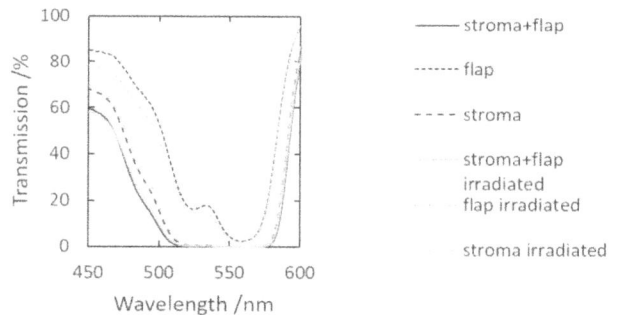

(c) RB concentration of 0.5%

Figure 4: Transmission spectra of cornea stained with different RB concentrations before and after irradiation, and emission spectrum of light source.

Taking the emission spectrum of the light source into account, the amount of light absorbed in the two examined layers were determined. Using a concentration of 0.02%, 25% of the total amount of light emitted by the light source was absorbed by the flap and 27% by the stroma. After treatment with a concentration of 0.1%, 52% was absorbed by the flap and 36% by the stroma and for a concentration of 0.5% an absorbance of 79% in the flap and 7% in the stroma occurred.

4 Discussion

A photo-mediated bonding between flap and stroma is possible using RB and green light. This result suggests that photo-bonding may at least partly reestablish the biomechanical properties of the cornea post-LASIK. From the data obtained in this study at least an increase in adhesion of a factor of 2 is possible.

Although RB has only been applied on the stroma bed light was also absorbed in the flap which means that after repositioning the flap, the dye partly diffused into the flap and caused absorption. In the 0.5% group, this diffusion was so strong that even after irradiation only a small amount of light was transmitted into the stroma. The absorbance was too high to be measurable so a bleaching effect could not be detected. In the 0.02% group most of the light was transmitted into the stroma and a bleaching effect was detected what indicates light absorption by RB and therefore crosslinking [14].

Also after treatment with a dye concentration of 0.1% detectible bleaching occurred. The transmission into the stroma was in average 35%, which seems to be the best solution for flap-bonding using photocrosslinking. A compromise between a strong but also comparable absorbance in flap and stroma was achieved to get a uniform light activation. Although not significant, this is confirmed by the separation experiments.

In summary, the significant effect of bonding LASIK-flap to the stroma bed by light activation of RB was experimentally verified and the dye concentration was optimized ex vivo. The procedure requires further parameter optimization, and evaluation in vivo to assess detailed biomechanical information about the photo-crosslinked cornea.

Acknowledgement

The work has been carried out at the Wellman Center for Photomedicine, Massachusetts General Hospital, Harvard Medical School, Boston, MA and supervised by the Institut für Biomedizinische Optik, Universität zu Lübeck. The authors thank Ziemer Ophthalmic Systems (Port, Switzerland) for the allocation of the Ziemer LDV Z6 femtosecond laser. In addition, we want to thank William Farinelli, Joe Musacchia and Theo Seiler for the inspiring discussions and their support during the experiments.

5 References

[1] I. G. Pallikaris, M. E. Papatzanaki, D. S. Siganos and M. K. Tsilimbaris,*A corneal flap technique for laser in situ keratomileusis. Human studies.* Arch Ophthalmol. 1991 Dec;109(12):1699-702

[2] A. Sugar, C. J. Rapuano, W. W. Culbertson, D. Huang, G. A. Varley, P. J. Agapitos, V. P. de Luise and D. D. KochLaser in situ keratomileusis for myopia and astigmatism: safety and efficacy. A report by the American Academy of Ophthalmology. Ophthalmology 2002;109:175–187

[3] T. Seiler and A. W. Quurke, Iatrogenic keratectasia after LASIK in a case of forme fruste keratoconus. J Cataract Refract Surg. 1998 Jul;24(7):1007-9.

[4] J. B. Randleman, M. Woodward, M. J. Lynn and R. D. Stulting,Risk assessment for ectasia after corneal refractive surgery. Ophthalmology. 2008 Jan;115(1):37-50.

[5] T. Seiler, E. Spoerl, M. Huhle and A. Kamouna,Conservative therapy of keratoconus by enhancement of collagen cross-links. Invest Ophthalmol Vis Sci. 1996;37:1017.

[6] G. Wollensak, E. Spoerl and T. Seiler, Riboflavin/ultraviolet-a-induced collagen crosslinking for the treatment of keratoconus. Am J Ophthalmol. 2003;135:620-7.

[7] A. J. Kanellopoulos, G. Asimellis and C. Karabatsas, Comparison of prophylactic higher fluence corneal cross-linking to control, in myopic LASIK, one year results. Clin Ophthalmol. 2014 Nov 27;8:2373-81.

[8] T. G. Seiler, I. Fischinger,T. Koller, V. Derhartunian and T. Seiler,Superficial corneal crosslinking during laser in situ keratomileusis. J Cataract Refract Surg. 2015 Oct;41(10):2165-70. doi: 10.1016/j.jcrs.2015.03.020.

[9] M. Yao, A. Yaroslavsky, F. P. Henry, R. W. Redmond and I. E. KochevarPhototoxicity is not associated with photochemical tissue bonding of skin. Lasers Surg Med . 2011; 42:123-131.

[10] E. E. Verter, T. E. Gisel, P. Yang, A. J. Johnson, R. W. Redmond and I. E. Kochevar,Light-initiated bonding of amniotic membrane to cornea. Invest Ophthalmol Vis Sci. 2011; 52:9470-9477.

[11] M. A. J. Rodgers, Picosecond fluorescence studies of rose bengal in aqueous micellar despersion. Chem Phys Lett. 1983; 78:509-514.

[12] H. R. Shen, J. D. Spikes, P. Kopecekova and J. Kopecek,Photodynamic crosslinking of proteins. I. Model studies using histidine- and lysine-containing N-(2-hydroxypropyl)methacrylamide copolymers. J Photochem Photobiol B. 1996; 34:203-210.

[13] S. Tsao, M. Yao, H. Tsao, et al. Light-activated tissue bonding for excisional wound closure: A split-lesion clinical trial. Br J Dermatol. 2012; 166:555-563.

[14] I. E. Kochevar and R. W. Redmond, Photosensitized production of singlet oxygen. Methods Enzymol. 2000; 319:20-28

Determination of quality and strength of animal pericardium
– Development of an optical device –

S. Hagen[1], J. Helfmann[2], and R. Huber[3]

[1] Medizinische Ingenieurwissenschaft, Universität zu Lübeck, simon.hagen@student.uni-luebeck.de
[2] Laser- und Medizin-Technologie GmbH, Berlin, j.helfmann@lmtb.de
[3] Institut für Biomedizinische Optik, Universität zu Lübeck, robert.huber@bmo.uni-luebeck.de

Abstract

Animal pericardium is an often used tissue in the production of biological, artificial heart valves. Its stability is based on the orientation of the collagen fibers, which needs to be quantified. Moreover the strength of the tissue is depending on the degree of fixation. This needs a testing method, too. To solve these problems, measurements using polarization and fluorescence were done. For the former a polarization microscope and for the second type a fluorescence spectrometer was used. As a result of these experiments it was shown, that it is possible to figure out the orientation of the collagen fibers using polarized light. The degree of fixation was determined by using the fluorescence of the tissue. Out of this knowledge a device has been developed, that combines both methods and should solve the problem of testing the quality in artificial heart valve production.

1　Introduction

In the field of artificial heart valves or expansion of arteries and veins, animal pericardium is an often used tissue. These heart valves or patches are made of porcine, equine or bovine pericardium, which consists mainly of collagen, so it has fluorescent and birefringent properties. For a safe and durable prosthesis made of pericardium it is necessary to know in which direction the collagen fibers are orientated. Moreover the stability is an important aspect. Often pericardium is fixated with glutaraldehyde, which makes the tissue quite hard and inflexible, because of the created cross-linkings between the collagen fibers. Nevertheless in an unfixed state it is way to flexible for being used as a heart valve. So it should be in between both states. To make a reliable statement about the degree of fixation, it needs to be measured. Figuring out the orientation of the fibers is typically done by mechanical tearing of the tissue, which destroys the used specimen. As mentioned above, pericardium has the property of being a birefringent medium, because of the different slices of collagen fibers. Via this, it is possible to quantify the orientation and retardation of the fibers, i.e. one can say whether they are packed and strictly orientated or not.

Furthermore the cross-linkings created by glutaraldehyde are fluorescent [1], which could make it possible to measure the degree of fixation by measuring the fluorescence.

On the basis of these different aspects, a device has to be built, that combines fluorescence and polarization imaging.

2　Material and Methods

To check the feasibility of the methods in case of pericardium, some preinvestigations have been done.

2.1　Samples

For this project different kinds of pericardium were used. They differed in thickness, as porcine was about $100 \mu m$, equine was about $200 - 250 \mu m$ and bovine was $> 350 \mu m$ thick and their state of fixation. While all three kinds of pericardium were available in the unfixed state, the fixated samples only were porcine with a thickness of about $210 \mu m$.

2.2　Polarization

The first experiments were done using a polarization microscope (DM2700P, Leica AG Wetzlar), with the aim to verify the quantification of the orientation and retardation of the collagen-fibers. For this the optical path difference has been measured with the help of a Berek-compensator. This instrument mainly consists of a tiltable birefringent plate, which compensates the optical path difference, if it's tilted in the correct angle. With the help of this angle, the difference between the ordinary and extraordinary rays can be calculated. A second motivation for these experiments was to find out how much influence the wavelength of the light and the thickness of the material have on the result. To get different wavelengths from a white-light LED, three filters were used. One bandpass-filter with transmission-maximum at 405nm (BG12 filter, Schott AG Jena) for blue light, a bandpass-filter for green light with transmission around 546nm (VSS 546 filter, Leica AG Wetzlar) and a longpass-filter (RG 630 filter, Schott AG Jena) with transmission starting at about 630nm. The thickness-depending measurements took place with all of the three different kinds of unfixed pericardium, that were available. In all of these experiments, the polarization microscope was used with transmitted light, so the light source was underneath and the camera above the sample.

2.3   Fluorescence

A fluorescence-spectrometer (LS55, Perkin Elmer Waltham, MA) was used for the fluorescence measurements. Thereby fixated and unfixed pericardium was put in the spectrometer to determine, whether the fluorescence of the collagen with glutaraldehyde induced cross-links is at the same wavelength as the pericardial fluorescence or not, which would make the two kinds of pericardium distinguishable. The excitation light shined with an angle of $45°$ on the samples leading to an emission, which is observed in reflection. For the measurements with unfixed samples porcine pericardium was used, because of its thinness to excite as much collagen as possible and not to loose too much of the emitted light due to scattering. Fixation of the tissue has the effect, that it is less flexible. This leads to a larger thickness, since there is no water or other liquid that can be pressed out of the specimen by the micrometer. Because both pericardium samples are stored in a solution made of sodium chloride solution and antibiotics, a measurement of this was done, too.

3   Results and Discussion

The results of the preinvestigations are shown in the next subsections followed by the corresponding requirements for the device and a concept for building it.

3.1   Polarization

The measurements using the Berek-compensator showed, that measuring the optical path difference of pericardium is possible with this method. In the literature a value of $\Delta n = 3 \cdot 10^{-3}$ is given as birefringence for strictly oriented and packed collagen[2]. In the measurements it was possible to get optical path differences of 474nm in 250 μm thick pericardium (see table 1). This leads to a birefringence of $\Delta n = \frac{0.474 \mu m}{250 \mu m} = 1.896 \cdot 10^{-3}$. The difference between these values is explainable by a non strict orientation in pericardial tissue. Nevertheless the Berek-compensator has the disadvantage that it has to be used manually and is not automatable, because the tilting-angles of the compensator have to be read. One possibility to solve this problem is to turn the specimen or the polarizers between $0°$ and $90°$ in some steps and take a picture at every step. Because of a $\frac{\pi}{2}$-periodicity the image turns dark and light every $90°$.

Table 1: Meanvalues of the optical path difference Γ and it's standard deviation measured with different wavelengths and on pericardium with various thicknesses.

	white light	green light	blue light	red light
	Thickness 0.25mm (side 1)			
$\bar{\Gamma}$ [nm]:	898.81	433.98	368.67	474.39
$\sigma(\Gamma)$ [nm]:	12.4067	8.9678	13.6029	16.2941
	Thickness 0.25mm (side 2)			
$\bar{\Gamma}$ [nm]:	384.385	422.505	248.47	405.42
$\sigma(\Gamma)$ [nm]:	6.5336	6.5768	7.736	7.9952
	Thickness 0.2mm (side 1)			
$\bar{\Gamma}$ [nm]:	690.37	658.605	504.185	385.6325
$\sigma(\Gamma)$[nm]:	12.3219	6.8074	7.2279	10.4337
	Thickness 0.2mm (side 2)			
$\bar{\Gamma}$ [nm]:	441.72	300.515	330.89	239.56
$\sigma(\Gamma)$ [nm]:	10.5738	20.4618	25.3808	5.6975
	Thickness 0.1mm (side 1)			
$\bar{\Gamma}$ [nm]:	1142.615	1269.6875	1774.8275	385.6325
$\sigma(\Gamma)$[nm]:	8.9766	156.8995	349.8114	10.4337
	Thickness 0.1mm (side 2)			
$\bar{\Gamma}$ [nm]:	1161.245	733.38	970.42	391.75
$\sigma(\Gamma)$ [nm]:	6.5444	7.0702	15.2921	11.4128

With the so measured intensity-values it is possible to calculate the phase and the retardation of the object, which gives evidence of the direction of the fibers. To make a first step in this direction an ImageJ-Plugin was written, which offers the possibility of reading images of the specimen belonging to one wavelength and then calculate in some steps the retardation and phase of each pixel of the image.

As mentioned in section 2.2 the influence of the wavelength and of the thickness of the sample had to be examined. With the bovine pericardium these examinations were not possible. A reason for this might be the circumstance of high scattering in this pericardium and more slices of collagen-fibers. The other two levels of thickness were unproblematic for the measurements and lead to different results in the optical path difference. The reason for this lies in the different number of collagen fibers. As the light has to overcome more birefringent slices, the optical path difference will grow.

The measurements using different wavelengths showed, that taking red light led to more reliable results and easier distinction between compensated and not-compensated optical path difference (see table 1). As it is well known from the literature [3], polarized light will be depolarized when it is transmitted through a scattering medium like pericardium. This effect grows with the thickness of the medium, leading to worse images. Moreover the scattering is wavelength dependent and has a larger effect at smaller wavelengths. For these reasons red or near-infrared light should be used.

For the manual measurements the problem of measuring in a higher order of birefringence occurred, meaning that not the first compensated transition was taken, but a later one, which led to a mismeasurement.

Calculated with the written plugin, which is based on a self-made algorithm for the analysis of a stack of images, the values of the optical path difference are often smaller than the measured ones. The reason for this phenomenon lies in the ambiguity of overlapping beams with an optical path difference, that differs in a multiple of the wavelength plus or minus the optical path difference. To solve this problem, two different wavelengths are needed, to compute the correct optical path difference [4]. Adding this information to the one about the wavelength, it results in two red or near-infrared wavelengths which are long enough to reduce part of the scattering. Nevertheless the spectral difference between them shouldn't be to high, so that the depolarisation at both wavelengths is as similar as possible.

3.2 Fluorescence

The results of the fluorescence measurements show, that there is one fluorescence-peak at 400nm in fixated (dots in fig.1) and unfixed (line-dots in fig.1) pericardium with excitation in the UV at about 340nm. But in the fixated with blue excitation another peak occurs at 520nm, which is a result of the cross-links caused by the glutaraldehyde (dashed line in fig.1) and can not be seen with the unfixed pericardium(full line in fig.1). This makes the two kinds of pericardium distinguishable. So a system with two excitation wavelengths (340nm and 470nm) and two wavelength bands for detection is needed to distinguish between fixated and unfixed pericardium. This leads to the consideration to excite a specimen at first with 340nm and then with 470nm taking images of the emission and then calculate the degree of fixation. The measurement of the sodium chloride antibi-

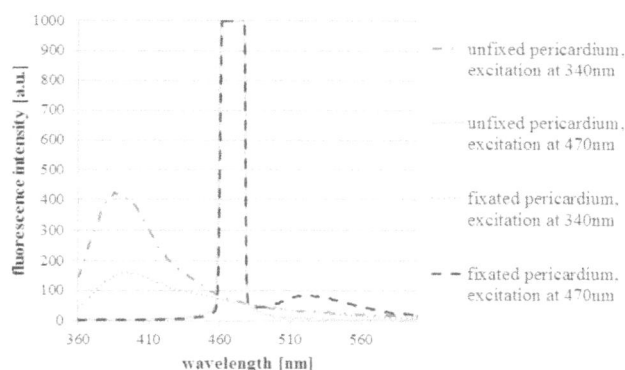

Figure 1: Comparison of the fluorescence between fixated and unfixed pericardium. Each of the tissues was excited at 340nm and 470nm. For the fixated pericardium a second fluorescence peak at 520nm can be seen. To prevent a to high signal, a 2 %-filter at excitation side was used.

otics solution showed, that it doesn't influence the result of the fluorescence measurements.

3.3 System Requirements

The device must include a fluorescence- and a polarization-beam path, but still shouldn't have a size of several meters, as it is for all day use in a company. Because the excitation of fluorescence normally is done from above, while polarization is done with transmitted light, the illumination paths

should be separated. The excitation of the fluorescence is done as mentioned with a 340nm and a 470nm light-source. To get the best results, these sources should have a sufficiently narrow spectral bandwidth, but shine on an area of about 10cm x 10cm.

Because for the excitation of fluorescence a wavelength in the UV is used, special optics with UV-transmission are necessary.

As the emission takes place at 400nm and 520nm, filters are required to prevent the excitation light, which might cover one of these wavelengths, from getting into the emission ray, where it might cause signal.

To avoid the problem of higher orders in the polarization-measurements, two wavelengths for this part are needed. Due to high scattering mainly red light transmits through the tissue, which leads to the use of red or near-infrared (NIR) light. After a simulation of the algorithm, that will be used for the calculation, 780nm and 880nm appeared as a well suitable wavelength-pair. Both are relatively long wavelengths compared to the visible, so the scattering of the light will be less than of other visible light of a smaller wavelength. Second the difference in depolarization between them is not as high as it will play a role in the analysis of the data.

Since the fluorescence light shouldn't be polarized, the polarizer on the emission side of the optical path should be movable. Moreover both polarizers should be rotatable for the acquisition of the different polarization images.

The last requirements are the needed camera and objective. The latter should image the whole specimen, so there is no need to move it. Furthermore the camera should cover a range of 400-900nm, have a big pixelsize and might allow binning. Moreover a high dynamic range is needed and temperature stabilization might be an option, depending on the signal-to-noise ratio and the reproducibility.

3.4 Concept

Based on the requirements a concept was developed. Because of simplicity and cost LEDs instead of lasers will be used. For the assembly of the LEDs in the particular beam paths different concepts would be possible.

3.4.1 Fluorescence excitation assembly

The light-sources for the excitation of fluorescence could have two different places. The first one would be sideways-above the specimen (see fig. 2a)). It has the disadvantage of a necessary dichroic-mirror, that must transmit the polarized light and the two emission-wavelengths but reflect the excitation-light or is removable, which will not be cost-effective. The second approach is to directly illuminate the sample from above (see fig. 2 b)). Thereby it is necessary not to shadow the emission beam and the camera imaging. Now the assembly of the light sources is possible in three different ways. The first possibility is to place both LEDs in an angle of 90° and collimate their light (see fig. 3 a)). Before both wavelengths will be overlayed by a dichroic mirror, the light with 470nm has to pass a short-pass filter,

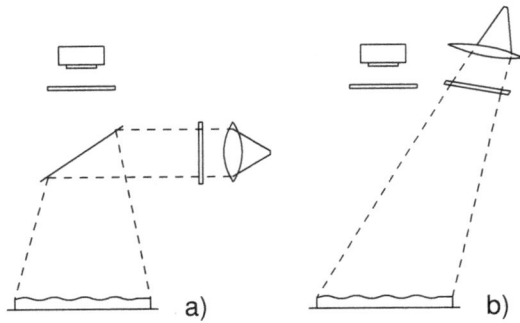

Figure 2: Schematic sketches of fluorescence excitation assemblies, a) illumination via a dichroic mirror, b) direct excitation from sideways above

so the amount of light in the wavelength range of the emission light is reduced to a minimum. For the other LED the dichroic mirror can work as a filter. After these optics, a lens system should enlarge the illuminated area to 10cm x 10cm and make the illumination as homogeneous as possible. In the second approach two LEDs would be placed next to each other or as close as possible (fig 3 b)). As in the first one both will be collimated, the unwanted part of light filtered out and homogeneous enlarged by a lens system. This possibility has the disadvantage, that the LEDs have to be very close to each other to illuminate the same area. For the third assembly LEDs should be changeable or on a wheel to change them as shown in fig. 3 c). For the corresponding filters a filterwheel is needed. The other parts would be the same like in the other approaches. The main problem of this is the price of an automatic filterwheel and the changeable LEDs.

For the mentioned reasons the filter stairs and an illumination from sideways above the specimen will be used.

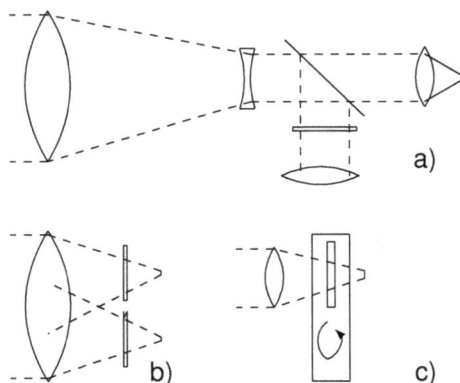

Figure 3: Schematic sketches of light source assemblies, a) using a dichroic mirror, b) location-depending filters, c) changing illumination

3.4.2   Polarization assembly

As mentioned above, for polarization imaging a transmitted light illumination of the sample will be used. Similar to

the fluorescence excitation assembly it is possible to use a dichroic mirror or to place both light sources close to each other. The second of both possibilities should give enough light for illumination of the specimen. Between the LEDs and the pericardium the first polarizer must be placed, to produce linear polarized light. After it has passed the sample, which should lie on a silica-glass plate on an x-y-stage, the now differently polarized light will have to get through a second polarizer, like in a common polarization microscope. This will be followed by a motorized filterwheel, that should let through all of the polarization light, but contains filters for the fluorescence light. Behind these filters the objective and camera will be placed.

4   Conclusion

As it has been shown in this work, it is possible to distinguish between fixated and unfixed pericardium by using fluorescence. Moreover the retardation and phase of the collagen fibers can be figured out with help of polarization. In theory all parts for a device, that combines both methods have been selected. In future work this device has to be built in reality. Furthermore verification tests have to be done, to see whether everything works as expected or if something has to be changed in the assembly.

Acknowledgement

The work has been carried out at Laser- and Medizin-Technologie GmbH, Berlin (LMTB) and supervised by the Institut für Biomedizinische Optik, Universität zu Lübeck. I would like to thank Dr. Jürgen Helfmann from the LMTB for his good mentoring of my internship and this project.

5   References

[1] J. S. Collins and T. H. Goldsmith *Spectral Properties of Fluorescence Induced by Glutaraldehyde Fixation.* In: The Journal of Histochemistry and Cytochemistry, Vol. 29, No. 3 1981, pp. 411-414

[2] D. D. Yakovlev et al., *Quantitative mapping of collagen fiber alignment in thick tissue samples using transmission polarized-light microscopy.* In: Journal of biomedical optics 21.7 (2016): 071111-071111

[3] V. V. Tuchin and V. Tuchin, *Tissue optics: light scattering methods and instruments for medical diagnosis.* Vol. 642. Bellingham: SPIE press, 2007.

[4] T. T. Tower and R. T. Tranquillo, *Alignment maps of tissues: I. Microscopic elliptical polarimetry..* In: Biophysical journal 81.5 (2001): 2954-2963.

[5] F. Massoumian, R. Juškaitis, M. A. A. Neil and T. Wilson *Quantitative polarized light microscopy.* In: Journal of Microscopy, Vol. 209, Pt1 January 2003, pp.13-22.

Dispersion Compensation for Fourier-Domain Mode-Locked Lasers

J. Klee [1], J. P. Kolb [2] and R. Huber [2]

[1] Medizinische Ingenieurwissenschaft, Universität zu Lübeck, julian.klee@student.uni-luebeck.de

[2] Institute of Biomedical Optics, Universität zu Lübeck, {kolb, robert.huber}@bmo.uni-luebeck.de

Abstract

Fourier-Domain-Mode-Locked (FDML) lasers are a new type of laser which allows for up to 140 nm sweep bandwidth or 5.2 MHz sweep rate. This is particularly useful in Optical Coherence Tomography, as the imaging speed is directly dependent on the sweep frequency. A major challenge of this laser type is dispersion, which means that different wavelength dependent delays are introduced in the fiber-based cavity. Compensating for this dispersion improves the overall laser performance: it enhances the coherence length and increases the phase-stability. In this paper it will be shown how the dispersion of such a laser has been reduced. With our approach the first second and third order of dispersion could be decreased by one order of magnitude. However, further improvements are still desired but the results are sufficient for further research.

1 Introduction

Fourier-Domain-Mode-Locked (FDML) lasers were first introduced in 2006 [1]. Those lasers are fiber-based, high repetition rate swept-source lasers which can reach up to 5.2 MHz sweep-rate [2].

An application for this laser is Raman-spectroscopy but most commonly it is used in Optical Coherence Tomography (OCT). OCT [3] is often described as ultrasound imaging using light since it is using the scattering of light waves instead of sound waves at distinct optical layers to reconstruct a depth image. OCT however needs an interferometric approach to resolve the resulting time delays. This imaging modality produces one depth-scan (A-scan) per sweep, via a rastering method utilizing one galvanometer scanner multiple A-scans can be composited to a 2D B-scan and a volume scan can be performed using a second galvanometer scanner. In opthalmologic applications the center wavelength is set around 1060 nm [4], because this wavelength is a local absorption minimum in the vitreous body. OCT has a typical penetration depth of up to 2 mm in scattering tissue at an axial resolution of 10 μm [5].

A fundamental problem of the FDML laser is the chromatic dispersion. This aspect describes the different round-trip times of the individual wavelengths within the laser cavity. For best performance the dispersion should be as low as possible to ensure synchronous behavior of all wavelengths. Dispersion compensation results in improved coherence-length and possibly better inter sweep phase-stability [4]. A phase-stable laser makes not only amplitude but also the phase information of the Fourier-Transform usable and provides additional data, which could be used for example for compensation of aberrations introduced in the lens the human eye [6].

2 Material and Methods

The setup used in this work is an FDML laser as shown in Fig.1, which has a center-wavelength of 1060 nm and a sweep range of 90 nm.

A tunable Fabry-Perot-Filter is driven by a 420 kHz sinusoidal voltage. The frequency f_{filter} at which the filter is driven is adjusted to the cavity length L based round-trip time. The theoretical round-trip time T can be calculated using (1)

$$T = \frac{n_i * L}{c_0} = \frac{1}{f_{\text{filter}}} \qquad (1)$$

where n_i denotes the refractive index of the fiber and c_0 the speed of light in a vacuum. Light of the whole sweep bandwidth is stored in the delay line, a spool comprised of SMF and LEAF fiber. Therefore lasing will not have to build up again, at each wavelength change of the filter. The laser gain medium is a semiconductor optical amplifier (SOA). Two optical isolators determine the light round-trip direction. Since optical fibers have a wavelength dependent dispersion coefficient, this has to be compensated for. This is largely solved by utilizing a chirped Fiber-Bragg-Grating (cFBG).

With optical buffering the 420 kHz are quadrupled to 1.68 MHz. The buffering is achieved by driving the laser-gain-medium with a duty-cycle of 25% synchronous to the filter-frequency. This shorter sweep is then split into

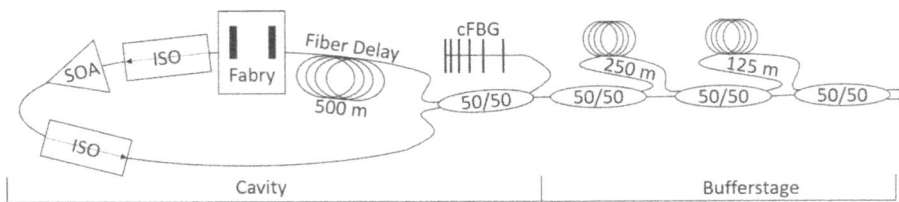

Figure 1: The basic FDML Laser setup as used for the purposes of this paper. The components are as follows: Semiconductor Optical Amplifier (SOA), Isolators (ISO), chirped Fiber-Bragg-Grating (cFBG), high frequency Fabry-Perot filter, Fiber-delays for cavity and buffer-stages and 50/50 couplers.

two by a fused fiber coupler, one half is delayed exactly by half the cavity round-trip time by using a fiber delay line of half the length of the cavity. This process is repeated and results in four identical sweeps at the desired frequency of 1.68 MHz.

2.1 Theoretical Dispersion

In fiber optics most components have a wavelength dependent refractive index and by using (1) it can easily be seen that a different refractive index leads to altered round-trip time for the individual wavelength. In order to calculate the dispersion of the fiber spool, which is the most significant component, a measurement was undertaken in a previous research project to asses the polynomial orders of dispersion. These have been Taylor expanded around the center wavelength λ_0 of 1060 nm given as

$$\sum_{n=0}^{3} d_n * (\lambda - \lambda_0)^n \qquad (2)$$

where d_n denotes the dispersion coefficient. Measurements of the dispersion coefficients for different fibers are shown in Table 1.This allows for a good estimate to use in the compensation process.

Table 1: Dispersion broken down into first and second order dispersion coefficients d_1 and d_2

Fiber type	d_1 / $\frac{ps}{nm*km}$	d_2 / $\frac{ps}{nm^2*km}$
HI1060	34.548±0.258	-0.0759±0.0030
LEAF	40.548±0,259	-0.0751±0.0078
SMF28e+	29.533±0.090	-0.0815±0.0071

2.2 Dispersion-Compensation

The different polynomial orders are compensated separately since the strategy differs for each and will be discussed in the following sections. In order to measure the current dispersion there are two possibilities. The first is capable of measuring large amount of dispersion but is in contrast to the second variant not very accurate. One observes the noise of the laser via a photo-detector coupled to an oscilloscope. This method relies on a frequency dependent

noise edge which marks the wavelength matching the current frequency. This noise edge can be moved via a frequency sweep of the tunable bandpass filter. By using (1) the round-trip time can be determined for every sample of the sweep. Utilizing another tunable wavelength filter and a spectrometer, the wavelength-sample-correlation can be determined. This results in a dispersion curve, which can be broken down into a linear and quadratic component and used to evaluate the compensation progress.

The second method utilizes the sweet-spot characteristics of the laser (see Fig.4 a,b for reference). By inducing a delay of 2 cm between the two interferometric arms of a Michelson interferometer, a very noisy signal can be observed on the photo-diode signal read out on the oscilloscope. In those sweet spots the noise dramatically decreases. This occurs when the round-trip time of the FDML laser is synchronized within ±100 fs to the rapidly tuned Fabry-Perot filter. The origin of this effect is not conclusively explained and further research is being done. By utilizing a frequency sweep, all wavelength of the swept laser will achieve their specific sweet spot. After determining a frequency range to sweet-spot correlation for the individual wavelengths, the central round-trip time T can be determined for each wavelength, again using (1) and resulting in a dispersion curve.

2.2.1 Dispersion compensation using cFBG

The main method of dispersion compensation is a cFBG. The cFBG used for the FDML laser in this setup is produced by TeraXion Inc at a center-wavelength of 1040 nm with a bandwidth of 200 nm. To compensate wavelengths that experience shorter delays, the cFBG reflects those later than wavelength that experience longer delays. This is achieved by the periodic refractive index modulation of the cFBG. Equation (1) shows that the resulting delay is approximately the inverse of the delay produced by the fiber delay line. Since there are variations in the fiber spool and other components, the cFBG cannot be designed exactly, further step are required which will be described in the following.

2.2.2 Compensation of the second polynomial order

The first step to minimize the dispersion is the compensation of the second polynomial order. This is done by evaluating the magnitude and orientation of the quadratic dispersion. Depending on whether it is positive or negative (orientation), fiber needs to be added or removed. The magnitude determines the amount of fiber. As shown in Table 1 all fibers have similar second order dispersion coefficients, therefore the choice of fiber type is not relevant.

2.2.3 Compensation of the first polynomial order

The most substantial influence on dispersion was in this case the linear component. As discussed in the previous subsection, magnitude and orientation are the two influencing factors. Since SMF and LEAF fibers have differing linear dispersion qualities (Table 1), an exchange of these can compensate for differing delays.

2.2.4 Compensation of the third polynomial order

In the last step of this process the third polynomial order needs to be compensated to such a degree that the maximum amplitude in dispersion is no larger than 100 fs. The main advantage of the following method is that it is capable of compensating all degrees of dispersion up to a maximum range of $\sim \pm 2\,$ps. As shown in Fig.2 a piece of shaped aluminum is used to create a specific temperature gradient in which the cFBG is embedded. According to

$$\frac{\Delta\lambda}{\lambda_0} = (1 - p_e) * \epsilon + (\alpha_\Lambda + \alpha_n) * \Delta T \qquad (3)$$

where λ denotes wavelength, p_e the strain-optic coefficient, ϵ the strain, α_Λ the thermal expansion coefficient and α_n the refractive index coefficient, the refractive index can be modified using temperature gradients. The use of strain is difficult since the fiber can be damaged easily by strain. The temperature gradient leads to a variable optical length in between the wavelength dependent reflective surfaces within the cFBG, thus allowing for modification of the delay characteristic of the cFBG. This is particularly useful, since imperfections due to the manufacturing process which have a magnitude of about $\Delta 200\,$ps, can theoretically be compensated. The shape of the aluminum plate is only modified in one dimension allowing for easier simulation and thus modeling. This was achieved by a Labview program based on

$$\dot{Q} = \lambda * A * \frac{\Delta T}{d} \qquad (4)$$

where $\dot{Q}$ denotes heat flux, λ the thermal conductivity, A the area of heat flux and d the thickness. The amount of inflection points in the dispersion curve determines the number of bases for temperature control, since the temperature course between two bases can only be monotonous. As a control and for production purposes the resulting aluminum shape was implemented and simulated in SolidWorks.

3 Results and Discussion

For the purposes of this paper two FDML lasers have been dispersion compensated to differing levels of accuracy. The first system, an FDML laser for use in clinical environment was only compensated for linear and quadratic dispersion.

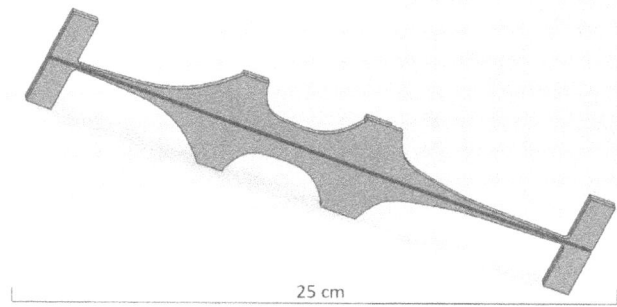

Figure 2: A 3D-Model of the CNC-made aluminium sheet for temperature gradient controlled optical length adjustment of the cFBG.

In the second system linear and quadratic dispersion where already mostly compensated for and only slight modifications had to be made. The cubic order however needed to be adjusted and compensated in the second system. Thus those two systems will be presented seperately.

3.1 Dispersion compensation of the first and second order polynomial

For the given system small iterative steps have been undertaken to compensate for dispersion slowly. As shown in Fig.3 the total linear dispersion was compensated down to six ps by exchanging LEAF for SMF fiber.

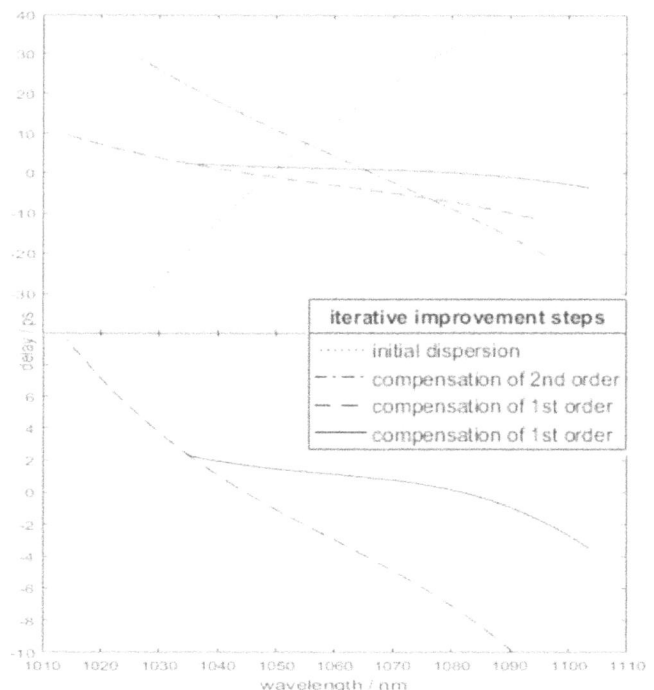

Figure 3: The improvement in quadratic and linear order are displayed progressively. The solid line represents the final state.

Although the linear dispersion is still quiet high, further improvements where not vital at this point, since the continuation of this project could further induce or reduce dispersion by adding further components. The quadratic dispersion however was successfully compensated for and only dispersion of third and higher degrees remains.

3.2 Dispersion compensation of the third order Polynomial

Dispersion reduction of the third order in the second system, proved to be more difficult, yet 90% of the sweep was successfully compensated within the sweet-spot-range as shown in Fig.4. Two imperfections in the manufactured cFBG could not be compensated for. Another critical point during the experiment was the exact alignment of the cFBG in the aluminum sheet. Possible solutions to these challenges would be an improved measurement of wavelength to cFBG depth relations and a new system with dynamic temperature gradients utilizing a higher density of bases. Measurement of the coherence length showed an improvement of the previous millimeter-range to a centimeter-range with the current system.

4 Conclusion

In this paper it was shown, that dispersion compensation of linear and quadratic order using the presented methods is a viable approach. Although further improvements to the cubic dispersion are desired, the current configuration is sufficient for the purposes of further research. This research will make use of the inter-phase stability gained to utilize the phase information of the Fourier transformed signal spectrum. A multi-base variant of the temperature gradient based method is currently in the design phase.

Acknowledgement

The work has been carried out at the Institute of Biomedical Optics at the Universität zu Lübeck.

5 References

[1] R. Huber, M. Wojtkowski and J. G. Fujimoto, *Fourier Domain Mode Locking (FDML): A new laser operating regime and applications for optical coherence tomography*. in: Opt. Express, OSA, Washington, pp. 59–62, 2006.

[2] W. Wieser, B. R. Biedermann, T. Klein and C. M. Eigenwillig, R. Huber, *Multi-Megahertz OCT: High quality 3D imaging at 20 million A-scans and 4.5 GVoxels per second*. in: Opt. Express, OSA, Washington, pp. 14685–14704, 2010.

[3] D. Huang et al., *Optical coherence tomography*. in: Science, American Association for the Advancement of Science, Washington, pp. 1178—1181, 1991.

[4] W. Wieser et al., *Extended Coherence Length Megahertz FDML and Its Application for Anterior Segment Imaging*. in: Biomedical Optics Express, OSA, Washington, pp. 2647—2657, 2012.

[5] J. P. Kolb et al. *Megahertz FDML laser with up to 143nm sweep range for ultrahigh resolution OCT at 1050nm*. in: Proc. SPIE 9697, SPIE, Washington , 2016.

[6] D. Hillmann et al., *Aberration-free volumetric high-speed imaging of in vivo retina*. in: Scientific Reports, Nature Publishing Group, London, pp. 1–11, 2016.

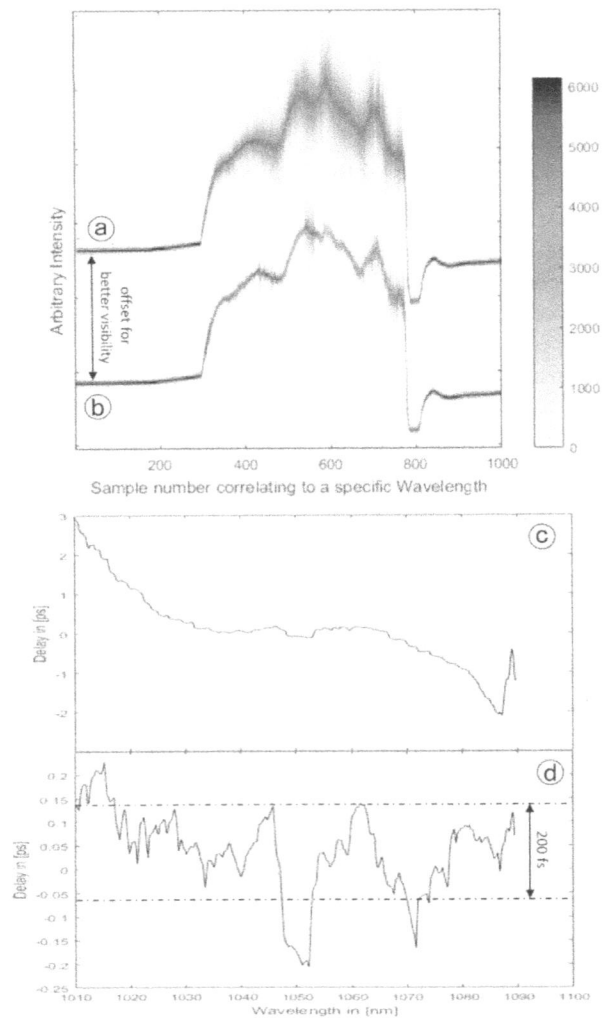

Figure 4: The signal of the photo-detector is displayed for the third order uncompensated (a) and compensated (b) dispersion. The width of the sweeps indicate the noise-level. As can be seen in (c) to (d) the remaining dispersion has been reduced from 5 ps to the sweet-spot range of below 200 fs for approximately 90% of the sweep.

Damage thresholds of retinal pigment epithelium (RPE) cells by laser irradiation (527 nm) in the microsecond time domain

D. Heinrich[1], J. Markmann[2], D. Theisen-Kunde[2], and R. Brinkmann[3]

[1] Medizinische Ingenieurwissenschaft, Universität zu Lübeck, devin.heinrich@student.uni-luebeck.de
[2] Medzinisches Laserzentrum Lübeck GmbH, {markmann,theisen-kunde}@mll.uni-luebeck.de
[3] Institute of Biomedical Optics, Universität zu Lübeck, brinkmann@bmo.uni-luebeck.de

Abstract

For treatment of retinal diseases selective retina therapy (SRT) was developed. It aims on damaging retinal pigment epithelium (RPE) cells selectively. Aim of this series of measurement is the investigation on thresholds of cell damage in the RPE depending on the laser pulse duration. Ex vivo enucleated porcine RPE explants were prepared and irradiated. Laser parameters used were pulse durations of 0,5 µs (15 - 131 µJ), 1,0 µs (23 - 208 µJ) and 1,5 µs (30 - 258 µJ). In order to investigate thresholds of cell damage on RPE layer the pulse energy for different pulse durations was varied. Viability assays were carried out on the irradiated RPE explants and evaluated in binary and ratio evaluation procedure via software by using an errorfunction. ED_{50}-values (effective dose) (162 - 383 mJ/cm^2 for binary, 186 - 492 mJ/cm^2 for ratio evaluation) varied depending on pulse duration and spot diameter. While binary evaluation is important for radiation protection, ratio evaluation is relevant for laser therapies.

1 Introduction

There are certain retinal diseases, which are conventionally treated by standard therapy of photocoagulation. By using laser irradiation in the millisecond time domain the tissue is damaged thermally. Proteins of the tissue degenerate and result in generation of scar tissue [1]. A negative aspect is that photoreceptors (Fig. 1 (a)) are destroyed and loss of vision occurs. Furthermore there is a chance of damaging the choroid (Fig. 1 (e)) which causes bleeding.

Selective retina therapy (SRT) is an alternative to photocoagulation for the treatment of diabetic macular edema (DME). Pulse durations below the thermal relaxation time are used which is necessary to damage retinal pigment epithelium (RPE) (Fig. 1 (c)) cells selectively. RPE cells damaged by irradiation result beneficially in migration and proliferation of surrounding cells which improve the metabolism [2]. Renewed RPE cells are thought of being a healing factor in multiple retinal diseases.

The mechanism and magnitude of cell damage in the RPE is dependent on the radiant exposure and pulse duration of the laser light. Around 40% of the energy of the light with a wavelength in the green part of the visible spectrum is absorbed in the RPE [3].

The thermal relaxation time defines the time after which the temperature peak in the center of a tissue is cooling down to a factor of e^{-1}. The adaption of the time of irradiation below the time domain, in which heat diffuses into nearby areas, allows the deposit of energy in a tissue without harming neighboring tissues like photoreceptors or the choroid (Fig. 1 (d)). Whether the RPE is damaged by thermal or thermomechanical effects depends on the pulse duration and en-

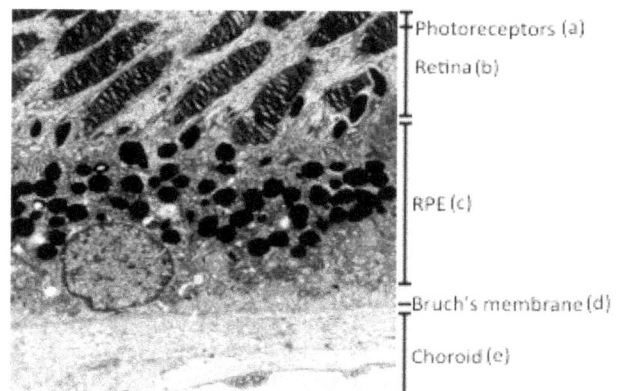

Figure 1: Arrangement of layers in the background of the eye. The innermost layer is the retina (b) followed by the RPE (c) consisting of a monocellular layer of RPE cells. Between the RPE and the choroid Bruch's (d) membrane is located. Bruch's membrane and the RPE layer build the blood-retina boundary. Behind it the choroid (e) builds the blood supplying layer (Adapted from [3]).

ergy. The thermal relaxation time of a melanosome is 400 ns, while for a complete RPE layer it is about 64 µs [4]. Using sufficient pulse energy the temperature peaks exceed the nucleation temperature of 145 °C on the surface of the melanin granule resulting in vaporization of the surrounding water and short disintegration of the melanosomes [4]. The expansion of the generated water vapor leads to the formation of microbubbles which damage the cell membrane mechanically due to expansion [5]. The lower the pulse duration the less energy is needed to exceed the nucleation temperature because the time domain gets too small for heat to diffuse.

Aim of this series of measurements is the investigation on thresholds of radiant exposure inducing cell damage in RPE cells by using single microsecond pulses of 0.5, 1.0 and 1.5 µs pulse duration.

2 Material and Methods

2.1 Setting and preparation

The frequency doubled laser light of a diode pumped solid state neodymium-doped yttrium lithium fluoride laser system was directed into a laser link, which was attached to a slit lamp, using a glass fiber. The slit slamp was positioned in such a way as the beam path was on axis to a contact lens attached to a cuvette in which a specimen holder was located.

After adjustment of the setting the pulse energies were measured with an energy detector head (J-10MB-LE, SN: 0049F12R, Coherent Inc., Santa Clara, California, USA). The detector head was fixated at the position of the sample inside the cuvette and the tip of the glass fiber was projected on its active area. Transmission of laser light inside the laser system could be varied by rotating a neutral density filter, which reflected percentage of the light. The reflected light was detected by a photo photodiode. Its signal was displayed via an oscilloscope (LeCroy WaveAce 2024, SN: LCRY2251C02055, Teledyne LeCroy GmbH, Heidelberg, Germany) to note reference values. By varying pulse durations and reference values the energies were measured and saved via the software EnergyMax PC (Coherent). Additionally spot diameters and spatial intesity variations were measured with a beam profile camera (BC106-VIS, SN: M00257581, Thorlabs Inc., Newton, New Jersey, USA). The camera was placed at the position of the cuvette such as the tip of the glass fiber was projected on its sensor area. By varying pulse durations and reference values multiple measurements were saved via the software Thorlabs Beam 4.1 with 12 bit resolution (Fig. 2). Settings of spot diameters could be noted by Thorlabs Beam 4.1 and were detected afterwards. The data was processed via software [MATLAB® R2016b]. Spatial intensity variations across the laser spot have to be quantified, which is done by the speckle factor. The definition of the speckle factor F is

$$F = \frac{H_{max}}{H_{mean}} \tag{1}$$

for a maximum radiant exposure H_{max} and a mean radiant exposure H_{mean} across the laser spot. H_{max} and H_{mean} were calculated for every pixel whose value is equal or superior to 13,53% of the maximum of all pixels (factor $\frac{1}{e^2}$). The speckle factor is used to correct the calculated ED_{50}-values (effective dose) - the radiant exposure for causing cell damage by a chance of 50% - to the corresponding maximum radiant exposure [6].

Samples of enucleated porcine eyes were prepared. By opening equatorially the front hemisphere was removed. A sample from the back hemisphere of a size of 1 cm^2 was cut out and placed in Ringer's solution. The retina was carefully removed with a pair of tweezers. The sample was placed in a specimen holder inside a cuvette filled with phosphate buffered saline (PBS).

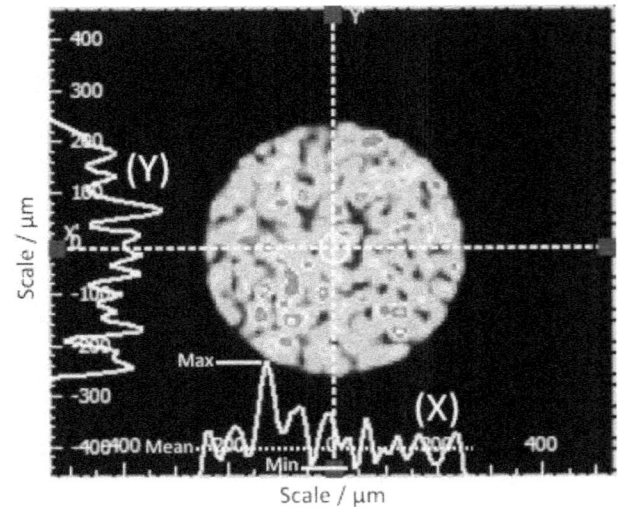

Figure 2: Measurement of a beam profile: Spatial intensity variation of the laser spot is shown for pulse duration of 0,5 µs with energy of 14 µJ. Additionally graphs of spatial intensity variations across the x-axis (X) and y-axis (Y) are shown. This beam profile shows H_{max} =21,7 mJ/cm^2, H_{mean} =7,6 mJ/cm^2 with a speckle factor of 2.8.

2.2 Irradiation and Viability Assay

Before irradiation marker lesions were applied on the tissue. Orientated on these markers pulses with specific energy and duration were applied. Either energy or duration was set constant to investigate dependency of cell damage.

After irradiation calcein-acetoxymethyl (calcein-AM) (C3100MP, Thermo Fisher Scientific Inc. Waltham, Massachusetts, USA, ratio of calcein-AM to PBS: 1:300) was added to the explant. Calcein-AM diffuses through cell membranes inside the RPE-cells because of its nonpolarity. Only vital cells contain esterases, which are able to process it to fluorescing calcein. These cells appear green on a fluorescence image. If cell membranes are damaged due to expansion of microbubbles, calcein-AM won't accumulate inside the cells and appears dark on a fluorescence image (Fig. 3) [6]. Fluorescence images were acquired with a fluorescence microscope (Eclipse Ti-E, SN: 10431294, Nikon Instruments GmbH, Düsseldorf, Germany) and saved via software NIS-Elements AR 4.13.05. The images were interpreted in two different procedures:

a) Binary evaluation: Every lesion was evaluated dichotomous. The value 1 means the damage of a cell compound of at least three cells. A lesion with a cell compound of less than three damaged and connected cells was assigned the value 0.

Table 1: The lower and upper 95% confidence bounds, $ED_{50,[1/0]}$-values determined by binary evaluation and $ED_{50,\%}$-values determined by ratio evaluation are shown for each combination of pulse duration and spot diameter used. Values for binary evaluation were multiplied to the specklefactor F (Eq. 1) before fitting the errorfunction for corresponding maximum radiant exposure. Values for ratio evaluation are mean radiant exposures H_{mean}.

Pulse duration	Spot diameter	Binary evaluation:			Ratio evaluation:		
		Lower 95% confidence bound	$ED_{50,[1/0]}$-value	Upper 95% confidence bound	Lower 95% confidence bound	$ED_{50,\%}$-value	Upper 95% confidence bound
[µs]	[µm]	[mJ/cm^2]	[mJ/cm^2]	[mJ/cm^2]	[mJ/cm^2]	[mJ/cm^2]	[mJ/cm^2]
1,5	200	203	219	235	298	318	337
1,5	150	278	383	468	477	492	509
1,0	150	269	317	352	380	391	404
0,5	200	133	162	188	177	186	193
0,5	150	172	194	216	249	255	261

b) Ratio evaluation: Every cell of every lesion was evaluated dichotomous and the ratio of damaged cells to the irradiated spot area was determined and evaluated.

Evaluations were realized via software Origin® 2016G: After plotting the dichotomous cell damage over the radiant exposures an error function $y = \frac{max}{2} \cdot (1 + erf(\frac{\sqrt{2}\cdot(x-l)}{w}))$ was fitted on the data to investigate on the transition point (Fig. 4). $Max = 1$ for the binary value of cell damage, $Max = 100$ for the ratio evaluation of cell damage. x are the radiant exposures used in the series of measurement. Using the *Least-Squares-Method* the values of l and w were calculated, for which the probability distribution approximates the data the best. The turning point l of the error function yields the ED_{50}-value and w provides information about the gradient at the turning point l. $ED_{50,[1/0]}$ of the binary evaluation procedure means the radiant exposure to apply lesions by a chance of 50% and $ED_{50,\%}$ of the ratio evaluation procedure means the radiant exposure to damage 50% of the cells in the irradiated area (Table 1).

Speckles in the laser spot caused changes of percentaged cell damage in the irradiated area (Fig. 1: Min to Max). If the percentaged amount of damaged cells was almost 0% the threshold of radiant exposure inducing cell damage was in the range of Max, because it covers the least area in the speckle field. The higher the amount of cells damaged the lower the threshold of radiant exposure in the speckle field. Therefore conversion factors were determined for every radiant exposure from Min to Max. Simultaneously the corresponding percentaged area in the speckle field was noted. In terms of a lookup table there was a conversion factor for every percentaged cell damage from 0% to 100%. Conversion factors were used to correct the mean radiant exposures used to the actual radiant exposure in the speckle field, which applied in a lesion, depending on the percentaged amount of cells damaged.

3 Results

Mean energies of multiple measurements were calculated for pulse durations of 0,5 µs (14 - 127 µJ), 1,0 µs (23 - 208

µJ) and 1,5 µs (30 - 258 µJ). Spatial intensity variations of the laser spot were quantified by the speckle factor, which was 3,2±0,3.

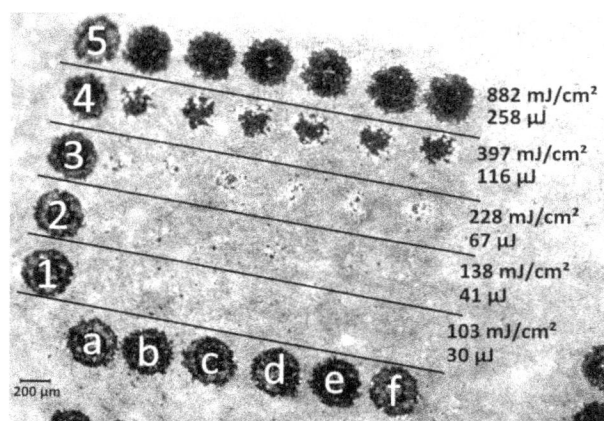

Figure 3: Example of a fluorescence image of RPE cells after irradiation. Marker lesions (Rows: 1-5, Columns: a-f) were applied. Six spots for each row were applied with same parameters. Pulse duration of 1,5 µs and radiant exposures of 104 to 882 mJ/cm^2 (row 1 to 5) were used. The radiant exposures are corrected in terms of variation of spot diameter, which was 193,1±0,4 µm.

In order to vary the radiant exposure per pulse duration the spot diameter (150, 200 µm) and pulse energy were changed. There was a total of 185 spots applied on the whole of three samples (as seen in Fig. 3) and evaluated to investigate thresholds of radiant exposure inducing cell damage (Fig. 4, Table 1). ED_{50}-values could be determined for both evaluation procedures. In the first procedure cell damage was count for a damaged cell compound of at least three cells which are 2,3% for 200 µm and 3,6% for 150 µm spot diameter.

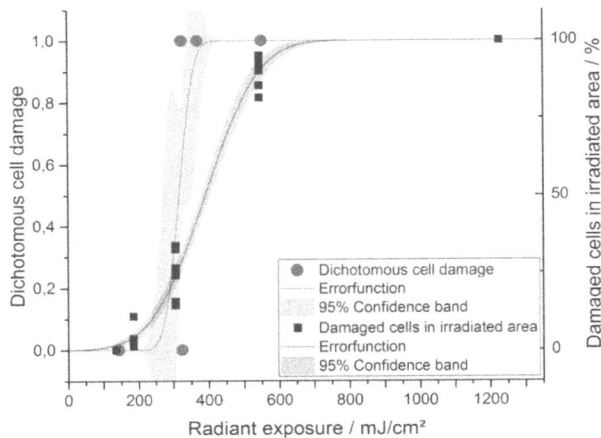

Figure 4: Plot of dichotomous cell damage per lesion and per cell over radiant exposure for 1,0 μs pulse duration and 150 μm spot diameter. Each is fitted with an errorfunction. Additionally the 95% confidence bands are shown. Binary cell damage is plotted over maximum radiant exposure $F \cdot H_{mean}$ (Eq. 1). Percentaged cell damage is plotted over mean radiant exposure H_{mean}.

4   Discussion

For the binary evaluation the speckle factor F concerned the $ED_{50,[1/0]}$-values. $F \cdot H_{mean}$ results in the maximum radiant exposure H_{max} in the laser spot which could have applied in a lesion while radiant exposures in the area around it did not damage RPE cells (e.g. Fig. 3 (2c)). H_{max} was determined before fitting the errorfunction to investigate on the $ED_{50,[1/0]}$-value. The ED_{50}-values rose for lower spot diameters because the heat arisen in the irradiated area could diffuse faster. The confidence bands could have been narrowed by investigation on radiant exposures in the domain near the ED_{50}-values.

To choose convenient parameters to apply lesions with certain magnitude of cell damage a top-hat beam profile would be best. The energy distribution across the laser spot would be constant resulting in a steeper fitting curve. Accordingly thresholds could be determined more accurate.

Determined thresholds are higher than in e.g. [7]. Damage of 50% of all cells occured at 140 mJ/cm^2 using 1,0 μs laser pulses with a speckle factor of 1,1 and spot diameter of 47 μm (which is 2,8 times lower than in this series of measurement (Table 1)). A reason could be that in this series of measurement no preselection of RPE samples was made. The explants varied in terms of pigmentation of the RPE cells which regulated the rate of absorbed energy at melanosomes.

5   Conclusion

For comparison of binary and ratio evaluation the speckle factor must be concerned. Irradiation can be performed with high speckle factors, but it should be included in thresholds of binary evaluation for corresponding maxi-

mum radiant exposure inducing cell damage. The outcome of this are different meanings of evaluation procedures: Binary evaluation is important for radiation protection, in which thresholds inducing minimal cell damage must be known. Ratio evaluation is more relevant for therapies, at which certain percentage of cells must be damaged.

5.1   Outlook

Prospective measurements will aim on testing longer pulse durations as this series of measurement covered a small part of low pulse durations.

Acknowledgement

The work has been carried out at Medizinisches Laserzentrum Lübeck GmbH.

6   References

[1] G. Schuele, M. Rumohr, G. Huettmann, R. Brinkmann, *RPE Damage Thresholds and Mechanisms for Laser Exposure in the Microsecond-to-Millisecond Time Regimen*, Investigative Ophthalmology & Visual Science, Vol. 46, No. 2, February 2005, p. 714

[2] J. Roider, N. A. Michaud, T. J. Flotte, R. Birngruber, *Response of the Retinal Pigment Epithelium to Selective Photocoagulation*, Archives of Ophtalmology, Vol. 110, December 1992, p. 1792

[3] V. P. Gabel, R. Birngruber, F. Hillenkamp, *Die Lichtabsorption am Augenhintergrund*, GSF-Report, Ludwig Maximilian University of Munich, 1974, p. 8 & 89

[4] R. Brinkmann, R. Birngruber, *Selektive Retina-Therapie (SRT)*, Zeitschrift für medizinische Physik, Vol. 17, March 2007, p. 54

[5] C. P. Lin, M. W. Kelly, S. A. B. Sibayan, M. A. Latina, R. R. Anderson, *Selective Cell Killing by Microparticle Absorption of Pulsed Laser Irradiation*, IEEE J. Select., Topics Quantum Electron., Vol. 5, Juli/August 1999, p. 963-968

[6] Schuele G., *Mechanismen und On-line Dosimetrie bei selektiver RPE Therapie*, PhD Thesis, University of Lübeck, 2002, p. 35

[7] R. Brinkmann, G. Huettmann, J. Roegener, J. Roider, R. Birngruber, C. P. Lin, *Origin of Retinal Pigment Epithelium Cell Damage by Pulsed Laser Irradiance in the Nanosecond to Microsecond Time Regimen*, Laser Surg. Med., Vol. 27, November 2000, p. 457

Development and Optical Simulation of a Needle Endoscope for Optical Coherence Tomography to visualize Adipocyte Cells during Coolsculpting

N. Meyer-Schell[1], H. Schulz-Hildebrandt[2], J. Rehra[2], M. Ahrens[3], M. Caspers[4], D. Manstein[4] and G. Hüttmann[2,3]

[1] Medizinische Ingenieurwissenschaft , Universität zu Lübeck, naja.meyerschell@student.uni-luebeck.de

[2] Institute of Biomedical Optics, Universität zu Lübeck, {schulz-hildebrandt,rehra,huettmann}@bmo.uni-luebeck.de

[3] Medical Laser Center Lübeck, Lübeck, Germany, ahrens@mll.uni-luebeck.de

[4] Wellman Center for Photomedicine, Massachusetts General Hospital, Harvard Medical School, {dmanstein,mjcasper}@mgh.harvard.edu

Abstract

Coolsculpting was developed to reduce fatty tissue by freezing the cells for a specific duration without any surgical intervention. The biological processes which happen during and after the therapy are not completely understood. OCT is an established diagnostic method in dermatology as it visualizes the dermis and epidermis well and might also be a good method to visualize adipose tissue. This paper investigates changes in extracted adipose tissue during cold exposure. A needle probe is designed and simulated–creating a method to investigate fatty tissue changes during coolsculpting.

1 Introduction

Fatty tissue is mainly composed of adipocyte cells that are classified as brown or white tissue. Fat in the human body is made predominantly of white tissue, which is located beneath the skin as subcutaneous fat, around organs and inter muscular. In healthy, non-overweight humans, white adipose tissue composes as much as $20\,\%$ of the body weight in men and $25\,\%$ of the body weight in women [1]. The main function of white adipose tissue is insulation because of its low heat conduction whilst a second important function of fatty tissue is energy storage. Although fat is essential for the human body, too much fat can lead to cosmetically problematic areas but also to hazardous health problems like diabetes or heart diseases [1].

Coolsculpting or cryolysis is a noninvasive method to reduce subcutaneous fat. It works by a controlled cool down of the adipocyte cells to $4\,°C$ through the skin. The extraction of the heat from the adipocytes leads to crystallization of the adipose tissue. This crystallization of the cytoplasmic lipids in adipocytes is already observed at $10\,°C$ which is well above the freezing point of tissue water [7]. The crystallization induces an inflammation response in the subcutaneous fat which last around a few months and leads to apoptosis of the adipose tissue. In its early inflammatory phase, panniculitis damages adipocytes whereas in its later phase, phagocytosis appears to account for removal of adipocytes and loss of fat tissue [3]. The great benefit, besides the noninvasive method of fat reduction with coolsculpting is that neither the dermis nor the epidermis is damaged, even though the cold is applied directly to the skin surface. The selectivity is caused by the specific inflammation response that is confined to adipose tissue, which has a greater sensitivity to cold than the dermis and epidermis [3]. Although good results have been achieved with coolsculpting, fat reduction was not always successful leading to the assumption that the biological processes underlying adipocyte cell death during cold exposure have to be further investigated to optimize coolsculpting treatment.

Optical coherence tomography (OCT) is an optical imaging modality that performs high resolution cross sectional imaging in biological tissue by interferometric measurement of backscattered light [6]. OCT was first developed to image the relatively transparent structures in the eye and later applied extensively to highly scattering tissue for morphological and functional imaging. As a medical imaging method, OCT is mainly used in opthalmology, cardiology and dermatology. OCT applications have been limited to the surfaces or lumina of organ systems, because the penetration depth of OCT is approximately $2\,mm$ in most tissues, hence it is not possible to image structures inside solid tissues or organs [8].

The objective of the presented paper is to visualize and investigate extracted adipose pig tissue during cold exposure with OCT and to design a side-viewing needle probe which enables visualization of the processes of cold exposure in fatty tissue in situ beneath the dermis. The key requirements for the needle endoscope are a diameter of $1\,mm$ to $2\,mm$, a free ranging length of at least $8\,mm$, a field of view of at least $75\,\%$ of the diameter and a cutting tip.

2 Material and Methods

In the following the approach of the cooling experiment is explained and the optical and mechanical considerations of the needle probe are introduced.

2.1 OCT Imaging System

The laboratory OCT imaging system (Telesto Thorlabs GmbH) was used for the measurement. With a center wavelength of 1325 nm and a bandwidth of 150 nm it provides deep image penetration in highly scattering sample that translate to 7.5 μm axial resolution at an imaging depth of 3 mm.

2.2 Cold Exposure

The back fat of a pig from a butcher was used to explore the feasibility of showing changes to adipose tissue by cold exposure. This fat was exposed to a cooling device that had as the cold applicator a rectangular stainless steel plate (40 mm × 40 mm). The rectangular plate was cooled with a cooling circulator (JULABO GmbH F10-HC) by circulating a cold antifreeze solution at $-10\,°C$ through the plate. The chiller was operating in constant temperature mode. To prevent heat exchange with the environment, the cooling applicator was covered with styrofoam and then grounded on the working table of the OCT system during the experiment as shown in Fig.1. Cold exposures were performed with 2 mm to 3 mm cut slices of fat which were first heated up to $30\,°C$ by a air heater to simulate body temperature. The fat was placed on the cooling plate and constantly cooled down until $4\,°C$.

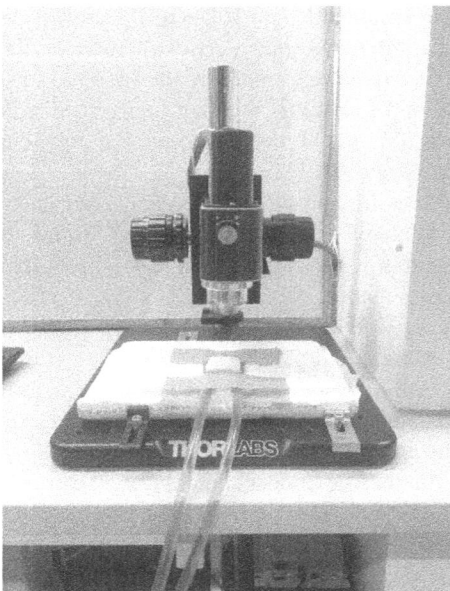

Figure 1: The setup of the cooling system

2.3 Optical Considerations of the Needle Probe

A GRIN-lens system implemented by a 4f-optic was used for the relay system as an optical consideration. GRIN lenses have a radial refractive index gradient with its maximum index at the center of the profile and, therefore, a plane surface which makes assembly easier [2]. Lenses with a diameter of 1 mm (key requirements) and a NA of 0.5 were chosen for the relay system. The refractive index gradient of the lenses were $n_0 = 1.616$ and a gradient constant $g = 0.687\,mm^{-1}$. The geometrical gradient constant g, characterizes the steepness of the index gradient with the lens length z_l. This together with the lens length is used to determine the focal length f and the working distance s, which can be calculated by (1) and (2). Right after the GRIN-lens system a prism is placed to deflect light with an angle of $90°$.

$$s = \frac{1}{n_0 \cdot g \cdot \tan(g \cdot z_l)} \quad (1)$$

$$f = \frac{1}{n_0 \cdot g \cdot \sin(g \cdot z_l)} \quad (2)$$

The optical simulation software OpticStudio™ 15.5, ZEMAX LLC was used for ray tracing to determine optical parameter.

2.4 Mechanical Considerations of the Needle Probe

2.4.1 Covering

The casing of the endoscope is made of a stainless steel pipe and covers the optical system in which the GRIN-lenses are mounted in. It should have at least a length of 1 mm and a diameter of 1 mm to 2 mm. Horizontally, at the top of the casing there is a notch in the pipe which serves as an optical window.

2.4.2 Needle Lace

The needle tip was designed to ensure a forceless penetration into the skin with minor tissue damage. Considering the thickness of the skin and therefore the minimal penetration depth for investigation of adipose tissue, a short needle tip of 1 mm was required. Two types of needle tops were introduced. The Lancet point which is part of the casing and looks like a hypodermic needle and the Trocar tip, a detached tip which is glued to the casing of the optical system (Fig.5).
Autodesk Inventor Professional® 2017 was used for modeling the design of the covering and the needle lace.

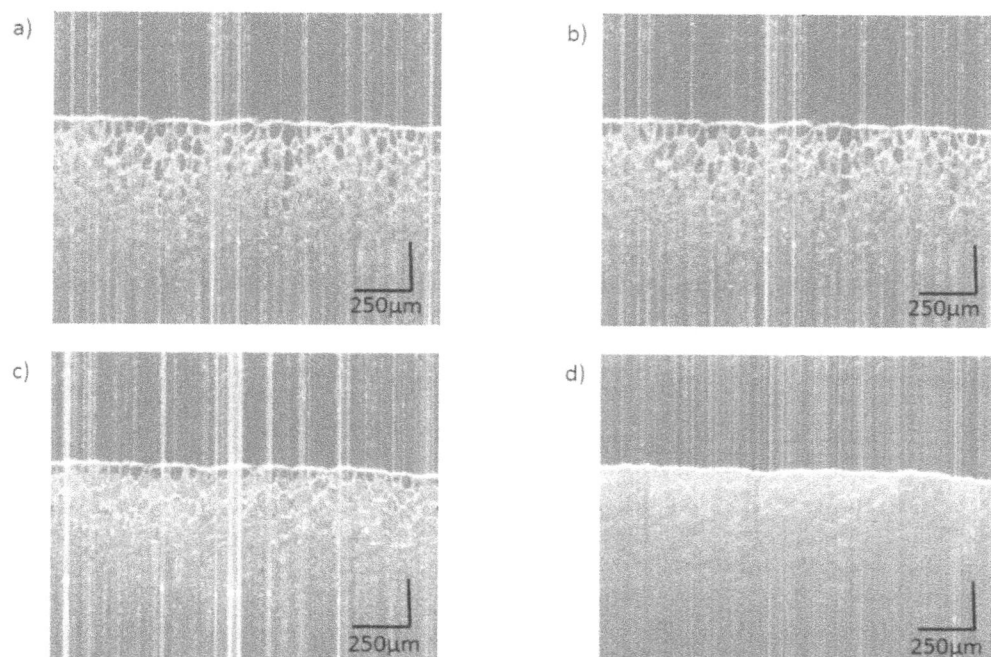

Figure 2: OCT images showing the process of crystallization, (a) Starting at $30\,°C$ the adipocytes are seen clearly (0th min), (b) $20\,°C$ adipocytes are still well seen (10th min), (c) $10\,°C$ starting of crystallization is observable (15th min), (d) $4\,°C$ fat cells are crystallized (25th min)

3 Results and Discussion

3.1 Cold Exposure

Cold exposure of the fat slices revealed noticeable changes in the adipose tissue. Fig.2 shows the results of the cooling procedure to four various points of time and temperature. As assumed, fat cells were seen clearly with the OCT (Fig.2a). It was observed that with temperature at around $10\,°C$ the fat bubbles are getting smaller and tighten together (Fig.2c). This process might show the crystallization of the adipose tissue, since it was described that crystallization of fatty tissue starts at temperature around $10\,°C$—significantly earlier than freezing of tissue water [7]. Fig.2d shows the completed crystallization process. Since this work was an ex vivo experiment, reduction of fatty tissue could not be observed. But since the crystallization is the decisive factor for the emergence of inflammation in adipocytes in vivo as described previously [6] [7], the results indicated that a needle probe might be a good method to visualize and investigate the process of cryolipolysis in vivo, as imaging through the highly scattering dermis is not possible. Therefore, this might be also a method to optimize the parameters (temperature and duration of treatment) of coolsculpting for better results of fat removal.

3.2 Optical Design

Considering and combining the key requirements and the optical considerations, an 8f-(four GRIN lenses) optical system was chosen for the optical design (Fig.3). The work-

ing distance of the needle endoscope should be between $0.5\,mm$ to $1\,mm$. The calculation of the working distance (1) of different lens lengths are shown in Table 1. Since the prism which is used for the deflection of the light is exactly placed right after the last GRIN lens, a lens length of $0.75\,mm$ were chosen to fulfill the working distance requirements. To ensure the required length of the relay system the considered 4f-system was extended to four GRIN lenses. The simulated optical system now has a total length of $14\,mm$, a FOV of $75\,\%$ of the lens diameter and an expected lateral resolution of $14\,\mu m$, which enables a good visualization of adipocytes, since adipocytes have a size of $100\,\mu m$. All key requirements are fulfilled.

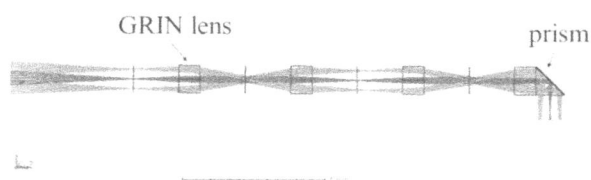

Figure 3: Ray Tracing

Table 1: Working Distance

Lens Lengths [mm]	Working Distance [mm]
0,5	2,518
0,75	1,590
1,0	1,098
1,25	0,179

3.3 Mechanical Design

The mechanical design of the endoscope shown in Fig.4 is composed of three parts. The inner part is made of three stainless steel pipes each with a length of 1.59 mm (working distance) placed between the lenses and act as spacers to keep the GRIN lenses in a constant distance. The spacers have the same outer diameter as the lenses and a wall thickness of 60 μm. The pipes will be glued to the surface of the GRIN lenses resulting in a fixed optical system. The outer part, the casing, is covering the optical system. This stainless steel pipe has a slightly greater diameter (1.1 mm, wall thickness 50 μm) than the lenses to ensure a push together of the lens system with the casing. The optical window lateral at the top of the casing has the dimensions of 1 mm × 0.5 mm .

Different needle tips are shown in Fig.5. Although it is not tested which of both tips is the better choice for the required considerations the Trokar tip (Fig.5a) (length 1 mm) was chosen for the needle probe. Since a typical hypodermic needle tip (Fig.5b) has an angle of 20° [5] and results in a too long tip (2.8 mm) for the considered requirements. Increasing the angle of the hypodermic needle for a shorter tip design would probably decrease the sharpness.

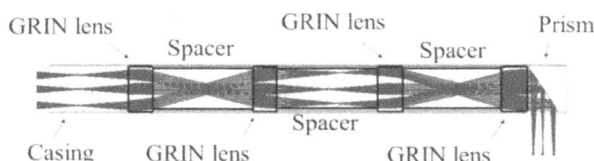

Figure 4: Schematic illustration of the mechanical components and the optical system

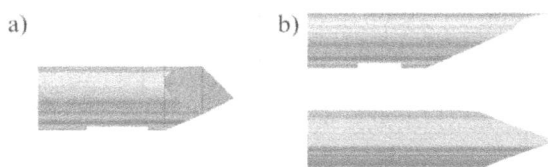

Figure 5: (a) Trokar tip (b) Hypodermic needle tip

4   Conclusion

In this paper it was demonstrated that OCT can be used to visualize changes of adipose tissue during coolsculpting. The crystallization of adipocytes was clearly observed. A needle endoscope was designed to investigate the processes of cryolipolysis with the future aim of optimizing coolsculpting. All optical and mechanical considerations meet the key requirements.

In future the design of the needle endoscope will be implemented and tested for animal use.

Acknowledgement

The work has been carried out at the Medizinisches Laserzentrum Lübeck GmbH, Lübeck, Germany.

5   References

[1] T. H. Schiebler, *Anatomie, Histologie, Entwicklungsgeschichte, makroskopische und mikroskopische Anatomie.* Springer, Berlin, Heidelberg, New York, 2005.

[2] H. Naumann, *Bauelemente der Optik.* Carl Hanser Verlag, München,Wien, 1992.

[3] B. Zelickson, B. M. Egbert and J. Preciado, *Cryolipolysis for Noninvasive Fat Cell Destruction: Initial Results from a Pig Model.* In: The American Society for Dermatologic Surgery, Wiley Periodicals, 35 pp. 1462–1470, 2009.

[4] X. Li, C. Chudoba and T. Ko, *Imaging needle for optical coherence tomography.* In: Optics Letters, Optical Society of America, Vol. 25, 2000.

[5] P. Han, D. Che and K. Pallav, *Models of the cutting edge geometry of medical needles with applications to needle design.* In: International Journal of Mechanical Sciences, Vol. 65 pp. 157–167, 2012.

[6] W. Drexler *Optical Coherence Tomography.* Springer, Berlin/Heidelberg, 2008.

[7] D. Manstein, H. Laubach and K. Watanabe, *Selective Cryolysis: A Novel Method of Non-Invasive Fat Removal.* In: Laser in Surgery and Medicine, 40 pp. 595–604, 2008.

[8] D. Lorenser, X. Yang and R. Kirk, *Ultrathin side-viewing needle probe for optical coherence tomography.* In: Optic Letters, Vol. 36, No. 19, pp. 3894-3896, 2011.

Design and Implementation of a Rigid Endoscope for Optical Coherence Tomography (OCT)

R. Tauscher [1], T. Pfeiffer [2] and R. Huber [2]

[1] Medizinische Ingenieurwissenschaft, Universität zu Lübeck, renke.tauscher@student.uni-luebeck.de

[2] Institute of Biomedical Optics, Universität zu Lübeck, {tom.pfeiffer,robert.huber}@bmo.uni-luebeck.de

Abstract

Optical coherence tomography is a non-invasive imaging modality which is mainly used in ophthalmology, early cancer diagnosis or dermatology. With the development of the ultra-fast swept-source OCT (SS-OCT) using Fourier Domain Mode Locked (FDML) lasers real-time 3D OCT imaging with multi-megahertz line rates was made possible [1]. A rigid endoscope for imaging with this OCT system was designed and tested. Therefore calculations and optical simulations had to be carried out in order to find lens combinations suited for endoscopic application. Two prototypes where built, one with 8 mm and one with 6 mm in diameter. The results show that the 6 mm endoscope performs significantly better almost achieving the diffraction limited resolution of 20 μm and continuously maintaining sharpness over the entire image field which could be suited for real-time 3D OCT imaging. The other prototype only achieves a lateral resolution of ca. 31 μm and ca. 60-70% of the outer part of the image is blurred.

1 Introduction

The Huber research group at the Institute of Biomedical Optics, Universität zu Lübeck has put emphasis on enhancing the ultra fast swept source FDML laser. One major research topic is its application for OCT. The OCT-system developed by this group is able to acquire, process and visualise 27 volumes per second. A volume is scaled into 320 x 320 depth scans where each of them has 400 pixels. This results in a volume rate of ca. 1 GigaVoxels per second which allows the fastest real-time 3D OCT imaging so far.

The FDML laser used for the OCT-system has a central wavelength of 1310 nm and a sweeping range of 120 nm. It can reach an A-scan rate of 3.2 MHz which means that the time necessary for one single A-Scan where the entire wavelength range is emitted is only around 0.3 μs [1].

The aim of this project is to develop and build a rigid endoscope that can be used to do real-time 3D OCT imaging since there are no such devices available yet. The extremely high A-scan rate demands the use of resonant galvanometer scan mirrors that operate at frequencies in the range of kilohertz.

For the design process of the endoscopic optics some considerations had to be made. One key feature of a rigid endoscope is its length. Inside its solid tube lies a relay-system that is crucial to transport the image towards its distal end onto an object. The length of the tube and the working distance are directly dependent on the focal lengths of the lenses of the relay-system. Two lens combinations that differ in focal length and diameter where chosen for the relay-system and their performance will be shown in this paper. Other important aspects of the scan optics, that had to be considered were, its image size or possible lateral resolution that is limited by the numerical aperture (NA) of the optics. These variables depend on the scan optics consisting of a lens collimating the beam and a scanning lens that focuses the collimated beam.

2 Material and Methods

The following section will give a closer understanding of the steps that were taken starting from preliminary considerations to the final construction of the endoscope prototypes.

2.1 Design of the Rigid Endoscope

For the design process of the rigid endoscope first the desirable characteristics had to be evaluated. The endoscope should have a diameter as small as possible but still allow for an image field as big as possible to be able to use existing body cavities and image a maximum area. The maximum endoscope diameter was set to 10 mm and an image field with a diameter between 4 mm and 6 mm was aimed for. Since the area of application of the endoscope was planned for the oral cavity an endoscope length between 70 mm and 100 mm was set to be adequate.

After specifying the desired parameters of the endoscope the next step was to find lens combinations with which they could be accomplished. Therefore it was necessary to take a deeper look into optical beam calculations and geometrical optics to estimate how the beam would propagate through the system.

The designed optics can be divided into three different parts.

First there is the scanning optics consisting of two lenses. In between them the second component is located which is the galvanometer mirrors that enable two directional scanning of a surface by moving the beam over an area. The third part is the relay-system which is necessary to transport the image over the desired distance.

The crucial parameter needed to start the calculations is the radius of the beam that is coupled out of the OCT-system through an optical fibre. It has a mode field radius of 9.2 µm for the central wavelength of 1.31 µm. This beam is collimated by the first lens of the scanning optics. The beam radius of the collimated beam can be calculated with the following equation:

$$\omega(z) = \omega_0 \sqrt{1 + \left(\frac{z\lambda_c}{\pi\omega_0^2}\right)^2} \quad (1)$$

$\omega(z)$ is the beam radius after the lens depending on the position z. ω_0 is the radius of the beam prior to the lens. λ_c is the central wavelength. With this equation it is possible to calculate the beam radius after every lens.

The collimated beam is scanned over the second lens of the scanning optics by the galvanometer mirrors.

The beam radius at the back focal point of the scanning optics can also be estimated by using (1). The numerical aperture (NA) of the system is calculated with

$$NA_2 = \frac{NA_1 f_c}{f_s} \quad (2)$$

where NA_1 is the numerical aperture of the fibre and f_c and f_s are the focal lengths of the collimation lens and the scan lens. Hence the NA depends on the two focal lengths NA_2 is needed to predict the diffraction limited lateral resolution of the system which can be assessed using the Rayleigh criterion.

$$D = \frac{1.22\lambda_c}{2NA_2} \quad (3)$$

where D is the minimal distance at which two points are still distinguishable. It can be seen that higher NA results in a better lateral resolution.

The diameter of the endoscope is estimated by

$$d_{min} = 2(\sin(\alpha)f_s + \omega) \quad (4)$$

where d_{min} is the minimal endoscope diameter, α is the half scan angle, f_s is the focal length of the scan lens and ω is the radius of the beam inside the relay-system.

The image field diameter d can be calculated using

$$d = 2\sin(\alpha)f_s \quad (5)$$

As can be seen through (1), (2), (3), (4) and (5) many of these parameters influence each other. So e.g. when a high lateral resolution is desired a small spot size and therefore a high NA of the system is needed. In (2) it can be seen that a high NA is achieved by choosing a long focal length for f_c and a short focal length for f_s. Unfortunately a decreasing value for f_s simultaneously results in a smaller field of view (5). That could be compensated by increasing the scanning

angle α but at the same time the endoscope diameter d_{min} could exceed the desired value.

A solution had to be found where all these correlation parameters would fit the predefined specifications.

2.2 Simulations

After the assessment of possible lens combinations for the scanning optics as well as the relay-system simulations which were carried out in order to get an even better perception on how the optics would behave. This was done with the help of the ray tracing software WinLens 3D Basic by QIOPTIQ, Paris, France.

This programme allows the user to test optical systems with several lens components ranging from simple lenses, mirrors to cemented triplets. WinLens 3D Basic features the so called Lens Editor where lens characteristics as surface radii, diameters, glass and spacings can be defined. That offered the possibility to simulate lenses from catalogues of different optical component manufacturing companies to estimate their performance before purchasing them. A variety of existing lenses could be simulated.

One way to evaluate the performance of the lenses was by using the full field spot pattern graph. Therefore the ray

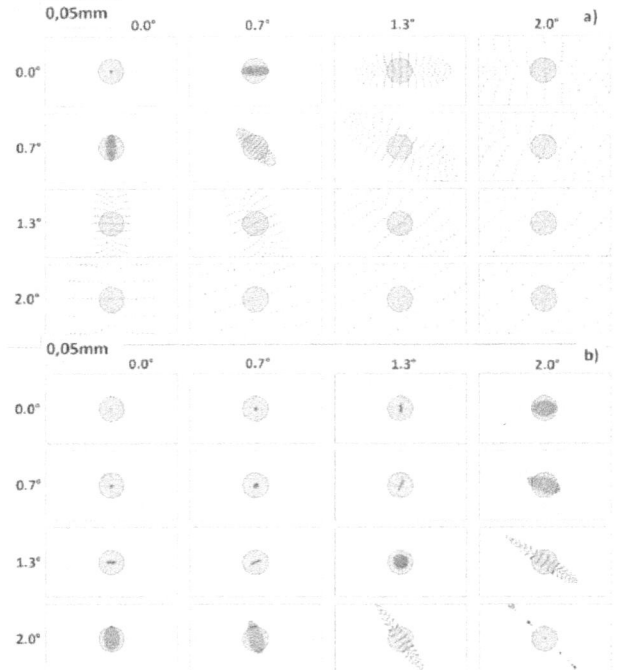

Figure 1: Full Field Spot Pattern of the a) 8 mm and b) 6 mm relay-system showing one quadrant of the image field each. The grey circle marks the diffraction limit of the system. It is the FWHM of the Gaussian intensity profile of the laser focus and calculated to be FWHM = 20 µm.

path was simulated with a definable object and a detector surface. The object emits of a selectable number of rays. The propagation of every ray and the influence that it experiences by the optical components is simulated. Especially the performance of the two different relay-systems were of interest. For that purpose a slightly simplified system was simulated leaving out the scan mirrors and the collimation lens. The incident object distance can be set to infinity and the radius can be defined according to a the calculated value resulting in a collimated source. The scan mirrors can be substituted by simply varying the object angle. Both can be done using the System Data Editor.

2.3 The Assembly of the System

After calculations and simulations were carried out and the final choice of the lenses has been made the system could be assembled.

An aspheric lens (C560TME-C) by Thorlabs, Inc. with a focal length of 13.86 mm was used as collimation lens. All lenses that were used can be found in the Thorlabs catalogue. In order to achieve collimation the end of the fibre must be in the frontal focal point of the lens. The collimated beam has a diameter of ca. 2.5 mm. (At this beam diameter intensity of the Gaussian beam has fallen to $(1/e^2)$ of its maximum value). It is scanned over a spheric scanning lens by the galvanometer mirror with a focal length of 50 mm. This choice of focal length results in a spot diameter of ca. 33 μm. The focal point of the scan lens is located between the two mirrors which is a compromise for not being able to image both pivot points of the mirrors at once with this system. This is the setup for the scanning optics. Two different relay-systems were connected to this system.

For the relay-systems two different combinations of achromatic doublets were chosen. They differ in their focal lengths and their diameters. In order to get a telecentric relay-system the correct spacing between the lenses has to be observed. By using matrix optics the cardinal points of the achromatic doublets could be calculated. Therefore the optical power of every one of the three surfaces of the doublet was calculated. The achromat was then assumed to consist of three thin lenses with the calculated optical powers separated by the thickness of the two components. In particular the front and back focal points (F/F') were of special interest because they determine the spacing in between the relay-lenses.

A relay-system in its simplest form can be built out of two identical lenses. When assuming thin lenses the length of the relay-optics is four times the focal length.

The first relay-optics was built out of two achromatic doublets (AC080-020-C) with F = -22.23 mm and F' = 15.90 mm for each lens. A relay length of ca. 72 mm can be achieved. The second relay-system is built out of four achromatic doublets (AC060-010-C) with F = - 8.53 mm and F' = 8.38 mm. Due to the shorter focal length using only two lenses would not be sufficient enough for the desired endoscope length. A relay length of 67 mm is achieved.

3 Results and Discussion

Imaging with both endoscopes was done using two linear galvanometer mirrors scanning over a half angle of 2.5° in horizontal and vertical direction (Fig.2 to Fig.4). The scanning angle is directly proportional to the given voltage. The lack of distortion, which would happen using resonant galvanometers allow for accurate measurement of the image field. For 3D real-time OCT imaging they can be easily replaced with resonant galvanometers.

The size of the image is limited by the chosen scanning angle and the radius of the relay-system. An image size of 5.6 mm and 5.4 mm were measured using the 8 mm and the 6 mm relay-system.

For determining the lateral resolution of both endoscopes a USAF Test Chart was used which can be seen in Fig.2.

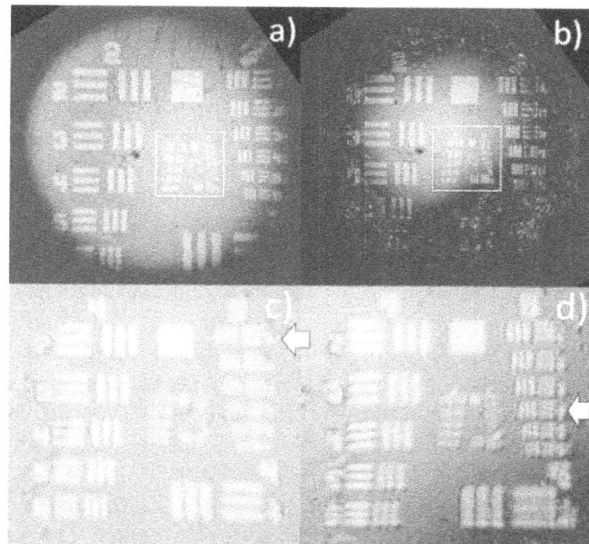

Figure 2: En-face images of a USAF resolution target a) using the 8 mm and b) using the 6 mm relay-optics. Underneath a) and b) zooms of the white boxes are shown in which the arrows mark where the lateral resolution was measured.

The USAF target is divided into six groups ranging from 2 to 7 and each group consists of six elements showing three horizontal and three vertical lines. Every combination of a group with an element determines a resolution. For the 8 mm endoscope a lateral resolution of ca. 31 μm was estimated by combining group 5 with element 1. The lateral resolution of ca. 22 μm was found out by choosing group 5 and element 4 (Fig.2c)).

The lateral resolution that was calculated is consistent with the one measured. The calculated lateral resolution on the basis of the lens combination for the scanning optics was ca. 20 μm using (2) and (3). Imaging with the two endoscopes shows that the lateral resolution of ca. 31 μm of the 8 mm endoscope does not reach the diffraction limited resolution. With the 6 mm endoscope a lateral resolution of ca. 22 μm is achieved which is close to the diffraction limit. In Fig.1a) and b) two spot patterns can be seen. They show a quadrant over 2° of the image fields. Although these plots were made using simulations on the basis of ray optics they help to evaluate the behaviour of Gaussian beams by manually adding the diffraction limited resolution described by the diameter of the grey disc. It is estimated by calculating the full width at half maximum (FWHM) of the Gaussian intensity distribution of the laser spot. The simulation predicted a increasing spot size exceeding the diffraction limit at ca. 0.7° for the 8 mm endoscope (Fig.1a)). Underneath is the spot pattern of the 6 mm endoscope. It shows a distinct difference in terms of ray distribution. Almost up until reaching the clear aperture of 2° it is within the range of the disc meaning that throughout the most part of the image the diffraction limited resolution can be maintained. The outcome of the simulations compare to the characteristics of the images which can be seen in Fig.4. There a) and b) show the en face z-projections of a finger done with the 8 mm and the 6 mm endoscope. They were acquired by the combination of the corresponding B-scans. One example for a B-scan image can be seen underneath for c) the 8 mm endoscope and d) the 6 mm endoscope. As predicted by the simulations (Fig.1a)) the lateral resolution of the 8 mm endoscope becomes increasingly worse towards the border. The blurring of the images can be identified easily. In the centre of Fig.4c) different structures can hardly be recognised. The images taken with the 6 mm endoscope (Fig.4c)) could also fit the simulations. The 2D-slice shows the maintaining of the diffraction limited resolution until the end of the image. The different tissue structures of the finger can be distinguished and even perspiratory glands can be seen more clearly. The surface of the finger shown on the 3D image also shows consistent sharpness throughout the entire image field.

The size of the area that can be imaged is ca. 200 µm greater in diameter using the 8 mm endoscope. Although a bigger area can be imaged the obvious drawback is the blurring of ca. 60-70% of the image emerging around the centre (Fig.3). The bigger diameter of the 8 mm relay lenses allows for a wider scanning angles of the galvanometer mirrors before the beam is clipping out of the optics which is responsible for the larger image size. The maximum half scanning angles that could be estimated by measuring the image data is ca. 2.1° for the 8 mm endoscope and ca. 1.9°

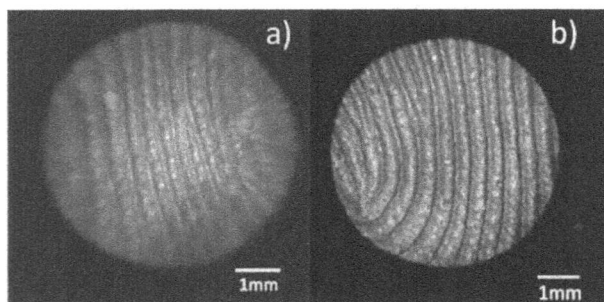

Figure 3: En-face image of a finger. a) 8 mm and b) 6 mm relay-system.

for the 6 mm endoscope. The lower right pattern of Fig.1b) shows that some rays clipped out of the optics.

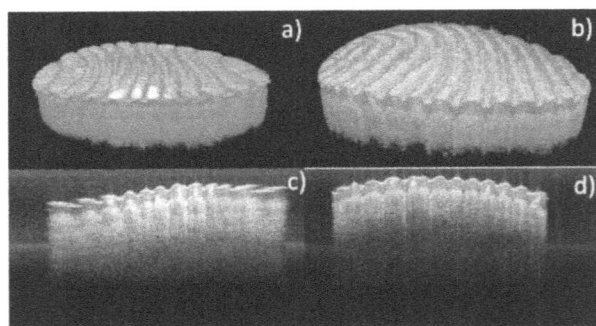

Figure 4: 3D images of a finger generated with the a) 8 mm and b) 6 mm relay-system. 2D slice (B-scan) through a finger is shown in c) for the 8 mm and d) for the 6 mm relay-system.

4   Conclusion

We were able to design and build two endoscope prototypes for OCT imaging. Preliminary considerations about the desired performance characteristics were made before sets of lens combinations could be simulated using ray tracing software. After simulating a variety of optical systems two lens setups were chosen for imaging. Both endoscopes fulfilled the desired requirements regarding image field size, endoscope diameter and length. However the 6 mm endoscope showed a significantly better performance regarding lateral resolution and image sharpness than the 8 mm endoscope.

Acknowledgement

The work has been carried out at Institut of Biomedical Optics, Universität zu Lübeck, Germany.

5   References

[1] W. Wieser, B. Biedermann, T. Klein, C. Eigenwillig, and R. Huber, "Multi-Megahertz OCT: High quality 3D imaging at 20 million A-scans and 4.5 GVoxels per second," Opt. Express 18, 14685-14704 (2010).

[2] R.Liang, *Optical Design for Biomedical Imaging*. SPIE Press, Bellingham, Washington USA, 2011.

[3] W. J. Smith, *Modern Optical Engineering*. McGraw-Hill, New York, USA, 2000.

[4] E. Hecht, *Optics. 4th Edition*. Addison Wesley Longman, Inc., Boston, USA, 2002.

[5] F. Pedrotti, L. Pedrotti, W. Bausch, H. Schmidt, *Optik für Ingenieure*. Springer-Verlag Berlin Heidelberg, Germany, 2008.

Optimization of a multi-wavelength stimulation unit for functional in-vivo full-field swept-source optical coherence tomography

Malte vom Endt[1], Clara Pfäffle[2,3], Dierck Hillmann[4], Hendrik Spahr[3], and Gereon Hüttmann[3,5,6]

[1] Medizinische Ingenieurwissenschaft, Universität zu Lübeck, Germany, malte.vomendt@student.uni-luebeck.de
[2] Graduiertenschule, Universität zu Lübeck, Germany, clara.pfaeffle@student.uni-luebeck.de
[3] Institut für Biomedizinische Optik, Universität zu Lübeck, Germany, {huettmann,spahr}@bmo.uni-luebeck.de
[4] Thorlabs GmbH, Germany, dhillmann@thorlabs.com
[5] Medical Laser Center Lübeck GmbH, Peter-Monnik-Weg 4, Lübeck, Germany
[6] Airway Research Center North (ARCN), Member of the German Center for Lung Research (DZL), Germany

Abstract

Intrinsic optical signals of single photoreceptors, detected by FF-SS-OCT showed characteristic intensity and exposure dependencies. These signals are assumed to be related to the visual phototransduction. To further assess the dynamics, there is need for high stimulation contrast and highly dynamic exposure, within safe ranges according to ophthalmic instrument standards. Additionally, multi-wavelength stimulation is desired, since it could be used to differentiate between rods and different types of cones. Therefore, an optic setup was built to combine the output of three narrow band LEDs (matched with the absorbances of s-, m- and l-cones) and project a pattern onto the retina of a test subject. Separate micro-controlled drivers were utilized in order to create a high degree of freedom regarding chosen LEDs, intensities and exposures. The built stimulation unit fulfills *DIN EN ISO 15004-2:2007*.

1 Introduction

Optical coherence tomography (OCT) is widely used by ophthalmologists to assess the health of human eyes [1]. And even though first functional applications like OCT-angiography were recently introduced, OCT is still mostly used to obtain only structural information [1]. Recently, it was shown that OCT can be used to obtain new in-vivo functional information in human retina [2]. These processes, which are related to light stimuli and thus the visual phototransduction, were visualized in humans for the first time. The promising results could be of high interest for a better understanding of the phototransduction itself and the detection of human retinopathies. Changes in the optical path length in the outer segments were measured by comparing phase data between the layer of the inner/outer segment of the photoreceptors and the retinal pigment epithelium (RPE). Since the observed changes in the optical path length were fast (ms regime) and small (nm regime), there are high technical demands on the used setup in order to obtain these signals. To fulfill the demands, a full-field swept-source optical coherence tomograph (FF-SS-OCT), based on [3], was used. The setup is visualized in Fig. 1. The former stimulation unit (SU) was implemented by a modified low-cost LED LCD projector (Tera UNIC UC40), which projected a pattern (PP) via a cold-mirror (CM) onto

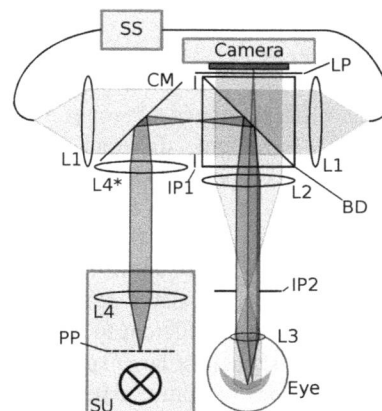

Figure 1: Mach-Zehnder type FF-SS-OCT setup with coupled stimulation unit (SU). L1 are collimation lenses for a fiber-coupled swept-source (SS). L4* couples the stimulation light via a cold-mirror (CM) into the sample arm of the OCT setup. A long-pass filter (LP) ensures that only IR light passes through onto the camera chip. With L3 acting as a model eye with $f = 20\,\mathrm{mm}$, L2 has a focal length of $f = 100\,\mathrm{mm}$.

the subjects retina. The low-cost SU suffered from lacking mechanical stability, suboptimal wavelengths at the color channels for photoreceptor stimulation and low stimulation

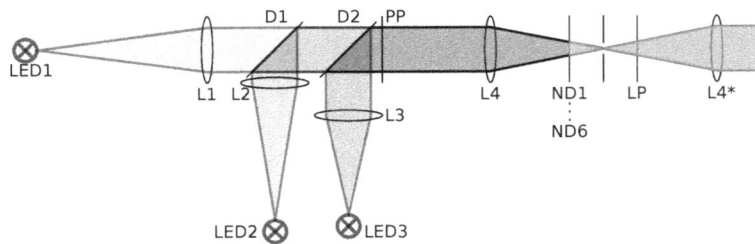

Figure 2: Optics of the stimulation setup: LEDs are projected via lenses (L1-L3) onto the pattern plate (PP) which is then projected into the OCT setup via lenses L4 and L4* (which corresponds to L4* in Fig. 1). LEDs 2 and 3 are coupled via dichroic filters (D1,D2). The output can be reduced by a neutral density filter-wheel (ND1-ND6). A low-pass filter (LP) removes wavelengths of 400 nm and below.

contrast. It is crucial to identify the different types of cones, which are responsible for vision at short-, middle- and long wavelengths. Therefore, stimulation with specific narrow-band wavelengths was implemented. Moreover there are demands regarding the quantification of the exposure and intensity dependency of the photoreceptor response. So there is a need for flexible exposure and intensity control of the SU, while maintaining secure intensity limitations based on ophthalmic standards. Lastly a mechanically stable integration into the FF-SS-OCT setup was required.

2 Material and Methods

The new SU is divided into three different components: Light sources, optics, and controllers.

2.1 Light Sources

There are three different types of cones within the human retina, which have three different absorption maxima [4]. Blue-sensitive cones (s-cones) at wavelength of 420 nm, green-sensitive cones (m-cones) at 534 nm, and red-sensitive cones (l-cones) at 564 nm. Rods were also shown with an absorption maximum at 496 nm, but will not be considered here, since the spatial resolution of the current FF-SS-OCT setup is insufficient to resolve them. The absorption curves in [4] suggest large spectral overlaps between the different types of cones, especially between m- and l-cones. Thus only narrow-band light sources were considered, in order to reduce cross-excitation of cone types. An additional demand was a narrow emission angle, so that the emitted light can be collected efficiently. In Table 1, the basic specifications of used LEDs are shown. Three of them have an emission angle of $2\Theta_{\frac{1}{2}} = 8°$, only B5B-433-B525 differs with $2\Theta_{\frac{1}{2}} = 15°$. The emitted spectrum is specified with $\lambda_{FWHM} = 30$ nm for two LEDs, while B5B-435-TL has $\lambda_{FWHM} = 20$ nm. B5B-430-JB is a white LED. While B5B-437-IX, B5B-433-B525 and B5B-435-TL are included for cone specific stimulation, B5B-430-JB is included to have a reference with former data which mainly used white light stimuli. All diodes were purchased from Roithner Lasertechnik GmbH, Austria.

Table 1: LEDs used in the stimulation unit.

LED	I_{ref}(mA)	λ_{peak}(nm)	(cd)
B5B-435-TL (LED1)	20	468	3.8
B5B-433-B525 (LED2)	30	528	10
B5B-437-IX (LED3)	50	630	13.5
B5B-430-JB (LED3)	30	450	7

2.2 Optics

The whole optical setup of the SU is displayed in Fig. 2. LED1 corresponds to B5B-435-TL (red), LED2 to B5B-433-B525 (green) and LED3 either to B5B-437-IX (blue) or B5B-430-JB (white). For B5B-430-JB (white) the dichroic filter D2 is replaceable by a mirror. Lenses L1, L2, and L3 collimate the LEDs onto the pattern plate (PP). The illuminated pattern is then projected via L4 and L4* into the FF-SS-OCT setup. L4 is the last lens that belongs to the SU. All lenses of the SU are Edmund Optics achromatic lenses, with focus lengths chosen, so that PP lies within the back focal planes of L1 to L3 and in the front focal planes of L4. The LEDs are all in the front focal plane of the corresponding lens. This was realized with 150 mm (L1), 125 mm (L2) and 88.9 mm (L3). L4 and L4* are both 100 mm lenses, so that the PP is imaged one-to-one into the intermediate image plane IP1, displayed in Fig. 1. By L2 in Fig. 1 which is 100 mm, it is again transferred one-to-one into IP2, and then due to the 20 mm focal length of the eye, imaged five times demagnified onto the retina. Dichroic long-pass filters were purchased from Edmund Optics and have 506 nm (D1 - Stock No. #67-080) and 562 nm (D2 - Stock No. #67-082) cut-off wavelengths. The long-pass filter LP (Schott GG400) blocks wavelengths of below 400 nm, in order to prevent ultraviolet hazards. The pattern plate was made of an aluminum plate (0.1 mm thick) with a milled-in pattern, which shows a cross.

2.3 Controller

A simplified block diagram of the controller is displayed in Fig. 4. The swept-source (Superlum Broadsweeper BS-840-1) and the camera (Photron FASTCAM SA-Z) require a trigger signal for each volume acquisition. This was realized by a microcontroller (Arduino Uno), which also con-

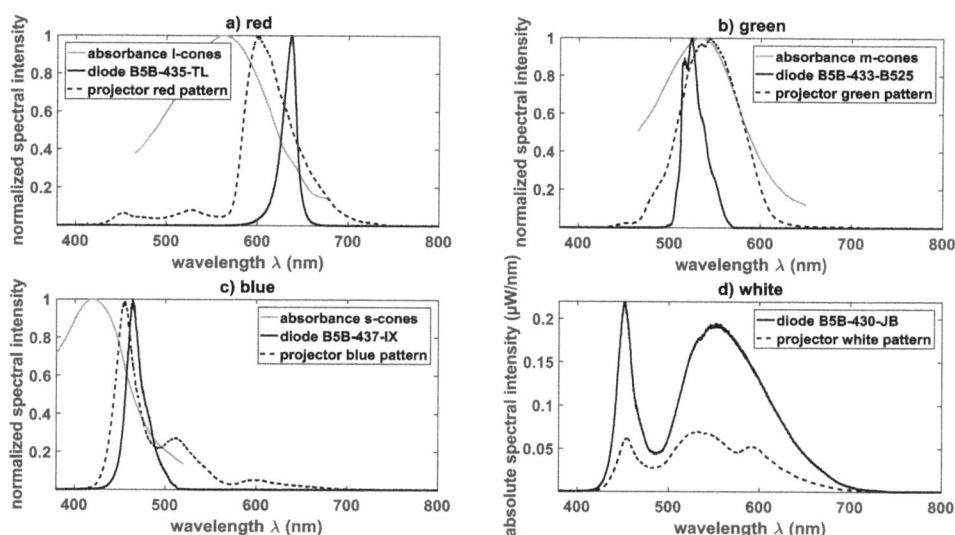

Figure 3: Spectral intensities of different light sources. Comparison between old and new light sources. a) to c) represent relative spectral intensities normalized to one. In addition, the absorbances of the corresponding type of cone is plotted in light gray. d) represents spectral intensities, obtained by spectrometric measurement, with integrals normalized to bolometrically measured powers. The white light diode displayed a bolometric output of $27.5\,\mu\mathrm{W}$ in contrast to $10\,\mu\mathrm{W}$ of the formerly used projector.

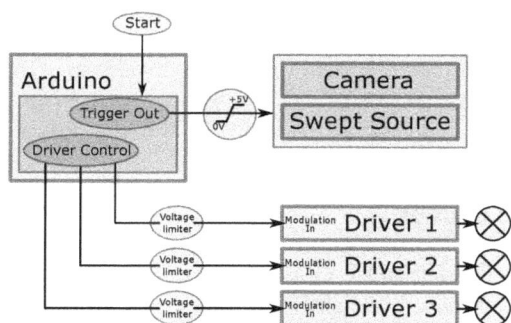

Figure 4: Simplified block diagram of the controller.

trolled the three light sources. To power the LEDs, three constant current laser-diode drivers (Lasertack SHS-2500) were used. These drivers have a modulation input and thus allow control of the diode-current within previously set limits. Since these drivers were built for currents between $500\,\mathrm{mA}$ and $2500\,\mathrm{mA}$, they were modified by replacing a built-in resistor. This resulted in a lower current range of approximately $5\,\mathrm{mA}$ to $500\,\mathrm{mA}$. According to the datasheet, this driver can be modulated with up to $250\,\mathrm{kHz}$ (TTL) or $500\,\mathrm{kHz}$ (analog). All electronic components were built into a housing with reverse polarity protection of the interfaces. Since the Arduino Uno does not provide analog outputs for driver modulation, the $5\,\mathrm{V}$ TTL signal from its digital outputs were connected to the modulation inputs of the drivers via potentiometers, acting as variable voltage dividers. This way, the desired maximum output can be defined, but not exceeded.

2.4 Ophthalmic instrument safety

To ensure a safe application, the output of the whole setup was assessed according to [5]. The LED drivers were calibrated, so that the maximum output is known and cannot be exceeded without opening and modifying the controller, or altering the optical setup. The maximum output of the setup was measured for each LED bolometrically with an Ophir Vega powermeter (P/N: 7Z01560) and spectra were measured with an Ocean Optics USB spectrometer (USB4000). The measured radiant flux were used to normalize the spectral integrals in order to get the correct spectral radiant flux for each LED. These could then be used to assess the output regarding [5]. Due to the $400\,\mathrm{nm}$ long-pass filter, there was no need to assess UV hazard. In order to fulfill the requirements of an ophthalmic instrument group 1 (no optical hazard by extended continuous wave exposure on the eye), the photochemical and the photothermal risks were assessed. Both risks were assessed by weighting the spectral radiant flux with the provided damage functions. The percentual usages of both limits have to be summed up and stay below $100\,\%$ in order to fulfill the standard.

3   Results and Discussion

Results are presented in Fig. 3 and Table 2. Table 2 gives an overview of possible output powers depending on diodes and diode currents. Since the driver had to be calibrated for safe operation, the outputs were measured for minimum (I_{min}) and reference currents (I_{max}). I_{exp} denote the currents that were used to calibrate the drivers. Photochemical hazards are the main limiting factor for visible light, based

Table 2: Measured powers in μW and resulting limit usage of DIN EN ISO 15004-2:2007 in percent for different diodes and diode currents.

	B5B-435-TL (red)	B5B-433-B525 (green)	B5B-437-IX (blue)	B5B-430-JB (white)
I_{max} (mA)	50	30	30	30
radiant flux (μW)	297	120	274	101.3
total limit usage (%)	4.71	10.89	415.37	46.09
I_{min} (mA)		6.2	3.8	3.8
radiant flux (μW)	not assessed	46.5	54.5	17.5
total limit usage (%)		3.84	78.33	7.4
I_{exp} (mA)	7.7	6.2	3.8	6.2
radiant flux (μW)	49.5	46.5	23.5 *(OD 0.4)*	27.5
total limit usage (%)	0.85	3.84	33.27	11.71

on [5]. Thus B5B-437-IX (blue) needed an additional neutral density filter with optical density of 0.4. Fig. 3 a to c show the measured relative spectral intensities for the three colored LEDs and compare them to the spectra that were obtained previously for the projector (colored pattern on black background). The spectra of the LEDs were measured and are displayed for I_{exp}, since a current dependency of the spectra was clearly visible. All colored LEDs provide more narrow-banded stimuli than the projector and have next to no spectral overlap. By comparing the very broad absorbances of the different types of cones, it can be seen that an approach is needed, that minimizes cross-excitation. Especially m-cones and l-cones have a large overlap. Even though B5B-435-TL is not ideal for stimulating l-cones, it should guarantee minimization of m-cone stimulation. Fig. 3 d compares the white LED with the projector (white pattern on black background). It can be seen that higher radiant fluxes are now technically possible and can be brought close to allowed limits according to [5]. Intensities emitted by each LED can be reduced individually by adding neutral density filters, or by dimming them down via the available potentiometers. Overall reduction of the output is possible via the ND-filter wheel (ND1-ND6). Stimulation contrast is expected to be much higher with the new PP compared to the LCD screen. The SU theoretically allows controlling the stimuli with 250 kHz, which would allow $4\,\mu s$ minimal exposure time. Driver modulation capability was verified by comparing the optical output, measured by a silicon photodiode (Thorlabs PDA100A-EC), to the TTL trigger signal with an oscilloscope (Tektronix TDS 210). The oscilloscope revealed a rising time of 15 us, whereof 8.75 us where caused by a constant delay. Falling time was measured to be considerably shorter with 2 us. Thus the current setup seems not to fulfill the specified 250 kHz, but is reasonably close. This provides headroom in terms of short excitation, since this is much shorter than the FF-SS-OCT maximum volume imaging rate of 167 Hz.

4   Conclusion

An versatile and safe ophthalmic stimulation unit - classifiable as ophthalmic instrument group 1, according to [5] - was created. Specific stimulation of the three different types of cones should now be possible. Exposure and intensities can be adapted for a large range. To further optimize the current setup, different LEDs could be considered. Goals could be to optimize the differentiation between s-cones and rods and m-cones and rods to provide rod-specific stimulation. The controller can be reused and calibrated for other LEDs and applications.

Acknowledgement

The work has been carried out and supervised at the Institute of Biomedical Optics, Universität zu Lübeck. Financial support of the BMBF (iCube, 98729873) and Thorlabs is gratefully acknowledged. We would like to thank all involved group members for providing valuable advice.

5   References

[1] W. Drexler and J. G. Fujimoto, *Optical Coherence Tomography*. Springer Berlin Heidelberg, 2008.

[2] D. Hillmann, H. Spahr, C. Pfäffle, H. Sudkamp, G. Franke, and G. Hüttmann, "In vivo optical imaging of physiological responses to photostimulation in human photoreceptors," *Proceedings of the National Academy of Sciences*, vol. 113, no. 46, pp. 13138–13143, 2016.

[3] T. Bonin, G. Franke, M. Hagen-Eggert, P. Koch, and G. Hüttmann, "In vivo Fourier-domain full-field OCT of the human retina with 1.5 million A-lines/s.," *Optics letters*, vol. 35, no. 20, pp. 3432–3434, 2010.

[4] J. K. Bowmaker and H. J. Dartnall, "Visual pigments of rods and cones in a human retina.," *The Journal of physiology*, vol. 298, no. 1, pp. 501–11, 1980.

[5] *Ophthalmic instruments - Fundamental requirements and test methods - Part 2: Light hazard protection (ISO 15004-2:2007); German version EN ISO 15004-2:2007*. Beuth, 2007.

Components towards fiber based Raman microspectroscopy

T. Eixmann[1], K. Karnowski[2], and D.D. Sampson[2,3]

[1] Medizinische Ingenieurwissenschaft Universität zu Lübeck, tim.eixmann@student.uni-luebeck.de

[2] Optical + Biomedical Engineering Laboratory, School of Electrical, Electronic and Computer Engineering, The University of Western Australia, Perth, WA 6009, Australia, {karol.karnowski,david.sampson}@uwa.edu.au

[3] Centre for Microscopy, Characterisation and Analysis, The University of Western Australia, Perth, WA 6009, Australia, david.sampson@uwa.edu.au

Abstract

The differentiation between different tissue types on-line in-vivo, e.g. cancer cells and the healthy surroundings, is a common goal of various optical techniques. Raman microspectroscopy is able to map specific molecule concentrations in a scanned sample but is limited to the penetration depth of light. There exists a strong demand to combine it with the practical ease of optical fibers to characterize samples inside the human body. Therefore a Raman spectroscopy system was built and published fiber probe designs were analyzed. Moreover, two basic concepts of Raman probes were considered. Possible noise sources were identified, the influence on the Raman signal characterized and ways to enhance the signal-to-noise ratio described. It is shown that single photonic crystal fiber based probes with gradient index fiber optics seem to be the best candidates for realizing high sensitivity small footprint fiber probes.

1 Introduction

Raman Spectroscopy is used to characterize specific molecules in a sample, to determine the concentration of a drug or even to distinguish between cancer tissue and the healthy surrounding. This originates from specific elastic scattering behavior of molecules which is based on vibrational energy levels. Raman spectroscopy techniques are non-invasive and marker free, which supersedes the use of in some cases toxic staining and enables tissue-conserving surgeries. For that reason, we used an existing spectrometer to build two kinds of setups. We characterized the system using a rectangular setup. Various noise sources were found, which decrease the signal-to-noise ratio (SNR) of the already weak Raman signal. We introduced optical fibers by changing the setup to a shared beam path.

1.1 Raman Spectroscopy

An incoming photon can be absorbed or scattered by a molecule, resulting in an oscillation of its charges with the frequency of the light (virtual energy state). Such an induced dipole emits light with the same frequency during relaxation into the ground state (elastic Rayleigh scattering) or a different frequency than the excitation photon: A higher frequency is a result of a relaxation from an excited rovibronic state onto the ground state (Anti-Stokes Raman scattering). A relaxation into a ground rovibronic state results in a lower frequency (Stokes Raman scattering) [1]. Because most molecules are in the ground-state at room temperature Stokes-Scattering has a far higher probability. Frequency-

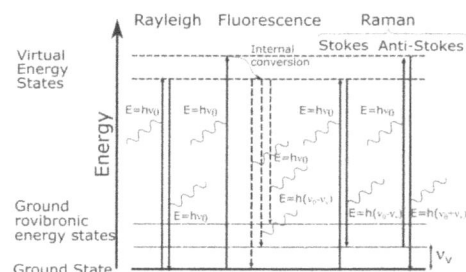

Figure 1: Jablonski Diagram with typical energy states of a molecule and the interaction with light during Rayleigh or Raman scattering process (adapted from [1]).

shifted light occurs usually 3 to 4 magnitudes less often than elastic scattered light and is therefore hard to detect. It is strongly wavelength dependent ($\approx \lambda^{-4}$). Also more probable than a frequency shift is a non-radiative transition of an absorbed photon into an excited rovibronic state to a singlet excited state. Afterwards the molecule relaxes back to the ground state by emission of a fluorescence photon [1]. Raman Spectroscopy is used to acquire the optical 'fingerprint' of molecules. This fingerprint is a result of characteristic vibration of a molecule or functional group. In combination with a microscope (Raman Microspectroscope) it provides label-free, non-invasive imaging with high chemical contrast [2]. The Raman signal is given as a Raman shift in wavenumbers (cm^{-1}) for normalization regarding the excitation wavelength. To overcome the limited penetration depth of the excitation light an optical fiber probe can be used [3].

1.2 Noise sources

As mentioned beforehand, the Rayleigh scattered light is predominant, but due to its known single wavelength it can easily be filtered out. In addition to the Rayleigh scattered light the autofluorescence background from the sample and also the Raman and fluorescence background signal generated by the fiber probe itself cover the weak Raman signal by magnitudes. Therefore, the optimal Raman fiber probe should generate an insignificant amount of Raman and fluorescence background not to decrease the already typical low SNR of the Raman signal.

2 Material and Methods

In the following, the basic components of the self-built Raman spectroscopy system and the different setups are explained.

2.1 Laser

To provide a high resolution a small wavelength is recommended, but these are predominantly absorbed by tissue resulting in higher fluorescence background and a lower SNR. Near-infrared light suppresses the fluorescence at the costs of a lower Raman signal. The best compromise between resolution and SNR is a 785 nm laser. We used an Integrated Optics 785L-41A high-power diode laser using volume Bragg gratings with a maximum output power of 500 mW and a central wavelength of 785 nm. The linewidth full width at half maximum (FWHM) is specified to be around 0.03 nm. The multiple transversal modes on the slow axis lead to a focusability of around 50 µm, preventing later experiments with single-mode fiber based systems (like hollow-core photonic crystal fiber based probes). To provide a variable laser power a simple attenuator based on a half-wave plate and a Glan-Taylor Polarizer was built. The intensity I_0 is attenuated to the intensity I by changing the angle ϕ between the polarizer and the half-wave plate.

$$I = I_0 \cos^2(2\phi) \tag{1}$$

2.2 Optical Filters

A bandpass filter was used after the attenuator to ensure that no other light source is coupled into the excitation path. Additionally, the Rayleigh scattered light and reflected excitation light needs to be filtered out before entering the spectrometer. For that reason, a long-pass edge filter (Semrock BLP01-785R-25) blocks light below 805 nm. For an excitation wavelength of 785 nm this edge corresponds to a Raman shift of 316.5 cm^{-1}, which is out of the fingerprint region of our interest. A dichroic mirror (Semrock LPD02-785RU-25) reflects the excitation light at 45° and passes the Raman-shifted wavelengths. The edge wavelength is 792.9 nm, which corresponds to a Raman shift of 126.92 cm^{-1} at 785 nm excitation wavelength. To further reduce the Rayleight scattered light the long-pass is still used in combination.

2.3 Spectrometer

In this study we used an automated, self calibrating imaging spectrometer (Jobin-Yvon/Horiba Triax 320). The spectral resolution with a 600 g /mm grating is 0.13 nm with a positioning accuracy of ±0.3 nm and a repeatability of ±0.06 nm. To ensure compatibility with the latest Microsoft Windows® OS a new software was programmed in National Instrument's LabVIEW™Version 2015 SP (32-bit). The Raman shift is calculated based on the assumption, that the gratings image the spectrum linearly onto the CCD chip, which is the case in the observed range. For calibration purpose the line of the excitation laser is imaged onto the CCD-Chip and the pixel position against the targeted wavelength monitored.

2.4 Setups

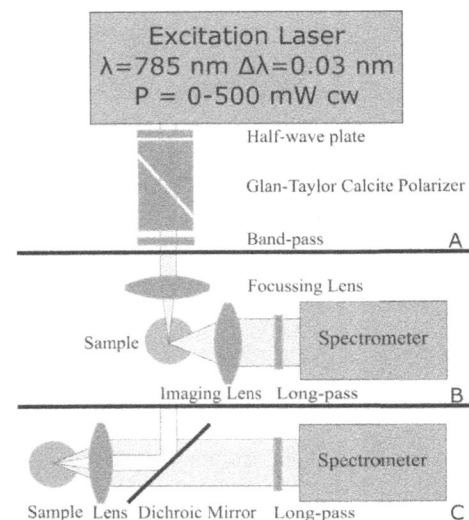

Figure 2: The schematic of both experimental setups used in this study. A) Excitation light source with adjustable power and notch filtering, B) Rectangular setup with two separate microscope objectives for illumination and Raman signal collection, C) Shared path setup with single path for excitation light and collection (can be used for both microscopic setup with microscope objective or with fibre probe).

The rectangular setup represents the standard setup for a Raman spectroscope [6]. The excitation laser is focused into the probe and the scattered light is collected with a second objective lens and guided to the spectrometer. This ensures that most of the excitation light is not coupled into the collection path and only the scattered light is acquired. An optical powermeter is used in transmission to adjust the laser power before it illuminates the sample. The basic system and component characterizations with liquids were done with this setup. The shared path setup [4] enables the analysis of surfaces of samples or the use of a single fiber probe for excitation and collection. Therefore, a wavelength depended beam splitter, a dichroic mirror, is used. This setup was compared to the rectangular setup and first experiments with single fiber probes were done.

2.5　Fiber probes

We evaluated two approaches for Raman spectroscopy fiber probes: a n-around-1 multimode fiber probe [2] [3] and a probe based on a single hollow-core photonic crystal fiber with additional optics [4] [6]. The first approach uses a single solid core multimode excitation fiber and in most cases n=6 multimode fibers around the central excitation fiber of the same type to collect the light. Additional filtering is necessary to enhance the Raman signal [2]. This probe type is restricted to a diameter of at least three fiber diameters and a significant amount of background is produced inside the fibers. The use of a low-dopant concentration in the fibers enhances the SNR by minimizing fluorescence light and Raman background. The smallest commercial available probes by EMvision LLC use on-tip filtering to ensure high signal quality [7]. Another approach is the use of single-mode hollow-core photonic crystal fibers (HC-PCF). Here the light is guided in air and only a negligible part is interacting with glass (down to less than 0.1 %). Therefore, Raman and fluorescence background signal produced within the probe itself is highly reduced. For future further reduction of the Raman background and to widen the transmission bands a new type of lattice needs to be used, the corresponding fiber is called Kagomé lattice hollow-core photonic crystal fiber [8]. Gradient index (GRIN)-lenses can be used as focusing optics, which can be made from GRIN-fibers with the same diameter spliced directly to the HC-PCF.

3　Results and Discussion

Isopropanol was used to compare the system with spectra given in literature. As shown in Fig.3, the acquired spectra were both comparable to the literature, but 0.6 nm shifted. This reveals an inaccurate calibration. The laser-line is not known exactly which may lead to this error, a calibration based on a known source such as a Mercury-Argon lamp is recommended.

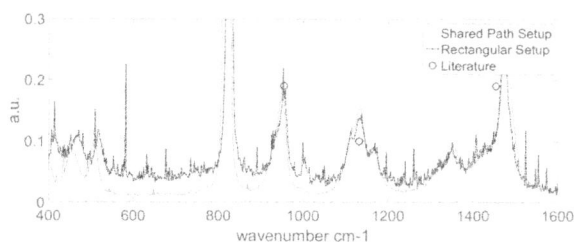

Figure 3: Insert of normalized Raman spectra acquired with the shared path setup (integration time 20 s, P=50 mW) and the rectangular setup (integration time 30 s, P=500 mW). For comparison characteristic peaks are taken from literature [9].

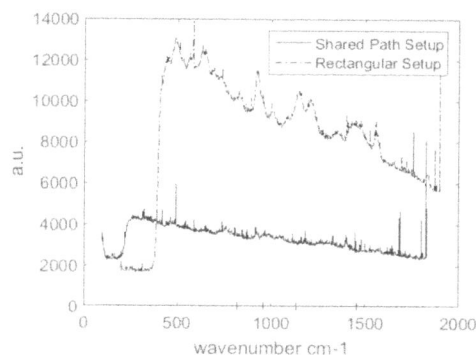

Figure 4: Raman spectra of a sugar solution made from 50 ml distilled water and 10 g brown sugar. The fluorescence signal is predominant. Shared path setup: integration time 20 s, P=50 mW, Rectangular beam setup: integration time 30 s, P=500 mW

3.1　Fluorescence

A major obstacle for tissue analysis is fluorescence. Water based solutions or contaminated samples exhibit the highest amount of fluorescence. The basic setups used in our experiment show, that the Raman signal of the fluorescing samples has an even lower SNR, which can lead to a total loss of the signal for too short integration times (Fig. 4).

A common approach to enhance the Raman signal is wavelength-modulated Raman spectroscopy (WMRS) [10]. Because the Raman light is a relative measurement, a minimal change in the excitation wavelength will change the Raman signal only. Since fluorescence signal will not change significantly for such small variations of excitations wavelength, it is possible with the use of principal component analysis to reduce fluorescence background and, as result, to enhance SNR of measured Raman spectra.

3.2　Fiber based Raman spectroscopy

The shared path setup was used to introduce fiber probes to the spectroscope. Therefore, the lens was replaced by a fiber port and a fiber pigtail. As expected, the fiber itself produced a high amount of Raman and fluorescence signal, which covered the whole sample signal. It was not possible to recover the signal by background subtraction. The background signal was analyzed and characterized. It is linearly depended on fiber length and on the used excitation power (Fig. 5). A gradient index fiber was used to focus the light into the sample. The Raman signal follows the gradient index fiber pitch with a sinusoidal behavior. Because the wavelength of the used laser cannot be modulated air-guiding optical fibers should be tested. Unfortunately, due to beam geometry of the laser, we were not able to couple in a sufficient amount of light into the hollow-core photonic crystal fiber.

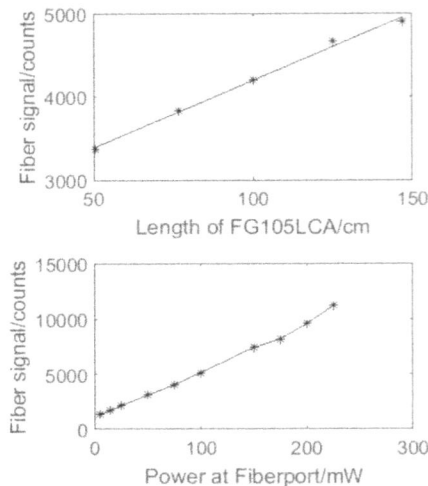

Figure 5: Mean of total spectrum counts with varying length of fiber (Thorlabs FG105LCA) acquired with 5 s integration time; P = 50 mW at sample and a fixed length of 33.7 cm with varying power acquired with 5 s integration time.

4   Conclusion

A Raman spectroscope was successfully built and the acquired spectra were comparable with literature. The goal to scan different samples with a fiber probe and acquire multiple spectra to distinguish between them was not accomplished, but the obstacles were characterized and possible ways to pass them are addressed. The fluorescence background prevented further investigation of tissue. To overcome this wavelength modulated Raman spectroscopy should be used in combination with a new wavelength modulatable laser [10]. With this technique new single lensed hollow-core photonic crystal fiber based probes with directly joined gradient index fiber sections may be tested. This should decrease the integration time and increase the SNR of the Raman signal. The background introduced by the multimode fibers covered the weak Raman signal by magnitudes, which wasn't solved by a decrease of excitation energy or fiber length. It has been shown that air-guiding probes increase the SNR significantly [6] and the use of Kagome hollow core photonic crystal fibers reduces the noise even further [4]. On-line in-vivo scanning Raman microspectroscopy would require an even shorter integration time (less than a second), which may not be reached with image comparison WMRS and the used spectrometer. The use of a locked-in amplified avalanche diode to scan only a specific wavelength or even a locked-in amplified line detector seems promising to acquire the differential signal directly.

Acknowledgement

The work has been carried out at Optical + Biomedical Engineering Laboratory, School of Electrical, Electronic and Computer Engineering, The University of Western Australia, Perth, WA 6009, Australia and supervised by the Institute of Biomedical Optics, Universität zu Lübeck

5   References

[1] B. E. A. Saleh and M. C. Teich, *Photons and Atoms*, pp. 423–459. John Wiley and Sons, Inc., 2001.

[2] I. Latka, S. Dochow, C. Krafft, B. Dietzek, H. Bartelt, and J. Popp, "Development of a fiber-based Raman probe for clinical diagnostics," in *European Conference on Biomedical Optics*, p. 80872D, Optical Society of America, 2011.

[3] I. E. I. Petterson, J. C. C. Day, L. M. Fullwood, B. Gardner, and N. Stone, "Characterisation of a fibre optic Raman probe within a hypodermic needle," *Analytical and bioanalytical chemistry*, vol. 407, no. 27, pp. 8311–8320, 2015.

[4] P. Ghenuche, S. Rammler, N. Y. Joly, M. Scharrer, M. Frosz, J. Wenger, P. S. J. Russell, and H. Rigneault, "Kagome hollow-core photonic crystal fiber probe for Raman spectroscopy," *Optics letters*, vol. 37, no. 21, pp. 4371–4373, 2012.

[5] S. Brustlein, P. Berto, R. Hostein, P. Ferrand, C. Billaudeau, D. Marguet, A. Muir, J. Knight, and H. Rigneault, "Double-clad hollow core photonic crystal fiber for coherent Raman endoscope," *Optics express*, vol. 19, no. 13, pp. 12562–12568, 2011.

[6] S. O. Konorov, C. J. Addison, H. G. Schulze, R. F. B. Turner, and M. W. Blades, "Hollow-core photonic crystal fiber-optic probes for Raman spectroscopy," *Optics letters*, vol. 31, no. 12, pp. 1911–1913, 2006.

[7] S. Dochow, D. Ma, I. Latka, T. Bocklitz, B. Hartl, J. Bec, H. Fatakdawala, E. Marple, K. Urmey, S. Wachsmann-Hogiu, and Others, "Combined fiber probe for fluorescence lifetime and Raman spectroscopy," *Analytical and bioanalytical chemistry*, vol. 407, no. 27, pp. 8291–8301, 2015.

[8] P. Russell, "Photonic crystal fibers," *science*, vol. 299, no. 5605, pp. 358–362, 2003.

[9] Spectral Database for Organic Compounds (SDBS), "Raman spectrum; SDBS No.: 2149; RM-01-00029:4880A.150M.Liquid." SDBSWeb http://sdbs.db.aist.go.jp. [date of access: 10.01.2017].

[10] B. B. Praveen, P. C. Ashok, M. Mazilu, A. Riches, S. Herrington, and K. Dholakia, "Fluorescence suppression using wavelength modulated Raman spectroscopy in fiber-probe-based tissue analysis," *Journal of Biomedical Optics*, vol. 17, no. 7, pp. 770061–770066, 2012.

Supercontinuum Generation with Sub-Nanosecond Pulses for Stimulated Raman Scattering Microscopy

T. Blömker [1], M. Eibl [2], and R. Huber [2]

[1] Medizinische Ingenieurwissenschaft, Universität zu Lübeck, torben.bloemker@student.uni-luebeck.de
[2] Institute of Biomedical Optics, Universität zu Lübeck, {robert.huber, eibl}@bmo.uni-luebeck.de

Abstract

Stimulated Raman scattering microscopy (SRSM) is a new label free imaging technology, which gained increasing attention in recent years. This paper presents a master oscillator power amplifier as a supercontinuum source used for SRSM. Narrowband 1064 nm pulses with tunable lengths and repetition rates are amplified in an all fiber based setup to peak powers of several kW. The amplified pulses generate nonlinear effects in standard single mode fiber. These effects are observed at slowly increasing pump currents at the multi-mode (MM) laser diode, to determine the spectral change, as well as the temporal evolution on the initial pulse at 1064 nm and the first Raman plateau at 1122 nm. The laser provided a high amount of power shift to longer wavelengths, wavelengths up to 1200 nm were obtained with low pump powers. Measurements have shown that the laser could be suitable as a versatile and affordable light source for SRSM.

1 Introduction

Raman spectroscopy enables label free identification of molecules. Each molecule possesses distinct vibrational and rotational energy states, which can be seen as a molecular fingerprint. Spontaneous Raman spectroscopy utilizes monochromatic light to excite the molecule into a virtual energy state. Following the excitation the molecule relaxes into an excited rotational/vibrational state. In this process photons with shifted frequencies occur, which yield information about the molecule. This technique provides information about the whole vibrational spectrum but suffers in signal strength due to its inherently low Raman cross section. Long acquisition times are required to achieve acceptable spectral resolution and signal-to-noise ratio [1],[2].

Stimulated Raman scattering (SRS) is a technique that enhances the Raman signal and thereby overcomes the limitation given by the low signal strength of spontaneous Raman scattering [2]. Besides the monochromatic excitation laser with the frequency ω_p a second laser with a frequency ω_s is utilized. The beams are referred to as pump and probe/Stokes beam. The sample gets irradiated with the probe and pump beam simultaneously. If a vibrational frequency ω_v matches $\omega_p - \omega_s$ the pump beam is attenuated and the probe beam gains power [3]. This gain is determined by the third order Raman susceptibility χ_R^3 [2]. The probe laser is either wavelength tunable or possesses a broad spectral bandwidth. The amount of frequencies ω_s determines the spectral scanning range covered by the system. Without a broad tunable laser *a priori* knowledge of the sample is required [3].

SRS is usable for label free microscopy. Most existing systems use a two laser setup with a monochromatic pump laser and a wavelength tunable probe laser [3],[4] or they use one laser but require a bulk-optical setup to broaden and shift the probe pulse [5].

Our goal was to setup and test an all fiber based laser, which provides both, the pump and the probe beam at the same time, without the need of a bulk-optical setup. Therefore a 250 ps 1064 nm pulse is amplified from 100 mW to some kW peak power in an all fiber based setup. The high power output pulse is emitted into standard single mode fiber, where nonlinear effects such as stimulated Raman scattering (SRS), self phase modulation (SPM), modulation instability (MI) and four wave mixing (FWM) occur. All effects combined should lead to a spectral broadening of the pulse and eventually to the generation of a supercontinuum. The optimal outcome would be a quasi monochromatic pulse at 1064 nm and broad spectral portions around the wavenumber where Raman bands are present.

2 Material and Methods

The setup is split into two main parts, the fiber based supercontinuum source and the measurement setup for the detection and characterization of the produced pulses.

2.1 Laser

The selected laser is an all fiber based master oscillator power amplifier (MOPA) (Fig. 1). In a MOPA setup a master oscillator creates a short optical pulse, which is amplified by power amplifiers to the desired levels.

An arbitrary waveform generator (AWG/TTi TGA12104) premodulates a semiconductor laser diode (Lumics

LU1064M200), which emits a spectrally narrow pulses around 1064 nm. Repetition rates range from 10 kHz to several MHz. The minimal produced pulse length of 10 ns however is too long, thus an electro-optical modulator (EOM/Photline NIR-MX-Ln-10) is utilized to shorten the pulse length to a minimum of 250 ps. The used EOM acts as a switch with an extinction ratio of up to 30 dB, operated by a second AWG (Waepond/DAX14000). At this point the peak power of the pulse is around 30 mW.

The first amplifier stage consists of 65 cm core pumped Ytterbium-doped fiber, which is backward pumped with a 975 nm laser diode (II-Vi LC96A74P-20R) delivering 400 mW power as a continuous wave (CW). The laser diode is coupled into the system with a wavelength division multiplexer (WDM), which allows the backward propagation of the 975 nm beam, while simultaneously protecting the 975 nm diode from the amplified 1064 nm pulse. Ytterbium pumped at 975 nm and seeded at 1064 nm behaves like a quasi four-level laser [6]. A problem of exciting Ytterbium at 975 nm is the strong spontaneous emission at 1030 nm [7], which further leads to more emission by 1030 nm. This effect is called amplified spontaneous emission (ASE). ASE limits the maximum achievable amplification within the fiber, however this limitation can be overcome by the use of a multistage amplification. A laser line filter after the first stage cuts out a 3 nm broad window around 1064 nm and suppresses everything else, hence the ASE does not get enhanced further.

The second stage resembles the first one with some small differences. The laser diode (Lumics LU0975M500-1002F10D) provides 500 mW at 975 nm to pump the Yb-fiber in backward direction. After the second stage an additional 1064 nm high power isolator is installed to protect the laser diodes from possible pulses in backward direction which could damage the laser diodes. The peak power of the pulse at the end of the second stage is about 40 W, which is not enough power to induce the desired nonlinear effects, thus more amplification is necessary.

The final amplification is done with a 5 m long double-clad (DC) Ytterbium fiber, which guides the pump light in the cladding. This allows higher pump powers without destroying the fiber. A high power 975 nm multimode (MM) laser diode (Lumics LU0975T090) with CW-powers up to 9 W is coupled counter-propagating to the pulse into the cladding with a DC beam combiner. The core with 10 μm diameter guides the 1064 nm pulse, thus the single mode character of the seed pulse is preserved. After the DC-stage several kW peak power can be achieved, depending on the current applied to the MM-diode.

These peak powers are high enough to induce nonlinear effects necessary for super continuum generation. SRS is one of the first observable effects inside the silica fiber, it generates a broad gain plateau around 1122 nm. An optional seed laser diode with 1122 nm (Innolume LD1122-FBG-400) is connected to the DC-stage. It can generate a pulse co-propagating to the 1064 nm pulse, which stimulates the Raman scattering in the silica fiber. A 1064 nm photon can lift a molecule into a virtual level, by interacting with an optical phonon. The depletion of this higher level by a 1122 nm photon leads to two 1122 nm photons. This stimulated emission changes the broad Raman plateau into narrow band pulses at 1122 nm.

To induce nonlinear effects standard several single mode fibers of different length are connected to the end of the DC-stage.

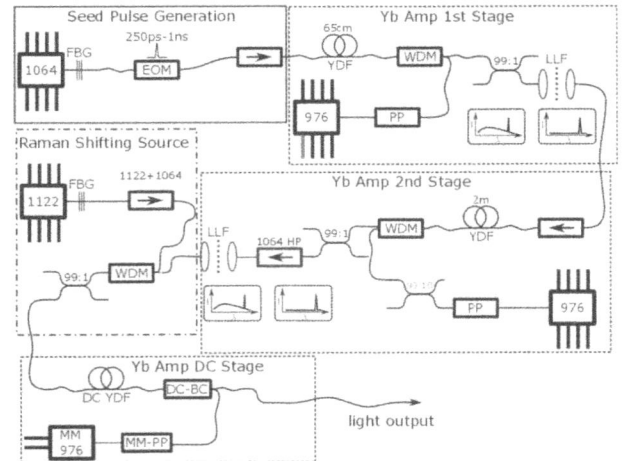

Figure 1: Setup of the used fiber based laser. A premodulated 1064 nm laser diode emits a 40 ns beam with 100 kHz repetition rate, which is reduced to a pulse length of 250 ps by an EOM. Each pulse gets pre-amplified by two single mode Ytterbium doped fiber amplifier stages, backward pumped at 976 nm. The occurring ASE background signal is blocked by a laser line filter after each stage. A DC-Ytterbium stage amplifies the pulse to the necessary level. The power of the MM-laser diode determines the strength of the occurring nonlinear effects.

2.2 Measurement

The measurement setup has to fulfill several requirements. The produced pulses with several kW peak power require attenuation to avoid instrument damage. This attenuation should be wavelength independent to ensure comparability between all wavelengths. Additionally the simultaneous measurement of CW-power, spectrum and noise at every occurring wavelength.

The incident beam is sent into a bulk optics setup via a collimator. A backside polished mirror splits it into a 96% and a 4% beam. The 96% are detected with a powermeter (Thorlabs PM100D) to constantly monitor the CW-power, while the 4% are guided to a mount for additional optical elements. These elements are either optical density filters to further attenuate the beam or spectral filters to observe a specific spectral region. A second collimator guides the attenuated beam into a single mode fiber, which is connected to an optical spectrum analyzer (OSA Yokagawa AQ6370).

A second bulk optics setup is utilized to split the pulse into single wavelength via a blaze grating with 600 lines/mm. The beam diameter and the grating parameter lead to a resolution of 1.4 nm. The single wavelength pulse is either coupled into the OSA, to determine the current wavelength or into a fast photodiode (Discovery Semiconductors DSC20H), which is connected to an Oscilloscope (Lecroy Labmaster 10zi 36GHz). This setup allows the determination of noise and peak power at each wavelength.

An existing problem of the bulk optics setup is the wavelength dependency of each collimation and the optical density filter. This problem is not totally avoidable but gets lessened by the attenuation with the backside polished mirror which is wavelength independent.

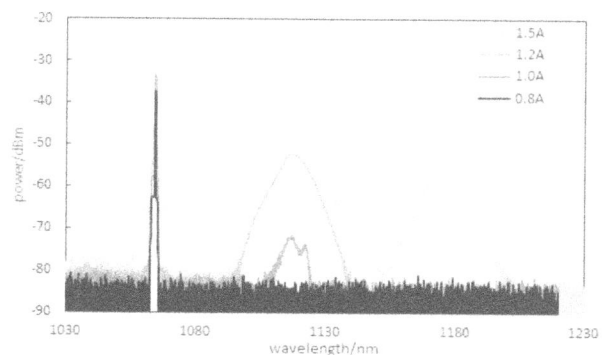

Figure 2: Spectral evolution with rising current at the MM-laser diode. Noticeable are the 3 nm socket around 1064 nm, caused by the ASE filtered with the 3 nm broad laser line filters, the rise and broadening of a first Raman plateau at 1122 nm (1.0/1.2 A), as well as the second raman plateau at 1180 nm (1.5 A). Additional side bands occur around 1064 nm at high peak powers.

3   Results and Discussion

Before generating a super-continuum the influence of the occurring nonlinear effects on the pulse at 1064 nm and the whole emerging spectrum are characterized. The development of the effects is observed with increasing current at the MM-laser diode and different fiber lengths.

3.1   Spectral Evolution

All measurements were performed with a repetition rate of 100 kHz and a pulse length of 250 ps. The current at the MM-diode is slowly increased in 100 mA steps. At each current the resulting peak power is estimated by factoring the CW-power and the inverse duty cycle.

$$\frac{1}{D_c} = \frac{1}{250ps \cdot 100kHz} = 40000 \quad (1)$$

Figure 2 shows the interesting steps in the produced spectra with 20 m Hi1060 fiber spliced to the DC-stage. At 0.8 A the 1064 nm pulse is amplified to about 0.44 kW peak power. The 3 nm broad socket around 1064 nm is the laser line filtered ASE background. The pictured proportion is misleading because of the CW-nature of the ASE and the integration measurement of the OSA. In comparison the socket should be lowered by factoring in the duty cycle (46dB). At 1.0 A (1.1 kW) the characteristic Raman gain of silica fiber arises around 1122 nm. After this point the rise of the peak power at 1064 nm is halted, more pumping rather leads to power shifting to longer wavelengths. The broad Raman gain rises and broadens with increasing peak power. At 1.5 A (3.9 kW) the spectrum shows a second broad plateau around 1180 nm. The frequency shift of around 14.4 THz corresponds to the first one from 1064 nm to 1122 nm, i.e. the power at 1122 nm reaches a point, where a second Raman band is generated. Modulation instability causes side bands around the narrow band 1064 nm pulse.

3.2   Temporal Pulse Evolution

The pulse development is observed at 1064 nm and 1122 nm, therefore the grating is rotated to couple the selected wavelength into the photodiode connected to the oscilloscope. The current powering the MM laser diode is slowly increased, while the pulses are monitored.

Figure 3. depicts the change of the pulse-shape. The pulse-shape at 1.0 A correlates to the shape present without activating the DC-stage. The front edge is rather steep, while the falling edge possesses a distinct shoulder, caused by the shape of the pulse operating the EOM. An increase of the current leads to an increase in power shift, as shown in Figure 2. At 1.06 A the power plateau of the 1064 nm pulse starts to decline. A further increase in pump power leads to a stronger decrease of the pulse, however both edges, especially the falling edge, retain power. This development might be caused by the initial pulse shape, while the power in the middle of the pulse is high enough to Raman-shift, the power in the edges is too small and retains its original wavelength. With increasing power the formerly low powered parts of the pulse-shoulder gain more power.

The pulse evolution at 1122 nm corresponds to the behavior of the 1064 nm pulse. When the decline starts, the pulse at 1122 nm arises. The first generated pulses possess a shorter pulse length, which correlates to the hole appearing in the 1064 nm pulse. An increase in the pump current leads to a pulse broadening up to the pulse length of the initial pulse but without the appearance of a shoulder at the falling edge. At 1.5 A the pulse-shape exhibits the same decline in the center as the 1064 nm pulse. In comparison to Figure 2 this marks the point, at which energy gets shifted into the next Raman plateau.

This behavior indicates, that there's a distinct threshold power, after which all the power would get shifted to the next higher level. Due to the non-rectangular pulse shape however, a portion of power remains at the initial wavelength.

3.3   Effect of Fiber Length

To characterize the effect of the fiber length on the occurring spectra, the Raman threshold was measured for the 20 m fiber (Figure 2 1.0 A). Afterwards the fiber was cut back to

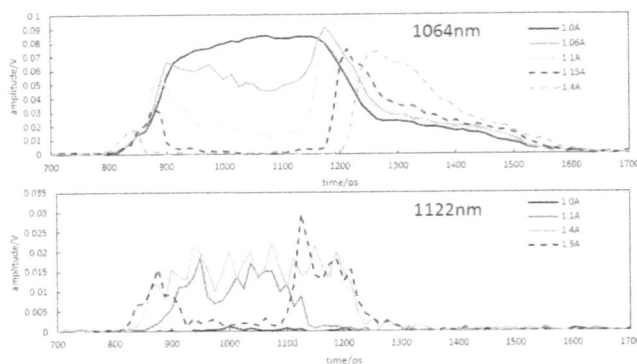

Figure 3: Plot of the pulse development at 1064 nm (top) and 1122 nm (bottom) with increasing current at the MM laser diode. The pulses correspond to the spectra shown in Figure 2.

10 m and 5 m and the Raman threshold was measured as well.

The Raman threshold for 20 m Hi1060 fiber is 1.12 kW (Figure 4). For 10 m fiber the threshold increases to 1.92 kW and for 5 m even further to 3.76 kW. The measurement indicates, that longer fiber lengths are beneficial for the generation of nonlinear effects. The Raman threshold shows a nearly inversely proportional relationship. The discrepancy could be caused by the nonlinear effects occurring in the short fiber, which connects the laser and the measurement setup. The influence of these effects rises if the initial examined fiber is shorter.

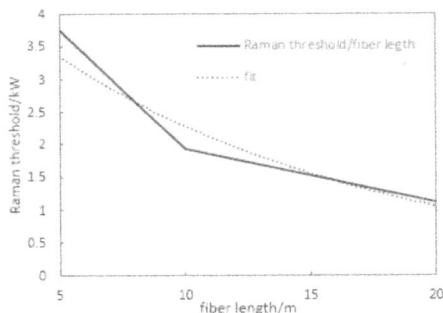

Figure 4: Raman threshold measured with different lengths of standard Hi1060 fiber. The threshold was defined as the point when the Raman plateau reaches a power of about -70 dBm with a sensitivity of -85 dBm and a resolution of 0.2 nm. A bisection of fiber length leads to a doubling of power needed to generate the Raman shifted wavelength.

4   Conclusion

The measurements show promising results for further tests regarding supercontinuum generation and stimulated Raman spectroscopy with the utilized laser source. The laser source is compact, highly tunable regarding pulse length and repetition rate and fairly low-cost in comparison to other comparable sources. The reached power even at rather low pump currents is sufficient to generate nonlinear effects in a standard single mode fiber. An even further increase in peak power could lead to more nonlinear effects and a further spectral extend.

The outcome shows, that a longer fiber length is beneficial, due to the faster initiation of the nonlinear effects, such as Raman shift. This allows more spectral broadening at lower peak powers. Wavelengths exceeding 1180 nm were reached with 20 m Hi1060 fiber at a peak power of about 4 kW over the whole spectrum.

Further tests will include the influence of different fiber parameters, such as zero dispersion wavelength and mode field diameter, the observation of the noise at single wavelength, as well as measurements with higher pump currents at the DC-stage, to stretch the spectrum even more.

Acknowledgement

The work has been carried out at the Institute of Biomedical Optics, University of Lübeck, Luebeck, Germany.

5   References

[1] F. Saltarelli et al. *Broadband stimulated Raman scattering spectroscopy by a photonic time stretcher*. In: Optics Express Vol.24, No.19, Optical Society of America, pp. 21264–21275, 2016.

[2] D. W. McCamant, P. Kukura and R. A. Mathies, *Femtosecond Time-Resolved Stimulated Raman Spectroscopy: Application to the Ultrafast Internal Conversion β-Carotene*. In: Journal of Physical Chemistry A, Vol 107, No.40, pp. 8208–8214, American Chemical Society, 2003.

[3] Y. Ozeki et al. *High-speed molecular spectral imaging of tissue with stimulated Raman scattering*. In: Nature Photonics, Vol. 6, pp. 845–851, 2012.

[4] C. W. Freudiger et al. *Label-Free Biomedical Imaging with High Sensitivity by Stimulated Raman Scattering Microscopy*. In: Science, vol. 322, pp. 1857–1861, 2008.

[5] E. Ploetz, S. Laimgruber, S. Berner, W. Zinth and P. Gilch, *Femtosecond stimulated Raman microscopy*. In: Appl. Phys. B 87, Springer-Verlag, pp. 389–393, 2007.

[6] H. M. Pask et al. *Ytterbium-Doped Silica Fiber Lasers: Versatile Sources for the 1-1.2 μm Region*. In: IEEE Journal of Selected Topics in Quantum Electronics Vol. 1, No. 1, pp. 2–13, 1995.

[7] R. Paschotta, J. Nilsson, A. C. Tropper and D. C. Hanna, *Ytterbium-Doped Fiber Amplifiers*. In: IEEE Journal of Selected Topics in Quantum Electronics, Vol. 33, No. 7, pp. 1049–1056, 1997.

Evaluation of filter-combinations within a DNA-Quant-Module for Next Generation Sequencing

B. Andersen[1], E. Sarofim[2],
[1] Medizinische Ingenieurwissenschaft, Universität zu Lübeck, bjarne.andersen@student.uni-luebeck.de
[2] Roche Diagnostics International Ltd, emad.sarofim@roche.com

Abstract

Next generation sequencing (NGS) describes a new technology of DNA-sequencing with the capability of multiple sequencing reactions at the same time. NGS requires a sample preparation and therefore a determination of the DNA concentration. For a DNA-Quant-Module, which is used for fluorescence imaging, different filter combinations for excitation and emission light are evaluated to fulfil the demands of a fast an accurate measurement. Using a workflow with two calibration steps, the DNA-Quant-Module can calculate the DNA concentration of an unknown sample. These samples are stored in a 96 multiwell-plate and the DNA concentration can be measured for each well individually. The filter evaluation considers the integration time of the camera, the limit of detection and the signal to concentration rate. Beside the optical parameters, the properties of different multi-well plates are also discussed.

1 Introduction

DNA-Sequencing describes a technique of determining the order of DNA nucleotides. For next generation Sequencing solutions, a library preparation is necessary for a successful DNA sequencing protocol [1]. This preparation may be done manually and time-consuming or fully automatically [2]. One essential step for sequencing is to determine the DNA concentration of a sample to ensure that enough material is available for the sequencing process [3] [4].
Therefore, a DNA-Quant-Module is currently in development which automatically measures fluorescence of samples within a 96-well multiwell-plate and calculates the DNA concentration afterwards. The module consists of a place for a multiwell-plate, a light source and a detection unit (Fig. 1). With the use of filters in front of the optics one can change the bandwidth of the excitation and emission light. The excitation light excites every sample in the

Figure 1: 3D rendering model of the DNA-Quant-Module. The multiwell-plate is located in the top left side.

multiwell-plate equally. After setting the integration time of the camera, the Quant-Module then measures the fluores-

cence light after excitation. Before measuring the concentration of an unknown sample, the module first needs to be calibrated. First, a "Dark"-plate with a DNA-concentration of 0ng/well is inserted in the Quant-Module. A second "Reference"-plate with a fix known DNA-concentration in each well is measured afterwards. The DNA-concentration of an unknown "sample"-plate can then be calculated using (1).

$$C_{DNA} = \frac{F_{Sample} - F_{Dark}}{F_{Reference} - F_{Dark}} \cdot C_{Ref} \qquad (1)$$

where C is the DNA-concentration and F is the measured fluorescence value.
The Quant-Module allows a total of 4 possible filter pairs for light transmission which are mounted on a filter wheel for automatic and fast change during runtime (Fig. 2). Finding a good default filter-combination for fluorescence measurements is the goal of this paper. Important parameters to look at are the integration time of the camera, a high signal to background ratio and a low detection limit. The properties of different multiwell-plates are also discussed.

2 Material and Methods

For the sample preparation, the Qubit® dsDNA HS Assay Kit was used. These reagents have a high photostability with <0.3% drop in fluorescence after 9 readings [5]. For an excitation wavelength of 450nm, the emission spectrum is shown in Fig. 3, as well as the excitation spectrum for an emission at 550nm. The first filter combination to test is used in a standard PCR amplification and detection instrument. This combination shall be compared with a new com-

Figure 2: Filter Wheel for light transmission. 4 possible filter combinations with 2 filters each results in a total of 8 filters.

Figure 4: Wavelength range for excitation and emission light for filter combination 1

bination of filters with a wider bandwith. The third choice is a combination of both the excitation filter of the standard instrument and the emission filter of the new filter combination. This results in a total of 3 different filter combinations which has to be tested.

The range of excitation / emission wavelengths for each filter combination is given in Table 1. Considering the white-LED lightsource of the Quant-Module, the excitation spectrum of the dye and the transmission of the different filter combinations, the excitation range for filter combination 1 is given in Fig. 4 and for filter combination 2 in Fig. 5. The emission range is also included considering the sensitivity of the CCD-camera, the emission spectrum of the dye and the emission of the filters.

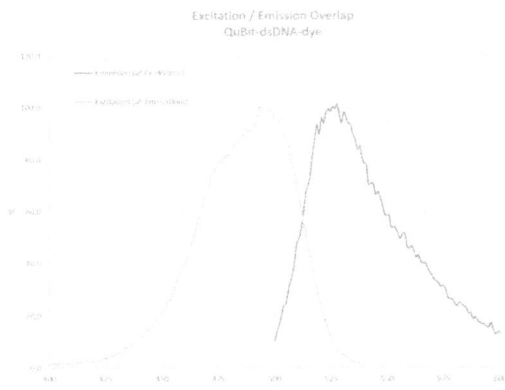

Figure 5: Wavelength range for excitation and emission light for filter combination 2

2.1 Experiment setup

For the test workflow, three black flat-bottom 96 multiwell-plates with a diameter of 6.39mm per well were used. Each well was filled manually with 100µl of the sample solution The first plate - referred as "dark"-plate - was filled with a sample solution with a DNA-concentration of 0ng DNA in each well. The second, "reference"-plate was filled with a DNA-concentration of 50ng DNA in each well. These two plates were used to calibrate the DNA-Quant-Module. The third and last plate was filled with a dilution series: The first three columns of the multiwell-plate were filled with a DNA-concentration of 0ng DNA in each well. Followed by a DNA-concentration of 3.28 ng per well for the next three columns. The following three columns with a DNA-concentration of 8.12ng per well and the last three columns with a DNA-concentration of 16.25ng per well.

Figure 3: Excitation and emission spectrum for two wavelengths

2.2 Experimental procedure

Each plate was measured using the same Quant-Module. First the reference-plate with the highest DNA-concentration was used to determine the maximum possible integration time, which does not lead to an overssaturation of the camera. The Quant-Module itself returns an saturation value, which is near one, if one or more pixels are oversaturated.

To make the test case interchangeable with other Quant-Modules, the saturation should be around 50% for the

Table 1: Range of excitation / emission wavelengths and bandwidth (nm)

Filter combination	Excitation (nm)	Emission (nm)
1	470/30	514/30
2	460/80	540/80
3	470/30	540/80

reference-plate for each filter combination. Therefore it is safe to use other Quant-Modules which might have a slightly different optics behavior, but using the same integration time without an oversaturation.

In order to measure the signal to background ratio for each filter combination, the arithmetic mean value over each well of the reference-plate was divided through the arithmetic mean value over each well of the dark-plate as shown in (2).

$$\text{Signal to background ratio} = \frac{\bar{x}_{Reference}}{\bar{x}_{Dark}} \quad (2)$$

where $\bar{x}$ is the arithmetic mean value.

The limit of detection was determined by using the dark-plate also as the sample-plate. Using (1) to calculate the DNA-concentration, the received data should theoretically be 0 for every well. Every deviation from that value will give information about the camera noise, which results to the limit of detection. We then defined the Limit of detection as 3 times the standard deviation of every well of the sample-plate, as shown in (3). The limit of detection was also measured with the sample-plate rotated by 180 degrees. This simulates a typical workflow with errors which might occur due to an inhomogeneous sample preparation.

$$\text{Limit of detection} = 3 \cdot \sigma_{Sample} \quad (3)$$

where σ is the standard deviation.

The dilution series was used to visualize the relation between signal and concentration. The axis intercept and the slope will be of interest.

3 Results and Discussion

The bandwidth of filter combination 1 is much smaller than the bandwidth of filter combination 2. Therefore, the integration time for the first combination is significantly higher to receive a camera saturation of 50%. As the last filter combination is a mixture of the other ones, the integration time is in between. For every filter-combination, the integration time is listed in Table 2. As the Quant-Module has a build in limit for the integration time at 5000ms, a lower value to reach the 50% saturation is an advantage.

Table 2: Camera integration time for a saturation of 50%.

Filter combination	Integration time (ms)
1	2500
2	300
3	900

Fig. 6 shows the result for the signal to background ratio. The first filter combination has the highest relation of 56.8. Followed by the third combination with 53.2 and the second combination with 41.8.
The results for the limit of detection is given in Fig. 7. The

filter combination 3 shows the best results in both the simple orientation with 0.021ng/well and the rotated orientation with 0.089ng/well.

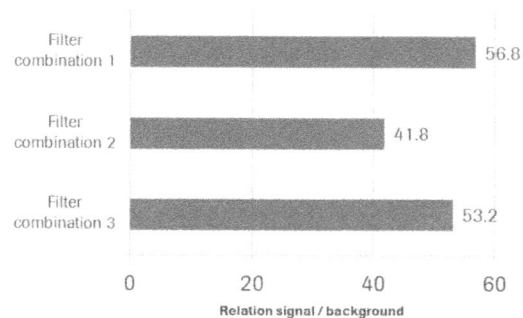

Figure 6: Relation between signal and background measurements using the dark-plate and reference-plate.

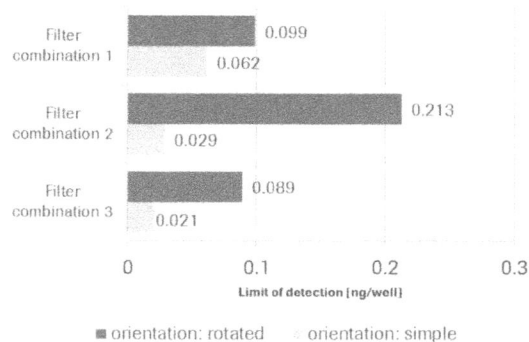

Figure 7: Limit of detection using the dark-plate as the sample-plate. We defined the limit of detection as 3 times the standard deviation of every well.

The first filter combination provides the best signal to background ratio whereas the third filter combination provides the best limit of detection. Looking at filter combination 1 and 3 there is not much of a difference in the test results except of the integration time, which is significantly lower using filter combination 3.
Possible errors may occur during the manual sample preparation, resulting in a contamination or different filling level in each well. Also the Quant-Module itself may have inhomogeneous fiberoptics which could lead to different results. As for now, there is only one prototype of the Quant-Module and the test case would be more accurate with more modules of the same type. Also an automatic sample preparation might be a better choice over the manual preparation, but calculating the mean value over all 96 measurements already reduces the influence of errors.
The shorter integration time for filter combination 3 and a good limit of detection were the crucial points of choosing this combination over filter combination 1 and 2.
The measurements for the dilution series are shown in Fig. 8. It shows the calculated concentration related to the measured signal for filter combination 3. On the axis intercept

one can see that the dark-sample generates 33 detections on average. This value is also an indication for the limit of detection as higher values will lower the contrast of the measurement.

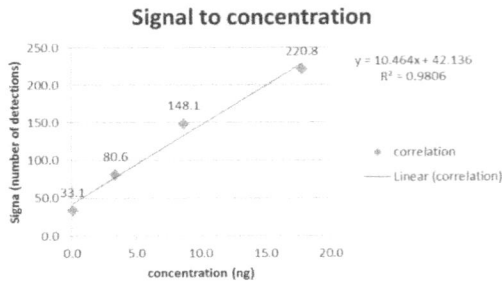

Figure 8: Signal to concentration ratio for filter combination 3

3.1 Influence of different multiwell-plates

Another improvement for the signal to noise ratio may be the use of 96-multiwell plates with a smaller diameter. Using filter combination 3 and a black 96-multiwell-plate with a diameter of 4.38mm instead of 6.58mm, the test case to receive the signal to noise ratio was repeated. The diameter of the plate was choosen to double the height of the head of the liquid.

As shown in Fig. 9, the ratio was increased once again from 41.8 to 89.1. Although a higher signal to background ratio is a benefit, the influence of certain errors may be higher. Contaminations may settle down on the bottom of the well and have a higher influence on the result when the diameter is smaller. Additionally an accurate positioning of the plate on the Quant-Module is more important with a smaller diameter. Inaccurate positioning may lead to blocked parts for the excitation-light. As the positioning will be also part of an automated solution, the multiwell-plate must fit perfectly into the enclosure of the Quant-Module.

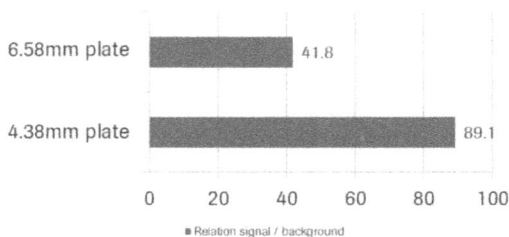

Figure 9: Relation between signal and background measurements using a plate with a smaller diameter.

4   Conclusion

In this paper, three different filter-combinations for a DNA-Quant-Module were evaluated, which will be used to au-

tomatically measure fluorescence and calculate the DNA-concentration of an unknown sample. The final filter-pair combines a short integration time with a good signal to background ration and a low limit of detection.

4.1   Outlook

As further shown, a multiwell-plate with smaller diameter might also improve the quality of the measurements. Further test cases which have yet to be done are tests for cross talk between two wells laying directly next to each other. The test case will include one well which is filled with a dark sample with no fluorescence material and surrounding wells with fluorescence material. Fluorescent light from one of the surrounding wells must not reach the dark well. This test can be combined with a positioning-test, where the multiwell-plate is slightly moved within the enclosure of the Quant-Module, if possible. Especially with regard to plates with a smaller diameter, the correct positioning of plates without blocking any light is important.

Acknowledgement

The work has been carried out at Roche Diagnostics International AG, and supervised by the Institute for Neuro- and Bioinformatics, Universität zu Lübeck.

5   References

[1] E. L. van Dijk, Y. Jaszczyszyn, and C. Thermes, "Library preparation methods for next-generation sequencing: tone down the bias," *Experimental cell research*, vol. 322, no. 1, pp. 12–20, 2014.

[2] M. D. Adams, C. Fields, and J. C. Venter, *Automated DNA sequencing and analysis*. Elsevier, 2012.

[3] J. D. Robin, A. T. Ludlow, R. LaRanger, W. E. Wright, and J. W. Shay, "Comparison of dna quantification methods for next generation sequencing," *Scientific reports*, vol. 6, 2016.

[4] T. A. Brown, *Gene cloning and DNA analysis: an introduction*. John Wiley & Sons, 7 ed., 2016.

[5] T. F. Scientific, "Qubit dsdna hs assay kits." https://tools.thermofisher.com/content/sfs/manuals/Qubit_dsDNA_HS_Assay_UG.pdf. Retrieved January 19, 2017.

Light-Induced Permeabiliziation of Liposomes

C. Malich[1], A. Link [2], and R. Rahmanzadeh[2]

[1] Medizinische Ingenieurwissenschaft, Universität zu Lübeck, carina.malich@student.uni-luebeck.de
[2] Institute of Biomedical Optics, Universität zu Lübeck, {link, rahmanzadeh}@bmo.uni-luebeck.de

Abstract

The delivery of therapeutic agents to a specific cellular site is a central challenge in the treatment of different diseases. Liposomes that can release their cargo upon an externally controlled trigger are attractive candidates for localized drug release. Light as external trigger can be controlled temporal and spatial with high precision. In this study, we investigated the potential of light sensitive liposomes with the photosensitizer 5,10-DiOH for light-induced release. To demonstrate permeabilization of the liposomes, we encapsulated calcein in high concentration inside the liposomes, so that the calcein fluorescence is quenched. If calcein is released from the liposome, quenching is avoided and the fluorescence increases. Liposomes with 5,10-DiOH were irradiated with light at 420 nm and the release was quantified through measurement of the fluorescence intensity. We demonstrated that liposomes including 5,10-DiOH release cargo effectively after irradiation. Further we observed through fractionated irradiation, that most of the calcein release took place during light irradiation, while the permeability of the liposome decreased shortly after light exposure.

1 Introduction

Liposomes have been developed to improve biodistribution and transport of various drugs across the cell membrane. They are artificial vesicles synthesized from non-toxic lipids [1]. The name liposome is derived from the Greek words lipos meaning fat and soma meaning body. Lipids have a hydrophilic head group and a hydrophobic tail group. Liposomes are made of one or more bilayers which are generally formed by self-assembling of the lipids in aqueous solution with the hydrophobic tails facing each other [1] [2].

Their attractiveness is based on their ability to load cargo in their hydrophilic core or within the hydrophobic region of the bilayer. Encapsulation of drugs into liposomes can reduce side effects and due to their greater size the circulation time in the blood can be extended [1].

Recently, liposomes have been developed that can release drugs on demand in a controlled and selective manner. This increases the efficacy of liposomal treatments by increasing the local drug concentration at a specific site of interest.

Several triggers have been demonstrated to release liposomal cargo. For light induced cargo release a photosensitizer is needed within the liposomal formulation.

When excited by light at a specific wavelength, the photosensitizer produces singlet oxygen [3] [4]. The excitation wavelength varies between different photosensitizers. The production of singlet oxygen and reactive oxygen species (ROS) causes damage in the membrane of liposomes, which induces cargo release [3].

In this study we investigated the potential of the photosensitizer 5,10-Di-(4-hydroxyphenyl)-15,20-diphenyl-21,23H-porphyrine (5,10-DiOH) for liposomal release. Therefore, we added 5,10-DiOH to the lipid formulation. As a test system, we encapsulated the fluorescent dye calcein in a high concentration inside the liposomes. As a result of close distance between the calcein molecules the fluorescence of calcein is quenched, meaning that the fluorescence is decreased to a minimum without destroying the fluorophore. When the liposomes are permeabilized, calcein is released from the liposomes and the fluorescence increases. The increase of fluorescence intensity was used to quantify the release. A scheme of the test system can be seen in Fig. 1.

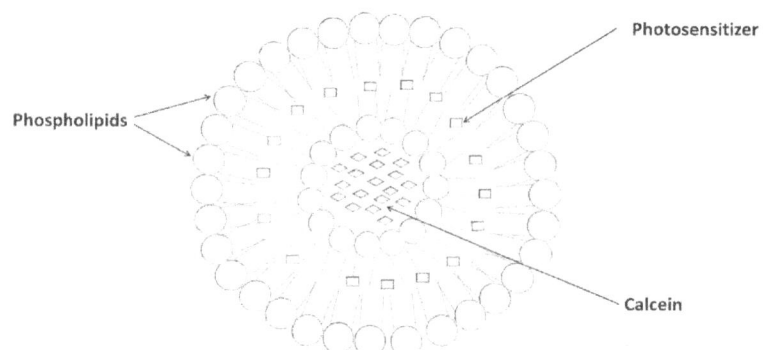

Figure 1: Scheme of light-sensitive liposomes.

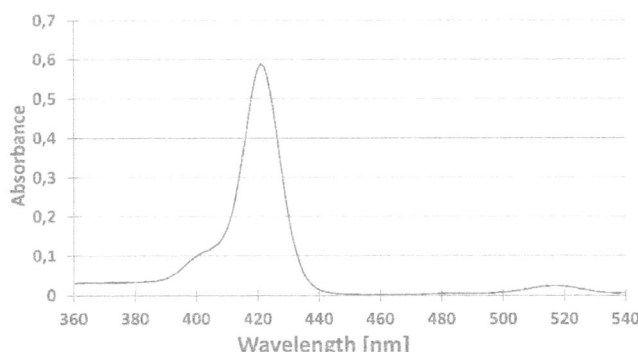

Figure 2: Absorption spectrum of 5,10-Di-(4-hydroxyphenyl)-15,20-diphenyl-21,23H-porphyrine in dimethyl sulfoxide.

2   Material and Methods

2.1   Lipids and Photosensitizer

To form L1-Liposomes, a specific liposome formulation, the lipids 1,2-Dioleoyl-3-trimethylammonium-propan (DOTAP), Dipalmitoylphosphatidylcholine (DPPC), 1,2-distearoyl-sn-glycero-3-phosphoethanolamine-N-[amino(polyethylene glycol)-2000] (DSPE2000-PEG) and Cholesterol were obtained from Avanti Polar Lipids, USA. Except for cholesterol all lipids were obtained as powders. Cholesterol was already solved in chloroform.
The amphiphilic porphyrine 5,10-Di-(4-hydroxyphenyl)-15,20-diphenyl-21,23H-porphyrine (5,10-DiOH) was used as photosensitizer. The porphyrine was obtained as a powder from Por-Lab, Porphyrin-Laboratories GmbH, Germany.

Absorption spectroscopy was performed to determine the optimal wavelength at which 5,10-DiOH produces ROS. Therefor 5,10-DiOH was dissolved in dimethyl sulfoxide (DMSO, Sigma Aldrich) with a concentration of 1 mM and then diluted in DMSO. The spectrum was acquired with a spectrometer (HITACHI U-2900 UV-VIS Double Beam Spectrometer). It is shown in Fig. 2 that 5,10-DiOH has an absorption maximum at 420 nm, therefore this wavelength is used for permeabilization.

2.2   Preparation and Characterization of Liposomes

The photosensitizer 5,10-DiOH was dissolved in DMSO, cholesterol, DOTAP, DSPE2000-PEG and DPPC were each seperatly dissolved in chloroform and then mixed in the ratio 1:1:1,6:3,9:7,9. Chloroform was evaporated under a stream of nitrogen [5] [6]. To remove chloroform residues the vial was placed overnight in an desiccator under vacuum. Calcein solution ($C_{Calcein} \approx 58\,\mu M$) was then added at 50 °C, a temperature greater than the transition temperature of the lipids [6]. The solution was incubated for 40 min at 50 °C in a waterbath and gently mixed every

8 min. Following this the solution was transferred on ice for 10 min and then incubated at 50 °C for another 10 min; this was repeated 6 times. The resulting dispersion consists of mulitlamellar vesicles. To form unilamellar vesicles the dispersion was extruded 13 times through a 100 nm-diameter membrane by using an extruder system (Avanti Polar Lipids, USA) [1] [2] [5] [6]. Non-encapsulated calcein was removed by gel filtration using a sepharose column (CL-4B, Sigma Aldrich) [5] [6] [7]. The concentration of encapsulated 5,10-DiOH was determined by absorption spectroscopy. Resulted liposomes were analyzed by dynamic light scattering (DLS) and transmission electron microscopy (TEM).

2.3   Permeabilization of Liposomes

Liposomes were permeabilized by light irradiation. To determine the parameters for liposomal irradiation different experiments were performed.
Liposomes were diluted to different concentrations of 5,10-DiOH (100 nM, 200 nM, 500 nM, 1000 nM, 2000 nM) in phosphate buffered saline (PBS, Sigma Aldrich) and then irradiated at 420 nm and 80 mW/cm^2 for 1 min. To evaluate different irradiances, liposomes were diluted to a 5,10-DiOH concentration of 200 nM. The samples were irradiated at 420 nm for 1 min with different irradiances (5 mW/cm^2, 10 mW/cm^2, 20 mW/cm^2, 40 mW/cm^2, 80 mW/cm^2). The third parameter that was investigated was the exposure time. Therefore liposomes were diluted to a 5,10-DiOH concentration of 200 nM and irradiated at 420 nm and 80 mW/cm^2 with different exposure durations. For all three experiments liposomes of the same 5,10-DiOH concentration were permeabilized with 4 % Triton (Triton X-100, SERVA Electrophoresis GmbH; diluted in PBS to a concentration of 4 %) served as control.

To determine the behaviour of L1-Calcein-5,10-DiOH after irradiation, liposomes were irradiated periodically. Liposomes were diluted in PBS to a total 5,10-DiOH concentration of 200 nM. As control liposomes of the same 5,10-DiOH concentration were permeabilized with 4 % Triton. One irradiation period consisted of irradiation at 420 nm and 10 mW/cm^2 for 1 min, followed by 5 min dark incubation; in total 7 periods were performed. Before and after every irradiation the fluorescence intensity was measured with a spectrophotometer (Spectramax M5, Molecular Devices, USA).

3   Results and Discussion

3.1   Characerization of Liposomes

The shape and size of the liposomes were characterized by TEM and DLS. The liposomes showed a spherical shape as can be seen in Fig. 3(A) with a median size of 186.56 nm. Fig. 3(B) shows the distribution density of liposomes relating to their size.

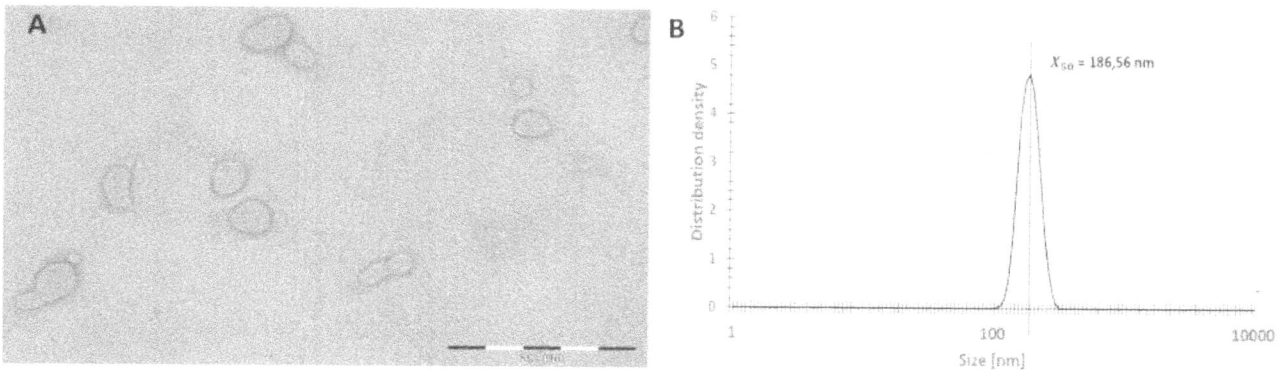

Figure 3: Characterization of Liposomes. (A) Transmission emission microscopy (TEM) image of L1-Calcein-5,10-DiOH showed a round shape of the liposomes. (B) Dynamic light scattering measurements (DLS) confirmed a median size for the liposomes of approximately 186 nm.

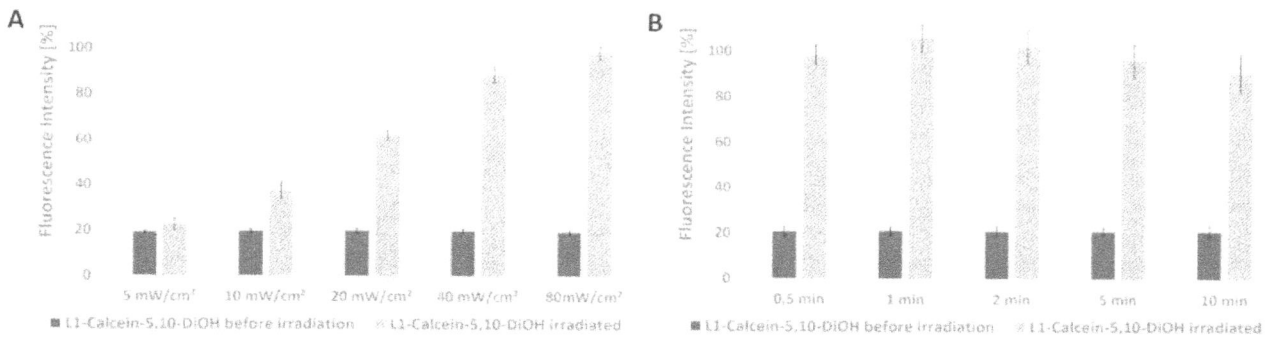

Figure 4: Light-induced permeabilization of L1-Calcein-5,10-DiOH by irradiation at 420 nm. (A) Irradiation for 1 min with different irradiances. (B) Irradiation at 80 mW/cm² with different exposure times. The fluorescence intensities are normalized to the fluorescence intensity of the control.

3.2 Permeabilization of Liposomes

Permeabilization of different 5,10-DiOH concentrations showed that a release of over 90 % can be already achieved with a concentration of 100 nM. With a concentration of 200 nM we achieved a release of 100 % and used this concentration for the following experiments.

Fig. 4(A) shows the release of calcein from liposomes after irradiation with different irradiances. With a low irradiance of 5 mW/cm² no significant increase in fluorescence can be observed. With increasing irradiance the fluorescence intensity increases as well. A release of 90 % can be achieved with 40 mW/cm² and nearly 100 % with 80 mW/cm².

The increase of the fluorescence intensity after irradiation with different exposure times can be seen in Fig. 4(B). With an irradiance of 80 mW/cm² nearly a complete release can be observed at an irradiation time of only 30 s. At longer exposure times such as 5 min or 10 min the fluorescence intensity decreases in comparison to shorter exposure times. This may be due to photobleaching.

To examine if the liposomes are able to reclose after permeabilization, liposomes were diluted in PBS and irradiated

several times at 420 nm. To visualize the behaviour of the liposomes after those irradiations the release rates were calculated after each irradiation.

$$Release\ Rate\ [\%] = \frac{F_i - F_0}{F_T - F_0} * 100 \qquad (1)$$

where F_i is the fluorescence intensity at the time of i, F_0 the background signal and F_T the intensity maximum of the control.

Fig. 5 shows light-induced release of calcein from L1-Calcein-5,10-DiOH. Depicted is the increase of fluorescence intensity over time. The slope is related to the release rate. The bars represent point and duration of irradiation. The black curve shows the time-dependent course of the release during periodic irradiation whereas the grey curve shows the release by only one irradiation.

During irradiation a significant increase of the black curve can be observed whereas a slight increase can be reported after dark incubation. The slope of the curve flattens and becomes saturated gradually.

As well as the black curve, the grey one increases signifi-

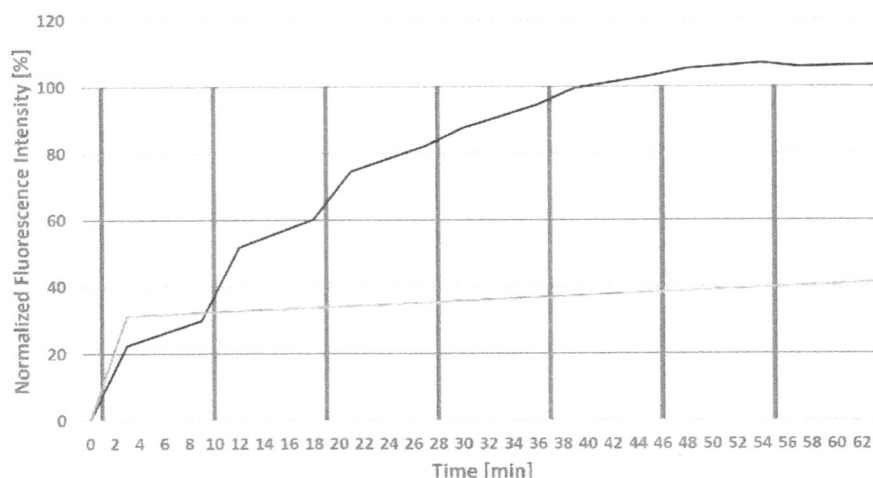

Figure 5: Light-induced permeabilization of L1-Calcein-5,10-DiOH at $420\,\mathrm{nm}$ and $10\,\mathrm{mW/cm^2}$. The black curve shows the progress of the release rate by periodic irradiation for 1 min each period. The grey curve shows the progress of the release rate after a single irradiation. The fluorescence intensities are normalized to the fluorescence intensity of the control.

cantly after first irradiation. After 1 h, the fluorescence intensity is measured again and an increase of $10\,\%$ can be observed.

It is evident that a single irradiation does not have the same effect on the liposomes as the periodically performed irradiation. The fluorescence intensity increased to $41\,\%$ 1 h after the single irradiation whereas the fluorescence intensity of the periodically irradiated sample increased to $100\,\%$ in the same time.

Nevertheless, the increase of $10\,\%$ indicates that L1-Calcein-5,10-DiOH liposomes do not reclose completely after irradiation.

The relatively flat increase of fluorescence intensity after a single irradiation may be due to the selected parameters. Relating to Fig. 4(A) the leakage after irradiation may be greater at higher irradiances.

4 Conclusion

We demonstrated that liposomes including the photosensitizer 5,10-DiOH release cargo effectively after light irradiation. The ability to release cargo on demand makes them attractive candidates for drug delivery.

Next step towards the use as drug delivery system would be the examination of in vitro efficacy. Therefore, the cytotoxicity of the porphyrine and the porphyrine containing liposomes on cells as well as their cellular uptake have to be determined.

Acknowledgement

The work has been carried out and supervised by the Institute of Biomedical Optics, Universität zu Lübeck.

5 References

[1] A. Akbarzadeh, R. Rezaei-Sadabady, S. Davaran, S. W. Joo, N. Zarghami, Y. Hanifehpour, M. Samiei, M. Kouhi, and K. Nejati-Koshki, "Liposome: classification, preparation, and applications.," *Nanoscale Res. Lett.*, vol. 8, p. 102, jan 2013.

[2] J.S. Dua, Prof. A. C. Rana, and Dr. A. K. Bhandari, "LIPOSOME : METHODS OF PREPARATION AND APPLICATIONS," *Int. J. Pharm. Stud. Res.*, vol. III, no. II, p. 7, 2012.

[3] D. Miranda and J. F. Lovell, "Mechanisms of light-induced liposome permeabilization," *Bioeng. Transl. Med.*, no. June, pp. 1–10, 2016.

[4] T. A. Debele, S. Peng, and H. C. Tsai, "Drug carrier for photodynamic cancer therapy," *Int. J. Mol. Sci.*, vol. 16, no. 9, pp. 22094–22136, 2015.

[5] R. Rahmanzadeh, P. Rai, J. P. Celli, I. Rizvi, B. Baron-Lühr, J. Gerdes, and T. Hasan, "Ki-67 as a molecular target for therapy in an in vitro three-dimensional model for ovarian cancer," *Cancer Res.*, vol. 70, no. 22, pp. 9234–9242, 2010.

[6] S. Wang, G. Hüttmann, Z. Zhang, A. Vogel, R. Birngruber, S. Tangutoori, T. Hasan, and R. Rahmanzadeh, "Light-Controlled Delivery of Monoclonal Antibodies for Targeted Photoinactivation of Ki-67," *Mol. Pharm.*, vol. 12, no. 9, pp. 3272–3281, 2015.

[7] S. Wang, G. Hüttmann, T. Scholzen, Z. Zhang, A. Vogel, T. Hasan, and R. Rahmanzadeh, "A light-controlled switch after dual targeting of proliferating tumor cells via the membrane receptor EGFR and the nuclear protein Ki-67.," *Sci. Rep.*, vol. 6, no. June, p. 27032, 2016.

2

Biochemical Physics

A SpIDA-derived method to analyse the *in vivo* GABA$_B$ receptor relocalization and oligomerization in chronic pain conditions

J.Bienzeisler [1], M. Landry [2], and A. Llorente [3]

[1] Medizinische Ingenieurwissenschaft, Universität zu Lübeck, jonas.bienzeisler@student.uni-luebeck.de
[2] BIC, CNRS UMS 3429, Université de Bordeaux, marc.landry@u-bordeaux.fr
[3] UPV/EHU, Universidad del Pais Vasco, alberto.llorente@ehu.es

Abstract

The response to the synaptic release of γ-Aminobutyric acid (GABA) in the central nervous system is mediated by the GABA$_B$ receptor, inter alia dependent on the formation of receptor oligomers. Spatial intensity distribution analysis (SpIDA) is an image analysis method measuring the oligomerization and density of proteins. It was used to quantify the oligomerization of GABA$_B$ receptors in the rodent spinal cord under conditions of chronic pain and inflammation. SpIDA reports a lowered density of tetrameric GABA$_B$ receptor entities in GABAergic neurons of the dorsal horn under these conditions compared to sham samples. The soma of neurons in the ventral horn is shown to be populated by single entities of the GABA$_B$ receptor. The findings underline the role of receptor oligomerization, as well as the membrane environment in regulating receptor function.

1 Introduction

The central nervous system (CNS) is formed by an information transmitting network of neurons, communicating mutually through the release and detection of neurotransmitters. Slight changes to how the system reacts to the release of these molecules have great physiological significance, but may also have patho-physiological implications. γ-Aminobutyric acid (GABA) is the most abundant inhibitory neurotransmitter in the mammalian CNS. The release is detected by the metabotropic GABA$_B$ and ionotropic GABA$_A$ receptor. Although the GABA$_B$ receptor (GBR) is suspected to be involved in neuronal malfunctions, the mechanisms regulating the function and distribution of the receptor are not yet well understood. The way of how many macromolecular complexes form the receptor, the oligomerization, is a promising candidate for playing a major role in this process [1]-[4]. Spatial Intensity Distribution Analysis (SpIDA) is a novel tool reporting the oligomerization and density of flurophores [5]. Combined with GBR labelled with enhanced green fluorescent protein (eGFP), SpIDA becomes a powerful approach to target investigating the function of this receptor.

1.1 Spatial Intensity Distribution Analysis

The number of observed photons per unit of time emitted from a single kind, excited, fluorophore and therefore also the number of photons in space by a number of this molecule fluctuates [6]. SpIDA is facilitated by the Poissonian behavior governing this process, thus empowering a super Poissonian fit to intensity histograms computed from confocal laser scanning microscopy images (CLSM) (Fig. 1) [5]. The intensity histogram of an imaged region of interest reports the number of pixels for each intensity value; it is fitted for the most probable number of molecules causing the data.

The probability of a detected fluorescence $\delta(k)$ of intensity k excited by the light intensity $I(\mathbf{r})$ by one fluorophore of the brightness ϵ can be calculated as

$$p^1(\epsilon; k) = \int \delta(\epsilon \cdot \mathbf{I}(\mathbf{r}) - k) d\mathbf{r}, \qquad (1)$$

and recursively for n molecules as

$$p^n(\epsilon; k) = p^1(\epsilon; k) * p^{n-1}(\epsilon; k). \qquad (2)$$

From these probabilities an expression describing the final histogram $H(\epsilon, N; k)$ can be derived, assuming a Poisson distribution:

$$H(\epsilon, N; k) = \sum_n p^n(\epsilon; k) \cdot poi(n, N) \qquad (3)$$

[5]. This equation can be fitted either for the number of fluorophores per excitation volume N, or the quantal brightness ϵ and can also be extended to describe the histogram of mixed populations of fluorophores. The probability distribution for the sum of two random variables is given by the convolution of the two individual distributions, therefore obtaining

$$H(\epsilon_1, N_1, \epsilon_2, N_2, A; k) = A \cdot H(\epsilon_1, N_1; k) * H(\epsilon_2, N_2; k) \qquad (4)$$

for two histograms with scalar prefactor A (Fig.1E). For known ϵ values these equations can be fitted for the particle densities, allowing the use of SpIDA for reporting the oligomerization and densities of proteins labelled with flurophores, as a labelled dimer has a monomeric brightness of $2\epsilon_0$. In practice the quantal brightness for a monomeric protein sample is determined with (1) and then protein densities can be assessed with (4).

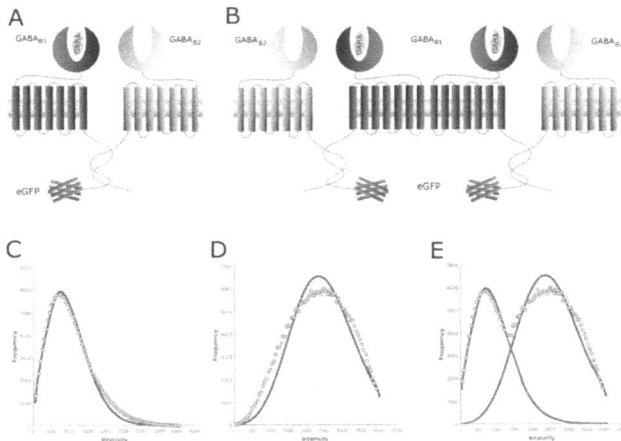

Figure 1: Detecting oligomerization of GBR by SpIDA. (A) A schematic heterodimeric GBR labelled with eGFP and (B), a tetraeric GBR labelled with two eGFP. (C) Intensity histogram (dots) and SpIDA fit (line) of a region of interest with a solely dimeric, (D) a solely tetraeric and (E), a mixed population of dimeric and tetraeric GBR.

1.2 The GABA$_B$ Receptor System

The GBR is a G-protein-coupled receptor mediating the response to the synaptic release of GABA. Affected by the sub-cellular localization, the GBR exerts distinct regulatory effects on the synaptic transmission making it a promising target for the treatment of psychiatric and neurological disorders [7]. Furthermore, GBRs are involved in the regulation and perception of pain [1]. The patho-physiological process of neuropathic pain is associated with neuronal sensitization in dorsal horn (DH) spinal networks, induced by a maladaptive GABA$_B$ inhibition of calcium-dependent intrinsic properties of DH neurons [3].

Albeit having diverse functionalities in the CNS, GBRs are invariably expressed as the same structure, being an obligate heterodimer of the two subunits GABA$_{B1}$ and GABA$_{B2}$ (Fig. 1A). Coexpression of the subunits is mandatory for the functional expression and proper binding of GABA to the receptor [7]. It is suspected that diversified function is achieved inter alia by the cell surface expression and spatial organization of the receptor, as multiple receptors form higher-order oligomers in association with themselves [2],[8]. Transient interactions among the receptors result in a functional cooperativity. In this fashion the formation of two GBR as a tetramer has been observed [2].

A key to understanding the physiological significance of the tetrameric organization and patho-physiological implications is studying the distribution of dimeric and tetrameric entities. If oligomerization is regulating receptor function, it can thus be postulated, that altered organization of GBR in the DH could play a role in the maladaptive GABA$_B$ inhibition under conditions of neuropathic pain in the CNS.

2 Material and Methods

To study the distribution of GBR and role in patho-physiological processes, spinal tissue of transgenic mice expressing eGFP-tagged GABA$_{B1}$ receptor-subunits under varying conditions was imaged and probed with SpIDA.

2.1 Fluorescence Microscopy

Samples were imaged with a true point-scanning, spectral confocal system (Leica, Mannheim) at a 63X magnification, NA 1.4. The eGFP was excited using a 488 nm diode laser at an intensity of 70 %. The fluorescence was detected by a PMT at a gain of 700 V. To calibrate the system for SpIDA and correct for error terms, the point spread function of the microscope was determined using 25 nm beads (Molecular Probes, Eugene). The slope of the linear correlation between the mean pixel value and the respective variance was asserted with a Argo-LM slide (Argolight, Pessac) [5],[9].

2.2 Measuring the monomeric brightness

Slides of viruses transfected with eGFP were prepared to measure ϵ_0 of eGFP under the given microscope setting. In these viruses the expression of eGFP can be assumed to be isolated and thus monomeric. SpIDA functions of the resulting images' intensity histograms were fit for ϵ_0.

2.3 Mouse Model

A total number of ten transgenic mice expressing GABA$_{B1}$ receptor-subunits fused to eGFP were sacrificed [10]. Among these, persistent neuropathic and inflammatory pain was evoked in two mice with Spared Nerve Injury (SNI), two mice were injected with Complete Freund's Adjuvant (CFA), two mice were injected with NaCl for sham conditons, four mice were left untreated. SNI was induced by a lesion of the tibial and common peroneal nerve branches causing a partial denervation of the sciatic nerve and hence pain-like behavior [11]. The intraplantar injection of CFA, triggering an immunoreaction, was applied to the mice's left paws. The experiments followed the ethical guidelines of the international Association for the Study of Pain and were approved by the local ethics committee in Bordeaux.

2.4 Sample Preparation

The mice were anesthetized with a 225 mg/kg lethal dose of sodium pentobarbital. After pinch response testing the animals were perfused through the left ventricle of the heart; two minutes with phosphate-buffered saline (PBS) 1x, pH

7.4 and two minutes with 4 % paraphormaldehyde, 0.1 % glutaraldehyde at pH 7.4 at room temperature. Brains and spines were extracted and postfixated for two hours at 4°C in 4 % paraformaldehyde, 0.1 % glutaraldehyde at pH 7.4. The samples were put in a cryoprotective solution of 15 % sucrose in PBS overnight, subsequently embedded in Tissue-Tek®(Sakura Finetek Europe, Leiden) and frozen in isopentan (-90°C). The samples were stored at -20°C and cut into 10 μm thick cryostat sections for mounting onto gelatin covered slides with Fluorescence Mounting Medium (Dako, Hamburg).

2.5 Image Analysis

The oligomerization of GBR (representatively the GABA$_{B1}$ subunit) was investigated in the ventral horn (VH) and the DH. The regions of interest for SpIDA were selected manually, considering a tradeoff between specificity (small ROI) and optimal sampling (large ROI) [5]. Background noise was empirically determined by measuring the mean intensity of image regions without fluorescence signal and substracted from the intensity histogram for final fitting with the SpIDA functions [5].

3 Results

Diverse oligomerization of the GBR system depending on the cellular location was observed under normal, neuropathic, inflammatory and sham conditions.

3.1 The GBR oligomerization in the spine of mice differs depending on the location

Since the function of the GBR-system depends on the subcellular region, SpIDA was applied to ROIs where these could be observed. In the soma of motoneurons innervating skeletal muscle located in the VH, high densities of the GBR are almost exclusively found in the dimeric state (Fig. 2B-D). Gaberic inhibitory interneurons, showing up as small regions with high fluorescence intensity in CLSM images, are found in the superficial DH (Fig. 2A). These regions with high densities of GBR display a very heterogenic distribution. The SpIDA results for these neurons show that the favorable oligomerization of the receptor ranges from dimeric over equally distributed to tetrameric. A tendency towards more tetrameric entities can be observed, as well as a linear correlation of the data (Fig. 2C). GBR in the surrounding DH laminae with low fluorescence signal and thus low receptor densities are favorably organized dimeric.

3.2 The GBR oligomerization in the Dorsal Horn is altered under Inflammatory and Neuropathic Pain Conditions

Weakened GABA$_B$ inhibition in the DH of the rodent spine is observed under neuropathic pain conditions. Concurrently the oligomerization and the density of the GBR is al-

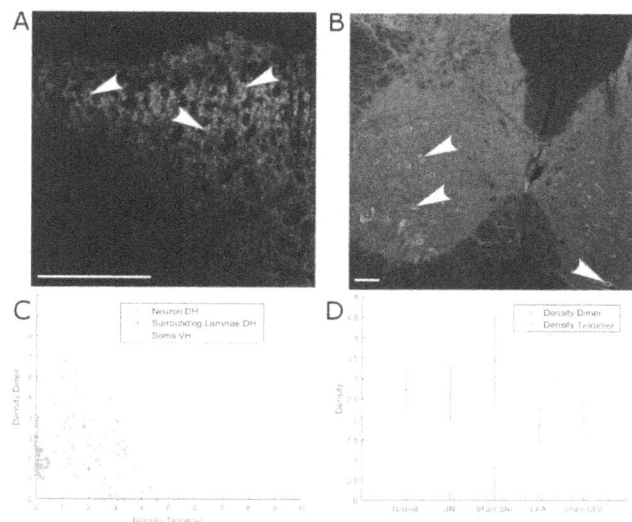

Figure 2: The Oligomerization of GBR in the DH and VH. (A) The superficial DH of a mouse expressing GABA$_B$. GABAergic inhibitory interneurons show high intensities of fluorescence (arrows). (B) The soma of motoneurons innervating skeletal muscle located in the VH (arrows). (C) The scatterplot of GBR densities in fluorescent molecules per laser beam-effective focal volume. Each data point represents one ROI, respectively one neuron (x), the soma of one nerve cell (circle) or surrounding laminae in the DH (plus). (D) The averaged oligomerization (with standard deviation) of the soma of nerve cells in the VH. Scale bar: 100 μm.

tered under neuropathic pain, inflammatory and sham conditions (Fig. 3). Under SNI, CFA and sham conditions high densities of almost exclusively isolated, dimeric GBR are found in the soma of VH motoneurons (Fig. 2D). The same is the case for receptors in the surrounding laminae of the DH with low fluorescence signal and thus low receptor densities, which are also favorably organized dimeric (Fig. 3D). Distinct changes in oligomerization were detected in the DH; the GABAergic inhibitory interneurons show definite alterations under neuropathic pain conditions. While under sham conditions a greater proportion of tetrameric organized GBR can be observed, under SNI conditions there are more neurons in which the GBRs are organized predominantly dimeric (Fig. 3B). The same is the case for CFA conditions, where a higher rate of neurons show rather dimeric GBR compared to sham conditions; however there is no obvious difference in data from the DH ipsilateral to the CFA injection (Fig. 3A). In all data a linear correlation of dimeric and tetrameric oligomerization can be observed.

4 Conclusion

Under all conditions there are two consistent observations: almost exclusively dimeric GBRs are found in the soma of VH motoneurons and all DH scatter plots show linearly correlated data of dimeric and tetrameric oligomerization. The

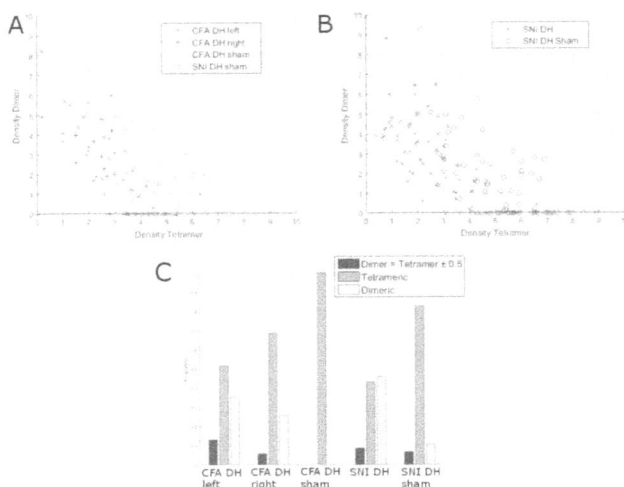

Figure 3: Oligomerization of GBR in pain and inflamation. (A) The scatterplot of GBR densities in fluorescent molecules per laser beam-effective focal volume in the DH of mice under CFA conditions and (B) under SNI conditions. (C) Summary of the data from (A) and (B) in the form of a box plot. The plots show the proportion of neurons (data points) that had a higher amount of dimeric, tetrameric GBR, and evenly distributed tetrameric or dimeric GBR.

former observation can be explained with the endocytotic mechanism that is involved in the control of cell surface expression of GBR. Surface trafficking of GBR means assembly of both the GABA$_B$ subunits [12]. This demonstrates that inactive GBR subunits are internalized in the soma of neurons isolated from each other, thus the SpIDA results report single receptors. Dissociation apparently occurs preferentially at the plasma membrane, which is in accordance with previous findings [4]. The linear correlations observed in the DH data indicate that the amount of receptors on the plasma membrane surface is limited and that the spatial configuration of receptors is altered under conditions of chronic pain or inflammation. Both observations suggest that the dimerization of the receptors is a dynamic process taking place on the membrane surface; changes in the membrane environment through lateral diffusion might also regulate function. It also hints that the assembly of GABA$_B$ subunits and endocytotic pathways might not be involved in altering the oligomerization of GBR on the membrane surface. Unequivocal is the physiological significance of the tetrameric organization and patho-physiological implications of changed distribution and hence the oligomerizaiton of GBR. Tetrameric organization as a dimer of two GBR apparently plays a role in regulating the function of the receptor. A lower proportion of tetrameric organization in the DH is seemingly intertwined with impeded GABA$_B$ inhibition and the pathological conditions of chronic pain or inflammation. These conclusions give further inside into the function of the GBR system and may help new ways of targeting neuropathical pain pharmacologically. However, the results need to be investigated on a greater scale *in vivo*,

as the conclusions drawn need to be further evidenced.

Acknowledgement

The work and microscopy has been carried out at Bordeaux Imaging Center, a service unit of the CNRS-INSERM and Bordeaux University, member of the national infrastructure France BioImaging. The help of Fabrice Cordelieres is acknowledged.

5   References

[1] C. Goudet et. al., *Metabotropic receptors for glutamate and GABA in pain* in Brain Res., vol. 60, no. 1, pp. 43-56, 2009.

[2] L. Comps-Agrar et. al., *The oligomeric state sets GABA(B) receptor signalling efficacy* in Embo J., vol. 30, no. 12, pp. 3239-3251, 2011.

[3] S. Laffray et. al., *Impairment of GABAB receptor dimer by endogenous 14-3-3ζ in chronic pain conditions* in Embo J., vol. 31, no. 15, pp. 3239-3251, 2012.

[4] S. Laffray et. al., *Dissociation and trafficking of rat GABAB receptor upon chronic capsaicin stimulation.* in Eur. J. Neurosci., vol. 25, no. 5, pp. 1402-1416, 2007.

[5] A. Godin et. al., *Revealing protein oligomerization and densities in situ using spatial intensity distribution analysis* in PNAS, vol. 108, no. 17, pp. 7010-7015, 2011.

[6] L. Mandel, *Fluctuations of Photon Beams and their Correlations* in Proc. of the Phys. Soc., vol. 72, no. 6, p.1037, 1958.

[7] D. Ulrich et. al., *GABA$_B$ receptor: Synaptic functions and mechanisms of diversity* in Curr. Opin. Neurol., vol. 17, no. 3, pp. 298-303, 2007.

[8] J. Schwenk et. al., *Native GABA$_B$ receptor are hetero-multimers with a family of auxiliary subunits* in Nature, vol. 465, no. 7299, pp. 231-235, 2010.

[9] C. Matthews and F. Cordelières, *MetroloJ: an ImageJ plugin to help monitor microscopes' health* in ImageJ User & Developer Conference 2010 proceedings, Luxemburg, 27-29 October 2010.

[10] E. Casanova et. al., *A mouse Model for Visualization of GABA$_B$ Receptors* in Genesis, vol. 47, no. 9, pp. 595-602, 2009.

[11] M. Pertin et. al., *The spared nerve injury model of neuropathic pain* in Methods Mol Biol., vol. 851, pp. 205-212, 2012.

[12] T. Grampp et. al., *Constitutive, agonist-accelerated, recycling and lysosomal degradation of GABA$_B$ receptor in cortical neurons* in Mol. Cell. Neurosci., vol. 39, no. 4, pp. 628–637, 2008.

Characterization of structural properties of bacterial membranes with atomic force microscopy

A. Crezelius[1], C. Nehls[2], and T. Gutsmann[2]

[1] Medizinische Ingenieurwissenschaft, Universität zu Lübeck, annette.crezelius@student.uni-luebeck.de

[2] Forschungsgruppe Biophysik, Forschungszentrum Borstel, {cnehls, tgutsmann}@fz-borstel.de

Abstract

This paper reports the use of atomic force microscopy (AFM) for a comparison between a natural system and an artificial model system. In the first part, the bacterial morphology and envelope structure was characterized by AFM. The bacteria samples were prepared with different protocols and in addition the influence of peptide interactions were investigated. For one peptide an impact on the bacterial envelope could be shown. In the second part the procedure for preparation of small unilamellar vesicles (SUVs) was systematically modified to get information on the influence of the different preparation steps. To determine the vesicle sizes distribution dynamic light scattering (DLS) was used. The SUVs will be used for the preparation of solid supported model membranes which will be comparatively investigated with AFM. Especially the parameters of ultrasonic treatment of the vesicles and the lipid composition had an influence on the size distribution.

1 Introduction

Biological membranes isolate the inside of cells, cell compartments and bacteria from the environment. To understand the influence of membrane active peptides on membranes it is an advantage to analyze the membrane structure. To gather more information about that, the membranes of natural systems and modell systems were compared.

The major component of a membrane are lipids. These lipids consist of a head group and fatty acid chains. They conform a bilayer with the head group pointing to the outside and the chains pointing to the inside. This happens in aqueos medium and can be explained by the amphipathicity of the lipids [2].

In this project especially reconstructed and natural membranes were investigated. Initially the envelope of *E.coli* was looked at with and without addition of membrane active peptides. The next step was the preparation of vesicles for solid supported membranes which are obtained by spreading them on the solid substrate. For that the preparation process was varied and the outcome was analyzed in terms of the vesicle size.

To examine the vesicle size dynamic light scattering (DLS) experiments were performed. DLS means that a laser beam of a certain wavelength illuminates a sample. In the sample the light gets scattered by the particles, in this case the vesicles. The light interferes on a detector which captures the intensity of the scattered light. Due to the Brownian motion of the particles intensities different time steps are considered which leads to a correlation function to compare these intensities [5]. The DLS provides data for the z-average which is the hydrodynamic vesicle diameter. Furthermore the Polydispersity Index (Pdi) is obtained. The Pdi is the width of the particle size distribution [5].

To perform measurements either of bacteria or of spread vesicles the atomic force microscopy (AFM) is a proper tool. AFM is a technique to study surfaces and properties of biological samples wich was introduced in 1986 [3], [4]. It is possible to reach a resolution less than 0.1 nm [1]. The AFM measures the force between a tip and a sample. The tip has a size between 10 to 30 nm and is usually made of silicon. It is fixed on a *cantilever*.

In general two modes exist to use an AFM - the *static* mode and the *dynamic* mode. In dynamic mode the cantilver oscillates close to the resonant frequency [4]. Depending on the repulsive or attractive force, the resonant frequency is shifted. Another way to measure is the static mode, in which the deflection of the cantilever is measured. This deflection is caused by the force affecting the cantilever which is proportional to the spring constant. One way to apply the static mode is the *contact* mode in which the tip of the cantilever is directly in contact with the sample. The tip scans systematic over the surface. If the height of the surface of the sample changes because of its structure the cantilever deflects, which is determined by the defelction of a laser beam [1]. Either the deflection of the cantilever is tracked or the z-position of the cantilver is adjusted with a feedback loop of the deflection resulting in a constant force.

2   Material and Methods

The following section describes on the one hand the AFM measurements of bacteria and on the other hand the size determination of the vesicles preparations by DLS.

2.1   Measurements of bacteria in air

For all measurements the bacterial strain *E.coli* ATCC 23716 was used. They were washed with HEPES and taken in buffer which contained of 20 mM HEPES and the pH 7.4. Five cups with each 200 μl of bacteria suspension were prepared. Two cups were combined with different peptides, peptide 1 and peptide 2. Afterwards all cups got incubated for 30 min at 37 °C with 150 rpm (revolutions per minute). From one cup 50 μl of the unfixed bacteria suspension were dried on a mica plate. The remainig bacteria in the four cups were fixed with PFA (4 g paraformaldehyd in 100 ml PBS (phosphate-buffered saline)). For that the cups were centrifuged and the pellet was picked up in 200 μl 4% PFA solution. Afterwards they were left 30 min and then washed with the buffer. In the end, the content of the 4 cups contained fixed bacteria in 200 μl buffer. 50 μl of the fixed bacteria only with *E.coli* were dried on mica. 100 μl of each the three remainig cups, two with *E.coli* with peptide and one only with *E.coli*, were centrifuged on mica by cytospin with 1200 rpm.

The surface properties were measured by the MFP-3D AFM (Asylum Research, Santa Barbara, CA, USA) and was handled with the software MFP3D 14.20.152, Igor Pro 6.37. The used cantilever was CSG11 by NT-MDT Co., Moskau, Russia. All pictures were taken in contact mode. The sample was oriented in the x,y-plain and can be moved by piezoelements. During measurements the deflection of the cantilever was kept constant by a feedback loop between the detector an the z-actuator.

2.2   Liposome preparation and characterization

The common process of small unilamellar vesicle (SUV) preparation was varied, the ratio of the lipids, the treatment with a tip sonicator and the temperature cycle.

For all different preparations 10 mg/ml stock solutions of the lipids were prepared in chloroform. The standard preparation process was to mix the lipids depending on the desired ratio in a vial. The utilised lipids were 48% of PE (Phosphatidylethanolamin) and 52% of PG (Phosphatidylglycerol) mimicking the lipid composition of bacteria. The chloroform was vaporised under nitrogen so that the lipids build a multilayer stack on the bottom of the vial. 1 ml of a buffer containing 100 mM of KCl (potassium chloride) and 5 mM of HEPES with pH 7.0 was added. In total the resulting lipid concentration was 1 mg/ml. To facilitate the budding of vesicles from the multilayer and to homogenize them ultrasonic waves were used. For that the Ultrasonic-Homogenizer (HTU SONI-130 by G. Heine-

mann, Schwäbisch Gmünd, Germany) was applied. Generally the vial was treated 4 min with the tip sonicator at an intensity of 30%. After the ulrasonic treatment a temperature cycle was performed by keeping the solution at 4 °C / 60 °C / 4 °C / 60 °C / 4 °C everything for at least 30 min and storage at 4 °C over night. The changes in the process that were made were at first the variation of the percentage of PG from 50%, 40%, 30%, 20% to 10%. Then the treatment with the tip sonicator was changed. One vial was not treated with ultrasonic waves and one was put into an ultrasonic bath (BANDELIN electronic GmbH & Co. KG, Bandelin Sonorex Rk 100, Berlin, Germany) for 30 minutes. Furthermore, the ultrasonic intensity was varied between 40%, 30% and 20%. In the end different durations of the tip sonicator treatment were compared. The exposed times were 15 sec, 30 sec, 1 min, 2 min, 3 min, 4 min, 6 min and 8 min.

All vesicles were characterized by DLS using the Zetasizer Nano-ZS90, Nano Series by Malvern Instruments Ltd., Worcestershire, UK with the zetasizer software for Nano, APS and μV Version 7.11. The vesicles were diluted to 50 μg/ml in the buffer to sample a volume of 1 ml in a cuvette. For each vesicle preparation three cuvettes were prepared. Each cuvette was measured 3 times and each measurment contained 10 runs. All acquired data was processed with the software Origin 8.5.

3   Results and Discussion

This section presents the results from the previous described measurements. At first the AFM measurements of *E.coli* bacteria are discussed. Afterwards vesicle size determination by the Zetasizer are presented.

3.1   Measurements of bacteria in air

Fig.1, shows two of the five bacteria samples. The bacteria suspension was dried on the mica plate. Both pictures reveal that this preparation approach is not applicable to observe bacterial surfaces properly beacause they are surrounded by crystals. Hence it is not possible to make any claim about the bacterial surfaces.

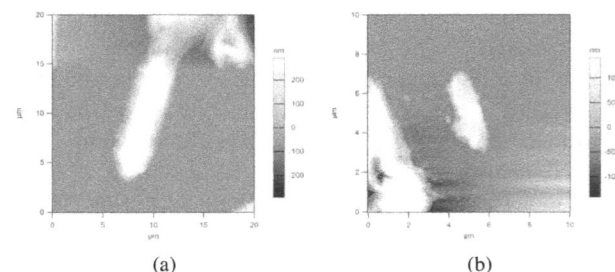

(a)   (b)

Figure 1: (a) 20 μm x 20 μm of *E.coli* fixed with PFA and dried on mica, (b) 10 μm x 10 μm of *E.coli* not fixed and dried on mica.

The remaining three samples prepared by cytospin were

better to observe the bacterial surfaces.

The three pictures (Fig.2) illustrate the bacteria without peptide (Fig.2(a)), with peptide 1 (Fig.2(b)), and with peptide 2 (Fig.2(c)).

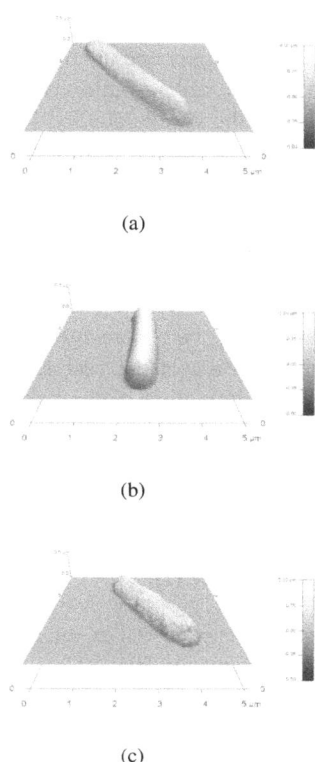

(a)

(b)

(c)

Figure 2: (a) 5 μm x 5 μm with a range of 1 μm, control group, *E.coli* fixed with PFA by cytospin centrifuged on mica,(b) 5 μm x 5 μm with a range of 1 μm, *E.coli* with peptide 1 fixed with PFA by cytospin centrifuged on mica, (c): 5 μm x 5 μm with a range of 1 μm, *E.coli* with peptide 2 fixed with PFA by cytospin centrifuged on mica.

The surfaces of the control group (Fig.2(a)) and the bacteria incubated with peptide 1 (Fig.2(b)) looked smoother than the bacterial surface combined with peptide 2 (Fig.2(c)). This was confirmed by the close-up pictures (Fig.3) and the raw image data of the different bacteria. The roughness of the image data were for the control group 4.416 nm, the *E.coli* with peptide 1 3.224 nm and for the *E.coli* with peptide 2 5.301 nm. Furthermore, the difference between the maximal and the minimal height were for the control group 24.116 nm, for *E.coli* with peptide 1 21.391 nm and for *E.coli* with peptide 2 29.887 nm. The effect of peptide 1 on the bacteria was not as visible as the effect with peptide 2. The image data however showed that the surface was even smoother than the surface of the control group. It could be possible to see an even bigger difference with a different sample preparation.

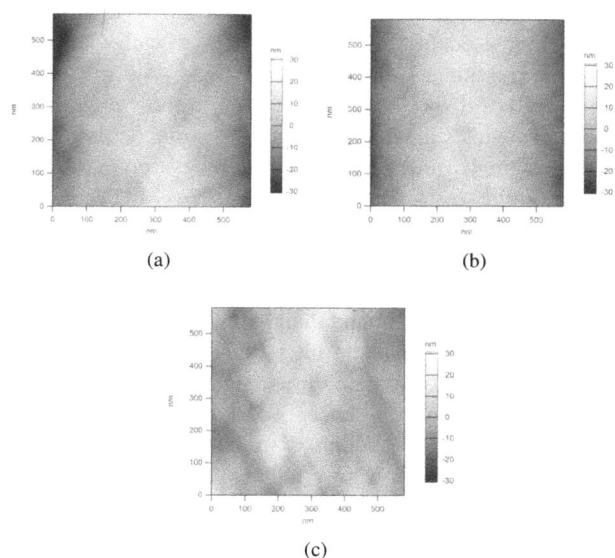

(a)

(b)

(c)

Figure 3: (a) 580 nm x 580 nm with a range of 61 nm, *E.coli* control group fixed with PFA by cytospin centrifuged on mica, (b) 580 nm x 580 nm with a range of 61 nm, *E.coli* plus peptide 1 fixed with PFA by cytospin centrifuged on mica, (c) 580 nm x 580 nm with a range of 61 nm, *E.coli* plus peptide 2 fixed with PFA by cytospin centrifuged on mica.

3.2 Liposome preparation and characterization

Some of the different preparation steps had effects on the vesicle size.

The least impact had the temperature cycle as one can see below in (Fig.4). Moreover for both measurements the Pdi was the same. This indicates that the width of the particle size distribution was the same as well. Refering to the quality of the vesicles it would be necessary to look at the solid supported membranes.

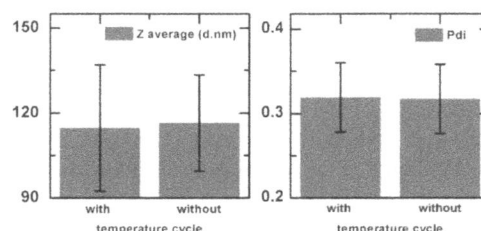

Figure 4: PE:PG vesicle size with and without temperature cycle.

Regarding (a) of Fig.5 the different time durations of the tip sonicator had a major impact. It appears that the z-average gets smaller by increasing exposure time up to 4 minutes but after that it gets larger again. Another interesting fact with longer ultrasonic treatment is that the standard deviation got much bigger for 6 min and 8 min than those from 30 sec up to 4 min. A similar evolution is visible for the Pdi, except that the width of the particle size ditribution already increased after 3 minutes. The vesicles without ultrasonic treatment (left bar Fig.5(a)) and with ultrasonic bath instead of tip sonicator were large. In the vial with the untreaded vesicles the lipds seemed to stay on the bottom of the vial so they were vortexed before measurement. After that the liquid was turbid. The Pdi of the vesicles treated with the ultrasonic bath had an average size of 748.5 nm with a standard deviation of 54.22 nm the Pdi was 0.782 with an standard deviation of 0.054. Moreover the values for z-average varied a lot and their standard deviations were big. This indicates that the ultrasonic bath had the same effect on the vesicles like no ultrasonic treatment.

In the next step the influence of the intensity of the tip sonicator was tested, which also had an influence on the size (Fig.5(b)). It showed that a 30% intensity had the smallest standard deviation of the z-average and the Pdi.

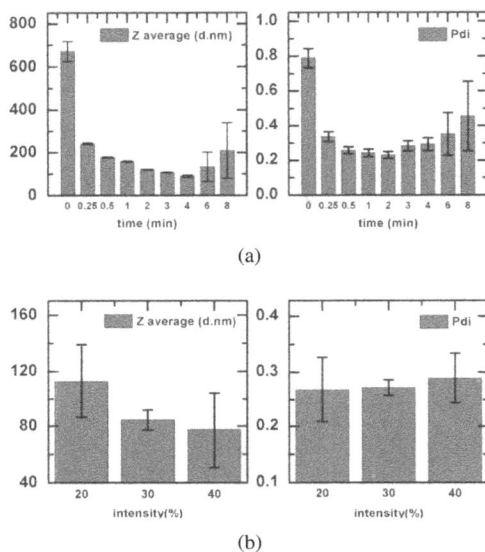

(a)

(b)

Figure 5: (a) PE:PG vesicle size with different exposure times to ultrasonic waves, (b) PE:PG vesicle size with different ultrasonic intensities

The concentration of PG had a big influence on the width of the particle size distribution (Fig.6). The vesicles with less than 30% PG had a smaller variation in size. Moreover, the standard deviation for the vesicle diameter got much bigger with a concentration higher than 30%.

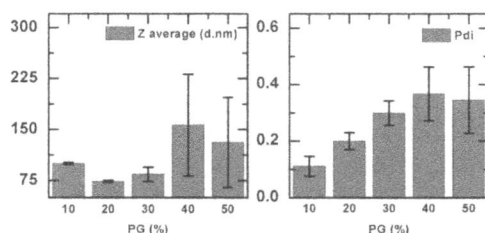

Figure 6: PE vesicle size with different amounts of PG.

4   Conclusion and Outlook

The peptides seemed to have an effect on the bacterial surfaces. It would be interesting to test which effect the fixation onto the bacteria had. Furthermore, we noticed the usefulness of the cytospin wich allowed to observe the bacteria without crystals around them.

We experiented that some steps of the preparation process have a specific influence on the size, especially the timing of the ultrasonic homogenizer and the lipid composition. Nevertheless the vesicle size distribution is not the only consideration to make for the vesicle quality. It would be possible that the effect, for example of the temperature cycle, is only visible in the ability of the vesicles to fuse on a solid support to form flat membranes. The results are only valuable for the used lipids - it would further be a benefit to test the size distribution for other lipid mixtures vesicle size distributions as well.

To continue the project two more systems should be tested. At first the investigation of the vesicle preparation should be expanded by examination of the solid supported membranes of the vesicles by AFM. Furthermore, it would be necessary to compare the described bacteria measurements with measurements taken in liquid.

Acknowledgement

The work has been carried out at the research group Biophysik, Forschungszentrum Borstel.

5   References

[1] C. Nehls *Charakterisierung der Wechselwirkung zwischen dem bakteriellen Protein VapA und der Phagosomenmembran an Modellmembransystemen.* Inauguraldissertation, Universität zu Lübeck, 2016.

[2] M. Eeman and M. Deleu *From biological membranes to biomimetic model membranes.* Univ.Liege, Gembloux Agro-Bio-Tech. Department of Biological Industrial Chemistry, 2009.

[3] G. Binning, C. F. Quate and Ch. Gerber *Atomic Force Microscope.* Stanford University, IBM San Jose Research Laboratory, California, 1986.

[4] E. Meyer *Atomic Force Microscopy.* Progress in Surface Science, Vol. 41, pp.3-49, University of Basel, 1992.

[5] F. G. Mayenfels *Fluorcarbonhaltige Nanoemulsionen zur Anwendung in der $^1H/^{19}F$-Magnetresonanztomographie.* pp.88-91, Inauguraldissertation, Albert-Ludwigs Universität, Freiburg, 2012.

How Antimicrobial Peptides Interact with Model Membranes and Affect the Lipid Mobility
– a FRAP-based Approach –

N. Ludolph [1], L. Paulowski [2], T. Gutsmann [2], and C. Hübner [3]

[1] Medizinische Ingenieurwissenschaft, Universität zu Lübeck, ludolph@physik.uni-luebeck.de

[2] Division of Biophysics, Priority Area Infections, Research Center Borstel, lpaulowski@fz-borstel.de, tgutsmann@fz-borstel.de

[3] Institute of Physics, Universität zu Lübeck, huebner@physik.uni-luebeck.de

Abstract

Antimicrobial peptides (AMPs) comprise a promising alternative to conventional antibiotics to circumvent the rising number of resistance against antibiotics. Within the scope of this paper, two major compounds are put in juxtaposition regarding their membrane activity and their interaction with model membranes. Therefore the reserve antibiotic polymxin B (PMB) and as representative for the AMPs, the α-helical peptide LL-32 were chosen. To prove the theory of the function of these peptides, different experiments have been made. Performed confocal images confirm the location of the peptides in combination with model membranes and FRAP experiments show the dynamic interaction of the peptide and the membranes. Model membranes were prepared at the University of Lübeck. The following experiments show several limitations of the used methods.

1 Introduction

Antibiotics are one of the biggest discovery in the 20th century for the human kind. They offer the possibility to fight against bacterial infections. After a century of success in the medication and prevention with antibiotics, the number of reports about antibiotic resistance is growing. To find alternatives to antibiotics is an important challenge of science and medicine.

Antimicrobial peptides (AMPs) are one part of the innate immunity in nearly every multicellular organism. A remarkable property about AMPs is that biological threats have not build a restistance against them. Until now, AMPs win the race of evolution and protect their host organism against several biological threats.

The interaction between AMPs and a biological membrane and the components of the membrane are still not understood in detail. On which component of the cell membrane does the AMPs bind and how does the AMPs perforate a membrane? To use AMPs as an alternative to antibiotics, it is mandatory to understand the mechanisms in a moleculare perspective.

To this end, Fluorescence Recovery After Photobleaching (FRAP) is one of the methods. With this technique the mobility of the labelled molecules is accessible. The standard fluorescence microscopy gives informations of the localization of substances in one specific area. To make the diffusion in the sample visible, we used FRAP.

2 Material and Methods

2.1 Membrane Compositions

In this paper, the focus is set on two membrane compositions: one reconstituted from the pure zwitterionic 1,2-dioleoyl-snglycero-3-phosphocholine (DOPC) modeling lipid bilayers without phase separation and a second composed of 1,2-dioleoyl-sn-glycero-3-phosphocholine (DOPC), N- (octadecanoyl)-sphing-4-enine-1-phosphocholine (SM), cholest-5-en-3β-ol (Chol) undergoing phase separation into liquid ordered and disordered domains below 25 degrees.

DOPC is a lipid of phosphatidylcholines, which is a class of phospholipids with cholin as the headgroup. DOPC is used to simulate an uncharged lipid bilayer. It does not show a functional unit called domain formation at room temperature and 37°C.

The composition of 1,2-dioleoyl-sn-glycero-3-phosphocholine, N- (octadecanoyl)-sphing-4-enine-1-phosphocholine, cholest-5-en-3β-ol has a lipid ratio of 2:2:1 and a molar ratio of 40:40:20. The difference to the DOPC lipid bilayers is that this solution builds large cholesterol-enriched domains.

The procedure of forming lipid bilayers from DOPC or DOPC/SM/Chol is exactly the same. First, the solution gets treated with ultrasound. Afterwards, a temperature cycle starts (three times 4°C and 60°C, 30 min). In the end, the sample is filled in LAB-TEK chambers (Thermo Fischer) or IBIDI μ-Side 8 Well (figure 1) and is spreaded over several hours. The process is schematical shown in figure 2.[1]

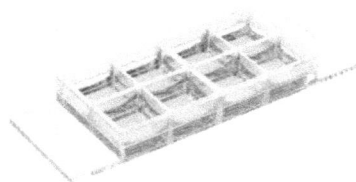

Figure 1: IBIDI μ-Side 8 Well [5]

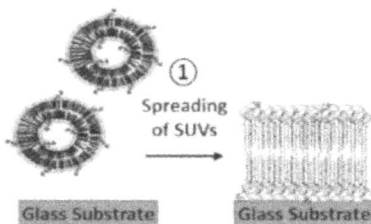

Figure 2: Spreading of small unilamelare vesicles (SUV's) out of DOPC or 1,2-dioleoyl-sn-glycero-3-phosphocholine, N-   (octadecanoyl)-sphing-4-enine-1-phosphocholine, cholest-5-en-3β-ol. [1]

2.2   Antibiotic Peptides

There are different kinds of antibiotic peptides and different models of the mode of action. The Shai-Matsuzaki-Huang model combines the most prominent aspects of the other models. The process which is distributed by the Shai-Matsuzaki-Huang model can be concluded in six steps. First, the antibiotic peptide (AMP) binds to the surface of the membrane. Secondly, the peptide self-inserts into the lipid membrane. At a critical peptide concentration, membrane lysis by pore formation is induced (step three). Now, lipid molecules and peptides can travel through the membrane into the cell (step four). The steps five and six represent alternatives or combinations of the acting of the AMPs. They can travel to intracellular targets or they can completely disintegrate of the target cell membrane.[2][3][4]

In this Paper, two different AMPs are used and compared, the human cathelicidin derivative LL32 and the peptide-antibiotic polymyxin B (PMB). In figure 3 the molecular structure of LL32 and Atto488 (the fluorescent label) are visualized and in figure 4 the structure of PMB is shown.

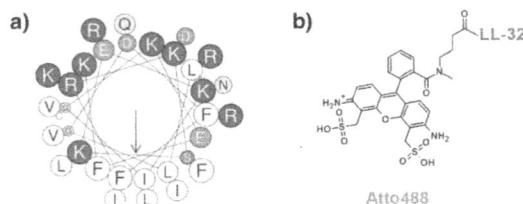

Figure 3: Illustration of the structure of AMP LL32. (a) α-helical wheel projection of the amino acid sequence of the LL32. (b) The structure of the labeling of Atto488-fluorescent to the N-terminus of LL-32. Fluorescent dye conjugate Atto488-carboxy introduced as fluorescence probe to the N-terminus of LL-32 [1]

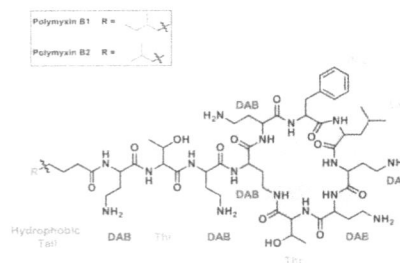

Figure 4: Illustration of the structure of polymyxin B.[1]

2.3   Fluorescence   Recovery   After   Photo-bleaching (FRAP)

The basic physical basis of FRAP is illustrated in figure 5-7. The region of interest (ROI) is located and irreversibly bleached. To bleach the sample, a high-power laser illumination is used (figure 5). After a specific time, fluorescence is detected at reduced laser power. Fluorescence molecules can diffuse into the ROI. The intensity of fluorescence grows again (figure 6-7).

There are two processes to explain the behavior of the rate of fluorescence recovery. At first, the fast diffusion of lipid-dye conjugates and fluorescently labeled peptides. The second slower process describes the binding between labeled and potential binding components in the ROI.[1]

Two model membrane types were prepared (1,2-dioleoyl-snglycero-3-phosphocholine (DOPC) and a composition of 1,2-dioleoyl-sn-glycero-3-phosphocholine (DOPC), N-(octadecanoyl)-sphing-4-enine-1-phosphocholine   (SM), cholest-5-en-3β-ol (Chol) (DOPC/SM/Chol)) and 150 μL

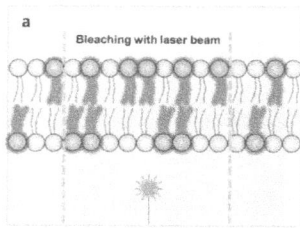

Figure 5: Bleaching with laser beam. The model membrane is labeled with Atto633-DOPE (0.05mol%, 1:1000, v/v) (darker lipids). A laser beam bleaches the sample in a defined area over a defined time (10 min). [1]

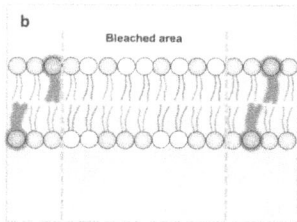

Figure 6: Bleached area. After the photobleaching, the fluorescence dye does not emit photons. The sample looks dark in a confocal microscope. [1]

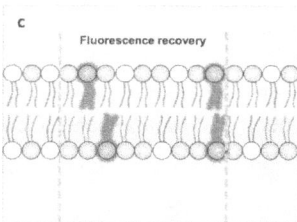

Figure 7: Fluorescence recovery. After a specific time, the fluorescence recovers in the bleached area. [1]

of this SUV's spread over night in a 8 well microscopy chamber (Nunc™ Lab-Tek™ II Chambered Coverglass, ThermoFisher Scientific, Roskilde, Denmark). The concentration of the applied lipid was varied in three steps: 100 nM, 500 nM and 1000 nM. The measurements were performed on a home-built confocal microscope. The objective in the confocal microscope was a water immersion objective with a N.A. of 1.2 (Plan Apochromat VC, NIKON, Melville, NY). [1][6]

The membrane was labeled with Atto633-DOPE (0.05mol%, 1:1000, v/v). A high performance laser system (OBIS™ 685 nm LX 40 mW, Edmund Optics, Barrington, NJ, USA) excited the sample at 630 nm wavelength. To use FRAP-effects, the laser system was set to maximum power (110%, 44 mW) for 10 min. The fluorescence recovery was detected in line scan mode (256 pixels, 20x20 μm, 2% laser power (1 mW), 8 ms per pixel integration time).

The excitation of the peptide-bound fluorophores

(BODIPY-, Atto488-) was realised with a solid-state continuum wave laser at 480 nm wavelength and an incoming power of 5 μW (Newport Spectra-Physics, Darmstadt, Germany).[1] To scan the sample, a combination of a piezo actuated objective stage and a x-y stage (PIfoc, PI, Karlsruhe, Germany) was used. [6]

3 Results and Discussion

The following figures show confocal images of the model membranes with antimicrobial peptides. The colour scale bars indicate the emitted photons. If there are a lot of events (emitted photons) at a point, this point is brighter than the background. At first, the SUV's (small unilamellar vesicles) and model membranes were prepared (in the laboratories of the institute of physics at the University of Lübeck). Figure 8 shows a confocal image of the spread membrane. The typical dapple pattern of a membrane is visible.

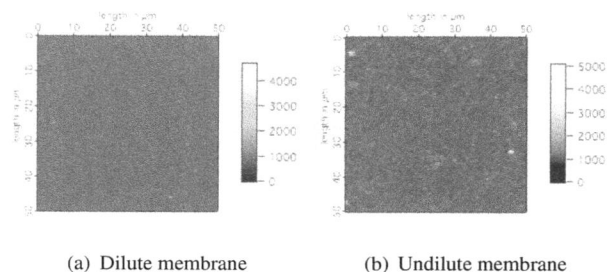

(a) Dilute membrane   (b) Undilute membrane

Figure 8: Comparison of dilute and undilute membrane. (a) shows the dilute membrane. Therefore pure DOPC and HEPES (pH 7.4) is used in a ratio of 1/10 (0.1mg/mL). The confocal microcope image shows a regular membrane. (b) shows the undilute membrane (1mg/mL). There are a few bright dots to see. This dots are unspread SUV's.

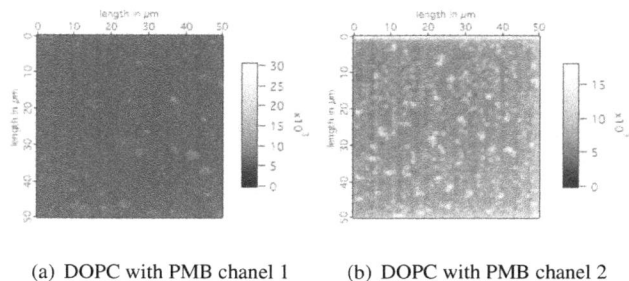

(a) DOPC with PMB chanel 1   (b) DOPC with PMB chanel 2

Figure 9: Confocal images of a sample illuminated at two different wavelengths. (a) Several bright dots in the DOPC membrane. (b) BODIPY-PMB was applied in a concentration of $1\mu M$.

At second, the two mentioned membrane types (pure DOPC and 2:2:1) are compared under the influence of PMB and LL32 (figures below). Previous experiments already showed the clustering behavior of PMB [1], that could be reproduced and confirmed herein (figures 9 and 11) and the membrane under the influence of LL32 shows no clusters of LL32 (figures 10 and 12), but in the figures 9-12 the membrane structure is not clear to see. The experiments should be repeated and optimizations to the system are ongoing.

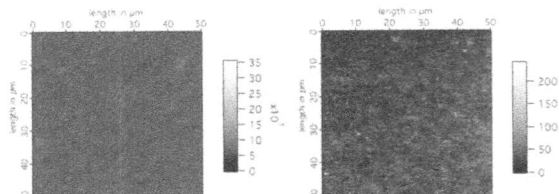

(a) DOPC with LL32 chanel 1   (b) DOPC with LL32 chanel 2

Figure 10: Confocal images of a sample illuminated with two different wavelengths. (a) A few bright dots in the DOPC membrane. (b) LL32 was applied in a concentration of 1000nM.

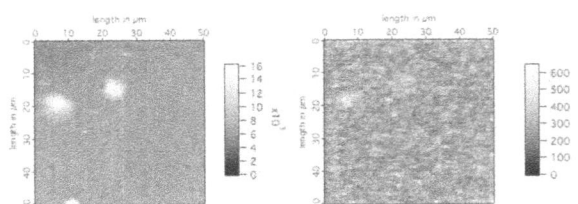

(a) DOPC/SM/CHOL with PMB chanel1   (b) DOPC/SM/CHOL with PMB chanel 2

Figure 11: Confocal images of the sample illuminated with two different wavelengths. (a) Several bright dots in the DOPC membrane. (b) BODIPY was applied in a concentration of 1000nM.

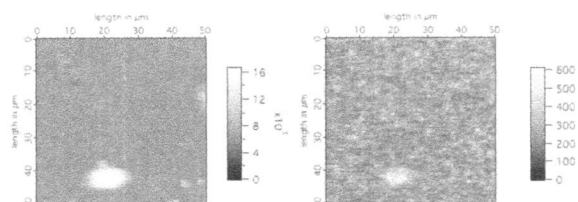

(a) DOPC/SM/CHOL with LL32 chanel 1   (b) DOPC/SM/CHOL with LL32 chanel 2

Figure 12: Confocal images of a sample illuminated with two different wavelengths. (a) Several bright dots in the DOPC membrane. (b) LL32 was applied in a concentration of 1000nM.

The bleached area in figure 13 is not clear to see because of the heavy fluorescence of peptid-cluster (bright dots). The experiments of this paper could not reach a FRAP image of a better quality. The hole FRAP method has to be adapted to the experimental requirements for biological and model membranes.

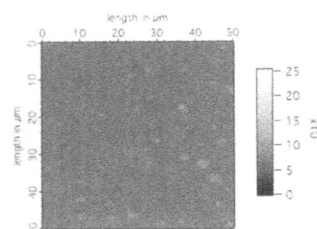

Figure 13: Fluorescence recovery by pure DOPC. After a specific time, the fluorescence recovers in the bleached area. This process is detected in a line scan mode with an integration time of 8 ms per pixel. Bodipy-PMB was applied in a final concentration of $1\mu M$. [1]

4 Conclusion

Model membranes were succesfully prepared with different equipment. The following experiments have to be overworked and should be measured again. Next, the experimental set up will be equipped with a temperature control, for enabling temperature-dependent measurements and the influence of domain formation on the lipid mobility and the membrane activity of AMPs.

5 Acknowledgement

The work has been carried out at the Institut of Physics, University of Lübeck and the Research Center Borstel (Division of Biophysics).

6 References

[1] L. Paulowski, *Anti-inflammatory Regulation of Immune Cells by Membrane Active Host Defense Peptides*. Universität zu Lübeck, Lübeck, 2016.

[2] L. Yang, T. M. Weiss, R. I. Lehrer and H. W. Huang *Crystallization of antimicrobial pores in membranes: magainin and protegrin*. Biophys J, 2000.

[3] K. Matsuzaki *Why and how are peptide-lipid interactions utilized for self-defense? Magainins and tachyplesins as archetypes*. Biochim Biophys Acta, 1999.

[4] Y. Shai *Mechanism of the binding, insertion and destabilization of phospholipid bilayer membranes by alpha-helical antimicrobial and cell non-selective membrane-lytic peptides*. Biochim Biophys Acta, 1999.

[5] Ibidi Labware, *μ-Slide 8 Well*. Available: http://ibidi.com/xtproducts/en/ibidi-Labware/Open-Slides-Dishes:-ibidi-Polymer-Coverslip/m-Slide-8-Well [last accessed on 11.01.2017].

[6] R. Borner, N. Ehrlich, J. Hohlbein, C. G. Huber, *Single Molecule 3D Orientation in Time and Space: A 6D Dynamic Study on Fluorescently Labeled Lipid Membranes*. J.Fluoresc, 2016.

Characterization of the interaction between the antimicrobial peptide LL-32 and a lipid monolayer using a film balance

J. Mertens[1], C. Nehls[2], and T. Gutsmann[2]

[1] Medizinische Ingenieurwissenschaft, Universität zu Lübeck, mertens@student.uni-luebeck.de
[2] Forschungsgruppe Biophysik, Forschungszentrum Borstel, Universität zu Lübeck, {cnehls, tgutsmann}@fz-borstel.de

Abstract

This paper features tension measurements by film balance in order to give information about the interaction of the antimicrobial peptide LL-32 to a reconstituted bacterial membrane. The work has been carried out by use of the KSV Nima (KN2001) to measure the lateral pressure and the mean molecular area between lipids. The intercalation depends on the lateral pressure and the state of the lipid-monolayer as well as the concentration of the used peptide, which is shown in the results. An approximation of the measured data can be made by a linear or polynomial model to gain comparable parameters. The reaction velocity can be described by the slope of the extracted line. Different intercalation rates for the same peptide concentrations can occur. When 30 mNm^{-1} is reached, the assumed maximum pressure for a still stable membrane, the experiments showed no intercalation reaction between the two substances.

1 Introduction

Every organism possesses a surrounding envelope, the biological membrane, that protects the living area of the organism from lifeless area surrounding it. Pathogens can attack the cell membranes of the host organism and can lead to severe diseases. If the immune system does not suffice to defeat a bacterial infection, antibiotics can be applied. During past decades, a considerable amount of bacteria developed resistances against this healing method. Research for alternatives to cope with these new pathogens has been going on for a long time. A possible option could show up in the form of antimicrobial peptides (AMPs). To investigate the effectiveness against such resistant bacteria, this paper is focused on a common model system for bacterial membranes [1].

Due to their amphiphilic character, lipids cause a higher order of the surrounding water molecules when in aqueous environment. This leads to a change in the entropy of the system and thus the lipids form structured aggregations. This process is called self-aggregation [2]. One method for measuring surfactant characteristics is the film balance or Langmuir-Trough. Langmuir discovered that fatty acids build monolayers when they get in between two phases such as gaseous and fluid phases [4]. Their hydrophobic part aligns itself in the direction of air. A film balance consists of a trough, a force transducer connected to a Wilhelmy-plate and movable barriers to control the film area.

2 Material and Methods

The measurements were performed by using the film balance KSV Nima (KN2001). Equation (1) describes the recording of the lateral pressure used by the film balance:

$$F = \varrho_p \cdot g \cdot l_p \cdot w_p \cdot t_p + 2\gamma \cdot (t_p + w_p)(cos\theta) - \varrho_i \cdot g \cdot t_i \cdot w_i \cdot h_i. \quad (1)$$

Here, ϱ_p stands for the density of the Wilhelmy-plate, g is the gravitational constant, l_p, w_p and t_p are the dimensions of the Wilhelmy-plate, γ is the surface pressure, θ the contact angle between plate and liquid, ϱ_i is the density of the liquid, t_i and w_i are the dimensions of the trough and lastly h_i stands for the immersion depth in the liquid.

Table 1: Overview of used substances

Substance	Molecular weight	Short
L-α-phosphatidylethanolamine	719.30	PE
L-α-phosphatidylglycerol	761.07	PG
Antimicrobial peptide LL-32	3922	LL-32
Chloroform	119.38	CHCl$_3$
2-[4-(2-hydroxyethyl)piperazin-1-yl]ethanesulfonic acid	238.31	HEPES
Potassium chloride	74.55	KCl
Potassium hydroxide	56.11	KOH
Ultrapure water	18.02	aqua ster.dest.

For the recording of the data a rectangular filter paper, as a Wilhelmy-plate with a perimeter of 20.6 mm was used. A water quench was used to stabilize the temperature at 25 °C throughout the experiments. The organic and inorganic materials that were used are shown in table 1. The injections were done with the syringe pump Symax (Spetec GmbH, Erding, Germany), the application of the lipids onto the liquid surface was done using Hamilton syringes with volumes of 20 and 50 μl.

A buffer solution with a pH-value of 7.0, 100 mM KCl and 5 mM HEPES was prepared. For the bacterial lipid mixture stock solutions of PE and PG in $CHCL_3$ were mixed to obtain a lipid ratio of PE:PG 1:1 M:M. The peptide LL-32 was used in a concentration of 1 mg/ml dissolved in buffer. The peptide is part of the human cathelizidin hCAP18 of the human immune system. It is a variation of the peptide LL-37 with a deletion of five C-terminal amino acids and is manufactured by Merrifield-synthesis in the Research Center Borstel [2], [3]. It operates for example during the phagosome maturation and possesses antimicrobial characteristics.

The approach for all experimental setups was as follows. The film balance trough was filled with 60 ml buffer solution. Afterwards, 7 μl of the PE:PG 1:1 M:M were applied to the surface by using the Hamilton syringe and the peptide solution was filled in the accompanying contraption of the syringe pump.

For the first series of experiments, when the target value of the lateral pressure was reached, the barriers were fixed at their position. Afterwards, the peptide was injected periodically with a constant volume and changes in the lateral pressure were recorded. The second series consists of reaching a high lateral pressure of 30 mNm^{-1} and subsequently injecting the peptide with a single injection volume. The lateral pressure was kept constant and the movement of the barriers was recorded while injecting the LL-32. After a two hour incubation time, an isotherm of the formed membrane-mix was measured. The film area is monotonically decreased, the lipid-membrane is compressed and then expanded again. Once the critical pressure is reached, the membrane gets unstable. A simultaneous change in the surface pressure and in the available area is reached through barrier movement [5]. For the third set, the lateral pressure was held constant while changes in the film area were recorded. By fitting a linear model to the measurement curves the extracted parameters slope and intercept of the lateral pressure curves or the film area curves, represent the reaction velocity and the starting pressure set for the experiments to illustrate reproducibility.

3   Results and Discussion

In Fig. 1 the pressure was held constant whereas the area varied. LL-32 was injected periodically for ten hours. The dashed line shows the average molecular area (MMA) per lipid at 20 mNm^{-1}, the solid line shows the MMA at 10 mNm^{-1} and the dotted line represents the control measurement with its median. The results outlined in Fig. 1

show the intercalation of the AMP into the lipid-monolayer. The control measurement shows no significant variation in the mean molecular area over the measured time period.

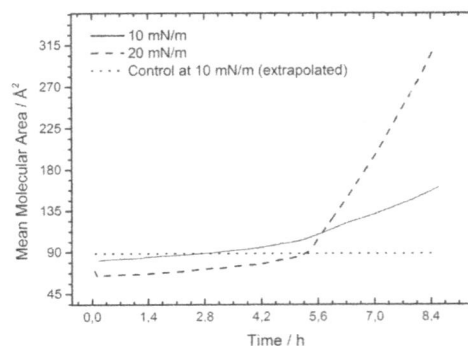

Figure 1: Results for the average of four sets of measurements with constant pressures and continuous peptide injection.

The buffer solution is thus disqualified to have caused the ascent of both averaged measurements at 10 mNm^{-1} and 20 mNm^{-1}. At approximately 5.5 hours, the increase of the MMA at 20 mNm^{-1} is 70 times higher than the rise at 10 mNm^{-1} - if the measurement would be prolonged one would see a continuous intercalation effect.

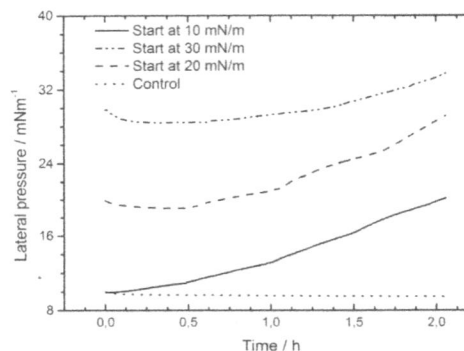

Figure 2: Lateral pressure for three measurements per average and continuous peptide injection with constant area.

The corresponding parameters for the intercalation rate for the linear fit of Fig. 1 are shown in Table 2. Fig. 2 shows the lateral pressure after peptide addition with a start value of 10 mNm^{-1} (solid), 20 mNm^{-1} (dashed) and 30 mNm^{-1} (dash dot), respectively. In this set of experiments, the barrier position was held constant while the peptide was injected. The recording lasted for two hours. Outlined in Fig. 2 is the intercalation of LL-32 as can be seen by the rise of the lateral pressure in contrast to the control measurement. For the measurements shown in Fig. 3 a start value of 10 mNm^{-1} (dashed), 20 mNm^{-1} (solid) and 30 mNm^{-1} (dash dot) have been used.

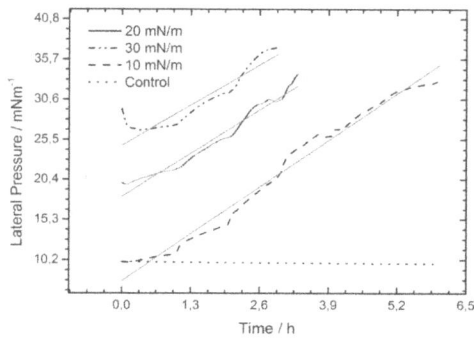

Figure 3: First phase of three averaged measurements per start-pressure with a peptide injection over ten hours and their linear fits.

In this setup, the barrier positions were held constant when the start value was reached. The dotted line represents the control measurement where buffer solution was injected. Fig. 3 shows the peptide intercalation also at different start pressures but with a fivefold increased incubation time in comparison to Fig. 2. The specified graph shows the measurements when they first reached their maximum pressure (in between 32-35 mNm^{-1}), the whole curve is divided in three characteristic parts (increase, stagnation and decrease) where two of them are presented in Figures 3 and 4. The incubation time between injections was doubled compared to those in Fig. 2. The curves and consequently the interaction between lipid and peptide are similar in Fig. 2 and Fig. 3. A saturation of peptide in the membrane is not indicated in these measurements.

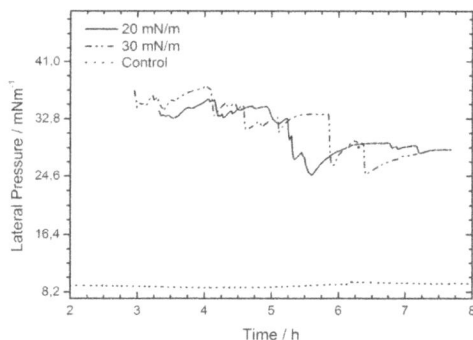

Figure 4: Second phase of three averaged measurements per start-pressure with a peptide injection over ten hours.

The solid line (20 mNm^{-1}) and dash dot line (30 mNm^{-1}) reveal a very distinct pattern for LL-32 at high pressures. This characteristic is repeatedly shown in all measurements. After reaching 30 mNm^{-1} or higher, it can be said that the system is in a meta-stable state where lipid and peptide compete for space at the surface. Fig. 5 shows the isotherm representing the behavior of the lipid membranes without any and with peptide, the lateral pressure is plotted against available area. The injection volume is equal to that in Fig. 3 and 4. The curves show the condition of the physical system. The monolayer is continuously compressed and expanded again.

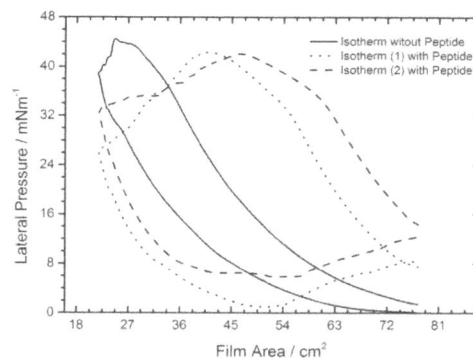

Figure 5: Isotherms after injection of LL-32 and a two hour incubation time.

Once the barriers are moved back to the opened position, the pressure decreases rapidly, indicating that the lipid-peptide-mix changes to a gaseous state. After the minimum pressure is reached, the lipids or peptides pushed down into the buffer begin to reintegrate into the monolayer as is indicated by the rise at around 54 cm^2. Also the meta-stable state is reached faster in the measurements with peptide at around 45 cm^2 compared to without peptide where the maximal pressure lies at 27 cm^2.

Table 2: Statistics of experiments with constant pressure

Lateral Pressure	Intercept	Slope
10 mNm^{-1}	66.798	0.003
20 mNm^{-1}	12.186	0.007
30 mNm^{-1}	130.519	-0.036

For comparison, also the extracted values from the averaged 30 mNm^{-1} measurements are shown. The corresponding curve is not shown in Fig. 1 due to a large scale difference. In order to evaluate statistically, the Residual Sum of Squares as well as the Pearson's R-coefficient are represented in Table 3. The Residual Sum of Squares is a quality criterion which describes how fatal an error affects the model whereas the Pearson's R-coefficient is a linear dependency between two variables, here the measurement data and the fit line.

Table 3: Statistics of experiments with constant pressure

Lateral Pressure	Residual Sum of Squares	Pearson's R
10 mNm^{-1}	$5.71 \cdot 10^4$	0.945
20 mNm^{-1}	$1.31 \cdot 10^6$	0.855
30 mNm^{-1}	$2.51 \cdot 10^5$	-0.997

The results show that except of the data obtained at 20 mNm^{-1}, the linear model is suitable for this kind of measurement.

Table 4 corresponds to Fig. 2, where the curves were linearly fitted as well.

Table 4: Statistics of experiments with constant area over two hours

Lateral Pressure	Intercept	Slope
10 mNm^{-1}	8.617	$1.42 \cdot 10^{-3}$
20 mNm^{-1}	17.315	$1.35 \cdot 10^{-3}$
30 mNm^{-1}	27.555	$7.85 \cdot 10^{-4}$
Control	9.978	$1.80 \cdot 10^{-5}$

Table 5: Statistics of experiments with constant area over two hours

Lateral Pressure	Residual Sum of Squares	Pearson's R
10 mNm^{-1}	$2.765 \cdot 10^2$	0.986
20 mNm^{-1}	$7.802 \cdot 10^2$	0.956
30 mNm^{-1}	$4.440 \cdot 10^2$	-0.898
Control	5.806	0.896

In Tables 6 and 7 the corresponding measurements with an incubation time over ten hours are evaluated.

Table 6: Statistics of experiments with constant area over ten hours, first phase

Lateral Pressure	Intercept	Slope
10 mNm^{-1}	7.597	$1.26 \cdot 10^{-3}$
20 mNm^{-1}	18.254	$1.17 \cdot 10^{-3}$
30 mNm^{-1}	24.765	$1.09 \cdot 10^{-3}$
Control	9.978	$1.80 \cdot 10^{-5}$

Table 7: Statistics of experiments constant area over ten hours, first phase

Lateral Pressure	Residual Sum of Squares	Pearson's R
10 mNm^{-1}	$4.76 \cdot 10^3$	0.991
20 mNm^{-1}	$1.12 \cdot 10^3$	0.895
30 mNm^{-1}	$3.38 \cdot 10^3$	0.945
Control	5.806	0.896

As can be seen, the values do not differ much from the computed values shown in Tables 4 and 5. In relation to reaching the critical pressure, it can be said that the short measurements of two hours are sufficient in their significance. In Table 3 the Residual Sum of Squares possesses a large error but the Pearson's R equals almost one, which means the model is in line with the measured data despite the model's error. Table 5 shows an error which is nearly one third of that in Table 3 and Pearson's coefficient is close to one. This is the expected result for a model that is suitable for the measurement data. The comparison between Table 4 and 6 gives a similar magnitude of the slope values despite their different incubation times, in Table 6 the incubation time was doubled.

4   Conclusion

As shown in the discussion, the AMP LL-32 intercalates into the lipid-monolayer as can be seen in either Fig. 1 or Fig. 2. A pressure of around 30-35 mNm^{-1} can be interpreted as the maximal possible lateral pressure for a PE:PG monolayer, when this point is reached the pressure decreases rapidly and increases again slowly as shown in Fig. 4. This behavior could be explained by the interaction of the peptide to the peptide-lipid-mix. The PE:PG monolayer could compete with the newly injected LL-32 for space on the surface. Once the minimal area per lipid in the monolayer is reached, lipid is forced out or they overlap and new peptide intercalates or the other way around. This course of the curve is a recurring event at the critical lateral pressure specific for this membrane composition. Different concentrations affect the intercalation - at higher concentrations the

intercalation is faster and the maximum pressure and consequently the unstable state is reached more quickly.

To determine the maximum insertion point of LL-32, the critical pressure at which the intercalation vanishes must be determined. We assumed this critical pressure to be near 30 mNm^{-1} but the experiments proved this not to be the case. The lipid membrane was not stable enough to hold this pressure constant for a given time interval, and further there was no intercalation of peptide into the monolayer at this pressure. It can be assumed that the peptide interacts with the monolayer in a way that the state of the membrane destabilizes.

Furthermore, in the experiments with constant area, the peptide-lipid-mix does not go into saturation. This aggravates the calculation of peptide in the monolayer so that the constant change in lateral pressure or area cannot be clearly attributed to a certain concentration of peptide. During the longterm measurements of over ten hours, the buffer solution crystallizes out on the plate mounting, the temperature of 25 °C does not seem to be suitable for long term measurements.

Acknowledgement

The work for this paper has been done in cooperation with the Research Center Borstel, especially in the division of biophysics led by Prof. Thomas Gutsmann. Furthermore, I would like to thank Dr. C. Nehls for his help and great support. I also would like to thank T. Gutsmann for the possibility to work in his group and his commitment in all biophysical questions.

5   References

[1] A. Hädicke, *Interactions of cationic peptides with anionic lipid bilayers and monolayers : influence of peptide and lipid modifications on bindin.* urn:nbn:de:gbv:3:4-17648, Martin-Luther-Universität Halle-Wittenberg, 2016.

[2] C. Nehls, *Charakterisierung der Wechselwirkung zwischen dem bakteriellen Protein VapA und der Phagosomenmembran an Modellmembransystemen.* Forschungszentrum Borstel, 2016.

[3] T. Gutsmann, S. Hagge, J. Larrick, U. Seydel, A. Wiese *Interaction of CAP18-derived peptides with membranes made from endotoxins or phospholipids.* Biophysical Journal 80 (2001), Nr. 6, pp. 2935-2945.

[4] S. Gromelski, *Wechselwirkung zwischen Lipiden und DNA - Auf dem Weg zum künstlichen Virus.* urn:nbn:de:kubv:517-opus-7629, Potsdam, 2006.

[5] E. Eugster-Meier, *Adsorptionsverhalten von Proteinen und niedermolekularen Lipiden der Milch an Phasengrenzflächen.* http://dx.doi.org/10.3929/ethz-a-004128810, Technische Wissenschaften ETH Zürich Nr. 14076, 2001.

Concepts for a process-safe application of a micro-precipitation method

R. Kaya[1], T. Guderjahn[2], C. Odefey [2], U. Eisenblätter[2] and K. Lüdtke-Buzug[3]

[1] Medizinische Ingenieurwissenschaft, Universität zu Lübeck, rafiye.kaya@student.uni-luebeck.de

[2] m-u-t GmbH, {tguderjahn,codefey,ueisenblaetter}@mut-group.com

[3] Institute of Medical Engineering, Universität zu Lübeck, luedtke-buzug@imt.uni-luebeck.de

Abstract

Nanotechnology, or device manufacture at the molecular level, is a multidisciplinary scientific field undergoing a rapidly progress. In view of the enormous global burden of infectious diseases it is expected to open application possibilities for the diagnostic, prevention and surveillance of living creatures and devices that can change over time due to the process with various bacteria. *Quantitative highest Sensitivity Particle measurement* (QSP) is a system which provides a first approach for the preventive measures in molecular diagnostic. The basis of the technology is a particle detector functioning as a molecular light barrier. In combination with magnetic nanoparticles, which get coated with antibodies, it is possible to determine size and quantity of particles. In close collaboration with interested companies and hospitals feasibility studies for the detection of the bacteria *Legionella* were carried out with promising results regarding prevention measures.

1 Introduction

Nanotechnology is the treatment of individual atoms, molecules, or compounds into structures to produce materials and devices with special properties [1], [4]. The ability to uncover the structure and function of Biosystems at the nanoscale, stimulates research leading to improvement in biotechnology [8]. The size of nanomaterials is similar to that of most biological molecules and structures; therefore, nanomaterials can be useful for both in vivo and in vitro biomedical research and applications [8]. In view of the 14.9 million deaths worldwide due to infectious diseases significant effort goes into developing new technologies as automatic measuring systems for the estimating size and quantity of particles [2]. This vision provides new standards regarding prevention approaches in biomolecular science. Moreover, it is cumulative crucial to process data from automatic systems in- and on-line as it arrives, both from the point of adaptation to changings of the system caused through bacteria. Because as well newly emerging as consisting bacteria still cause scientific uncertainties due to their affiliation to the most common causes of death and disability worldwide [1]. As a highlight one of the feasibility studies based on the technology Quantitative highest Sensitivity Particle measurement (QSP) for the detection of Legionella is executed in detail. The results promise a highly specific and automatic solution approach in flow cytometry for the real-time measurement and surveillance of various bacteria.

1.1 Legionella spp.

Legionella (Fig. 1) is a rod-shaped, gram-negative bacteria [10]. It can be found in freshwater and soil worldwide and tend to contaminate man-made water systems [10]. Currently, about 57 different legionella species are known which comprise at least 79 different serogroups [10]. All species are classified as potentially human pathogens [10]. At temperatures between 25 °C and 45 °C the bacteria has the best settings to multiply very quickly biofilm-associated [10].

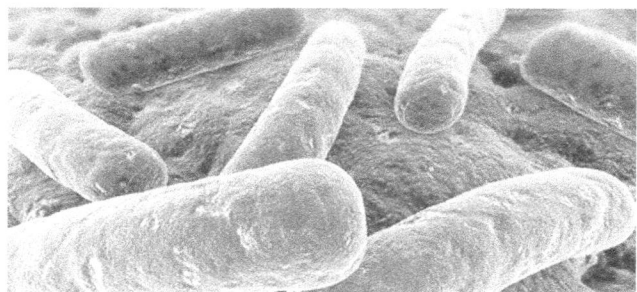

Figure 1: The rod-shaped, gram-negative bacteria Legionella is figured [12].

The most important specie for causing disease is Legionella pneumophila [10]. The disease is classically described as a severe pneumonia that may be accompanied by systemic symptoms such as fever or diarrhoea for example and has to be medicated with antibiotics [10]. A transmission from person to person is not proved. Generally it is caused through inhalation of contaminated aerosols or aspiration of contaminated water [10]. Despite recent advances in molecular methods, the official, internationally accepted detection method for Legionella in water samples (ISO 11371) is still based on cultivation [10]. This method has major disadvantages such as a long assay time of 10 days and the detection of cultivable cells only by surface-counting [10]. Due to the importance of water for the mankind, a cultivation-independent, sensitive, fast detection method is necessary for the surveillance of water systems to prevent disease outbreaks.

2 Material and Methods

2.1 Experimental setup

As seen on Fig. 2 the technology consists of three main units: a laser, a measuring chamber and a detector. In principle, this technology is a scattered-light measuring device, which is able to detect particle aggregations or particle growth quickly and much more sensitively with a very low detection limit of 10 fg/l (one hundred billion grams per liter).

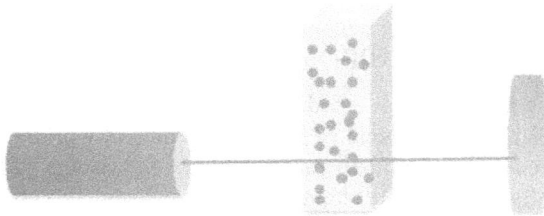

Figure 2: The measuring system of QSP is illustrated. The technology consists of three essential units: a laser, a measuring chamber and a detector.

The functionality is strained by the classical process particle measurement technology, which is widely used in laboratory measurement instruments. Any particle that passes through the laser beam produces a signal. Larger particles scatter the light more than smaller particles. Particles may pass anywhere along the length of the exposed laser beam. The number of signals corresponds to the statistical distribution of the particles in the measuring chamber. If several particles are simultaneously in the laser beam, only the largest is measured. On the basis of the different velocities at which the particles pass the laser beam, the respective particle size can be determined from 20 nm to 5 μm (Stokes-Einstein equation). The cleaner the solution, the more likely it is that particles are particular in the laser beam. The more polluted the solution is, the more the signals superimpose, resulting in reduced sensitivity. Therefore QSP decides from other analyzers by giving more information by diluting strongly (millions or even millions of times in order not to exceed the device) than by concentrating.

2.2 Measuring principle

QSP is a device that evaluates real-time measurement of micro precipitations of antigen-antibody reaction products based on the irregular particle signal resulting from the Brownian motion. For a cross linking reaction to take place, the particles contained in the sample liquid must have at least two antibody binding sites so as to be able to function as an antigen. All body-borne substances (bacteria, viruses, toxins) are antigens. When cross linking antibodies with antigens, larger particles are formed than before. This is exactly the reaction the technology makes use of. However, this is only possible with the *magnetic nanoparticles* (MNPs).

Magnetic Nanoparticles (Fig. 3) are aggregates of atoms or molecules that have controllable sizes ranging from a few nanometers up to tens of nanometers, which places them at dimensions that are smaller than or comparable to those of a cell (10–100 μm), a virus (20–450 nm), a protein (5–50 nm) or a gene (2 nm wide and 10–100 nm long) [5]-[6].

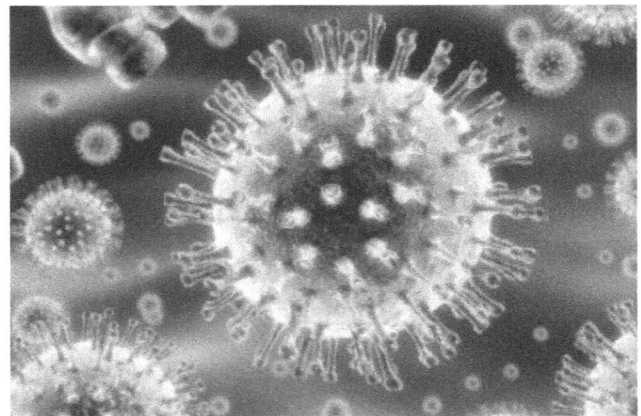

Figure 3: Magnetic Particles [5].

In general they represent a new class of particles due to their unique size, whereby they can get close to a biological entity of interest [5], [7]. They can be coated with biological molecules force them to interact or tag to a biological entity. Thus the MNPs build the coupling surface for efficient protein immobilization. With this background and the possibility of a high magnetization value of the particles, they can be manipulated by an external magnetic field, which is enabling the detection the entities in question at a distance [5]-[7]. In Fig. 4 the measuring principle of QSP is illustrated. To prepare the detection cocktail, the reaction vessel is first coated with the corresponding antibodies for a specific antigen. The components of the lysed bacteria function as antigens. The capture antibody can now cross link with the corresponding epitopes of the antigen. After the magnetic nanobeads are added, they perform the same reaction and bind to the corresponding epitopes of the antigens (path A) of Fig. 4). The antigens thus act as a binding member between the MNPs and the antibodys. The liquid sample can now be analysed by QSP, which is able to determine the specific bindings of the MNPs with a high sensitivity. If the liquid sample does not contain any associated antigen,

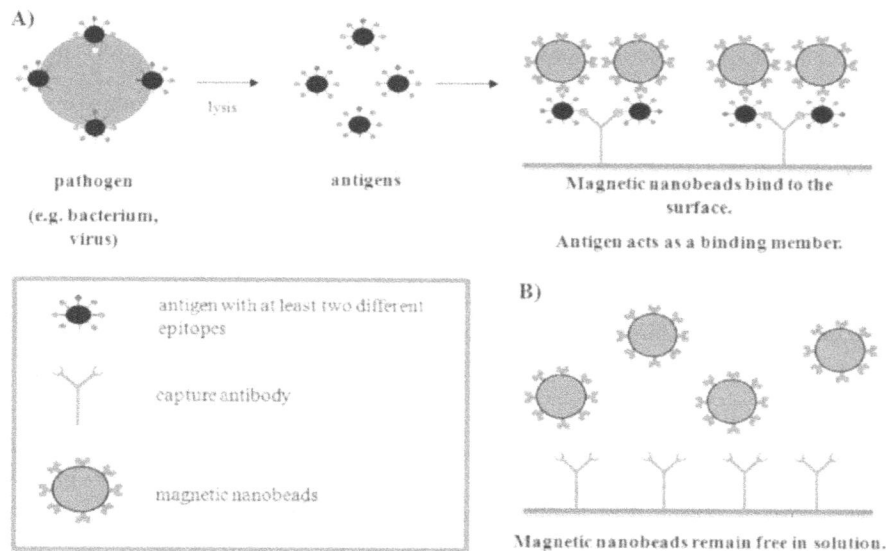

Figure 4: The measuring principle of QSP is illustrated. The first path demonstrates a succesful tagging of the MNPs with the entities of interest. Path B) shows the negative result.

the test will be negative(path B), Fig. 4).

3 Results and Discussion

3.1 Possible application-field: *Detection of Legionella*

For a first feasibility study an unknown liquid sample was provided by a company. In view of the various legionella species, a polyclonal antibody was used which could bind to epitopes of antigens of different legionella species. Table 1 shows an extract of the legionella studies.

Table 1: An extract of the measurement results of the legionella detection. One sensor level (sl) corresponds to ten thousand bindings of the MNPs.

Nr.	Zero value in sl	Sample value in sl	Amount of bound particles
1	25	43	180.000
2	25	44	190.000
3	25	37	120.000
4	25	49	240.000
5	25	51	270.000

Five results of the measurements are listed by way of example, with 50 μl of the eluate (= detection cocktail) used in each case. The values were always given in sensor levels (sl), one sensor level corresponding to 10,000 bound MNPs. Before the actual detection of the sample began, pure water was measured, which has functioned as zero value. The 25 sl for water is an approximate value, which is always in this range because of the uniformly scattered light quantity. The device has shown a display value for the samples in the range of 37 to 51. The first measurement delivered

for example a sample value of 43 sl. However, it is important to note the difference of 18 sl to the zero value. This would mean that 180,000 bound particles are present in the eluate. In order to ensure that only the bounds nanobeads have been measured a crosscheck must be carried out. This can be specified by the magnetic property of the nanobeads. The application of a magnetic filter opens the opportunity to eliminate false positive results.

Table 2: The results after the application of a magnetic filter are listed. In relation to the amount of bound particles (Table 1)the filtration enables the detection of the amount of the real bound nanobeads.

Nr.	Sample value in sl	Magnetic filter value in sl	Real amount of bound beads
1	43	27	160.000
2	44	29	150.000
3	37	25	120.000
4	49	27	220.000
5	51	25	250.000

As seen on Table 2 the displayed value after the application of a magnetic filter was 16 sl with a difference to the bond particles of 2 sl. The real amount of bound MNPs is thus 160.000. The difference of 20.000 bound particles is an indication for nonspecific bonds. Possible reasons could be contaminations while diluting the eluate or inaccurate work. For the third measurement for example, which demonstrates a best case, no difference between the amount of particles in direct comparison to the real amount of bound MNPs could be determined. In both cases the technology detected an amount of 120.000 bindings. In this case no contamination could be recorded. The fifth measurement demonstrates with 27 sl the eluate with most bond particles. The real amount is however 25 sl.

Fig. 4 shows the contrast of the amount of bound particles

compared to real amount of bound beads. At first sight, it is noticeable that the difference for all five measurements does not exceed 50,000 bound particles. The third measurement is the only measurement where no difference could be determined after the application of a magnetic filter. For the fifth measurement, 10,000 nonspecific bindings were calculated. A difference of 20,000 bonds was recorded for the first and fourth measurements. With 40,000 bonds, the largest difference of the five measurements was found for the second measurement. In summary, four out of five measurements with a maximum of 20,000 nonspecific bindings have been reported.

4   Conclusion

The successful validation of the system is promising a huge potential to bring benefits in the biomolecular science. In consideration of the nanotechnology which encompass a range of techniques rather than a single discipline and stretch across the whole spectrum of science, physics, engineering and chemistry, QSP is an important future-oriented device [9]. The device is characterized by a very low detection limit of 10 fg/l and is able to determine the quantity and sizes of particles in the range of 20 nm to 5 μm. One of the special advantages is the acquisition of information by diluting, which enables the detection with a small amount of eluate. One the one hand the use of magnetic nanobeads expands the flexibility of the device and opens wide-ranging application possibilities on the other hand it increases the selectivity. The feasibility study relating to the legionella detection supports this hypothesis. The bacterium was detected in all five cases. A crosscheck by filtering the bound particles also made it possible to exclude false-positive results and to sensitify the quantification of the real amount of bound nanoparticles. Four of five measurements had very low nonspecific bound particles, which could originate due to inaccurate work. In further studies the detailed functions of an automatic detection system for bacteria will be studied.

Acknowledgement

The work has been carried out at m-u-t GmbH and supervised by the Institute of Medical Engineering, Universität zu Lübeck. Note: The elaboration contains confidential data. In this context not all research results are shown.

5   References

[1] D. F. Emerich and C. G. Thanos *Nanotechnology and medicine*. In: Expert Opinion on Biological Therapy, vol. 3, pp. 663–665, 2003.

[2] D. M. Morens, G. K. Folkers and A. S. Fauci, *The challenge of emerging and re-emerging infectious diseases*. In: Nature, vol. 430, pp. 242–249, 2004.

[3] K. E. Jones et al. *Global trends in emerging infectious diseases*. In: Nature, vol. 451, pp. 990-993, 2007.

[4] A. P. Nikalje, *Nanotechnology and its Applications in Medicine*. In: Medicinal chemistry, pp. 81–89, 2015.

[5] G. Schmid, *Nanoparticles: From Theory to Application*. Wiley-VCH Verlag GmbH & Co. KGaA, 2006.

[6] C. Wilhelm, C. Billotey, J. Roger, J. N. Pons, J. - C. Bacri and F. Gazeau, *Intracellular uptake of anionic superparamagnetic nanoparticles as a function of their surface coating*. In: Elsevier, 2002.

[7] L. Steinke, A. Vetter and S. Ripperger *Die Partikelzählung in Flüssigkeiten unter Berücksichtigung der Rückführbarkeit der Messgrößen*. In: Filtrieren und Separieren, Jahrgang 28, Nr.6, 2014.

[8] M. Singh, S. Singh, S. Prasad and S. Gambhier *Nanotechnology in Medicine and antibacterial effect of silver nanoparticles*. In: Digest Journal of Nanomaterials and Biostructures, vol. 3, no. 3, pp. 115–122, 2008.

[9] D. Karunaratne, *Nanotechnology in Medicine*. In: J. Natn.Sci.Foundation Sri Lanka, vol. 53, no.3, pp. 149–152, 2007.

[10] D. Karunaratne, *Nanotechnology in Medicine*. In: J. Natn.Sci.Foundation Sri Lanka, vol. 53, no.3, pp. 149–152, 2007.

[11] ECDC Surveillance Report, *Legionnaires' disease in Europe 2014*. Available: http://ecdc.europa.eu/en/publications/Publications /legionnares-disease-europe-2014.pdf [last accessed on 2016-12-30].

[12] All Seasons Hire, *Legionella: What it is, why it Matters and How to Prevent it.*. Available: http://allseasonshire.eu/blog/post/legionella-matters-prevent/ [last accessed on 2017-02-07].

Characterization of nanofluidics devices for high-throughput single-molecule fluorescence detection

T. Wenzel[1,2], C. Fijen[2], J. Hohlbein[2], and Christian Hübner[3]

[1] Medizinische Ingenieurwissenschaft, Universität zu Lübeck, timo.wenzel@student.uni-luebeck.de
[2] Laboratory of Biophysics, Wageningen University & Research, contact: johannes.hohlbein@wur.nl
[3] Institut für Physik, Universität zu Lübeck, contact: huebner@physik.uni-luebeck.de

Abstract

The continuous monitoring of enzymatic reactions in real time and at the single–molecule level is hugely challenging. Therefore, the Hohlbein group developed novel nanofluidic devices that combine the advantages of confocal microscopy in solution, namely the high temporal resolution and high sensitivity, with that of TIRF–microscopy, namely the ability of monitoring many individual fluorescent molecules in parallel. The nanochannels offer a height smaller than the height of the detection focus of a microscope objective such that the molecules stay in focus as they flow through the field of view (FOV). This special device geometry allows long–time observations detecting many molecules in parallel. Whereas the overall project aim is to monitor conformational changes of DNA polymerases during the DNA synthesis without quick depletion of a fluorescently labelled DNA polymerisation sensor, our task was to optimise parameters for the detection of single molecules within the devices.

1 Introduction

In the last 20 years, techniques based on the observation of single–molecule fluorescence enabled the determination of sub–populations in heterogeneous biological samples, the possibility to detect rare events and the observation of the temporal behaviour of protein-DNA interactions, protein folding and membrane proteins [1].

Single molecule detection depends strongly on the high sensitivity of the used optics and the technical realisation remains challenging. There are two major schemes for single-molecule-fluorescence detection: diffusion-based confocal microscopy in solution and TIRF-microscopy. In a confocal microscope, molecules diffuse freely through a femtoliter-sized detection volume. In TIRF-microscopy, hundreds of surface-immobilised molecules are imaged simultaneously with a sensitive camera. For the purpose of studying the dynamics and enzymatic reactions on the single-molecule level and in real time, the Hohlbein group developed novel nanofluidic devices that combine the strengths of both detection schemes: the ability to observe fast conformational changes of many single molecules (sm) in parallel with high time resolution. Through the specially designed geometry of the nanofluidics devices (Fig. 1), the inserted biomolecules flow inside a restricted volume, in which they are exposed to the evanescent field generated by the TIRF-setup (Fig. 2). By using a labelled DNA sensor changing its FRET signal after binding of polymerases, conclusions about the interactions and the conformational changes of the involved DNA can be drawn.

1.1 Nanofluidic devices

The rise of microfluidics in the 1990s and further advancements in nanotechnology, nowadays allow to manufacture devices with structural features smaller than 1 µm. In order to provide alternatives to utilise immobilised DNA molecules, the nanofluidic devices (Fig. 1) were developed and designed in collaboration with Klaus Mathwig (University of Twente, The Netherlands) and manufactured by Micronit Microfluidics (Enschede, The Netherlands) via wet etching. The nanochannels are flanked by a microchannel to ensure the required flow velocities of 10 to 1.000 µm inside a NC.

2 Material and Methods

2.1 Total Internal Reflection Fluorescence (TIRF) Microscopy

In this work, experiments were performed using a TIRF-microscope and stroboscopic laser alternating laser excitation, in which the donor and the acceptor fluorophore are excited consecutively and with an illumination time of approximately 1 to 2 ms during each camera frame of 10 to 50 ms [2]. A sufficiently high signal-to-noise ratio is required to analyse biomolecular structures and interactions with sm-FRET and to detect a diffusing or immobilised fluorescent molecule against the solvent background.

As shown in Fig. 2, the TIRF setup for single-molecule detection uses excitation laser light of up to four different

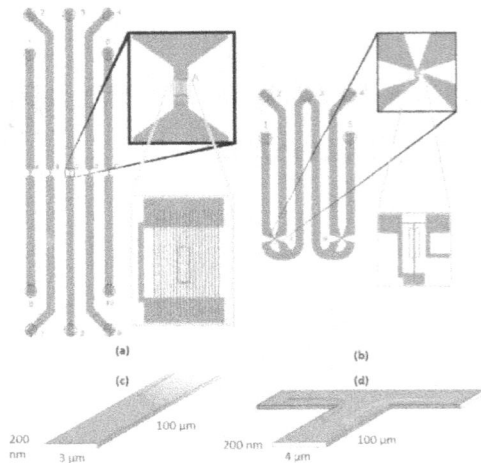

Figure 1: Schematic overview over the two different design types of the nanochannel slides. (a) A device contains five independent units with single inlets and single outlets and their respective array of nanochannels (pNC). (b) Schematic view showing two devices with a mixing nanochannel (mNC) geometry featuring two inlets and one outlet per unit. Labelled molecules (e.g. ssDNA) can be inserted from different sides by programmable syringe pumps, followed by mixing in the T-shape-NC. This configuration provides the possibility to observe non equilibrium studies as well as pre steady-state kinetic in the detection area. (c+d) Dimensions of the NCs. (Figure obtained from [3]).

colours, which is then focused into the back focal plane of the objective hitting the glass/sample-interface at a higher-than-critical angle ($> 61°$). Based on *Snell's law of refraction*, 100 % of the incident light is then reflected (i.e. total internal reflection) at the glass/water-interface and, hereby, an evanescent field is created, which decays exponentially within around 100 nm (thinning lines) illuminating just a tiny volume inside the sample. Hence, only in close proximity to the glass-interface labelled molecules can be excited and the area of excitation is roughly 50 µm by 50 µm. Due to its longer wavelength, the emitted fluorescent signal passes the polychroic mirror and is spatially filtered by an aperture. The fluorescence light splitted by dichroic mirrors is captured in three spectrally separated channels by an enhanced cooled charge-coupled device (emCCD) camera for further analysis.

2.2 Labelled DNA constructs

The used DNA samples consist of single stranded DNA oligonucleotides with 30 to 80 base pairs (bp), purchased from IBA, Germany (Table 3.1). The labelling and annealing protocols are reported in the Electronic Supplementary Material (ESI) of [2]. In order to monitor FRET, complementary DNA strands are annealed forming double-strand DNA (Fig 3). The acceptor–labelled linker strand JH004 and the complementary, donor-labelled cover strand JH024 were annealed to form a small DNA Hairpin showing two FRET states, one closed and one opened state with differ-

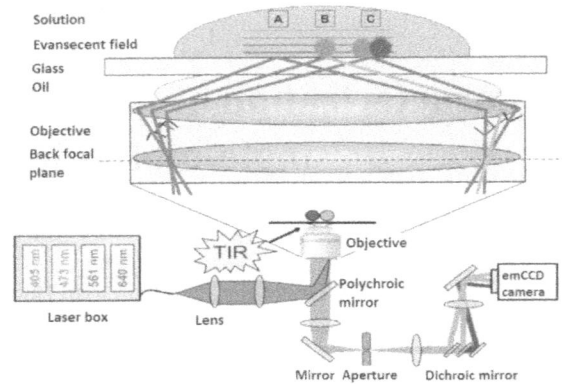

Figure 2: Schematic overview of the total internal reflection fluorescence microscope suitable for sm detection.

ent transfer efficiencies ($E = 0.75$ or $E = 0.35$), dependent on the salt concentration of the buffer medium. Due to the biotinylation on the $5'$-end of the linker strand, the sequences JH004–024 were immobilised on NC–surface coated with biotin–PEG (section 2.4) [3]. By annealing ssDNA JH004 and JH025, DNA polymerisation sensor with an expected raw FRET efficiency ($E = 0.5$) was applied in solutions with varying glycerol concentrations (section 2.3) [4] [6].

ID	Sequence
JH004	(biotin)-$5'$-CCT CAT TCT TCG TCC CAT T̲AC CAT ACA TCC$_H$-$3'$
JH024	$5'$-**C**CC AAA AAA AAA AAA AAA AAA AAA AAA AAA AAA AAT TGG GCT GGA TGT ATG GTA ATG GGA CGA AGA ATG AGG-$3'$
JH025	$5'$-CCA CGA AGC AGG C**X**C TAC TCT CTA AGG ATG TAT GGT AAT GGG ACG AAG AAT GAG G-$3'$

Table 1: Sequence of the labelled, single-strand DNA (ssDNA). In all constructs, the donor dye Cy3B in bold type ($\lambda_{ex} = 559$ nm, $\lambda_{em} = 570$ nm) and the acceptor dye Atto647N ($\lambda_{ex} = 646$ nm, $\lambda_{em} = 664$ nm), underlined within the table, were used as complementary FRET pair.

We further used fluorescently labelled DNA polymerase β (POLB, kindly provided by the Sweasy lab, Yale University) to investigate the biotin-PEG-treatment of the mixing nanochannels. POLB is a human DNA polymerase and performs base excision repair required for DNA maintenance in eukaryotic cells by filling single nucleotide gaps in DNA. For sample preparation, the DNA constructs were diluted in buffer, either PBS 1x or TE-buffer, to achieve the required, low concentrations of 1 to 10 nM for JH024-004, 62.5 pM to 1 nM for JH004-025 and 1 nM for POLB.

2.3 Molecule diffusion in glycerol

During this experiment, the effect on the diffusion of molecules by adding glycerol was studied. As glycerol is usually a weak protein binder and has much higher viscosity than water ($\eta_{glycerol} = 1,480\frac{\text{N s}}{\text{m}^2}$ to $\eta_{water} = 1\frac{\text{N s}}{\text{m}^2}$), a decrease of the diffusion velocity due to the increase of the solution viscosity was expected.

In publication [5], which provided the scientific basis, dye Rhodamine 6G (R6G) was resolved in water-glycerol-mixtures of the glycerol concentration ranging from 0 to 39 % and characterised with FCS (Fluorescence Correlation Spectroscopy). This research results show, that the Stokes-Einstein relation between the measured diffusion coefficient D and the solution viscosity η holds even in highly concentrated glycerol:

$$D = \frac{k_B T}{6\pi\eta r},\qquad(1)$$

whereas k_B is the Boltzmann constant. r defines the radius of the molecules solved in a liquid at temperature T.

In order to reproduce the experiments with comparable results, DNA polymerisation sensor JH004-025 was diluted in a broad range of glycerol-water-solutions (from 0 to even 100 % glycerol (Sigma-Aldrich, USA)) for observation of its diffusion behaviour in CultureWells^{TM} Gaskets (Sigma-Aldrich). For evaluation of the measurement data, recorded movies of 1,000 frames were subsequently evaluated by two lab-written *Matlab*-software packages (Algorithm 1: *gSGUI* & Alogrithm 2: *CDFJH2*), suitable for further analysis of molecule localisation and traces tracking.

2.4   Biotin-PEGylation of mNC

Since DNA polymerases are known for non-specific adsorption to glass surface, former experiments often resulted in blocking of the nanochannels.

A common problem-solving approach is the surface treatment and biocompatible passivation using polyethylene glycol (NHS-ester mPEG, Laysan Bio Inc., USA). An initial study of the group with immobilised polymerase, testing a Vectabonding-biotin-PEGylation protocol (Appendix 2., [3]), used Ibidi-slides (Ibidi USA Inc., USA) and was effective in preventing sticking and blocking. This process prevents protein adsorption to the surface and leaves enough space for biomolecules to flow through the NCs. However, for an effective application of the well-working method on mNC, the protocol had to be adapted due to their special geometry. Because biotin-PEGylation is highly time-consuming, the slides were treated on the day before their usage. A positive side effect, however, is, that overnight incubation leads to a higher quality of the PEGylation.

To confirm the success of the process, an experiment for surface immobilisation of biotinylated DNA was performed on the coated mNC afterwards. NeutrAvidin with a concentration of 0.25 $\frac{\text{mg}}{\text{ml}}$ was pumped into the NC and incubated for 5 minutes.

Since NeutrAvidin has an extremely strong affinity for biotin and the avidin-biotin complex is the strongest known non-covalent interaction (dissociation constant $K_d = 10^{-15}$M), it attaches to the biotin-end of the PEG (Fig. 3). If biomolecules with an attached biotin-group then flow through the channels, they bind to the NeutrAvidin (biotin-NeutrAvidin-biotin-linkage (b-N-b)) leading to their immo-

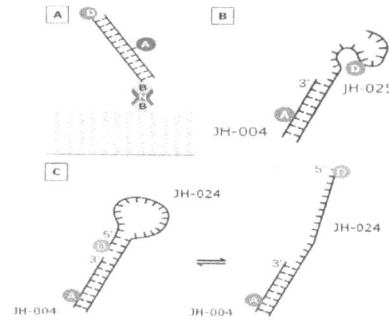

Figure 3: (A) Scheme of labelled, biotinylated DNA attached to the biotin-PEG-coated NC-surface through biotin-NeutrAvidin-biotin-linkage. (B) DNA polymerisation sensor JH004-025. (C) DNA Hairpin JH004-024. (adapted from [6]).

bilisation to the biotin-PEG covered surface [4]. Subsequently, their fluorescent signal can be observed with the TIRF-setup. Such an molecule example is the utilised small DNA Hairpin, which contains a biotin-group on the 5'-end of the acceptor-strand. On that account, biotinylated JH004-024 and, for contrast, non-biotinylated Polymerase β (examined for its expected adsorption capacity) were diluted in PBS 1x to 10 nM and then inserted into biotin-treated mNC to test whether they immobilise or not [3].

3   Results and Discussion

3.1   Molecule diffusion in glycerol

Both analysis software packages calculated the diffusion coefficient D for JH004-025, dissolved in differently concentrated glycerol-solutions. Due to different calculation algorithm and their dependencies on manually adjustable localisation and tracking parameter, the computed values between DgSGUI and D$_{\text{CDFJH2}}$ differ by ± 5 %. Diffusion coeffiecents D_{Article}'s of R6G in differently concentrated glycerol were also calculated with estimated mean diffusion times of a molecule through the observation volume τ_D's (Fig. 1 of [5]). For comparison, they were normalised by unity-based normalisation, subsequently.

ϕ_{glycerol}	D'_{gSGUI}	D'_{CDFJH2}	ϕ_{glycerol}	D'_{Article}
0	100.00	100.00	0	100.00
10	93.94	92.76	4	89.89
20	78.89	75.43	8	84.21
30	69.90	67.33	16	61.54
35	65.28	61.43	27	40.00
40	46.54	44.35	35	29.63
50	33.87	32.75		
80	14.79	8.69		
100	1.49	0.85		

Table 2:   Normalised diffusion coefficients D'_{gSGUI} & D'_{CDFJH2} of JH004-024 and D'_{Article} of R6G, solved in various glycerol concentrations ϕ_{glycerol} (in %).

The results prove that the flow speed and the diffusion of biomolecules can be slowed down by adding glycerol: The higher the glycerol concentration of the solution, the slower the diluted molecules. However, comparing the calculated values for D'_{gSGUI}/D'_{CDFJH2} with $D'_{Article}$, the discrepancy between the slowing effects is significant. Whereas $D'_{Article}$ slopes with gradient $m_{Article} \approx -2$ per additional glycerol percentage, D'_{gSGUI}/D'_{CDFJH2} fall off by only half of this value ($m_{gSGUI}/m_{CDFJH2} \approx -1[\frac{1}{\%}]$).

Possible reasons for these differences are the different behaviour patterns of R6G and the used JH004-025 in glycerol solutions, as they differ in characteristic parameters such as the hydrodynamic radii R_H or their interaction with the surrounding. In addition, the measurement (e.g. TIRF/FCS) and ensuring analysis procedures are performed under different conditions. Furthermore, the difficulties during this work with mixing the desired glycerol concentration as precisely as possible should not be neglected. For this reason, another preparation procedure than using pipettes for mixing, especially for high glycerol concentrations, should be considered. Additionally, the NC-preparation should not be performed in the cold room, as suggested in [3], since the low temperature provoked excessive leaking of the inlets as well as adoption difficulties of the slides (e.g. breakage).

3.2 Biotin-PEGylation of mNC

The evaluations of all recorded movies, whether in- or outside the mixing NCs, showed similar results. In Frame (I) of Fig. 4, the JH004-025 sample, pumped into the mNC with a flow speed of 0.5 $\frac{\mu l}{h}$, was excited with green laser light. While in the green detection channel the molecules flow constantly in- and outside, the red detection channel showed no motion of the acceptor-labelled and biotinylated DNA indicating successful immobilisation of the inserted DNA through biotin-NeutrAvidin-biotin-linkage. The detected flow within the green channel also indicates that the equal concentrated strands JH004 and JH025, usually forming the small hairpin, do not anneal in some cases within the short passage time, but rather are present in their single-stranded forms.

For further confirmation of an effective biotin-PEGylation, POLB was also tested in coated mNC after NeutrAvidin-incubation. Due to the lack of attached biotin for immobilisation, the expected free flow through the FOV was observed. Only a motionless fluorescent signal in front of the two mNC-entries was noted, pointing out that the polymerase sticked temporarily in this restricted area before entering the nanochannels.

4 Conclusion

According to the results, using glycerol concentrations of 30 to 40 % are worthwhile to achieve longer observation times of the DNA constructs within the nanofluidic devices. Furthermore, the successful application of biotin-PEGylation protocol on the mixing nanochannels was verified. However, the procedure remains challenging, due to

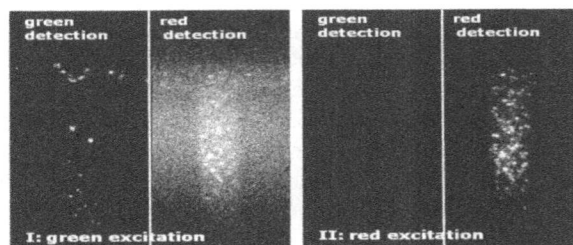

Figure 4: Two consecutive frames showing the immobilisation of the small hairpin on biotin-PEG-coated mNC.

complications arising during surface treatments and the absolute necessity that the channels are free of any molecules at the end of the procedure to obtain only fluorescence of the actual probe as well as a high PEG-coating-density. Moreover, complete prudence and tidiness during the NC-treatment and sample preparation has to be ensured, since other PEGylation-experiments during this work failed and suffered from background artefacts caused by contaminated chemicals and already used, but unmarked tubing/syringes.

Acknowledgement

The work has been carried out in the Single Molecule Group at the Laboratory of Biophysics (Wageningen University & Research, The Netherlands) and supervised by the Institute of Physics, Universität zu Lübeck.

5 References

[1] A. N. Kapanidis and T. Strick., *Biology, one molecule at a time.*, In: Trends in biochemical sciences, vol. 34, no. 5, pp. 234-43, 2009.

[2] S. Farooq and J. Hohlbein, *Camera-based single-molecule FRET detection with improved time resolution.* In: Physical Chemistry Chemical Physics, vol. 41, no. 17, pp. 27862-72, 2015.

[3] E. Acun, *The study of fluorescently labelled biomolecules in nanofluidic devices.* Bachelor thesis, Wageningen University & Research/Hogeschool Van Hall Larenstein Leeuwarden, 2016.

[4] E. Fijen, A. M. Silva, A. Hochkeppler and J. Hohlbein, *A single-molecule FRET sensor for monitoring DNA synthesis in real time.* In: Physical Chemistry Chemical Physics, 2017.

[5] E. Sherman et al., *Using Fluorescence Correlation Spectroscopy to Study Conformational Changes in Denatured Proteins.* In: Biophysical Journal, vol. 94, pp. 4819-4827, 2008.

[6] M. Fontana, *Characterization of nanofluidics devices for high-throughput single molecule fluorescence detection.* Master thesis, Wageningen University & Research/Universita di Bologna, 2015.

Design and implementation of a gel documentation system for multi-wavelength fluorescence detection for Biosafety Level S2 laboratories

J. Weiler[1], C. G. Hübner[2], Y-H. Song[2]

[1] Medical Engineering Science, Universität zu Lübeck, joana.weiler@student.uni-luebeck.de
[2] Institute of Physics, Universität zu Lübeck, {huebner, song}@physik.uni-luebeck.de

Abstract

Single molecular fluorescence detection is a powerful method for studying the structure and function of biomolecules under nearly native conditions. Prerequisite to employing such a method are the fluorescence protein samples, which can be either natural fluorescence proteins or chemically labelled proteins with synthetic fluorophores. The quality control of the sample is of prime importance and the standard method of sample characterization is gel electrophoresis. For this purpose, a fluorescence gel documentation system with a multi-wavelength detection is needed. The aim of this project is to design an optimized versatile gel documentation system for laboratories with a Biosafety Level S2. Here, the identified specification and a 3D-construction model of the instrument are discussed. Furthermore, a few preliminary results are presented. The resulting apparatus has a robust and simple design, with a multi-wavelength fluorescence detection. The realization of the construction design enables that it can be operated as an open access instrument.

1 Introduction

Biophysical studies on the level of single molecules have experienced tremendous development in the last two decades. In particular, single molecular fluorescence (smF) detection and imaging methods are gaining popularity in a wide research field [1]. Single-molecule fluorescence resonance energy transfer (smFRET) can be employed to study conformational dynamics and molecular interactions of proteins [2]. To examine proteins using smFRET *in-vitro*, they are purified and either they have an intrinsically fluorescence property, e.g. fluorescence proteins such as GFP or mCherry or they have to be chemically labelled with fluorescent dyes. Apart from spectrometry and mass spectrometry, SDS-PAGE (sodium dodecyl polyacrylamide gel electrophoresis) is routinely employed for quality control. Depending on the nature of the proteins, they can be visualized either via staining with coomassie brilliant blue or using fluorescence detection methods. For the visualization of proteins labelled with fluorophores, a gel documentation system with the capability of excitation of the fluorophores and detection of the emitted fluorescent photons is needed. In biochemical and molecular biology laboratories, a gel documentation system is commonly used. It is generally equipped with a camera and a transilluminator, consisting of white visible and UV-light sources. In the context of sample preparation for smF, most commercial instruments need an adaptation. Furthermore, they are massive, heavy-weighted and expensive. Widely used fluorophores in smFRET are e.g. the green-fluorescent dye Alexa Fluor® 488 (AF488)

and the red-fluorescent dye Alexa Fluor® 647 (AF647). They have a high photostability and suitable spectral properties, to be used with the light emitting diodes (LEDs) as light source. AF488 has the excitation maximum at 490 nm and that of emission at 525 nm [4]. The excitation maximum of AF647 is located at the wavelength of 650 nm and that of emission at 665 nm [4]. The work presented here shows a constructional design of a gel documentation system with a robust, water impermeable and easy to clean surface material, compatible for laboratories of Biosafety Level S2. The chosen LEDs are suited to bring the variety of fluorophores to their excited energy states and the chosen camera is capable of detecting the fluorescence emission. In addition, the design of the illumination box of the system ensures a homogeneous light coupling for the exposure of the samples.

2 Material and Methods

2.1 Fluorescence excitation with RGB LEDs

To illuminate the protein bands, and thereby exciting the fluorophores, RGB (red, green, blue) LEDs (RevoArt, Borsdorf, DE) were used. The chosen LEDs have a wavelength window between $\lambda = 435$ nm and $\lambda = 650$ nm (Fig. 1). This allows an excitation of a broad variety of fluorophores, suitable for samples with multi-fluorescence labels.

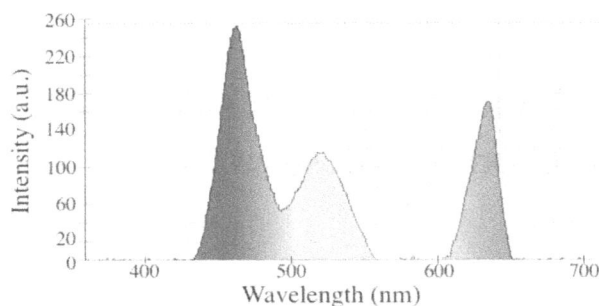

Figure 1: RGB LED spectrum. Figure adapted from [5].

2.2 Homogeneous light decoupling for optimized fluorescence excitation

To achieve a homogeneous illumination for the excitation of the fluorescence samples embedded in a gel and a subsequent detection of the emitted fluorescence photons, the LED light has been coupled into an optical plate made of a PMMA material (polymethyl metharylate). A special LED PLEXIGLAS® (LED-PLEXI) was selected (EVONIK Industries AG, Essen, DE). LED-PLEXI contains some embedded scattering particles, which should ensure optimal coupling of the LED light at the edges of the plate (information of the manufacture, Fig. 2). The incident beams are fed over the polished edges of the transparent plate and decoupled homogeneously over the detection area (Fig. 2) [6]. In addition, PMMA has a suitable chemical property: It is resistant to most inorganic compounds, allowing a direct contact to the polyacrylamide gel [6].

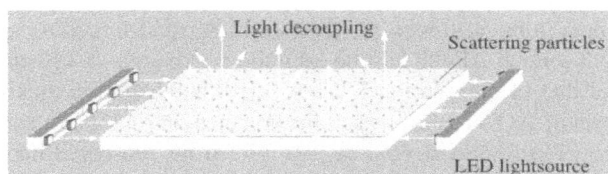

Figure 2: Scheme of the LED light coupling over the edges of the optical LED-PLEXI. Figure adapted from [6].

2.3 Camera requirements

The protein bands to be imaged are inherently with low contrast and without sharp borders. In addition, the fluorescent protein bands are often of limited brightness. However, imaging those bands with fuzzy borders is of great importance to assess the quality of the fluorescence labelling reactions. To detect multiply labelled protein bands and compare the images to each other by superposition, the images should be taken from the same gel in an exact same positioning of the camera. Therefore, a camera is chosen with the capability of manual focus and manually varied exposure time: PowerShot SX230 HS (Canon inc., Tokyo Metropolis, Japan).

2.4 Band-pass filters

Specific optical band-pass filters are used to distinguish the chosen fluorescence detection and eliminate photons orig-

inating from the incident beam (excitation wave). In addition, the band-pass filters increase the spectral brightness of fluorescence due to the reduced background. The selected filters belong to the class of interference filters: For AF488 the band-pass filter HQ532/70 and for AF647, the filter AT685/70 (Chroma Technology Corporation, Vermont, USA) are chosen.

2.5 Optimized selection and characterization of the material for the construction of the outer body

A careful evaluation of diverse materials on the market has been carried out in the decision making process for the selection of the material for constructing the gel documentation system. The material characteristics are defined as follows: (a) light weighted ($\varrho = 1.19\,\mathrm{g\,cm^{-3}}$), (b) highly shatter resistant (tensile strength = 80 MPa, elastic modulus = 3300 MPa), (c) optically tight, (d) easy to clean and (e) water impermeable. The chosen material, a black dyed PMMA material with an one-sided velvet surface, has all the above-mentioned properties (PLEXIGLAS®, satinice, EVONIK Industries AG, Essen, DE). Furthermore, it is resistant to inorganic compounds, acids and bases [6]. The material is completely non-tranmissable for visible and UV-light [6]. The design of the system allows a firm closure, therefore no external light can overlay the desired excitation wavelength.

3 Results and Discussion

The prototype of a gel documentation system with the desired purposes has been optimized, while several improvements in the construction materials and design have been achieved.

3.1 Fluorescence detection of labelled proteins

To accomplish an optimized homogeneous illumination of the detection area, the LED light has been coupled into the LED-PLEXI. All photographs of the samples and the labelled protein gels have been manually focused, photographed in presence of band-pass filters and the exposure time has been adapted for an optimal dynamic range. Illumination studies have been carried out after the new design and compared with the existing prototype. In Fig. 3a, b the imaged fluorescing samples are proteins from *Streptococcus agalactiae* (SAG), labelled with the synthetic fluorophore AF488 and illuminated with blue light, respectively. The photographs show, that the illumination of the detection area using LED-PLEXI is homogeneous (Fig. 3a). Still, the fluorescence image has high background scattering due to the decoupled incident light from the LED-PLEXI. In Fig. 3b without using LED-PLEXI, the fluorescence of the sample is visually dominant and the background scatter is almost eliminated.

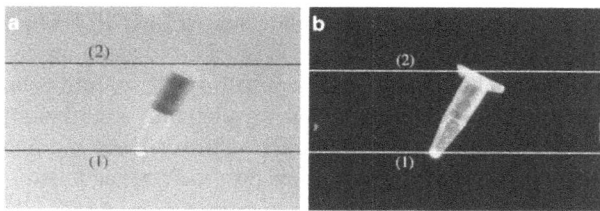

Figure 3: Photographs of AF488 fluorescence detected using the band-pass filter HQ532/70; exposure time: $1/10$ s. (a) Light coupled with using LED-PLEXI. (b) Without LED-PLEXI. The two pixel slices taken for the quantification are shown as two lines and annotated as (1) and (2).

By plotting a brightness profile along the horizontal pixel slices within the detected fluorescence area, the brightness ratio between the sample and the environment of the detection area can be made unambiguously visible (Fig. 3 and 4). The plot in the Fig. 4a corresponds to the brightness profile of the photograph in the Fig. 3a and that of the Fig. 4b corresponds to the Fig. 3b. In direct comparison of the plots, the brightness ratio of the illuminated fluorophore in Fig. 4 appears higher, than that of the image taken using the LED-PLEXI (Fig. 4a). Furthermore, Fig. 4a shows that the background scattering of the LED-PLEXI is higher, than that of the image taken without LED-PLEXI (Fig. 4b). These results are taken into account for realization of the optimized design.

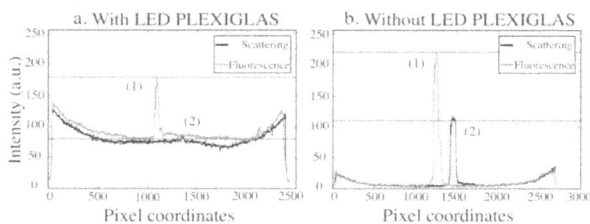

Figure 4: Brightness intensity profiles along the pixel slices denoted as (1) and (2). (a) With LED-PLEXI. (b) Without LED-PLEXI.

To evaluate the fluorescence detection tests with the authentic molecular biology samples, a gel with CFP-10 (culture filtrate protein; mycobacterium tuberculosis antigen) and hDDP3 (human dipeptidyl peptidase 3) protein bands has been prepared. The proteins were doubly labelled with AF488 and AF647. The fluorophores in the gel have been excited with blue LED light for AF488 (Fig. 5) and with red LED light for AF647 (Fig. 6). The brightness of the fluorescence of AF488 and AF647 are restricted by the different quantum yields of the fluorophores ($QY_{AF488} = 0.92$, $QY_{AF647} = 0.33$). To achieve a satisfying differentiation of the fluorescence image, the exposure time of the camera has been chosen far shorter for the photographs with LED-PLEXI, than that of without it. The results of the tests with the labelled SAG protein sample are consistent with the images taken from the gel with the labelled CFP-10 and hDPP3 proteins. The backgrounds of the images are slightly higher with the usage of LED-PLEXI in both cases,

although the illuminated sample area looks more homogeneous (Fig. 5b, Fig. 6b). Protein bands can be clearly detected and differentiated from one another. Photographs taken without using the homogenizing illumination (without LED-PLEXI) exhibited both high scattering at the borders of the gel and at enclosed air bubbles (white arrow). This is the drawback of the images taken without LED-PLEXI, although this method show less background (Fig. 5a, Fig. 6a).

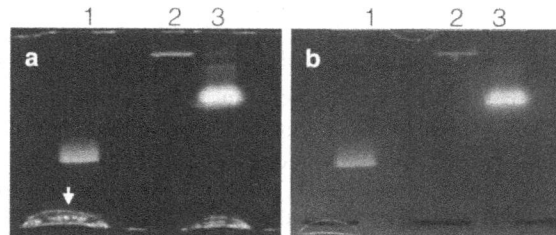

Figure 5: Photographs of fluorescence of a gel containing AF488 labelled in CFP-10 (1), hDPP3 (2) and of GFP (3), using the band-pass filter HQ532/70. Fluorescence detection without LED-PLEXI (a) and fluorescence detection with LED-PLEXI (b). Exposure times: a) $1/10$ s, b) $1/50$ s.

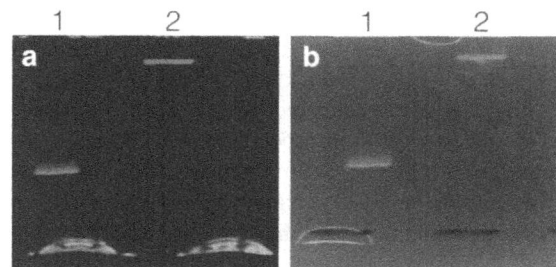

Figure 6: Photographs of fluorescence of a gel containing AF647 labelled in CFP-10 (1) and hDPP3 (2), using the band-pass filter ET685/70. Fluorescence detection without LED-PLEXI (a) and fluorescence detection with LED-PLEXI (b). Exposure times: a) 2 s, b) $1/10$ s.

Furthermore, a naturally occuring green fluorescence protein (GFP) has also been loaded on the same gel. It can also be excited with the wavelength spectrum of the RGB LEDs. GFP has the following spectral properties: $\lambda_{GFP,ex.max} = 488$ nm, $\lambda_{GFP,em.max} = 510$ nm [7]. At the laboratory of the Institute of Physics, also other natural fluorescence proteins are employed, such as mCherry, which has the spectral properties: $\lambda_{mCherry,ex.max} = 587$ nm, $\lambda_{mCherry,em.max} = 610$ nm [7]. It can also be excited and detected with the gel documentation system presented in this publication. These preliminary tests show the benefit of the multi-wavelength gel documentation system, to detect proteins with double fluorescence labels. It can be used as a method for the quality control prior the studies of smFRET or other smF techniques. The presented results show the limitation of the usage of LED-PLEXI due to the high background, but that can be compensated by short exposure time.

3.2 Design of the gel documentation system

The 3D design of the multi-wavelength gel documentation system is based on a prototype. The computational design is carried out using the CAD Autodesk® software Fusion 360™ (Autodesk Inc., San Rafael, USA). Improvements of the construction have been implemented according to the requirements needed for the laboratory setting on a Biosafety Level S2. The device is composed of three main parts and has a robust cubic design (Fig. 7).

The electronics of the system should be placed below the illumination and detection chamber, which is protected from the exterior working area (Fig. 7A). This design ensures any damages of the electronics caused by moisture. Four RGB LED strips for the fluorescence excitation should be attached to the four inner edges of the upper chamber, next to the detection area (Fig. 7B, C). An over-exposure of the LED photons by direct excitation light is prevented by adding a removable shielding (Fig. 7D). The removal of the shielding is necessary for maintenance purpose. To protect the detection area from external daylight, the coverage box is placed on the detection chamber. Nevertheless, the front and back panels can be slided to one side, to allow a fast access to the sample chamber and day light (white light) photographs, which is important for caputring images from coomassie brilliant blue stained gels (Fig. 7E). To facilitate the fluorescence detection of double labelled protein bands, requiring two kinds of band-pass filters, a slidable filter holder is mounted on the top of the coverage (Fig. 7F). According to the specification of the equidistant photographs, a removable camera mount is also integrated. It allows a fixed camera position, therefore the overlays of perfectly matching photographs of stained and fluorescence labelled protein bands are ensured (Fig. 7G).

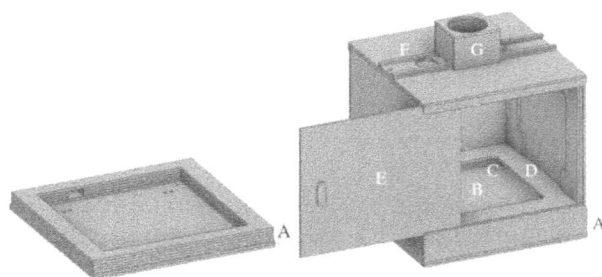

Figure 7: CAD drawing of the gel documentation system. Left: A. Bottom chamber for stowing electronics. Right: The complete system. B. Sample and detection area. C. RGB LED strips location. D. LED shielding. E. Slidable panel. F. Filter holder. G. Camera mount.

4 Conclusion and Outlook

In this project, the implementation of the improved gel documentation system is presented and first preliminary encouraging results are included. For the computer aided 3D design, subsequent improvements and steps are shown.

Scattered photons from the homogenizing LED-PLEXI and from other scattering sources can be reduced using modern and high quality band-pass filters with small spectral bandwidths in the desired detection wavelength. Additionally, digital signal processing and filtering of scattered and fluorescent photons can be performed post to image acquisition, to achieve a better signal-to-noise ratio.

The usage of LED-PLEXI improves the homogeneity of excitation illumination, but reduces the intensity of the fluorescence and enhances scattering background. Nevertheless, differentiation between scattering non-fluorescent photons and that to be detected fluorescence can be improved. Furthermore, the control of the LEDs will be realized with an Arduino® board, which has an integrated microcontroller circuit board. Therefore the selection of the desired wavelength range can be chosen using a dial and the chosen wavelength will be displayed on an LCD-window (EA W164B-NLW, Electronic Assembly GmbH, Gilching, DE). All controlling devices will be mounted easily accessible at the front of the gel documentation system. The realization of these preliminary results into the construction of the new designed multi-wavelength gel documentation system is currently ongoing.

Acknowledgement

The work has been carried out at the Institute of Physics at the Universität zu Lübeck. We are grateful to Till Zickmantel and Christian Magana for the support and Dennis Pohle at the FabLab Lübeck for his advice and help.

5 References

[1] W. E. Moerner, *Microscopy beyond the diffraction limit using actively controlled single molecules.* J Microsc, 2012.

[2] K. A. Henzler-Wildman, V. Thai M. Lei., M. Ott, M. Wolf-Watz.,T. Fenn et al., *Intrinsic motions along an enzymatic reaction trajectory.* Nature, pp. 450, 2007.

[3] H. Schägger, H. Aquila, G. VonJagow, *Coomassie blue-sodium dodecyl sulfate-polyacrylamide gel electrophoresis for direct visualization of polypeptides during electrophoresis.* Anal Biochem, 1988.

[4] V. Hirschfeld, H. Paulsen, C. G. Hübner, *The spectroscopic ruler revisited at 77K.* Phys Chem 15, 2013.

[5] L. Reiner, *Design und Konstruktion eines Do-it-yourself Geldokumentationssystems mit Multi-Wellenlängen Fluoreszenz-Detektion.* Bachelorarbeit, Universität zu Lübeck, 2016.

[6] PLEXIGLAS®, Available: https://www.plexiglas.de [last accessed on 2017-01-21].

[7] R. Y. Tsien, *Constructing and Exploiting the Fluorescent Protein Paintbox (Nobel Lecture.* Angew. Chem.48, 2009.

Investigation of the correlation between the natural moisturizing factor and non-invasive parameters of the skin barrier function in vivo

T. R. Stilla[1], I. Angelova-Fischer[2], D. Zillikens[2], and T. W. Fischer[2]

[1] Medizinische Ingenieurwissenschaft, Universität zu Lübeck, tasja.stilla@student.uni-luebeck.de
[2] Department of Dermatology, Universität zu Lübeck, {Irena.Angelova-Fischer,Detlef.Zillikens,Tobias.Fischer}@uksh.de

Abstract

Irritant contact dermatitis (ICD) is a leading cause of occupational skin disease worldwide. Exposure to the detergent sodium lauryl sulphate (SLS) is a common workplace factor for ICD. Repeated exposure of SLS leads to decreased natural moisturizing factor (NMF) levels and thus decreased stratum corneum hydration. Additionally SLS disrupts the skin barrier causing an increase in transepidermal water loss (TEWL). The Corneometer based on the measurement of capacitance is the most commonly used device to non-invasively assess stratum corneum (SC) hydration. The aim of the study was to investigate the correlation between the NMF levels and the Corneometer values in healthy and irritant-exposed skin. A positive correlation between the Corneometer values and the baseline NMF levels of the volunteers and also after SLS exposure was found. These results suggest that the Corneometer might be able to indirectly assess the NMF levels without the necessity of tape stripping and laboratory analysis.

1 Introduction

Chronic irritant contact dermatitis (ICD) is one of the most common occupational diseases worldwide. ICD is characterized by dryness, scaling and inflammation. Irritants damage the stratum corneum (SC) function and may consequently negatively influence the water balance of the skin, determined by the skin lipids and the natural moisturizing factor (NMF) levels. The outermost layer of the skin, the SC, is a barrier to the environment protecting the skin against irritant damage. Besides the barrier characteristics, the SC fulfills a water binding function. NMFs in corneocytes are able to bind water whereby the intercellular lamellar lipids preventing them from leaching out thus these two biophysical mechanisms providing an effective barrier maintaining the water balance. There are several of non-invasive bioengineering devices available to assess the SC hydration. The most commonly used device is the Corneometer that is based on the capacitance principle. Additionally the non-invasive measurement of the transepidermal water loss (TEWL) enables the assessment of the skin barrier function. [1]

The aim of this study was to investigate the correlation between the capacitance measurement, TEWL and the NMF levels in the skin under physiological conditions and after skin damage by irritants.

2 Material and Methods

In the following chapters the used material and methods will be described.

2.1 Study population

65 healthy individuals aged between 20 and 65 years (47 females and 18 males; mean age 28.0 years) without any systemic or skin disease in the medical history were included in the study. Intensive UV-exposure six weeks prior to inclusion and during the study, pregnancy and lactation were defined as criterions for exclusion. Prior to the study the volunteers gave written informed consent. The protocol was approved by the ethics committee of the University of Lübeck.

2.2 Irritants and mode of exposure

43 out of the 65 volunteers underwent repeated irritation with 0.5% aqueous solution of sodium lauryl sulphate (SLS; 99.0% purity; Sigma-Aldrich, Steinheim, Germany). The irritant was applied on four consecutive days (D1-D4) for 30 minutes twice daily at the same time ($\pm$ 1 hour) using large Finn chambers (12 mm diameter, SmartPractice, Reinbek, Germany). Untreated skin served as control. The volunteers were allowed to take a shower as usual but were not allowed to use any skin care products and to be exposured to UV radiation in the test area during the five days of the study.

2.3 Bioengineering of the assessment of the skin irritant response

Transepidermal water loss (TEWL) and skin hydration (capacitance) were used to monitor the irritant response of the skin. The assessments were performed before the first application of the irritants on day one (0h), after 48h and after 96h prior to tape stripping. The open chamber system was used to measure TEWL (Tewameter TM300; Courage and Khazaka Electronics, Cologne, Germany) and the capacitance principle was used to assess the skin hydration (Corneometer CM825; Courage and Khazaka Electronics, Cologne, Germany). Two consecutive measurements per field were performed under controlled ambient conditions (temperature $21 \pm 1°$ C; mean relative humidity 38–40%) according to the published guidelines. [2], [3]

2.4 Natural moisturizing factor analysis

On day 5 (96h) six corneocyte samples per test field of each volunteer were taken with the standardized tape stripping method. The samples were taken with commercially available 14mm D-Squame discs (CuDerm Corp., Dallas, TX, USA) and stored in 1.5 ml Eppendorf tubes (Eppendorf, Hamburg, Germany) at $-80°$ C until analysis. The NMF components of the stratum corneum (histidine, 2-pyrrolidone5-carboxylic acid (PCA), trans- and cis-urocanic acid (t-UCA, c-UCA)) on the tapes were extracted with 400 μl of 25% (w/w) ammonia solution, evaporated to dryness and reconstituted in 200 μl pure water before high-performance liquid chromatography (HPLC-UV) analysis. The NMF levels were corrected for the amount of protein extracted from the tapes and expressed as mmol NMF/g protein.

2.5 Statistical analysis

GraphPrism Version 6 (GraphPad Software Inc., San Diego, CA, USA) was used to perform the statistical analysis. The level of significance was $p<0.05$. Repeated analysis of variance (ANOVA) or Friedmann tests were used to evaluate changes for the respective field over time. The data in the respective tables and figures are represented as mean and standard error (SEM) with the exception of NMF levels that are represented as median. The correlation between the assessment of the skin hydration and the natural moisturizing factor levels was analyzed by the nonparametric Spearman correlation.

3 Results and Discussion

At baseline there were no significant differences between the values of the test and control fields of the assessed parameters TEWL and capacitance. On D5, the TEWL values were significantly increased at the SLS exposed test fields. Repeated SLS exposure resulted in significantly lower capacitance values. The results are shown in Fig. 1 and Fig. 2. The NMF levels and also the levels of the single NMF

components of the SLS exposed skin were significantly decreased compared to the untreated (control) skin. A proportional relationship between capacitance and the total NMF levels as well as components was found. In the untreated skin (non SLS-exposed) the capacitance and the total NMF levels and also the single components except of t-UCA were significantly correlated. In contrast, except of t-UCA and c-UCA the capacitance and NMF levels as (total NMF and single components) of the SLS exposed skin were not significantly correlated. In SLS-treated skin the capacitance, t-UCA and c-UCA were poorly correlated. However, altogether the capacitance, NMF levels and its single components tend to decrease after repeated SLS exposure. The results are shown in Table 1 and Table 2. An inverse relationship between TEWL and the NMF-levels was found in healthy, untreated skin and also healthy SLS exposed skin. Except of histidine and c-UCA, the NMF levels and TEWL were significantly correlated. After SLS exposure only the total NMF and TEWL were significantly correlated.

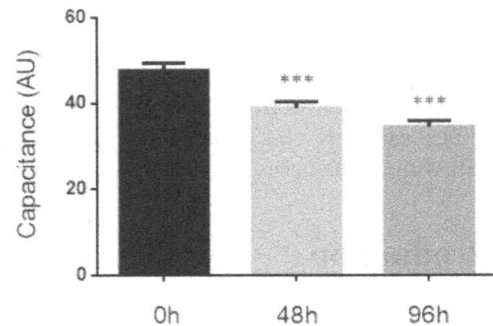

Figure 1: Repeated irritation test (RIT) with 0,5% sodium lauryl sulphate (SLS). Comparison of the Capacitance decrease with the time shown as mean ± standard error of mean (SEM); level of significance < 0.05, *p < 0.05, **p < 0.01, ***p < 0.001 compared with the non-exposed (normal skin) control field.

Table 1: Correlation between the capacitance values of the non-irritant exposed (healthy skin) and NMF, measured in tape strips by UV-HPLC.

	Spearman r	p-value
total NMF	0,29	< 0,05
Histidine	0,34	< 0,001
PCA	0,01	< 0,05
t-UCA	0,13	ns
c-UCA	0,45	<0,0001

The aim of the study was to investigate correlations between the measured Corneometer values and the NMF levels. There was a significant correlation between these two parameters and thus the Corneometer might be suitable to non-invasively and indirectly provide information on the

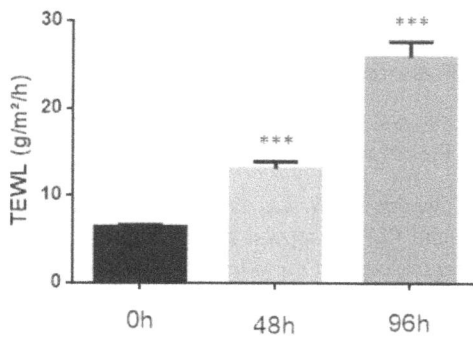

Figure 2: Repeated irritation test (RIT) with 0,5% sodium lauryl sulphate (SLS). Comparison of the transepidermal waterloss (TEWL) increase with the time shown as mean $\pm$ standard error of mean (SEM); level of significance < 0.05, *p < 0.05, **p < 0.01, ***p < 0.001 compared with the non-exposed (normal skin) control field.

Table 2: Correlation between the capacitance values of the SLS exposed skin and NMF, measured in tape strips by UV-HPLC.

	Spearman r	p-value
total NMF	0,27	ns
Histidine	0,30	ns
PCA	0,24	ns
t-UCA	0,30	< 0,05
c-UCA	0,33	< 0,05

NMF levels in vivo. In addition a correlation between TEWL, the parameter for the assessment of the skin barrier function and NMF levels were found.

There are several commercially available non-invasive bioengineering methods to assess SC hydration in vivo that are based on electrical measurements. Commonly used devices are the Corneometer (Courage and Khazaka Electronics, Cologne, Germany) that is based on measuring capacitance and the Skicon (IBS Company, Hamamatsu, Japan) that is based on conductance. The Corneometer has a worldwide popularity. It allows measurements in vivo and in vitro and it is established in indirect measuring the SC hydration in healthy and also diseased skin. The reasons for the high popularity in dermatological research are the relatively low costs and the easy, fast (1s) and reproducible measurement technique. [4]

Environmental influences like the room temperature and humidity influence the skin hydration and thus the measurement of SC hydration. As a consequence it is necessary to assess the SC hydration in temperature and humidity controlled rooms, according to the published guidelines. [2] This was also the case in the present study.

In addition to the electrical methods, there are also a few non-invasive spectroscopic techniques available to assess

the SC hydration such as infrared, Raman and photoacoustic spectroscopy. Besides the possibility to directly measure the water content of SC the Raman spectroscopy is also able to assess the NMF levels in vivo. The Raman spectroscopy is useful to non-invasively investigate the epidermal water gradients and water handling ability in vivo however its use in dermatological research and practice is still limited because of substantial costs and necessity of well-trained personnel for the measurements and interpretation of the results. [5], [6]

ICD is an inflammatory skin disease that is caused by external stimuli and has a nonimmunologic mechanism. Different intrinsic and extrinsic factors influence its development. The skin barrier function is an important intrinsic factor for the outcome of the interaction of an irritant with the skin. Exposure to common irritants like the anionic detergent SLS is a well-known extrinsic damaging factors contributing to ICD. [1]

The damaging effect depends among other factors on the duration of the exposure and the concentration of the irritant. Aramaki et al. reported a positive correlation between the concentration of the irritant, the duration of the irritant exposure and the skin response using the patch test method and SLS concentrations in a range from 0,125% up to 2,0% for different periods of time (3-48 hours). In the study the skin irritant response was assessed by TEWL measurements and laser Doppler flowmetry. [7]

The sum of the damaging effects as well as the repair capacity of the skin is critical for the outcome of interaction of an irritant with the skin. TEWL, capacitance and measurement of erythema are the most commonly used in vivo parameters for studying the skin irritant response under experimental exposure and daily life conditions *in vivo*. [1] The application of these methods might help to examine and to understand the interactions between different irritants and the skin barrier as well as its components.

The influence of SLS on the SC hydration and also on other parameters such as TEWL and erythema has been examined in many in vivo studies in human skin. Decreased SC hydration as well as decreased NMF levels after single and repeated SLS exposure has been observed in several studies that used different models with various concentrations of SLS and durations of exposure. Irritants like SLS have damaging effects on the skin barrier and influence the water balance by interaction with both the lipid and NMF components. [8]

Increased TEWL values and erythema along with reduced NMF levels after cumulative dermal SLS exposure have been recently reported by Angelova-Fischer et al. [9]

A recently published occlusion modified tandem repeated irritation test confirmed these findings whereby occlusion enhanced the skin damaging effects of SLS. Additionally repeated SLS exposure leads to decreased Corneometer values and an increased blood flow that is shown as erythema as a sign for inflammation. Erythema was assessed by Colorimeter and visual score based on Kligman and Frosch. [10]

NMFs are present in the corneocytes in the SC. Reduced

NMF levels lead to skin dryness while skin dryness is a common finding in ICD. Therefore the irritant-induced reduction of NMF could be a major contributing factor to the skin dryness observed in ICD. The NMF is a mixture of moisture absorbing compounds including amino and organic acids, urea, as well as inorganic ions that primarily serve as humectants. [11] Additionally, NMFs maintain the healthy, acidic pH value of the skin [12].

In this study histidine, t-UCA, c-UCA and PCA were measured.

The source of the investigated NMF components is the histidine-rich protein filaggrin that is formed by phosphorilization of its precursor protein profilaggrin. UCA is derived from the amino acid histidine and absorbs UV-light that leads to isomerization of the naturally occurring trans isomer to the cis isomer. UCA has a photoprotective function in the skin. PCA derives from the amino acid glutamine. [11]

4   Conclusion

In conclusion, in the present study we found a correlation between capacitance and the NMF levels so that the capacitance measurement may possibly provide indirect information about the concentration of NMF components that determine the SC hydration. Although not all components of the NMF and capacitance values were significantly correlated after SLS exposure there was always a positive correlation between them. It might be useful to investigate more volunteers to get reliable information about the observed correlations and the possibility to non-invasively measure NMF levels with the Corneometer.

Acknowledgement

The work has been carried out in the Laboratory of Skin Physiology at the Department of Dermatology, University of Lübeck. The authors would like to thank Sanja Kezic (Coronel Institute of Occupational Health, Academic Medical Centre, University of Amsterdam) for the technical support of the NMF level analysis and the COST Network StanDerm for support.

5   References

[1] A. Chew and H. I. Maibach, *Irritant Dermatitis.* Springer, Berlin/Heidelberg, 2006.

[2] E. Berardesca, *EEMCO guidance for the assessment of stratum corneum hydration: electrical methods.* Skin Res Technol, 3:126–132, 1997.

[3] V. Rogiers and the EEMCO Group, *EEMCO guidance for the assessment of transepidermal water loss in cosmetic sciences.* Skin Pharmacol Appl Skin Physiol, 14: 117–128, 2001.

[4] P. Elsner, E. Berardesca and H. I. Maibach, *Bioengineering of the Skin: Water and the stratum corneum.* CRS Press, Inc, Florida, 1994.

[5] R. Marks and P. A. Payne, *Bioengineering and the skin.* MTP Press Limited, Lancaster, 1981.

[6] P. J. Caspers, G. W. Lucassen, E. A. Carte, H. A. Bruining and G. J. Puppels, *In vivo confocal Raman microspectroscopy of the skin: noninvasive determination of molecular concentration profiles.* J Invest Dermatol, 116(3):434-42, 2001.

[7] J.Aramaki, C. Löffler, S. Kawana, I. Effendy, R. Happle and H. Löffler, *Irritant patch testing with sodium lauryl sulphate: interrelation between concentration and exposure time.* Br J Dermatol,145(5):704-8, 2001.

[8] T. Agner, *Skin barrier function.* Karger, Basel, 2016.

[9] I. Angelova-Fischer, I. Dapic, A. K. Hoek, I. Jakasa, T. W. Fischer, D. Zillikens and S. Kezic, *Skin barrier integrity and natural moisturising factor levels after cumulative dermal exposure to alkaline agents in atopic dermatitis.* Acta Derm Venereol, 94(6):640-4, 2014.

[10] I. Angelova-Fischer, T. Stilla, S. Kezic, T. W. Fischer and D. Zillikens, *Barrier function and natural moisturizing factor levels after cumulative exposure to short-chain aliphatic alcohols and detergents: Results of occlusion-modified Tandem Repeated Irritation Test.* Acta Derm Venereol, 96(7):880-884, 2016.

[11] J. J. Leyden and A. V. Rawlings, *Skin moisturization.* Marcel Dekker, Inc, New York, 2002.

[12] J. Levin, S. F. Freidlander and J. Q. Del Rosso, *Atopic dermatitis and the stratum corneum: Part 1: The role of filaggrin in the stratum corneum barrier and atopic skin.* J Clin Aesthet Dermatol, 6(10): 16–22, 2013.

3

Safety and Quality

Extension of a medical engineering laboratory course in consideration of the research environment at HAWK and UMG

R. Kutlu[1], Ch. Rußmann[2], and St. Wieneke[2]

[1] Medical Engineering Science, University of Lübeck, kutlu.resul@gmail.com
[2] Faculty of natural science and technology, HAWK Göttingen, {christoph.russmann, stephan.wieneke}@hawk-hhg.de

Abstract

Two experiments for a medical engineering laboratory course have been newly designed. These experiments augment the already existing experiments, which altogether form the laboratory course that will be part of a new degree program at the University of Applied Science and Art (HAWK) in Göttingen. The first experiment demonstrates the connection between the frequency of light and the strength of absorption. In addition the phenomenons of diffraction and interference are used to illustrate the wave nature of light as well as to measure the thickness of a human hair and to determine the wavelength of laser light. The second experiment includes the dissection of a pig's eye and biometric measurements on human eyes. Several requirements had to be observed during the design of the two experiments: The experiments had to fit to the curriculum of the new degree program, they had to complement the already existing experiments and they had to fit in the research environment of the universities that host the new degree program.

1 Introduction

The University of Applied Science and Art in Hildesheim, Holzminden and Göttingen (HAWK) offers studies in the field of engineering sciences and is complementing the range of courses offered by the Georg-August-University in Göttingen [6]. As part of the reestablishment of the "Gesundheitscampus Göttingen", a new degree program called "Medical Engineering" will be offered in collaboration with the University Medical Center Göttingen (UMG), starting from the winter term 2017. Within this program a new technical module—called „Praktikum zur Medizintechnik" (laboratory course for medical engineering)—shall link specialized know-how with practical experiences. This laboratory course should convey in an illustrative manner:

- basics of medicine and medical technology

- practical knowledge in the analysis of biomedical data.

The task of the project described in this article was to augment the already existing four demonstration experiments of this new laboratory course (sonography, human physiology, hematology and electrical physiology) with new experiments that should give to the students an overview about the main research of the two collaborating universities in the fields of medicine and of medical technology:

- Laser and plasma medicine, medical photonics with a Focus on ophtalmonological and intraoperative medical imaging (HAWK),

- neurology, cardiology and oncology (University Medical Center Göttingen).

Table 1: Experiments in the laboratory course medical technology. Existing experiments are printed in normal roman letters, experiments that are currently under design are printed in italics. The experiments described in this article are emphasized by bold letters.

Medical basics	Medical technology basics
Human physiology	Sonography
Hermatology	**Optical measurement of hair (HAWK)**
Electraphysiology and cardiography (HAWK)	**Biometry (HAWK)**
Neurology (HAWK)	*Slit lamp (HAWK)*
Oncology	*Further experiments (currently under discussion)*

A compilation of the already existing demonstration experiments as well as of the experiments designed in this project is given in Table 1.

The remaining part of this article is organized as follows: In the next section the newly designed experiments are described in detail. The subsequent section contains a critical appraisal of the new experiments in the context of the new degree program and it is discussed in how far the educational objectives have been reached. Conclusions drawn from this analysis and an outlook close this article

Figure 1: Colored balloons are irradiated with laser light.

2 Material and Methods

Two new demonstration experiments were designed, the first one demonstrating optical basics, the second one applying optics to investigate structure and function of eyes.

2.1 Properties of light

Some basic properties of waves, in this case of visible electromagnetic waves, are illustrated using simple laser pointers with different colors. In the first part of this experiment the student should learn the connection between the frequency (color) of light and its absorption by matter. In the second part the wave-like nature of light and its consequences like interference and diffraction are studied at the example of a hair illuminated by laser light. At the same time the students learn to investigate the anatomy of hair using coherent light.

2.1.1 Colored balloons absorb laser light

The students have to inflate red and green balloons and irradiate them with the same red laser pointer. While the red balloon will burst, nothing is going to happen to the green one. Afterwards a red balloon is inserted into a white balloon and both balloons are inflated with air (see Fig. 1). Irradiation with red laser light lets burst the inner red balloon while the outer balloon is staying untouched.

The students should discuss, first, the reason for the burst and, second, the connection between the color of the laser, the color of the balloon and its susceptibility to be destroyed by the laser. This discussion process should help the students to recollect the terms frequency and absorption and understand the meaning of an absorption spectrum.

2.1.2 Diffraction of light by a hair

In a dark room a human hair with unknown diameter d is vertically fixed in a clamp and illuminated by green light from a laser pointer with a wavelength of $\lambda = 520$ nm. On a distant screen a horizontal pattern of green points is visible. In a first step the students are asked to give a qualitative explanation for this observation. This should lead to a recollection of the terms diffraction and interference. In a second step the observation has to be explained quantitatively. To do so, a number of quantities have to be measured: the

distance b between hair and screen, for each green point (maximum number n) the distance x_n to the central maximum. From the measured distances the angle α_n between the optical axis and the rays of light directed to maximum number n can be determined according to equation (1):

$$\tan \alpha_n = \frac{x_n}{b} \, . \tag{1}$$

The angle for maximum number n can be inserted into Bragg's law,

$$\lambda = \frac{d \sin \alpha_n}{n} \, , \tag{2}$$

which can be easily rearranged in order to obtain the diameter d of the hair. In this way each maximum that is visible on the screen can be used to yield an estimate for the diameter of the hair. In order to minimize the influence of measurement errors the mean value of these estimates and their standard deviation are calculated. The students are then expected to give a critical analysis of the accuracy of the diameter of the hair determined in this way.

The procedure described above is repeated with a second laser pointer with different color and unknown wavelength. Since the diameter of the hair has been determined, Bragg's law can now be used to measure the wavelength of the laser light:

$$d = \frac{n\lambda}{\sin \alpha_n} \, . \tag{3}$$

Again this has to be done for each observed maximum and mean value and standard deviation have to be calculated.

2.2 Optical Biometry

The second experiment is divided into two parts. In the first part an eye of a pig is dissected in order to convey a basic knowledge of the anatomic structure of the eye. In the second part biometric measurements of human eyes are performed in vivo. The students shortly reiterate the meaning of interference and the setup of a Michelson interferometer. The principles of optical coherence tomography and the anatomy of the eye should be studied by the students as a preparation to this experiment.

2.2.1 Dissection of a pig's eye

The eyes of pigs are very similar to human eyes and for this reason well suited for an introductory anatomic study of the human eye [8]. During the dissection special emphasis is given to the lens, the sclera, the choroid, the retina and to the vitreous body (see Fig. 2 for details). After the dissection the students are asked to recapitulate the constituent parts of the eye as a preparation for the biometric measurements in the second part of this experiment.

2.2.2 Biometry

This part of the experiment is supervised by a medical doctor of the university hospital in Göttingen. At the outset the

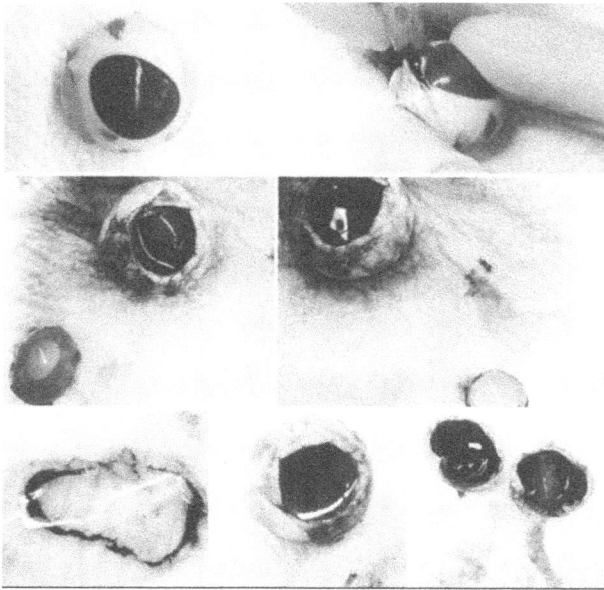

Figure 2: Dissection of a pig eye.

Table 2: Measured quantities in the biometry experiment.

Abbrev.	Quantity
RC	Radius of the cornea
ref	Refraction of the target
d	Depth of the anterior chamber
L	Length of the axis

students are divided into groups of two. Afterwards they perform biometric measurements on each other's eyes by means of the optical biometer IOLMASTER500. The quantities to be measured by the students are given in Table 2. The (fictive) refractive index of the cornea ($nC = 1.3315$) and of the intraocular fluid and the vitreous body ($n = 1.336$) are given to the students, as well as the vertex distance between cornea and glasses ($dBC = 12$ mm). By means of the Haigis formula [2,4] these quantities can be used to calculate the refringence DL of the intraocular lens (IOL) [1-3]:

$$DL = \frac{n}{L - d} - \frac{n}{(n/z) - d} \quad (4)$$

with

$$z = DC + \frac{ref}{1 - ref + dBC} \quad (5)$$

and

$$DC = \frac{nC - 1}{RC} . \quad (6)$$

When these measurements have been repeated five times for each of the students, average, median and standard deviation of the calculated IOL refringence are derived. These results together with minima and maxima have to be displayed in a box whisker plot.

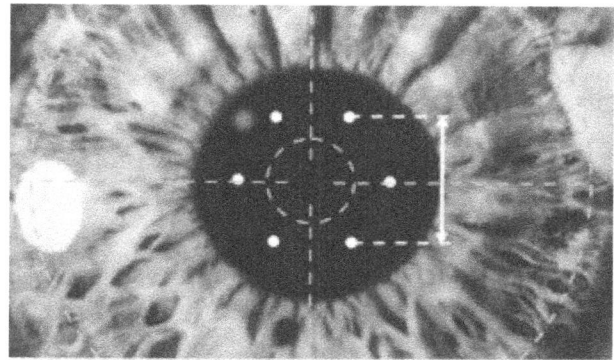

Figure 3: This image shows the measurement with IOL-MASTER500.

3 Results and Discussion

The objective of this project was the extension of an already existing laboratory course for medical engineering. The new experiments that were included into this laboratory course had to fulfill certain boundary conditions: (I) they should complement the existing four experiments of this course, (II) they should transfer basic knowledge in medical technology and biometry, (III) they should fit to the research field of the universities responsible for the degree program for which this laboratory course is designed.

The experiments presented in the previous section obviously fulfill condition I as can be seen for instance in Table 1. In order to test whether objective II has been reached, the new experiments have been presented to selected teachers and students of related study programs. Interviews with teachers and students confirmed that these experiments are based on the basic lectures and extend the knowledge of the students in the field of medicine and medical technology. For the first experiment lasers were selected, because this topic is taught to the students in the lecture Physics 2, and because lasers in medicine is one of the research focuses of the HAWK. The individual parts of the first experiments have a degree of difficulty that should be adequate to all students including those with average proficiency in physics. For the future a test run of the experiments with a larger group of students is planned.

4 Conclusion

The example of the two experiments designed in this project illustrate the interdependencies between the curriculum of the degree program, the research projects of the hosting universities and the design of the laboratory course. The procedure for designing new experiments for the medical engineering laboratory course established in this project will be used for the design of additional experiments in order to complete the course. While during this project the suitability of the experiments has been evaluated qualitatively, it is intended to combine the design of further experiments with a quantitative evaluation together with tests with a larger number of students.

Acknowledgement

This work has been conducted at the HAWK University of applied science and art and their location in Göttingen.

5   References

[1] T. Carl Zeiss Meditec AG, Iolmaster500 – Broschüre 2015.

[2] W. Haigis, *Präoperative Berechnung der Stärke intraokularer Linsen bei Problemaugen* Z. Med. Phys. 2007, H. 17, S.45-54.

[3] MacLaren RE, Natkunarajah M, Riaz Y, Bourne RR, Restori M, Allan BD. *Biometry and formula accuracy with intraocular lenses used for cataract surgery in extreme hyperopia. American Journal of Opthalmology*, 2007;Band 143, S. 920-931.

[4] Wang JK, Hu CY, Chang SW., *Intraocular lens power calculation using the IOLMaster and various formulas in eyes with long axial length*, Journal of Cataract and Refractive Surgery, 2008,H. 34, S. 262-267.

[5] ULB Sachsen-Anhalt, *Optimierung der Linsenkonstanten für die SOFLEX2*. Available: https://www.sundoc.bibliothek.uni-halle.de/diss-online/05/05H047/t4.pdf; [last accessed on February 08, 2017].

[6] HAWK, *Physics 2*. Available: http://modul.hawk-n.de/modulblattn.php?name=Physik2 [last accessed on January 24, 2017].

[7] Ernst, Dava Eva, *wave optics - diffraction* Available: http://www.exphys.jku.at/Skripten/Unterricht/Stoffgebiete/Wellenoptik/33_Wellenoptik_Beugung_oberstufe_ernst.pdf [last accessed on February 07, 2017]

[8] Schmitz Kerstin, *Dissection of pig eye's* Available: https://www.yumpu.com/de/document/view/38574850/anleitung-schweineaugen-praparation [last accessed on February 06, 2017].

Price-performance optimized hardware for molecular simulations

A. Flammiger[1] and H. Paulsen[2],

[1] Medizinische Ingenieurwissenschaft, Universität zu Lübeck, andreas.flammiger@student.uni-luebeck.de
[2] Institut für Physik, Universität zu Lübeck, paulsen@physik.uni-luebeck.de

Abstract

Improvements in the implementation of GPUs (graphics processing units) for parallel computations of molecular dynamics simulations have made CPU-only servers obsolete over the last few years in terms of molecular simulations, not only for reasons of cost but also for performance. In this field, a market analysis of existing processors and graphics cards was carried out and evaluated from a price-performance perspective. Not only one-off acquisition costs were assessed but also current costs. Based on benchmarks, the hardware was selected and compiled for a price- and performance-optimized server system. Optimum selection of components and subsequent optimization of the software environment can be used to build up workstations that stand out due to low acquisition costs and low operating costs.

1 Introduction

Since the early 1990s, GROMACS (Groningen Machine for Chemical Simulations) has been proven to be one of the most efficient programs for the parallel calculation of molecular dynamics.[1] With the use of GNU General Public Licenses, this has resulted in widespread dissemination within academia at universities and research facilities. By compiling the kernel on the target hardware, an optimal performance can be achieved by the hardware, which is not possible with precompiled software. Since the emergence of graphics cards (GPU) with programmable kernels at the beginning of the 2000s, it is possible to influence the pipeline on the softwares side and to use the massive performance of parallel computing power for scientific calculations. Multi-core CPUs these days have a number of processors in the 1-32 range, while GPUs have more than 3000 processors, making them ideal for any algorithm that can be parallelised.

Hardware companies such as NVIDIA and AMD have recognized this market at an early stage and have provided drivers and programming environments to enable customers to implement their products widely. In 2007, NVIDIA published "CUDA" (Compute Unified Device Architecture) a programming environment specifically for the use of its GPUs for the calculation of highly parallel tasks such as climate research, financial calculation or molecular simulations. [3]

Since GROMACS version 4.6 it is possible to outsource large parts of the molecular dynamics calculations to CUDA on GPUs. Fig. 1 provides an overview of which parts of the simulation are transferred to the GPU and how the data flow can be processed in parallel.

Fig. 1 shows the load distribution between CPU and GPU. Since the calculations on the CPU and the GPU have to

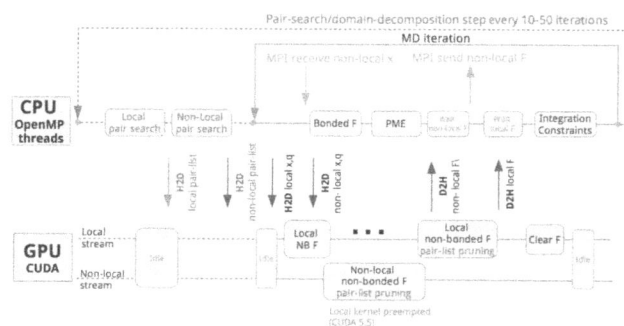

Figure 1: Flow diagram of the distribution of GROMACS computing operations between GPU and CPU.[4]

be calculated in parallel, the calculations must be available at certain points in time and should be terminated in the ideal case at the same time. In order to design a price-performance optimized system for molecular dynamics calculations, it is of utmost importance to coordinate a ratio between the performance of the processor and the graphics card. Thus, an unnecessarily high investment in performance is avoided, which can not be retrieved due to the lack of balance.

2 Material and Methods

For the evaluation of the performance with regard to molecular dynamics simulations, two components of a workstation are the most important power carriers: the processor and the graphics card. While in the area of the processors essentially AMD and Intel are prevailing, are the dominant players in the area of the graphics cards NVIDIA and AMD.

Table 1: Current Intel CPU generations

CPU	CPUMark	Price [€]	P-P Ratio
i7-6700K	11043	350	36.81
i5-6600K	7862	240	35.75
i7-5775C	11055	380	29.09
i5-5675C	8033	300	26.77
E3-1275v5	10254	350	29.30
E3-1285v4	11170	470	23.77
E5-1650v4	14353	680	21.11
E5-1680v4	17060	1890	9.03
E5-2650v4	15939	1220	13.06

Table 2: Current Intel AMD generations

CPU	CPUMark	Price [€]	P-P Ratio
FX 9590	10245	200	51.23
FX 9370	9501	190	50.01
FX 8350	8940	150	59.6
FX 8150	7619	130	58.61
A10 7850K	5525	100	55.25
A10 7700K	5190	80	64.88
Opteron 6380	10082	1150	8.77
Opteron 6376	9414	750	12.55

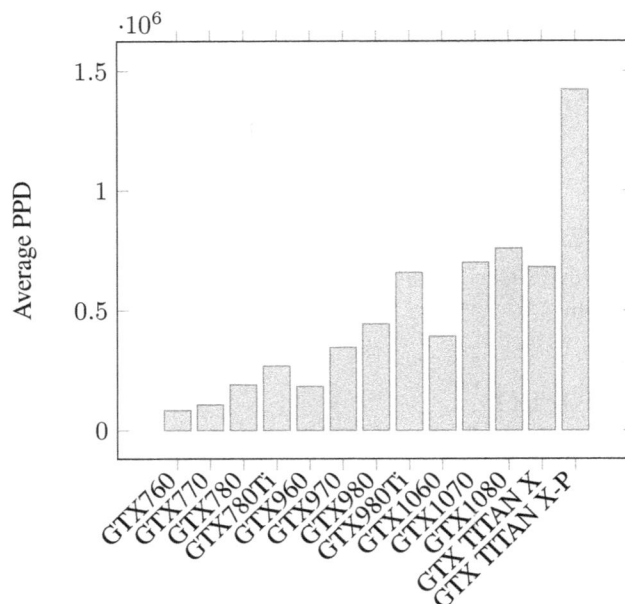

Figure 2: Development of the PPD performance for NVIDIA graphics cards

2.1 CPUs

The latest generation of processors from Intel is marketed under the name Kaby Lake and is the 7th generation Core model series. Like the predecessor generation, Broadwell and Skylake, the Kaby Lake chips are manufactured in a 14 nm FinFET process.

Consumer CPUs are equipped with a maximum of 4 cores on the Intel side. If more CPU cores are required, you must switch to server CPUs with the name Xeon. These are produced with up to 22 cores, which have a lot more power, but also at a significantly higher price.

In Table 1, the two consumer CPUs, as well as the Xeon E3 and E5, models of the last two generations are listed with the i5 and i7 processors. The program *PassMark PerformanceTest* was used as benchmark program and the performance data were compiled with the current store prices in a price - performance ratio. (A higher mark is better than a lower mark) [5]

AMD introduced the FX series in 2014 and had problems with manufacturing technologies until 2016, so that the next generation, named Ryzen, was not be announced until 2017.

2.1.1 FPGA

In recent years FPGAs (Field Programmable Gate Arrays) have become more affordable and powerful. A successful implementation of FPGAs as processors for MD (molecular dynamics) simulations is reported in *FPGA Based Accelerator for Molecular Dynamics*. [2]

In most cases, an OpenCL environment is simulated in order to use existing software parts. Compared with a CPU-only simulation an acceleration with a factor of 4 to 18

could be achieved. [7]

But due to the acceleration through GPUs and their cost-effective cost factor, as well as a complicated integration to GROMACS, it is currently still ineffective to use FPGAs. In this case, basic programming principles must still be given in order to be able to offer pure users of MD programs a real advantage.

2.2 GPUs

NVIDIA has not only been able to increase the available computing power over the last few years, but also through its excellent software support for various MD programs, simplifying the implementation and minimizing the CPU-to-GPU interface through very hardware-near programming.

Fig. 2 shows the development of the computational or convolution performance of the last NVIDIA generations. Folding@home, [6], is a software package managed by the Stanford University, which provides rendered computing power from home users to simulate research projects in the shortest possible time. Users will receive PPD (points per day) for the rendered computing power. As the global interest in this project is continually high, benchmark information is available, which can also be obtained from GROMACS, as Folding@home uses the GROMACS simulation core.

Fig. 3 shows the development of computing performance in the context of the Folding@home project. On one hand, AMD was able to offer cheap hardware, but the poorly optimized implementation of OpenCL, in contrast to CUDA, still significantly hinders the full development of the simulation performance.

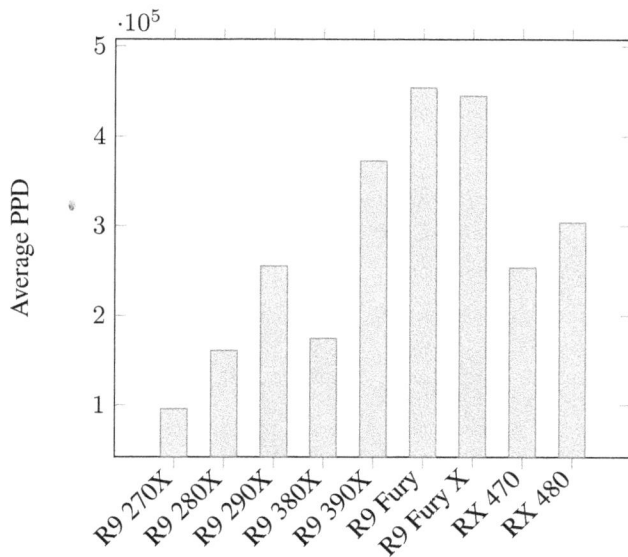

Figure 3: Development of the PPD performance for AMD graphics cards

Table 3: Comparison of current GPU generations

GPU	Powerlimit [W]	PPD/Watt	Price [€]
GTX 1060	120	3376	252
GTX 1070	150	4581	399
GTX 1080	180	4238	599
TITAN X	250	2840	750
TITAN X-P	250	5690	1299
R9 380X	190	920	209
R9 390X	275	1354	350
R9 Fury X	275	1599	399
RX 470	120	2114	179
RX 480	150	1996	199

Table 4: Comparison of current GPU generations

GPU	PPD/Price	Price/Perfromance
GTX 1060	1608	4984
GTX 1070	1722	6303
GTX 1080	1274	5512
TITAN X	947	3787
TITAN X-P	1095	6785
R9 380X	836	1756
R9 390X	1064	2419
R9 Fury X	1102	2701
RX 470	1417	3531
RX 480	1505	3501

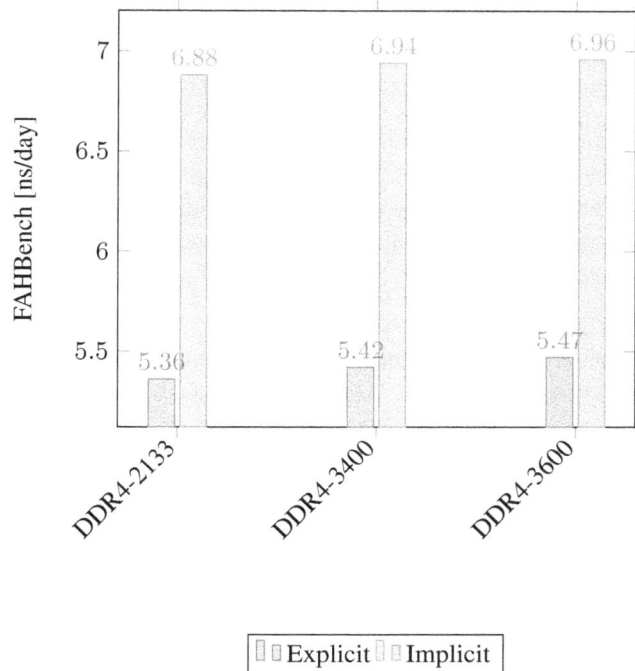

Figure 4: Comparison of different RAM speeds

2.3 RAM

Memory plays only a subordinate role with regard to performance in MD simulations. While the size of the memory should be sufficient to represent the MD system as a whole, the speed as Figure 4 shows, is almost irrelevant for the simulation speed. Significant costs can be saved by paying attention to standard speeds.

3 Results and Discussion

Through the market analysis presented in the preceding chapters, it is now possible to easily select the hardware for a price-performance-optimized system based on a set price frame for the individual computing station. While CPU performance in the past few years was only in the single-digit performance gain per year, the performance of the graphics cards grew significantly in the high double-digit range per year. The high-performance models of all manufacturers are also characterized by increased costs and thus lower the cost-benefit ratio, so that should be placed on mid-range models.

With the ratio of PPD in the middle / price set up, future graphics card genotypes can be quickly assessed with regard to their performance for GROMACS. While a couple of years ago, dual GPU systems were worthwhile, a somewhat different picture emerges today: due to the dramatically increased performance of the individual graphics cards, the balance between CPU and GPU is tilted to the point that a single processor can hardly support more than one graphics card with calculations and commands. Dual CPU mainboards are significantly more expensive and you should evaluate very closely whether you need the total power for a single simulation or rather two workstations to work parallel to several simulations and still cost-saving. Due to the rapidly changing technologies in the field of graphics cards, it should be noted that it is more worthwhile to procure a workstation for a very limited time of 1 or 2 years and not to plan for a longer period of time over 2-3 years. Improvements in the OpenCL implementation can also make AMD more competitive again in the coming years and allow additional cost savings for the user.

4 Conclusion

Due to the ever-improving implementation of graphics cards by CUDA and OpenCL in the GROMACS computing core, simulation calculations could be accelerated many times. This makes the CPU-only workstation completely obsolete in a price-performance-oriented environment.

By means of the relationship PPD / price a quick evaluation can be made and a CPU can be selected to find a matching hardware. In doing so, care must be taken to find a composition which does not unnecessarily give away performance, which results in uneven waiting for the simulation process.

The initial exclusive calculation of Van der Waal's forces on the graphics card has now been expanded to the Coulomb interactions. Thus, still bound interactions and the PME (particle mesh Ewald) load are calculated by means of the CPU. If further parts of the arithmetic load can be transferred to the GPU in the next GROMACS versions, a higher non-uniformity with regard to CPU and GPU performance can be tolerated.[8]

At the time of the publication, the 7th generation of Intel's processors, Kaby Lake, was introduced. There was a similar situation with AMD. Ryzen will only be available in the second quarter of 2017. Both product lines were therefore neither available, nor were valid benchmark data available. [9],[10]

Acknowledgement

The work has been carried out at the Institut für Physik, Universität zu Lübeck.

5 References

[1] GROMACS (Groningen Machine for Chemical Simulations), *GROMACS*. Available: `http://www.gromacs.org` [last accessed on 2017-01-08].

[2] Nicolae Goga, Mihaela Malita, David Mihaita, Gheorghe M. Stefan, *FPGA Based Accelerator for Molecular Dynamics*,2016.

[3] CUDA Zone, *NVIDIA accerelerated computing*. Available: `https://developer.nvidia.com/cuda-zone` [last accessed on 2017-01-09].

[4] GPU acceleration flow, *GPU acceleration*. Available: `http://www.gromacs.org/@api/deki/files/229/=gmx_4.6_gpu_acceleration_flow_parallel.png` [last accessed on 2017-01-09].

[5] PassMark Performance Test, *PassMark Software Solutions*. Available: `https://www.passmark.com/products/pt.htm` [last accessed on 2017-01-12].

[6] Folding@home, *Folding@home - Protein Folding*. Available: `https://folding.stanford.edu/home/faq/faq-gromacs/` [last accessed on 2017-01-13].

[7] The open standard for parallel programming of heterogeneous systems, *OpenCL*. Available: `https://www.khronos.org/opencl/` [last accessed on 2017-01-15].

[8] Ulrich Essmann, Lalith Perera, and Max L. Berkowitz Tom Darden, Hsing Lee, and Lee G. Pedersen, *A smooth particle mesh Ewald method*. in: The Journal of Chemical Physics, vol. 103,no 19, 1995.

[9] The "Zen" Core Architecture, *RYZEN*. Available: `http://www.amd.com/en-us/innovations/software-technologies/zen-cpu` [last accessed on 2017-02-02].

[10] Intel Core i7-7xxx Processor Series, *Kaby Lake*. Available: `https://ark.intel.com/products/codename/82879/Kaby-Lake#@All` [last accessed on 2017-02-02].

Evaluation and comparison of two different types of motor control for BiPAP and CPAP mode in non-invasive mechanical ventilation

F. Dietzel[1], W. Hanssen[2], and P. Rostalski[3]

[1] Medizinische Ingenieurwissenschaft, Universität zu Lübeck, frank.dietzel@student.uni-luebeck.de
[2] Löwenstein Medical Technology GmbH + Co. KG, Hamburg, wolfgang.hanssen@loewensteinmedical.de
[3] Institute for Electrical Engineering in Medicine, Universität zu Lübeck, philipp.rostalski@uni-luebeck.de

Abstract

Due to the widespread occurrence of lung diseases and the associated necessity of mechanical ventilation, the development of ventilation systems is always a topical issue. This paper deals with the comparison of two types of control for electric motors and the respective advantages of ventilation in CPAP- or BiPAP-mode. Some technical and medical backgrounds were discussed and the two motor controls were presented. The two types of commutation, sinus and block commutation, are tested on the basis of appropriately optimized example hardware and the results are compared. A theoretical discussion follows. The power drain and temperature behavior are examined. The result of this paper is that no type of control can be favored for the cases examined. Further ideas for experiments on this topic are mentioned in the conclusion.

1 Introduction

A versatile used ventilation mode in clinical and home-care ventilation is the CPAP-mode. CPAP stands for "Continuous Positive Airway Pressure", which means that the ventilation machine continuously induces the same pressure at all times (PEEP, Positive End Expiratory Pressure). One advancement of the CPAP-mode is called the BiPAP-mode (Biphasic Positive Airway Pressure). The main difference between the CPAP- and the BiPAP-mode is that the BiPAP-mode uses a low level of airway pressure during the expiration and a higher second level for inspiration. The BiPAP-mode reduces the need of respiratory work this way. CPAP ventilation, as well as BiPAP ventilation, is used for the treatment and therapy of atelectasis as well as for the reduction of the respiratory work [1].

1.1 Area of application

These types of mechanical ventilation are used for various diseases. In particular, they are used for obstructive sleep apnea syndrome (OSAS) [2]. In the course of this disease patients will suffer phases of sleep apnea, which can have a significant impact on a person's health and life expectancy [3].
They are also used for the treatment of COPD-Patients (chronic obstructive pulmonary disease). This respiratory obstruction results in an intensified inflammatory reaction of the upper respiratory tract and the lung. It is characterised by a variable combination of obstructive bronchiolitis and pulmonary emphysema [3]. COPD is one of the main causes for morbidity and mortality worldwide [3]. Statistically it is the third main cause for death in the world (in Germany: fifth place [3]). A rising trend is expected for the next decades [3].

1.2 Ventilation machines

Two types of ventilation machines are produced: clinical- and home-care-use. The home-care variation is optimised for certain requirements such as battery life-time and wearing comfort for example. The clinical one is more focused on monitoring and loud alarms for example.
The ventilation machine consists of the following parts:

- Blower: motor-driven impeller to produce gauge pressure
- Respiratory humidifier: humidification of the respiratory air
- Electronics and measurement systems

1.3 Blower

In essence, three drives have been established for the generation of overpressure during overpressure breathing [4]:

- Pneumatic bellows
- Electric piston drive
- Electrical blower

In a radial blower, the respiratory air is compressed

by centrifugal forces. Such blowers can produce the required pressures without additional devices. Due to the low inertia of the impeller, high speed changes can be achieved in a short time. This feature is especially interesting for BiPAP mode where step changes of pressure need to be realised in every breathing cycle. The impeller is an essential part of the blower system. Its design is significant for the creation of flow and pressure.

The structure and the principle of operation of brushless DC-motors (BLDC) are shown in figure 1. The brushless DC-motor has six coils (stator) which induce a magnetic field that interacts with the magnetic field of the rotor and causes motion. For energy-efficient use of a motor at the best possible performance, the choice of the appropriate control method is decisive. This includes the current-signals for the coils and the determination of the position of the rotor. The injection of the phase currents can be effected essentially by a block or a sinus commutation [5].

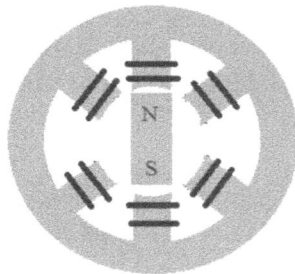

Figure 1: Schematic representation of the BLDC operating principle: A permanent magnetic rotor is driven by a precisely controlled external field.

1.4 Block commutation

The six transistors of the motor driver, which are needed to switch between the different coils, are defined as three "high-side" and three "low-side" transistors. For block commutation always one "high-side" and one "low-side" transistor of different phases are supplied with current. "High-side" gets a PWM-signal with a fixed pulse duty factor, "low-side" gets a not modulated signal for the time of the period.

In order to bring current into the correct phase at all time, the position of the rotor must constantly be determined. The position is usually determined by a control magnet and three Hall-Sensors staggered by 120°. During a commutation cycle, six different switching combinations are cycled through. In relation to the rotor, this means switching over every 60°. Since the phase current within these 60° always remains constant, the rotor-field does not change in this time period, but the stator-field does. Thus, only in the middle of the commutation cycle, the rotor-field is orthogonal to the stator-field to produce the maximum torque. At the begin-

ning and at the end of a commutation cycle, the torque decreases by up to 13.4 % [5]. A type of control which always keeps the two fields orthogonal to each other may improve the efficiency. This is the main idea of the sinus commutation.

1.5 Sinus commutation

All three phases of the motor are supplied with different PWM-Signals. The pulse duty factors are sinusoidal modulated and thus create a continuously rotating magnetic field with constant amplitude [5].

For control with a sinus commutation, it is suitable to use a FOC (field oriented control). In this case, the sinusoidal alternating variables are not directly controlled in their temporal instantaneous value, but in a current value corrected by the phase angle within the period. The resulting magnetic field rotates with the rotational speed of the motor and is always at an angle of 90° to the rotor magnetic field. Thus the maximum torque is produced [5].

A motor can be optimized for the control with a sinus commutation or a block commutation via the arrangement and construction of the coils. Controlling a motor which is optimised for sinus commutation with a block commutation could damage the motor.

The aim of this paper is to compare the different motor controls and find out which one is more suitable for controlling a motor of a blower in a mechanical ventilation system for CPAP and BiPAP ventilation.

2   Material and Methods

The blower mainly consists of three parts:

- motor
- cooling unit
- impeller

Its design has to meet certain specifications, which give highest priority to performance and efficiency. In this paper only those parameters are considered which can be influenced by commutation. It is possible to reach the needed speed and dynamics with both commutations. To this end, correspondingly powerful electronics are required. The noise level is determined by the design of the blower. The balancing of the impeller is very important, since an unbalanced impeller can cause vibrations and noise at high speeds. The required electronic power for the achievement of certain speeds and the temperature behavior of the engine as well as the respiratory air are analysed. This gives information about how efficient a motor, which is optimised for a certain commutation, works under control of this commutation in comparison to the other one. The heating of the motor means power loss. Although there might be power loss in other parts of the system, e.g. the motor driver. The heating of the motor is very important for the generation of heat in the whole ventilation machine. So the following experiments compare two systems which are optimised for sinus commutation and block commutation. The following requirements must be applied to the blower:

- Dynamics (change of speed per time): >250.000 rpm/s (the dynamics are mostly important for the BiPAP-mode to quickly switch between the pressure for the inspiration and the pressure for the expiration).
- Maximum speed: 70.000 rpm (the maximum speed of the motor is responsible for the maximum pressure that is producible).
- Maximum permissible temperature: 43°C (according to DIN EN ISO 60601-1 section 11, it is assumed that the respiratory air adopts the temperature of the motor and the mask is heated to this temperature).
- Noise level: A low noise level is necessary, as the patient is sleeping right next to the device.
- Efficiency: For reasons of battery life and heat development the efficiency should be as high as possible.

2.1 Test setup

Two blowers are compared. The motor of one blower is optimised for the control with block commutation, the other one for the control with sinus commutation. Both impellers and the diameters of the outputs are identical to ensure comparability.

The blowers are tested over a speed range from 10.000 rpm to 50.000 rpm and the input power is calculated by $P = V \cdot I$, where I is the input current and V is the input voltage. Both of the blowers have Hall-sensors included. As described above, they are only used in the control board for block commutation to determine the position of the rotor. But for these experiments they were also used to calculate the speed value. The magnetic field of the rotor creates Hall-impulses every revolution. By multiplying the frequency of one Hall-signal with 60 s, the speed value is calculated in rpm. The Hall-signals were measured with an oscilloscope.

To compare the temperature of the two motors, a temperature-sensible resistor (NTC) mounted in the motor is measured after the motor has been running for 5 minutes at 50.000 rpm. The characteristic curve of the resistor was used to convert the resistance value into temperature. The temperature of the respiratory air is measured by a temperature-sensor 15 cm away from the blower.

The sinus commutation with a FOC is generated by an evaluation board of the company Microsemi. The board is called SmartFusion2. It is controlled by a software from the company Microsemi. The control is realized by a microcontroller and an FPGA-IC.

The block commutation is realized by the electronics of a ventilation device from the company Löwenstein Medical Technology. This board is controlled via the communication terminal "hTerm".

2.2 Theoretical discussion

The sinus commutation requires a much more accurate position determination of the rotor in comparison to the block commutation. The resolution of three Hall-sensors is sufficient for block commutation, but resolutions up to 0.01° are used for sinus commutation [6]. The more efficient motion of the motor controlled by a sinus commutation leads to lower torque ripple. For ventilation, this means a more constant flow and pressure than in block commutation.

This smooth motion with sinus commutation only works for low speed levels. At higher speed levels the frequency of the sinusoidal current signals increase and due to the controllers a lag between control of the stator-field and rotor-field results. This restriction can be eliminated by using a FOC. The FOC makes a more stable motion at higher speed levels possible [7].

Theoretically, sinus commutation is more energy efficient than block commutation and produces a more stable motion at low speed levels. Block commutation, on the other hand, can be more easily implemented and thus also more cost-effectively and is relatively efficient in the high speed range. But the sinus commutation with a FOC combines a stable motion at low speeds and an efficient running at high speeds. With both commutations powerful driver units are required for the use in ventilation machines because of the high speeds and dynamics that are necessary [8].

3 Results and Discussion

Fig. 2 shows that the graphs approximately follow an exponential function. At 10.000 rpm, less than 5 W are needed for both commutations, whereas powers of about 50 W are needed at speeds of about 50.000 rpm. In this measuring range, the power consumption with the sinus commutation is always smaller than with the block commutation. The maximum difference in power consumption is 2.42 W at 50.000 rpm. The difference in power consumption can be assessed as insignificant. It is not possible to determine an advantage of the sinus commutation in comparison to the block commutation with respect to the power consumption. However, based on the graphs, it can be assumed that the sinus commutation is found to be appreciably advantageous at higher speeds.

The temperature of the motor increases in both commutation modes to over 55°C within 5 minutes of measuring, with a tendency to rise shown in fig. 3. However, the temperature rise is almost the same for both commutations.

The temperature of the respiratory air has differences up to 6°C. It is always higher when controlling with block commutation than with sinus commutation. However, there is no correlation between the temperature graph of the motor and the temperature graph of the respiratory air. Therefore, it is not clear that the temperature difference of the respiratory air results from the temperature of the motor and thus from the commutation mode. The reason for that could be the different design of the two motors but not the commutation. Thus, no advantage of a commutation mode can be

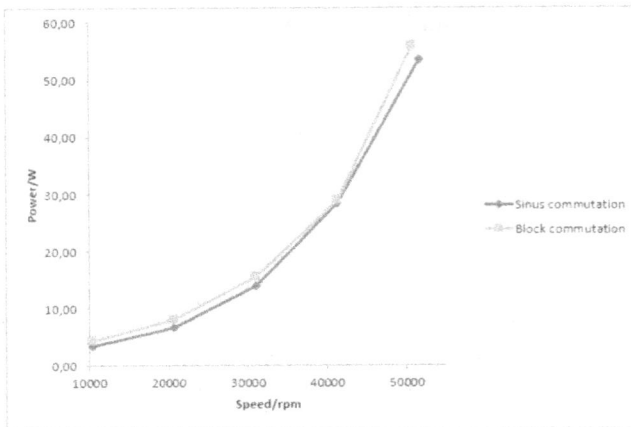

Figure 2: Power input for certain speed levels between 10.000 rpm and 50.000 rpm. Five values have been measured with a test distance of about 10.000 rpm.

derived from this finding. The temperature of the respiratory air is also influenced by the design of the blower and the ambient temperature.

It should also be noted that efficiency does not solely depend on commutation. The entire electronics as well as the design of the motor are also crucial.

It should be mentioned that this level of stress for the motor and therefore its temperature are not part of the normal use in a ventilation machine.

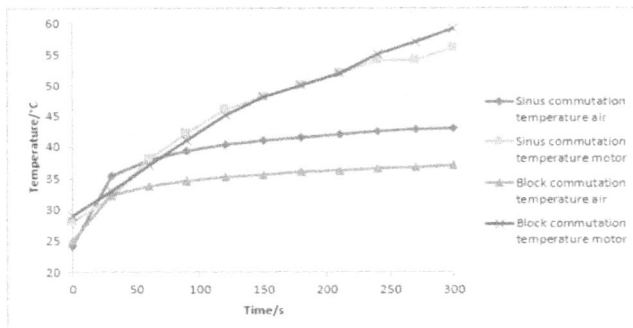

Figure 3: Results of the temperature experiments. Both systems where examined for 5 minutes. For each system the motor temperature and the temperature of the respiratory air were measured.

4 Conclusion

In summary, no type of control can be favored for the considered speed range and operating time, as no decisive advantages of one commutation has been noticed in any experiment.

From the theoretical point of view, the sinus commutation is more energy-efficient and thus particularly advantageous for battery-powered ventilation devices. The sinus commutation with a FOC also produces a more stable motion.

Further studies could include higher speeds, other work-loads on the engine, researches on the dynamics and a volume test. To improve the comparability both of the commutations should be performed with the same board and motor driver.

Due to the fact that every blower is different, series of measurements with many motors could be done. Furthermore, the physiological effects of the torque ripples and the ripples themselves could be examined. If these ripples and the resulting differences in the produced pressure have negative physiological effects, the sinus commutation would be favored for the mentioned reasons.

The drivers for block commutation are easier to realise and therefore usually less expansive than for sinus commutation.

Acknowledgement

This work has been carried out at Löwenstein Medical Technology GmbH + Co. KG, Kronsaalsweg 40, 22525 Hamburg and supervised by the Institute for Electrical Engineering in Medicine, Universität zu Lübeck.

5 References

[1] H. Lang, *Beatmung für Einsteiger*. Springer, Berlin/Heidelberg, 2016.

[2] L. Ziegenfuss, *Beatmung*. Springer, Berlin/Heidelberg, 2013.

[3] D. Bösch, *Lunge und Atemwege*. Springer, Berlin/Heidelberg, 2014.

[4] S. Leonhardt, *Medizinische Systeme*. Springer Vieweg, Berlin/Heidelberg, 2016.

[5] all-electronics.de, *Feldorientierte Regelung für EC-Motoren*. Available: http://www.all-electronics.de/stabil-drehen [last accessed on 2017-01-14].

[6] E. Hering, *Handbuch der elektrischen Anlagen und Maschinen*. Springer, Berlin/Heidelberg, 1999.

[7] motioncontroltips.com, *What's the difference between field oriented control and sinusoidal commutation?*. Available: http://www.motioncontroltips.com/faq-whats-the-difference-between-field-oriented-control-and-sinusoidal-commutation [last accessed on 2017-01-17].

[8] S. Lee, T. Lemley, *A comparison study of the commutation methods for the three-phase permanent magnet brushless DC motor*.

Advancement of a sensor-system simulation for sleep therapy- and ventilation test benches

J. Stubbe[1], S. Fricke[2], and P. Rostalski[3]

[1] Medizinische Ingenieurwissenschaft, Universität zu Lübeck, jasmin.stubbe@student.uni-luebeck.de
[2] Löwenstein Medical Technology, Hamburg, sebastian.fricke@loewensteinmedical.de
[3] Institute for Electrical Engineering in Medicine, Universität zu Lübeck, philipp.rostalski@uni-luebeck.de

Abstract

The final test of the final assembly of sleep therapy devices and respirators in the field of medical technology is one of the decisive steps before a new market launch. It is carried out using a test system based on a test PC with special test software and an electronic and a pneumatic module. Due to the high importance of these test systems, they are subject to the same requirements as the newly developed devices. For this reason, the hardware, as well as the software of the systems must be checked and updated regularly. Since the new test software or test software updates must also be tested even when a test bench is not available at the moment, a test bench simulation has been developed, which can be used for this purpose. The problem is, however, that the test scripts are too extensive and too inflexible for tests in endurance runs. Therefore, an expansion of the software has been made, which provides a new mode to fulfill exactly this purpose and to create a platform with which the new test software can be tested in endurance runs. Another part of the work was that new test scripts were developed to test all existing sleep therapy devices and respirators with the test bench simulation.

1 Introduction

The device groups considered here are sleep therapy devices and respirators, which belong to the group of medical devices because of their intended use. When a new device of this type of product is introduced into the market, special requirements relating to quality and quantity are required under Directive 93/42/EEC concerning medical devices [4]. These relate not only to the device itself, but also to the device test, and thus the test methods and the test systems used. For this reason, the final inspection of the equipment must also be assessed on the basis of this directive. The system used for the final inspection is a test bench consisting of a PC with installed test software, as well as three modules, a power supply module, an electronic module and a pneumatic module. The modules supply the device under test (DUT) both electrically and pneumatically. Furthermore, the DUTs are measured from the test bench during the device test and subsequently evaluated by the test software. The test of the test bench consists of individual test parts, which are partly automated and partly controlled by the subjective assessment of the examiner. The quality management provides a 100% coverage with regard to the sample size in order to keep the error rate as low as possible from the beginning. In order to achieve such an audit quality, it is indispensable that, on the one hand, the sensor system has a sufficiently high accuracy and is regularly calibrated, on the other hand, the test software is constantly being further developed and kept up to date. In order to validate and verify this, a test bench is required, which can test a DUT with the current version of the test software. The problem at this point is, that the existing test benches are located in the production lines and thus are not available locally or in time. For this reason, an alternative of verifying the test software needed to be implemented. This should be achieved in the form of a test bench simulation. The work is based on an already existing system and for its extension, requirements have been defined which correspond to the wishes of the later user:

1. Create additional test scripts for all device types

2. New software mode which only intervenes in the test if necessary (e.g. simulate voltage, send manipulated command)

This should ensure that the final test simulator has three different modes.

1. Bypass Mode: only serves to allow all received and sent commands to pass through without active intervention in the test

2. Simulation Mode: allows the user to manipulate every possible command and therefore to arrange the test exactly according to his ideas

3. Endurance Run Mode: only intervenes in the test at specific points in order to simulate, e.g. a voltage or a keypress, if necessary

2   Material and Methods

A bachelor's thesis [1], which has been preceded by this work, already implemented a system solution that allowed to adjust the system group of the test bench and to specify a simulation sequence using an XML script containing individual commands of the device command structure. XML stands for Extensible Markup Language and is a language used to represent hierarchically structured data in form of text files [3]. The simulation software is able to simulate all relevant flows and pressures in the pneumatic system and some voltages and currents in the electronic system. With the created software and the electronic module, it was already possible to test the current test software by performing a test with completely positive results or individual part tests by manually created XML files. For the development of test benches it is interesting to carry out tests with completely positive or completely negative results in order to check, whether the test bench shows all positive, but also every possible negative result, correctly [5]. The goal of this thesis was to create additional XML files for two other device types and to extend the part tests of the already existing device types so that a single script is available for each individual part. Furthermore, a new mode should be implemented, which allows the software to intervene only at the points where it is really necessary, to create a mode that is capable of checking the test software in endurance runs. In addition, the communication between one of the device types that had not yet been implemented and the test system had to be created. The real test system consists of a test PC with a built-in NI measurement card and three modules. An electronic and a pneumatic module and a power supply module, which contains the mains voltages. The test bench can be seen in fig. 1. The power supply module is responsible for the power supply of all components as well as the DUTs. Via the pneumatic module, the pressures and flows are adjusted and measured. The DUT is connected to the module via a breathing hose. The entire information exchange takes place via the electronic module and the LabVIEW module cards. These are used to itemize the channels of the measurement card so that they can be further processed and forwarded by the hardware. The contained relay cards process the digital signals which are used to switch voltages and signals on and off. In addition, the analog inputs are combined here and all measurements are passed on to the test software with current and voltage sensors on additional module cards. Some of the values must also be scaled to 0-10 volts. This is implemented with the help of voltage dividers and a conversion within test software. There are three different types of tests. The production test, the repair test and the part test. The production test and the repair test comprise the entire test parts. The production test is used to carry out the final inspection of the newly produced devices. Each device must have successfully passed it once before it can enter the market. The repair test serves to carry out a final inspection of the devices, which were returned due to errors. These are then inspected and repaired in the service and must be checked

Figure 1: Front view of the test bench

again before they are returned to the owner. Each individual test part can be selected and tested separately during the part test. This makes it possible to check the test points individually and to make troubleshooting easier. This test mode is usually used in the service if a faulty device is being sent in and only the defective component, such as the hose heater, should be tested. This can be carried out with the help of the part test before the repaired device has to pass the repair test at the end of the repair. The various tests of the test bench consist of individual test points. These can differ from device to device, even if the basic test sequence is the same. However, there are differences in the setpoints of individual tests as well as for certain components to be tested. The hose heating, for example, is not available with every DUT. It serves to prevent condensation in the breathing hose, or to get moisture into the interior of the device.

The test software is written in LabVIEW, which is a development environment with a graphical programming interface designed specifically for engineers and measurement and control systems [2]. The program was written object-oriented. The various device types and test sequences with the individual test parts are stored in different classes. This makes it easier to customize the test software if certain test sequences, device features or thresholds should change. The simulation software is also written in LabVIEW to ensure the best possible communication between the two programs.

The existing test bench simulation consists of a standard test PC with the current test software, a simulation PC with the simulation software and an electronic module, via which voltages and currents can be simulated [1]. With this module pneumatic values such as pressure and flow, or currents and voltages are simulated. These are entered in the XML file, converted into a voltage value between 0 and 10 volts within the software and then simulated via the voltage sensors of the electronic module so that the test software can measure the correct values. The existing simulation software had already two modes. The first mode, the simulation mode, passes through a test script in the XML format, which covers the entire scope of testing of the respective devices and allows the user to manipulate each checkpoint if he wants to. The XML file is read in and checked for errors. It consists of individual checkpoints, each of them is divided into individual commands. It is indexed by two "For loops" and then the individual commands are executed. The second mode, the bypass mode, is used only for testing as livetracking and does not actively intervene in it by simulating values via an XML file. The problem with this implementation was that on the one hand, XML files for two other device types were missing to go through the entire test and to manipulate and on the other hand, the previous main simulation mode was too inflexibel for its actual purpose. The goal was to have a third simulation mode running through an XML file that only intervenes at the points where it is really required and values had to be simulated or commands be manipulated. The previous state required very large XML files and the goal was that these should become much clearer, by the simulation software only at those places at which really values are simulated or commands must be manipulated and everything else in the test is left unaffected. This has to be done at the points where, due to the lack of sensors, e.g. pressure and flow sensors, no real signals are available and because of that these values have to be simulated.

In order to meet all these requirements, first step was to put the existing system into operation. The Windows operating system had changed from Windows 7 to Windows 10 and thus the communication between the two computers had to be reset with their software programs and the DUT. Here, a first change was made to the existing software, in which the assignment of the COM ports in the software was set flexibly and no longer hardcoded as before. As a next step, new XML files were created for the two missing device types, and they were made to run with the existing simulation software. As a last step at this point, XML files were created and tested for each individual part test. After completion of this block, the development of the new mode and the software expansion started. The new mode should combine the two already existing, by acting as an observer on the one hand, to pass the test that had been performed through live tracking, without influencing it and, on the other hand, to intervene at the points where a simulation of values or a manipulation of a command was necessary. This mode should also run with an XML script. A major difference, however, is that the XML contained only the commands that were actually required for the test. The flowchart of the new mode is shown in fig. 2. In the flowchart, it can be seen that the XML file is read as usual and then waited for a trigger command. If the program sees this trigger, the corresponding point in the XML is to be processed. If the program sees no trigger, it does nothing, except continuing its live tracking. If the command of the test software matches the trigger of the XML file, the tree structure of the file is processed further. The script specifies whether a command is to be

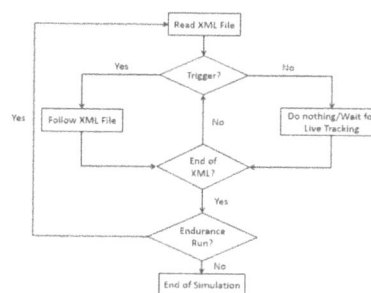

Figure 2: Flow chart of the new mode

replaced by another, for example, to manipulate sensor values or to simulate a specific current or voltage value. The XML is processed until the incoming command no longer matches the next XML trigger, then it is returned to Livetracking mode. In addition, the user should have the option to specify whether he wants to run the mode in the endurance run or not.

Working with XML files is easy because the LabVIEW programming interface has pre-built blocks (VIs) to read and process. This structure is particularly suitable for creating a script for the communication between simulation software and the DUT. The script consists of the individual checkpoints, which in turn are divided into different commands, which consist of 14 subpoints. The structure can be seen in fig. 3.

3   Results and Discussion

First of all the test bench simulation was made workable for the still missing device types. For this purpose, XML files had to be written for both device types, which serve as a script template and thus as the basis for the already existing simulation mode. Problems arose because of the different command structures of the various device types and possible changes that could occur during the course of device or firmware development. The smallest change causes that the XML file cannot be executed correctly. An example of an instruction from the XML is shown in fig. 3. In this example, the first open command shows how the implementation of the manipulation of a command is performed. The second open command shows the simulation of a voltage value. The analog and digital ports are also addressed here and the required voltage of 12.8 V is connected via the relay. When creating the XML file for the newest of the devices, there were some problems to specify the correct commands as triggers in the XML. The command structure was inconsistent in some places and made it difficult to create a script template.

Second part was to implement a new mode in LabVIEW, which should make it possible to carry out endurance runs to test new device firmware. This mode should also work with XML files, but these should not be as extensive. This should make the simulation more flexible. To achieve this, another event case was created. The configuration and the

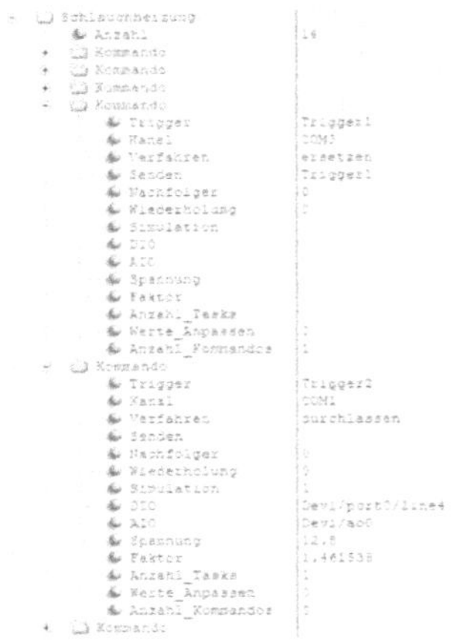

Figure 3: Extract from XML file command structure

import of the XML file take place as in the already implemented simulation mode. The new mode consists of a while loop with a case structure. The case structure has two cases whose calls are based on the match between XML trigger and the read command. If they match, the true case, to be seen in fig. 4, is traversed and the XML command executed in the same way as in the previous simulation mode. The status of the trigger is constantly interrogated by a notifier.

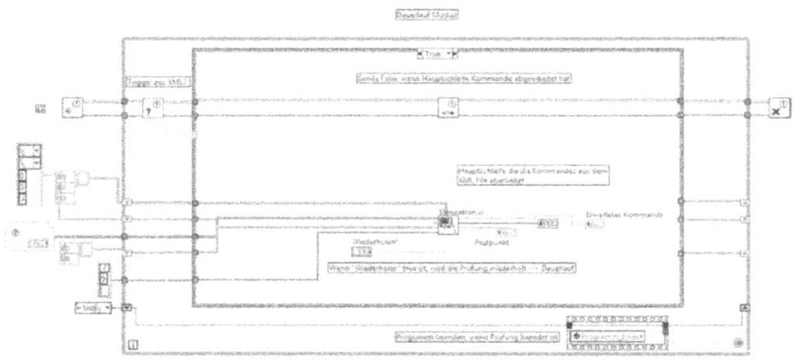

Figure 4: True case of endurance run mode

If this no longer matches, the false case, seen in fig. 5, is traversed and only a live tracking of the transmitted and received commands without active intervention of the software is performed.

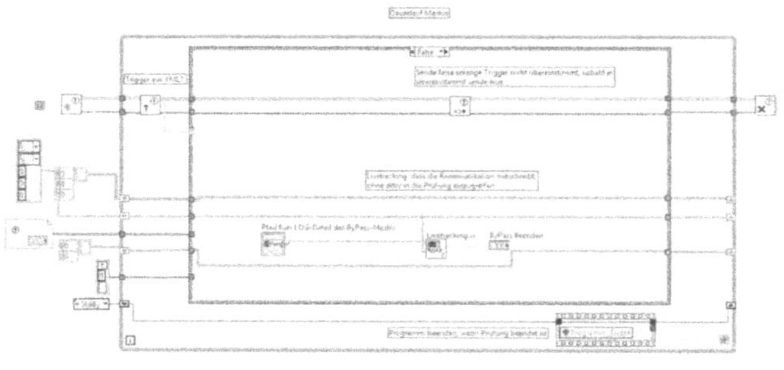

Figure 5: False case of endurance run mode

After completion of the test, it is checked whether the user has selected the endurance run mode or not. If he does not,

the simulation is terminated. If the run button is activated, the mouse clicks are simulated via property nodes, which are required to restart the simulation. This means that the configuration mode is first selected, the XML file is re-read and then the simulation is restarted. In this mode the program remains until the termination button is selected from the outside and the simulation is terminated.

4 Conclusion

In conclusion, the newly created XML files of the two missing devices, as well as the extensions of the XML files of the already existing devices work as expected. Thus, for all existing sleep therapy devices and respirators, a test with completely positive results, as well as each individual part test can be performed with the simulation.

Furthermore, the new mode has been successfully implemented and tested to enable endurance runs with the simulation and to work with less extensive XML files. The main problem, however, remains communication and timing between simulation software, test software and DUT. Due to the "looping" of the commands from the simulation computer through the test computer to the DUT and the other way round, timing problems can arise which prevent a real-time examination of the test. One of the main problems remains the interplay between LabVIEW and Windows 10.

Acknowledgement

The work has been carried out at Löwenstein Medical Technology, Hamburg and supervised by the Institute for Electrical Engineering in Medicine, Universität zu Lübeck.

5 References

[1] N. G. Poor, *Entwicklung einer Sensor-System-Simulation für Beatmungs- und Schlaftherapieprüfstände*, Hamburg, 2016.

[2] Georgi, Wolfgang; Metin, Ergun: *Einführung in LabVIEW.*, 5. Auflage, München, Carl Hanser Verlag, 2012

[3] Friesen, Jeff: Java XML and JSON., 1. Auflage, Dauphin (CA), Apress, 2016

[4] Böckmann, Rolf; Frankenberger, Horst: *MPG & Co.: Eine Vorschriftensammlung zum Medizinprodukterecht mit Fachwörterbuch.*, 7. Auflage, Köln, TÜV Media GmbH TÜV Rheinland Group, 2015

[5] Chlebek, Paul: *Praxis der User Interface-Entwicklung - Informationsstrukturen, Designpatterns, Vorgehensmuster.*, 1. Auflage, Wiesbaden, Vieweg+Teubner, 2011

4

Biomedical Engineering

An Improved Procedure for Automated Testing of Dialysis Machines

T. Gano[1], A. Schrörs[2], and P. Rostalski[3]

[1] Medizinische Ingenieurwissenschaft, Universität zu Lübeck, theresa.gano@student.uni-luebeck.de
[2] Fresenius Medical Care, Bad Homburg v.d.H., alexander.schroers@fmc-ag.com
[3] Institute for Electrical Engineering in Medicine, Universität zu Lübeck, philipp.rostalski@uni-luebeck.de

Abstract

MultiFiltratePro is a dialysis machine for extracorporeal blood purification. During the development process multiFiltratePro devices are subjected to automated trials that consist of endurance tests in order to detect errors in the device's hardware or software. Automation of endurance tests requires the simulation of the actual patient's treatment. Automated bag changes that refill empty bags of the multiFiltratePro device by using the testing's setup are integrated into this automation. An improved algorithm for automated bag change is to be implemented. The idea is to run a certain number of checks before an automated bag change begins in order to determine if clamps on tubes attached to bags are open or closed. Additionally, a monitoring of the waste bag weight is being integrated. The checks allow the detection of an obstacle in the flow path. The automation stops and the user receives a message that a clamp is shut and needs to be opened.

1 Introduction

A general and basic requirement of medical devices is that they must be designed in such a way to allow resistance to ordinary operating conditions without impairing their characteristics. No user interaction shall lead to endangerment of people's health or security [1]. Therefore, system tests are performed in order to detect possible errors in hardware or software as early as possible. The more critical the systems are, the more important these tests become, leading to early error correction during the development process [2]. Acute dialysis is a method for extracorporeal blood purification that is used for treatments of acute kidney failure. Frequent cause of acute kidney failure include multiorgan failure, high blood loss or even trauma, so the acute dialysis is carried out in the intensive care unit [3]. In contrast to chronic dialysis, acute kidney failure is treated using lower blood and dialysis flow rates over a longer period of time that can take up to several days. In addition, dialysate is supplied using bags, which is not the case in chronic dialysis [4]. It is important that the machines continue to function correctly even after several days of use. Therefore, the endurance tests used during system tests are essential.

At intensive care units where multiFiltratePro machines are usually used, is it not possible for dialysis machines to mix their dialysis fluids themselves because the required connections for ultrapure water are not installed. Therefore, multiFiltratPro machines possess a scales system where the weight of the bags containing dialysate, substitute fluid and filtrate (waste fluid) are placed and constantly monitored. During the development process of a dialysis machine, its

software is continuously refined. One goal of system tests is to detect and document errors and to feed this information back into the software development process. The system is influenced by all its components, which are also interconnected and interact with each other [5]. Therefore all components and their interactions must be tested. Software errors and their possible consequences need to be discovered early in the process in order to be fixed and to avoid consequential errors. Therefore, patient treatments are simulated using automated permanent machine runs. This automatised testing has the advantage that it can be conducted in a predefined sequence. Furthermore, documentation of the trials is included since all the results are directly transcribed into a log-file. In order to avoid manual intervention during the automated test runs, used dialysis fluids are refilled by an automated bag change setup. The software that runs the automated test runs uses the message displayed on the screen of the multiFiltratePro devices for a bag change. Specifications are performed in response to different messages. As soon as a dialysis bag is empty, a certain message appears on the screen and fluid from the waste bag is pumped up into the dialysate bag.

There are clamps on the tubes and on the waste bag. These clamps, once they are shut, do not allow any fluid to pass through. During the permanent runs, users replace used tubing sets at the machines on a weekly basis. After these replacements, a possible fault condition is the remaining of closed clamps on the bags and tubes. When clamps on the connecting tubes between the machine and the setup are shut or when a machine is standing on a tube, detachment of tubes, damage to the setup or leaking of fluids may occur.

At present, the automation does not yet provide an appropriate fault-detection mechanism. Closed clamps are not detected and the pump continues to pump fluid until a user intervenes manually and stops the process. It is therefore desirable to develop a possibility to supervise and control the bag refill process in order to avoid these errors and to increase the availability of permanent machine runs.

In addition, a monitoring of waste bag weight that avoids the pump to get filled with air has to be developed. When the pump draws too much air, its correct functioning is impaired and proper bag change cannot occur either.

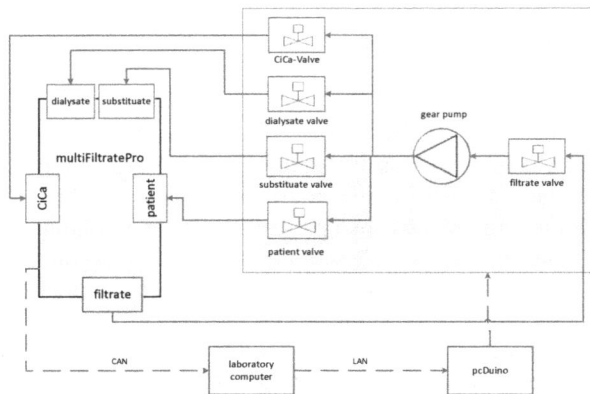

Figure 1: schematic test setup for the automated multiFiltratePro testing with the main devices

2   Material and Methods

Permanent runs can be performed due to a special setup; the test setup is shown in figure 1. In the following, the necessary components of the setup will be described. Concerning the multiFilratePro machine a CAN-Bus (Controller Area Network) is used to exchange data between the machine and a laboratory computer which, in turn, commands a pcDuino that controls the test setup. The pcDuino is a single board computer [6]. The software used for automation is written in the coding language Python 3. Python 3 is a modern coding language that is also commonly applied in science [7]. The programme for automation consists of two software packages, one for the laboratory computer and one for the pcDuino. They communicate with each other via a network connection. These software modules include different tasks such as control of programme runs, network communication and logging functions.

The test setup consists of several hardware components such as valves, an amplifier and a gear pump. The valves used are solenoid valves. There are five solenoid valves per test setup. In the construction a gear pump is being used for the transport of fluids from the bags. The process of transporting fluids is generated by filling and squeezing out the gaps of the rotor. The pump can rotate forward and backward and the speed of the rotor is set to a fixed value. An advantage of the test setup is that the number of revolutions

is constant and that pressure pulses can be avoided. A disadvantage of the gear pump is that the pump's power immediately decreases once it is filled with air. For the endurance tests it is necessary that the gear pump is filled with liquid. With multiFiltratePro regional anticoagulation with citrate can be performed. This is an alternative to classical anticoagulation with heparin. An advantage of the citrate anticoagulation is the fact that it only affects the extracorporeal blood circuit and thus, it is also suitable for patients with an increased risk of hemorrhage. Before the reinfusion of the treated blood into the patient, the necessary amount of a suitable calcium solution is infused [8].

In the test setup the calcium and citrate fluid is represented as a fluid bag so-called "CiCa"bag. The real patient, who is normally connected to the multiFiltrate in hospital is represented in the test setup by a filled liquid bag which is called "patient".

3   Results and Discussion

A monitoring procedure is created which allows the process of filling and emptying the bags to occur properly and that interrupts it as soon as an error is detected. At first, the idea of monitoring bag changes was based on a modification of the algorithm. Primarily, the functioning of the current algorithm for automated bag change shall be defined, followed by explanations concerning the latest innovations.

A bag change is made as soon as the machine displays the message «Bag Empty» on its GUI. This means that the dialysate or substitute bag does not contain enough fluid to continue the treatment. During treatments at the hospital, empty bags are replaced with full bags by the nursing staff. During the permanent runs, this is done with the help of an automated bag change using the setup.

The automation software is divided into different classes and functions. The function "commandPCDuino" is used to fill and empty bags. In this function, the message numbers of multiFiltratePro indicating when bag change is necessary are listed. When a bag change is needed, the software sends commands to the pcDuino that the bags must be refilled via the setup. The process of bag change starts by opening the necessary valves of the setup. These valves are grouped according to different virtual flow paths. By measuring weight at the scale where the bag to be refilled is placed, its initial weight is determined. While the pump works continuously, the weight of the bag being refilled is constantly monitored until the weight reaches a pre-set weight of 1500 g. As soon as the final weight is obtained, the pump stops and the valves are closed.

The scale weight is only taken into consideration to determine when bag refilling or emptying is to be finished, but it is not analysed for continuous changes in weight. Therefore, the pump would work infinitely if tubes were permanently obstructed by a closed clamp or a machine that stood on top of a tube. The refilling or emptying process could not be correctly executed and detachment of tubes from the setup would occur. Thus, the integration of a newly developed function named "check_DiaSub" was designed to detect such obstructions before a disruption could occur. To this end, a procedure is incorporated into the automation software that checks whether all flow paths are open. This control runs before starting the process of bag change as described above, meaning before fluid transport is initiated. At the beginning, the initial weight of the waste bag and the weight of the bag to be refilled are measured. The valves determining the flow path are switched and the pump starts pumping backwards. The backward movement has explicitly been chosen because it allows the pump to be filled with fluid even before filling of the bags which is necessary for the correct operation of this kind of pump. After three seconds of liquid transport, the weight of the waste bag is rescanned. This rescanning takes place after three seconds because after this interval, there is already enough fluid being pumped to detect an increase in waste bag weight and a decrease in the bag to be refilled. Plus, this time span is still short enough, in case of an obstacle, for the pressure in the tubes not to build up sufficiently to cause tube detachment. This interval has been determined experimentally for one setup, by running through the new function using different time slices. Three seconds is the time where all criteria are being met. By looking at the scale readings, a difference in weight before pumping and after three seconds of pumping can be detected. If the weight has increased by more than 1.5 g, a correct pumping into the waste bag can be confirmed. If the weight has stayed the same or decreased by less than 1.5 g, the clamp on the bag appears to be shut. A decrease in weight up to 1.5 g can still be detected because the scales of multiFiltratePro are very sensitive and react to the slightest disturbing influences from outside, such as air blasts. By calculating the weight difference, the software can detect closed clamps. As a consequence, the automation and the pump are stopped. The user receives a message that a clamp is shut and needs to be opened. Using a weight scan of the bag to refill after three seconds of pumping, a weight difference can also be detected in this case. If the weight of the bag has decreased, the clamp is open. If the weight has remained the same or decreased by less than 1.5 g, the clamp is shut. The pump and the automation are stopped and a message is sent, asking the user to check for closed clamps. When the user confirms the message, the function "check_DiaSub" is run again, starting from the beginning, checking if all the clamps are open now. The function is repeated until correct fluid transport is established. The check for closed clamps is only done before the bag filling process because it is not likely that users close clamps again after having opened them beforehand. Therefore, no continuous control for shut clamps is necessary during the filling process.

Figures 2 and 3 show scale signals that have been recorded during the automated bag change with the newly implemented function "check_DiaSub". Figure 2 shows the weight of the waste bag (filtrate). Figure 3 represents the weight of the dialysate bag. Both weights were recorded during the same treatment and at the same time. The clamp on the tube leading to the dialysate bag was shut intentionally in order to test the monitoring procedure. The function

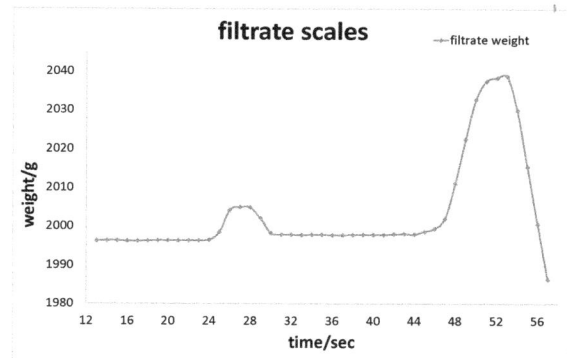

Figure 2: Signal of the scale where the waste bag (filtrate bag) is suspended during automated bag change with the newly implemented functions.

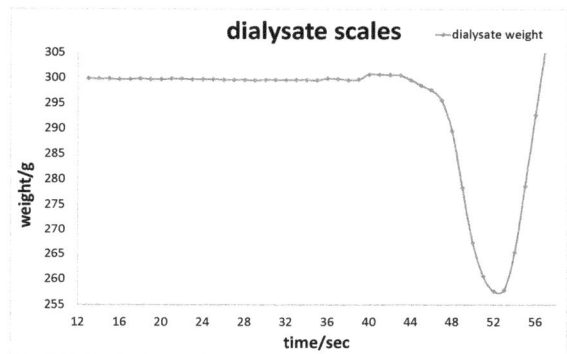

Figure 3: Signal of the scale on which the dialysate bag is placed during automated bag change with the newly implemented functions.

"check_DiaSub" passes all the steps described earlier which can be seen when examining the scale signals. After sending the message «Bag Empty», the function checks if the clamps are shut.

At second 24 in figure 2 and 3, the pump is pumping backwards. The weight of the waste bag increases. The weight of the dialysate bag stays the same, except for slight variations which can be seen in figure 3. Thus, there appears to be an obstacle between the dialysate bag and the gear pump. This obstacle is detected by the software which then stops the automatic procedure and sends a message to the user. The obstacle is manually removed and after an entry by the user, the function starts again from the beginning. At second 44 in figures 2 and 3, it can be seen that the valves are opened and that the pump is pumping backwards again. The weight of the waste bag increases and the weight of the dialysate bag decreases. Due to this weight difference, the automation recognises that there is no more obstacle. The function "check_DiaSub" is now finished and the filling of the dialysate bag is performed as usual by the setup. When stopping the pump after having obtained the target weight of 1500 g, some overfilling occurs which does not cause any problems, since the bag has a capacity of 5 l and there is no risk for the bag to burst.

After refilling the empty bag, it is checked whether fluid has to be refilled into the "CiCa" or "patient" bag. The levels of "CiCa" and "patient" are calculated by the multiFiltratePro. The necessary values for the calculation are taken into account and are related to each other, depending on the treatment. If a specific volume has been used up since the last bag change, these bags are refilled. The newly developed process consists of a newly integrated function named "check_CiPat" that checks before bag filling if there is an obstacle on the tubes leading to the bags. The process allowing identification of an obstacle is identical to the function "check_DiaSub". However, only the weight of waste bag is taken into account. All the weight measurements are done using this scale. The "CiCa" and "patient" bags do not have their own scales. The clamps are open if the weight on the scale increases by more than 2 g. The "CiCa" or "patient" bag is filled until the target weight is reached on the scale. The resulting overfilling is weighed and subtracted from the target volume during the next filling process. In addition to the current process algorithm for filling of "CiCa" and "patient" bags, a monitoring of the waste bag weight has been implemented. It stops the pumping process when the bag weight is less than 1500 g. The weight of 1500 g in the waste bag must not be undershot because otherwise, there will not be enough fluid in the bag for a refill. If there is not enough fluid in the bag, it may happen that the pump is not wetted with enough fluid, either which impairs correct functioning. Filling is cancelled when the limit is undershot and the process of bag change is declared finished.

Another idea of how to achieve monitoring of obstacles during automated bag change would be the incorporation of pressure sensors that would allow pressure measurements in the tubes of the test setup. This would require the integration of the pressure sensors into the hardware and the software of the setup. The integration of a new element into the test setup requires modifications of its construction which may disturb the performance of the already existing setup. An advantage would be a permanent monitoring of the pressure in the tubes. Possible obstacles and errors that have not been uncovered yet might be detected which could avoid possible damage by an earlier stopping of the automation.

4   Conclusion

The integration of the functions "check_DiaSub" and "check_CiPat" allows the detection of the presence or absence of an obstacle in the flow path by the software before the bag filling procedure. Thus, errors that occur when the flow is disturbed are avoided because the process stops beforehand. Additionally, a monitoring of the filling of "CiCa" and "patient" bags has been incorporated that en-

sures that the weight of the waste bag of 1500 g is not undershot so that the bag is not drained completely and the pump does not draw air.

Since the performance of the programme has only been tested using one setup, it is now important to integrate the refined process into the software of the other three setups as well. As the interval after which the weight during backward pumping is measured depends on the tube length and the tube characteristics, it needs to be determined experimentally for each setup. Another criterion for proper monitoring is the necessity of a specifically built connection piece between the tubes on the bag to be refilled. If all these criteria are met, the modified algorithm for automated bag change can be integrated into and used by all four setups in order to improve the availability of endurance tests and to promote further system tests on the multiFitratePro machine.

Acknowledgement

The work has been carried out at Fresenius Medical Care, 61346 Bad Homburg v.d.H., in the department Research and Development and supervised by the Insitute for Electrical Engineering in Medicine, Universität zu Lübeck.

5   References

[1] N. Leitgeb, *Sicherheit von Medizingeräten, Recht-Risiko-Chancen*. Springer Vieweg, Berlin/Heidelberg, p.91, 2015.

[2] H. M. Sneed, M. Baumgarnter and R. Seidl, *Der Systemtest, Von den Anforderungen zum Qualitätsnachweis*. Carl Hanser Verlag, München, p.8, 2012.

[3] A. Khwaja, *KDIGO Clinical Practice Guidelines for Acute Kidney Injury*. In: Nephron Clinical Practice, Karger AG, Basel, pp.179-180, 2012.

[4] J. T. Daugirdas, P. G. Blake and T. S. Ing, *Handbook of Dialysis, Third Edition*. Lippincott Williams & Wilkins, Philadelphia, pp.102-103, 2001.

[5] R-D. Böckmann and H. Frankenberger, *MPG and Co*. Loseblattwerk, TÜV Media GmbH, Köln, p.833, 2015.

[6] LinkSprite: pcDuino Family, Available:http://www.linksprite.com/pcduino-family/ [last accessed on 2017-01-16].

[7] S. Gerlach, *Computerphysik; Einführung, Beispiele und Anwendungen*. Springer-Spektrum, Berlin/Heidelberg, pp.55-56, 2016.

[8] A. Davenport and A. Tolwani, *Citrate anticoagulation for continuous renal replacement therapy (CRRT) in patients with acute kidney injury admitted to intensive care unit*. In: NDT Plus Journal, p.439, 2009.

Determination of intradialytic electrolyte shifts in dialyzers

F. Rubin-Schwarz [1] and W. Janik [2]

[1] Medizinische Ingenieurwissenschaften, Universität zu Lübeck, Friederike.Rubinschwarz@student.uni-luebeck.de

[2] B.Braun Melsungen AG, Melsungen, Waldemar.Janik@bbraun.com

Abstract

During dialysis the dialyzer takes on the task of the kidneys of renal patients. But the behavior of certain electrolytes, like potassium and calcium, in the dialyzer during dialysis is not investigated, yet. In this research article the impact of the Gibbs-Donnan effect (GDE) on these cations during dialysis is inspected. Different experiments were performed during ongoing dialysis treatment with blood substitutes. GDE is visible for potassium and calcium, during dialysis for protein solutions, but the effect is in such a small range, that it has no impact on the dialysis treatment. What means if the concentration of the cations of interest is supposed to be constant before and after the treatment, the dialysis fluid used should have the same concentration as the blood of the patient. In Case of needed withdrawal, the concentration of these electrolytes in the dialysis fluid has to be adjusted accordingly.

1 Introduction

Approximately 2.5 million patients suffer from renal insufficiency per year. During the dialysis treatment no individual dialysis can be performed [1]. This may be critical in terms of regulating the electrolyte concentration in the patient's blood, especially for potassium and calcium. There is hardly any information on the distribution of these cations during the dialysis treatment in the dialyzer in the scientific literature. Normally, the concentration of electrolytes is controlled by filtration in the kidneys. This procedure has to be carried out during a patient's dialysis treatment in the dialyzer, three or four times a week. If the various electrolyte concentrations in the body are too high or too low, severe side effects and consequences may occur during and after the dialysis treatment [2]. Ensuring the required electrolyte concentration in the blood of the patient should be of utmost importance during dialysis therapy. How the cations within the dialyzer behave during the dialysis therapy is of first interest. Especially the Gibbs-Donnan-effect (GDE) is being investigated, because it has already been proven to have an impact on sodium, the most frequent electrolyte in the blood. What is the impact of the GDE on sodium during dialysis? The higher the blood's albumin level in the dialyzer, the greater the difference of the electrostatic potential on the dialyzer blood membrane layer. Albumin is too large to pass through the membrane pores and is also negatively charged. Therefore, more cations must accumulate on the layer which consequently are no longer available to the diffusion. In order to equalize the concentration, more sodium ions have to be exchanged between blood and dialysis fluid. As a consequence the sodium concentration of the blood changes [3]. If the GDE also has an impact on the cations potassium and calcium, is of interest.

In this study, the impact of the GDE on potassium and calcium during dialysis treatment will be investigated. Thereby the behavior of the cations can better be predicted during dialysis treatment and the blood can be filtered, according to the needs of the patient, to ease the already unpleasant treatment.

2 Material and Methods

In order to gain new insights, different experiments were performed. The goal is to determine the changes in the potassium or calcium concentration during the dialysis for the respective composed blood substitute.

2.1 Methods

The experiments are carried out according to the scheme shown in Fig. 1. Dialysis fluid flows past the blood substitute in the counter-current principle to allow substance to exchange. The blood substitute is dialyzed with usual clinical parameters. Conductivity meters (CDM) are placed at the inlet and outlet of the dialysis fluid to ensure a continuous conductivity measurement. With this, unregulated changes in the electrolyte concentration of dialysis fluid and blood are immediately registered and can be remedied. For a more detailed analysis of the electrolytes, samples are taken from the blood inlet (BE), the blood outlet (BA), the dialysis fluid inlet (DE) and the dialysate outlet (DA) and the various electrolyte concentrations are checked. A blood gas analyzer system (RAPIDLab 348EX, Siemens), operating with ion probes, determines the electrolyte concentration of BE, BA, DE and DA during the procedure.

In order to compare the measurements, a special method was used. A potassium concentration as well as a calcium concentration was searched, that ensures that the concen-

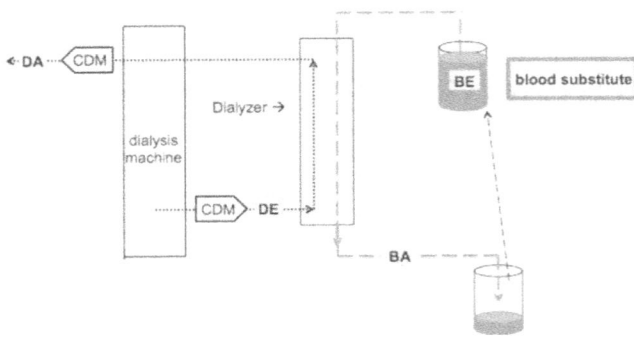

Figure 1: In dialysis therapy, a blood substitute is dialysed with a dialyzer to replace the function of the kidneys. Dialysis fluid flows past the blood substitute in the countercurrent principle. For the continuous conductivity measurement, conductivity meters (CDM) are placed at the inlet and outlet of the dialysis fluid into or out of the dialyzer. For a more detailed analysis of the electrolytes, samples are taken from the blood inlet (BE), the blood outlet (BA), the dialysis fluid inlet (DE) and the dialysate outlet (DA).

tration of the cations in the blood of the patient is constant during dialysis. That is to say no potassium or calcium may be diffusely added or withdrawn to the patient during therapy. This phenomenon is called a *constant dialysis* in this work. In order to ensure this, the dialysis fluid must contain a certain concentration of the cations. The searched-for concentration depends on the electrolyte concentration of the patient's blood.

2.2 Used blood substitutes

In order to simulate realistic dialysis conditions, various blood substitutes are used.

Protein solution

The tests are performed with a self-made protein solution. As a basis serves an electrolyte solution containing the relevant electrolytes of the human blood. Dialysis fluid can be used, since it contains all the necessary electrolytes. In order to produce a more complex blood substitute, the protein *albumin* is added to the electrolyte solution. For this purpose, bovine serum albumin (BSA) was enclosed. Due to only minor structural differences, BSA can be used as a cost-effective replacement for human serum albumin (HSA) in research [4].

Bovine blood

Bovine blood is used as another blood substitute. It is very similar to the human blood in the individual components and is often used equivalently for laboratory purposes. Some values of bovine blood and human blood, which are of interest to this report, are presented in [5]. The albumin concentration in bovine blood is with about about 30 g/l to

39 g/l lower than in human blood. The electrolytes relevant for this report are available in similar concentrations for both types of blood. potassium is present in the range of 3,9 - 5,2 mmol/l and calcium has got a value of 2,00 - 2,54 mmol/l [5].

2.3 Experiments

First, the experiments with protein solution are carried out. During each cycle, different conductivities were set for the dialysis fluid. This changes the electrolyte composition and as a result the potassium and calcium concentrations. After measuring the electrolyte concentration, the test was performed until the concentrations of the cations of the blood sample are equivalent to the concentration of blood inlet (c_{BE}) and blood output (c_{BA}). It can be written as $c_{BE}=c_{BA}$. This means no potassium or calcium has been added or removed from the blood, equals the constant dialysis.

If a concentration of the dialysis fluid was found for a constant dialysis, the difference between the concentration of potassium at the blood inlet and dialysis fluid inlet (c_{DE}) was calculated, such as

$$DIFF = c_{DE} - c_{BE}. \qquad (1)$$

The same was calculated for calcium. As a result, it was possible to determine which concentration had to be adjusted in the dialysis fluid so that no cations diffuse while establishing a potential equilibrium.

In order to test whether GDE plays a role for the behavior of the cations in the dialyzer, the same experiment was accomplished with various amounts of BSA, added to the blood substitute before dialyzing the substitute again. For a healthy adult, the serum albumin concentration is in the range of 35 g/l - 50 g/l [4]. Therefore the investigations were carried out in these boundaries. However, a serum albumin concentration <35 g/l is common for dialysis patients, too. To account for this, serum albumin concentrations below 35 g/l were also investigated. A total of eight measurements are performed between a serum albumin concentration of 12.3 g/l and 43 g/l.

To gain further insights, the experiments were carried out with bovine blood as well. First of all it was tested again whether a constant dialysis is feasible and recognizable. Subsequently, an influence of GDE was checked. For this purpose, 5, 10 and 15 g/l albumin were added to the blood and then dialyzed according to the scheme described above. More albumin than 15 g/l was not added to the bovine blood, since this does not correspond to physiologically realistic values.

3 Results and Discussion

The results of the previously presented methods are illustrated in the following section. Special emphasis is given to the measurements regarding the electrolyte shifts during the performed dialysis.

3.1 Measurements with electrolyte solution to determine electrolyte shifts during dialysis

It had to be shown whether a potassium or calcium concentration of the dialysis fluid can be determined for carrying out a constant dialysis under given boundary conditions. The first experiment was accomplished with electrolyte solution, without added albumin.

The sought-after DIFF, is the concentration, allowing a constant dialysis, which is needed in order to compare the experiments. Based on the experimental measurements a DIFF of 0.0033 mmol/l can be observed for potassium. Resulting in the following equation where $c(K^+)_{DE}$ corresponds to the potassium concentration of the dialysis fluid inlet and $c(K^+)_{BE}$ to the K^+ concentration of the blood inlet: $c(K^+)_{DE}= c(K^+)_{BE} + 0.0033$ mmol/l. The dialysis fluid should contain 0.0033 mmol/l more potassium than the blood used which can be rounded down to 0 mmol/l. This means the potassium concentration of the dialysis fluid should correspond to the potassium concentration of the blood substitute in order to perform a constant dialysis under these conditions. Thus $c(K^+)_{DE}= c(K^+)_{BE}$ applies.

The DIFF concerning calcium results to -0.0039 mmol/l which can be rounded up to 0 mmol/l as well and leads to the following equation $c(Ca^{++})_{DE}= c(Ca^{++})_{BE}$. Thus, there is a possibility to carry out a constant dialysis with respect to the potassium and calcium ions and to observe and compare the behavior of the cations in the dialyzer.

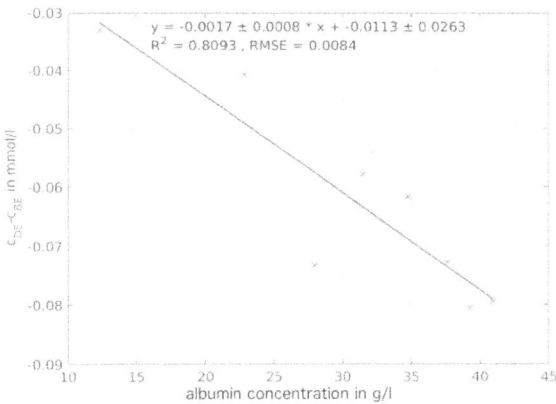

Figure 2: The DIFFs of the concentration of potassium are shown in this graphic, compared to the used albumin concentration of the protein solution. The negative slope of the line indicates that the higher the albumin level, the lower should be the potassium level in the dialysis fluid, in order to perform a constant dialysis.

3.2 Measurements with protein solution to determine electrolyte shifts during dialysis

In order to investigate the GDE, the DIFFs of the individual measurements are plotted against the albumin concentration

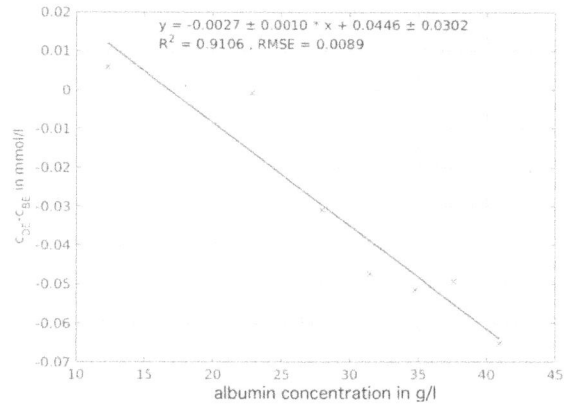

Figure 3: The DIFFs of calcium are shown in this graphic, compared to the used albumin concentration of the protein solution. A negative correlation indicates a higher albumin level leads to a lower calcium level in the dialysis fluid, in order to perform a constant dialysis.

of the protein solution used, in Fig. 2 for potassium and in Fig. 3 for calcium. There is a negative trend for both electrolytes. What means the higher the albumin concentration of the protein solution, the greater the DIFFs. This leads to the conclusion, the higher the albumin concentration of the blood substitute, the less potassium or calcium must be present in the dialysis fluid compared to the protein solution in order to perform a constant dialysis. The high R^2 value of these features indicates a strong linear relationship.

3.3 Measurements with bovine blood to determine electrolyte shifts during dialysis

In order to assess the electrolyte shifts of the potassium and calcium molecules in bovine blood, the experiments are examined accordingly.

Whether the observed dependence on the albumin concen-

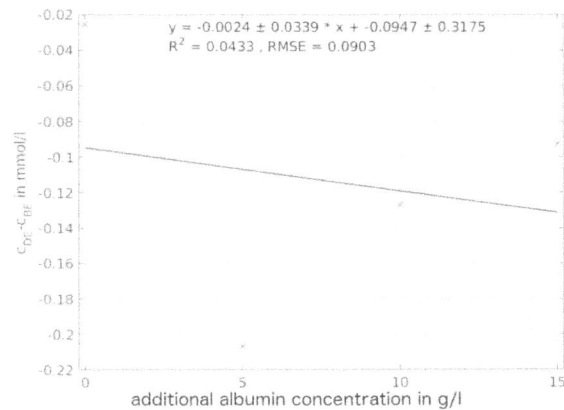

Figure 4: The DIFFs of potassium in bovine blood are shown, compared to the additional added albumin concentration of the bovine blood.

tration of protein solutions can also be expected for bovine blood, which consists of additional particles and substances

is examined in this section. For each concentration, however, the DIFF of the potassium and calcium concentration is determined. The result for potassium is shown in Fig. 4 and for calcium in Fig. 5.

The additional albumin concentration is visible on the x-axis, since the bovine blood already contains a certain quantity of albumin. Furthermore, the differences in the corresponding albumin concentrations can be seen. For both electrolytes, the negative trend, which appears in the measurements with protein solutions, can not be determined unequivocally. This can be explained by the more comlex composition of the bovine blood, probably leading to different biological reactions. At the same time, insufficient data are available to counteract the results so far. Due to the small quantity, the data is not as meaningful as the data for the protein solutions.

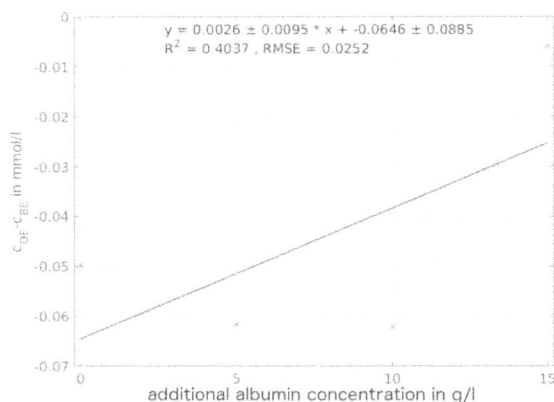

Figure 5: The DIFFs of calcium in bovine blood are shown, compared to the additional added albumin concentration of the bovine blood.

4   Conclusion

The main component of this report involves the determination of the impact of the GDE for the electrolytes potassium and calcium, in order to clearly classify the behavior of the cations during dialysis. For this purpose a constant dialysis was searched-after, in order to compare the different measurements. By means of this, it was shown that a GDE exists in the dialysis for the cations potassium and calcium. This is especially true for protein solutions. The GDE is caused by charged proteins (albumin) in the blood plasma of the patient. It manifests itself to the extent that the higher the albumin concentration of the protein solution, the more negative is the DIFF of the concentration for the constant dialysis. For bovine blood these findings are not recognizable at first. On the one hand, this can be attributed to the more complex structure of bovine blood, on the other hand to the small number of measurements. Furthermore, a secondary membrane can be formed during dialysis with human blood or bovine blood by depositing substances, which results in a worsening of the cleaning performance of certain substances. In addition, clotting may occur in the dia-

lyzer. So it would be interesting if the secondary membrane or the clotting can be observed during further experiments. For example by changing the ultrafiltration rate, what can lead to a higher water removal of the blood during dialysis.. With this, the particles are drawn stronger to the membrane and could produce such a secondary membrane. In summary, it can be said that unintended electrolyte shifts of potassium and calcium during dialysis can be observed. However, the concentration ranges that influence the GDE are so low that they do not have to be considered in practice during dialysis therapy. In order to prevent unwanted diffusion of potassium and calcium, the respective concentration in the dialysis fluid should be adapted to the concentration of the blood of the patient. To withdraw cations, the concentration has to be adjusted accordingly, what means the concentration in the dialysis fluid must be lower than the concentration in the blood.

Acknowledgement

The work has been carried out at B. Braun Melsungen AG and was supervised by the Institute of physics, Universität zu Lübeck.

5   References

[1] Fresenius, *Geschäftsbericht 2013*. Der Dialysemarkt, 2013.

[2] G. Breuch, *Fachpflege Nephrologie und Dialyse*. Elsevier Verlag, 2008.

[3] F. Donnan, *Die Membrangleichgewichte*. Colloid and Polymer Science vol. 61, no. 2, pp. 160 - 167, 1932.

[4] T. Peters Jr, *All about albumin: biochemistry, genetics, and medical applications*. Academic press, 1995.

[5] Medizinische Tierklinik, *U. L. M. Referenzbereiche - rind, 2016*. Available: http://www.vetmed.uni-leipzig.de/ik/wmedizin/labor/diagnostik/referenzwerte/rind.htmCallforPapers.aspx [last accessed on 2017-06-01].

Tests with a new optical sensor for hematocrit and oxygen saturation at extracorporeal life support systems

M. L. Severin[1], U. J. Netz[2] and R. Huber[3]

[1] Medizinische Ingenieurwissenschaft, Universität zu Lübeck, mara.severin@student.uni-luebeck.de

[2] Laser- und Medizin- Technologie GmbH, D-14195 Berlin, u.netz@lmtb.de

[3] Institut für Biomedizinische Optik, Universität zu Lübeck, robert.huber@bmo.uni-luebeck.de

Abstract

Monitoring hematocrit and oxygen saturation during cardiac surgery could reduce the amount of blood samples needed and as a consequence the risk of an operation. Blood gas analyses are done in constant intervals to control the quality of blood during the usage of extracorporeal circulation like heart-lung machines or extracorporeal membrane oxygenation. This paper describes the situation of an existing optical system based on spatially resolved reflectance (SRR) that is developed by the Laser- und Medizin- Technologie GmbH Berlin and its progress. For further development measurements with the sensor are done and the signals are analysed to evaluate the influence of different properties on the signals of the sensor.

1 Introduction

Oxygen saturation (sO2; measured in %saturation) describes the percentage of oxygen-saturated hemoglobin relative to total amount of hemoglobin in blood. Hematocrit (hct; in vol.%) is the volume percentage of red blood cells in blood. Both parameters are important for the assessment of the patient's oxygen supply [1]. Monitoring of sO2 and hct is important during cardiac surgery when using a cardiopulmonary bypass or an extracorporeal membrane oxygenation (ECMO). The continuous monitoring of blood parameters in extracorporeal circulation allows the optimisation of perfusion and anesthesia without delay. As the accuracy of commercial non-invasive optical blood parameter monitoring systems is not comparable with the accuracy of blood gas analysis (BGA) for small tubings, a BGA is done every 15-20 minutes while using cardiopulmonary bypasses. Taking a blood sample can be problematic, particular in neonatal cardiac surgery. With a higher accuracy of the monitoring systems the amount of blood gas analyses and as a result the required blood samples could be reduced to an absolute minimum. Especially during the use of ECMO it would be helpful to reduce the required blood samples because over the time of application up to 10 days with a blood gas analysis taken every hour the risk of infection increases. The measurement of the oxygen saturation during ECMO is important for controlling the function of the oxygenator. Most monitoring systems are pre-calibrated and a recalibration with a blood gas analysis needs to be done at the start of a cardiopulmonary bypass procedure. Additionally most systems need a cuvette that represents a flow resistance and increases costs for disposables. These cuvettes do not have a long time approval for what reason these systems are not licensed for long time applications. Laser- und Medizin-Technologie GmbH Berlin has developed an optical in-line sensor that only needs to be calibrated once in manufacture without human blood. The system is adapted for the use of a clear transparent plastic cuvette that is incorporated into the venous side of the cardiopulmonary bypass [1]. In previous measurements it was observed that the sensor measures accurate signals directly on the tubing without a cuvette. So the aim is to develop the existing sensor further to measure directly on the tubing reliably and independently from properties of the tubing like diameter, wall thickness, colored lines or annotations and place it on the market. All these parameters should be eliminated and observed in the calculation of oxygen saturation and hematocrit.

2 Material and Methods

In this chapter the used material and methods are described.

2.1 Analysis of monitoring systems for sO$_2$ and hct

Existing monitoring systems for sO$_2$ and hct were analysed. There are eight systems from different manufacturers that can measure sO$_2$ and hct in-line during cardiac surgery. Some sensors can measure more parameters than sO$_2$ and hct but according to information of perfusionists these are the most important parameters to measure. The range of measurement is between 15% and 50% for hematocrit and between 40% and 100% for oxygen saturation. The measurement deviation of oxygen saturation for the sensors lies between 1.9% and 6.3%, for hematocrit between 1% and

5% [2]. The sensor of Laser- und Medizin- Technologie GmbH already offers values with an accuracy of ±1.1% on a cuvette and ±1.4% on the tubing for hematocrit and an accuracy of ±1.2% on a cuvette and ±2.3% on the tubing for oxygen saturation. This existing prototype of the sensor with the mentioned accuracy was used for measurements on different tubings to investigate the accuracy.

2.2 Optical sensor concept

The sensor uses the principle of spatially resolved reflectance (SRR) in the near-infrared range [4]-[5]. The principle of spatially resolved reflectance can be seen in fig. 1. Light is emitted by four light emitting diodes (LEDs) at three different wavelengths between 730nm and 845nm to irradiate the blood from one position. Because of the available luminance efficiency for one wavelength two LEDs are used. The optical properties of the blood are dependent from the percentage of hematocrit and oxygen saturation [3]. Using photodiodes (PDs) that are arranged in a line with the LEDs the diffuse reflected light is detected at different distances from the light source of the light emitting diodes. The distances are depending on the wavelength. An additional reference photo diode that monitors the output of the LEDs is used to correct the optical signal. This correction is done by measuring the dark signal and substracting it from the reflected light and dividing the result by the monitor diode. Every measurement is averaged over 6 seconds. The intensity and spectral distribution of the LED output is stabilised by controlling the temperature. Independently from ambient temperature the temperature of the flowing blood in the tubing can change between 38°C and 15°C within a few minutes. These changes need to be compensated to pretend that they influence the result.
A firmware calculates the values of oxygen saturation and hematocrit from the signals that are measured.

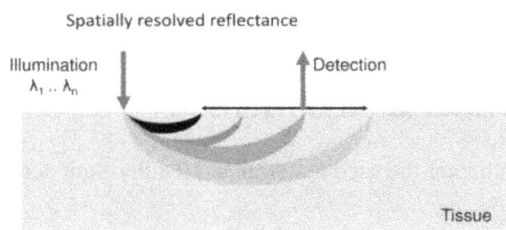

Figure 1: Principle of spatially resolved reflectance spectroscopy; backscattered light is detected at given but varying distances from the light source at different wavelengths λ_i (i = 1...n) [1].

2.3 Measurements on blood circulation system

Following the setup and the execution are described.

2.3.1 Experimental setup

To analyse the influence of different properties of the tubing an experimental setup of a blood circulation system with a simulated heart-lung machine is built. The simulated heart-lung machine consists of an oxygenator, a reservoir, a water quench for tempering and a multiflow roller pump. Five tubings of 3/8" of different manufacturers and types are used. The first tubing in the circulation is tested at the end of the circulation for a second time so it can be proven that there are no changes in the signal strength that result from the flow. The setup consists of a circulation that is built with six tubings connected to a reservoir and an oxygenator. A roller pump is used to adjust the volume blood flow. Additional to the monitoring sensors of Laser- und Medizin-Technologie GmbH a temperature sensor is installed in the circulation to measure the temperature of the oxygenator, the reservoir, the water quench and the room. Another sensor is installed measuring the oxygen saturation and the temperature of the blood in real time. A further sensor is used to control the oxygen saturation. A gas supply is used to adjust the saturation values. The setup can be seen in fig. 2.

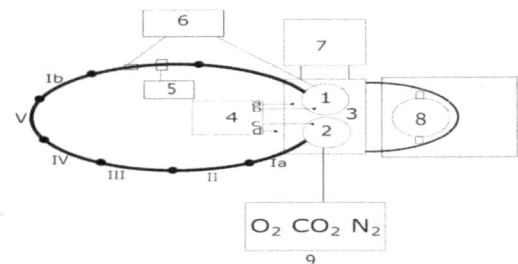

Figure 2: The experimental setup. 1: reservoir, 2: oxygenator, 3: water quench, 4: heat meter a: water quench, b: reservoir, c: oxygenator, d: room, 5: Baxter OxySat-Meter, 6: Fibox oxygen and temperatur sensor (temperature of the blood in the reservoir), 7: heater-cooler unit, 8: Stöckert Multiflow roller pump, 9: gas supply, Tubings: Ia+Ib: Raumedic noDOP Bloodline, II: Dideco, III: Sorin, IV: Raumedic Implantat, V: X Coating.

2.3.2 Test execution

The system is run with red cell concentrate. After setup a flow test was made to control whether the circulation is consistent and the required amount of red cell concentrate to run the circulation without air inside could be calculated. The measurements are performed at specific values of hematocrit and oxygen saturation. At the beginning a specific value of hematocrit is adjusted by diluting the red cell concentrate with phosphate-buffered saline (PBS) whose osmolarity and ion concentration match those of the human body (isotonic), it is innoxious for cells and buffers at pH=7.4 [6]. Oxygen saturation can be adjusted with gas connections (oxygen, carbon dioxide and azote). As the circulation only represents a model of a patient there are no sudden pathologic changes in hematocrit and oxygen saturation, so three values of sO2 and hct are simulated – a pathologic too low, a normal value and a pathologic too high value to describe the normal range. sO2=60/80/99%, hct=25/30/35%. Three values improve the precision of the calibration model. After every adjustment of the values a blood sample is taken to control oxygen saturation and hematocrit. Oxygen saturation is controlled by a blood gas analysis, hematocrit is controlled with a Hettich-centrifuge. For every adjustment of the parameters the sensor is set on the different tubings. Every measurement takes 30 seconds with a readout interval of six seconds. The temperature of the hypothermia machine is set on 22°C. The experiments should give information about the influence of turbulent flow at high volume flow rate and emerging changes in the flow pattern in cross-section. For this information the velocity of the flow and consequently the shear velocity is modified by changing the speed of the roller pump. The adjusted volume flows should cover all shear-velocities that occur in the use of extracorporeal circulations. This depends from the diameter of the tubings and the weight of the patient. The volume flow in neonates is different compared to the volume flow in adults. The measurements with different shear velocities are done for a low and a high value of hematocrit and the oxygen saturation is set on 100%. Further the spectral characteristic is analysed for wavelengths of 730nm, 807nm and 845nm for a fixed hematocrit in dependence of the oxygen saturation and the type of the tubing e.g. different manufacturers and different turbidities. The measurements need to be done within one day as the properties of the blood alter due to haemolysis. For this reason the measurement of the influence of markers or labels was left out. Every measurement is done with two sensors measuring directly on the tubing, the prototype of the sensor and a sensor with a modified sensor head to improve the connection between sensor and tubing. As a reference a sensor is set on a cuvette and not changed over all the time measuring hct and oxygen saturation. Fig. 3 shows the three sensors during the measurements. After the measurements the data of the sensors were used for a subsequent calibration. For every wavelength (730nm, 807nm, 845nm) the signals of the four installed photodiodes are normalised to a monitor photodiode that monitors the irradiated LED-output per LED. The reference values for hematocrit and oxygen saturation of the laboratory apparatuses (hct: Hettich-centrifuge, sO2: Eschweiler combi line blood gas analyser) are assigned to the specific times of measurements.

2.4 Multivariate data analysis

Multivariate data analysis is used to analyse the content of many characteristics that are measured by a number of objects and a target figure. To analyse the data measured by the sensor partial least square regression (PLS) is used. The measurement gives us an N × M matrix **X**, where N is the number of measurements and M is the amount of photodiodes (M=12). **Y** is a matrix of N × K, with K the amount of target figures. In this case **Y** is a vector $\vec{y}$ with K=1, $\vec{y}$

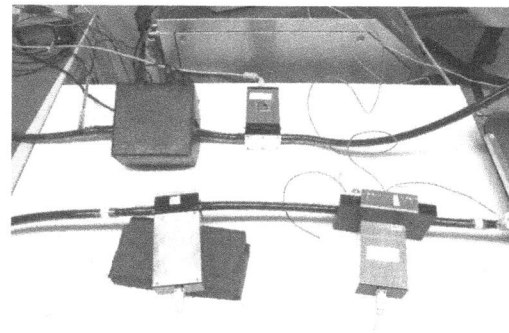

Figure 3: Bottom left: Prototype of sensor, bottom right: sensor with modified sensor head, both measuring directly on the tubing, top: sensor for reference measurement on cuvette.

is the vector of reference values of HCT or sO_2. PLS is the most commonly used algorithm of regression for multivariate regression. It generalises principal component analysis (PCA) and multiple linear regression. The goal is to predict dependent variables Y ($SatO_2$, HCT) from independent variables X (measurements of the photodiodes for different wavelength) [7]. Fig. 4 illustrates the principle of PLS regression.

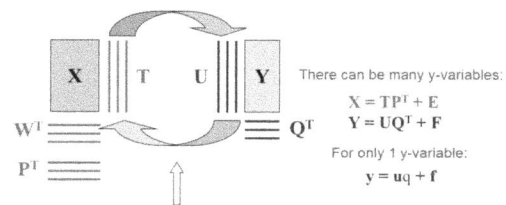

Figure 4: Schematic diagram of PLS and involved matrices. Data of X influences the PCA of the data of Y and data of Y influences the PCA of the data of X [8].

For the analysis of the measured data the program The Unscrambler X 10.2 by Camo Software AS was used.

3 Results and Discussion

Due to limited time only the signals of two examplary tubings were taken for the evaluation of the results and analysed. So in further analysis a more precise inspection of the results and propagation of uncertainty needs to be done. Tubing number one was a tubing by Raumedic (noDOP bloodline) and tubing number two was a tubing by Dideco. In fig. 5 the signals for 730nm on the tube of Dideco can be seen. For every photo diode the signals of every photo diode are shown for the same sensor with the normal measurement and the measurement with the modification. One can see that the modification of the sensor influences the strength of the signal as the signals of the modified sensors are different to those of the normal sensor with the same blood. Fig 6 shows the signals for the tubing by Raumedic. The relation between the signals of the modified measurement and the normal measurement is the same but it is visible that the strength of the signals belongs to the properties of the tubings. The signals of the measurement on the tubing by Raumedic are higher than the signals of the measurement on the tubing by Dideco for the same blood properties. It is visible that the strength of the signals of photo diode number one increases with the modification while the strength of the signals of the other photo diodes decreases.

In further analyses it needs to be analysed why the trend of the signals of photo diode number one is contrary to the signals of the other photo diodes. Additionally the differences between the different tubings need to be evaluated and to be taken into account for the calibration of the sensor.

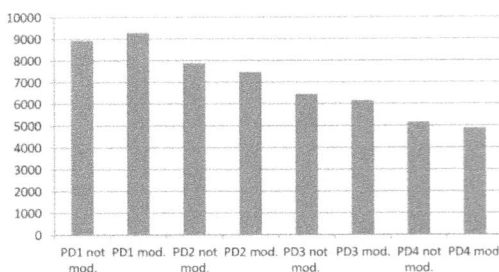

Figure 5: Signals for 730nm on tubing two (by dideco) of the normal sensor and the modified sensor for every photo diode, signals in the unit of the A-D converter

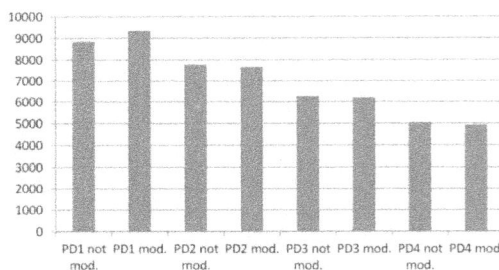

Figure 6: Signals for 730nm on tubing number one (by Raumedic) of the normal sensor and the modified sensor for every photo diode, signals in the unit of the A-D converter

4   Conclusion

The aim is to develop the sensor further so that it can measure oxygen saturation and hematocrit reliably without a cuvette. The sensor already offers a high accuracy which is competitive to similar sensors of other manufacturers. Measuring without a cuvette offers the possibility to enter the market of monitoring in extracorporeal membrane oxygenation. So far monitoring is not established here as EC-MOs can take up to ten days and cuvettes have no long-time approval. Especially during ECMOs the amount of blood samples is high as a control of the parameters is important here, so monitoring could reduce the number of blood gas analysis and lower the risk of an ECMO. In addition the number of disposables can be reduced. Therefore the cost of an ECMO decreases. A further advantage of the senor of LMTB is that it does not need to be calibrated on the patient's blood before every use but is calibrated once during production. In summary one can say that there is a high potential in the manufacturing of the sensor that offers reliable values of hct and oxygen saturation measuring directly on the tubing. For developing the sensor further and establishing it on the market different aspects need to be analysed more precisely and to be taken into account when calibrating the sensor to make the measurements less sensitive to errors.

Acknowledgement

The work has been carried out at Laser- und Medizin- Technologie GmbH, Berlin. Thanks to my colleague Uwe Netz for supervising me and to my colleague Enno Ott for doing the measurements with me. The work has been supervised by Institut für Biomedizinische Optik, Universität zu Lübeck.

5   References

[1] U. J. Netz, M. Friebel and J Helfmann, *In-line optical monitoring of oxygen saturation and hematocrit for cardiopulmonary bypass: Adjustment-free and bloodless calibration*. In: Photon Lasers Med, vol. 4, Iss. 2, pp. 187–192, 2008.

[2] H. Weise et al., *Aktuelle Herz-Lungen-Maschinen im Vergleich*. In: Kardiotechnik vol. 3, pp. 74–79, 2006.

[3] S. M. Daly, *Biophotonics for blood analysis*. In: Biophotonics for Medical Application, pp. 243–299.

[4] S. Andree, C. Reble, J. Helfmann, I. Gersonde and G. Illig, *Evaluation of a novel noncontact spectrally and spatially resolved reflectance setup with continuously variable source-detector separation using silicone phantoms*. In: Journal of Biomedical Optics 15(6), pp. 067009-1–067009-12., 2010.

[5] M. Meinke et al., *Clinical feasability tests on a novel optical on-line blood monitoring sensor for cardiopulmonary systems*. In: Medical Laser Application vol. 22, Iss. 4, pp. 248–255, 2008.

[6] Sigma-Aldrich, *Product Specification of Phosphate buffered saline*.

[7] H. Abdi, *Partial least squares regression and projection on latent structure regression (PLS Regression)*. In: Wiley Interdisciplinary Reviews: Computational Statistics, vol. 2, Iss. 1, pp. 97–106, 2012.

[8] W. Kessler, *Multivariate Datenanalyse für die Pharma-, Bio- und Prozessanalytik Ein Lehrbuch*. Wiley-VCH, Weinheim, 2007.

Endoscopic Suction
–Evaluation and Comparison of Specialized Pump Technologies–

L. Schulz[1], H. Macher[2], D. Braun[3] and P. Rostalski[4]

[1] Medizinische Ingenieurwissenschaft, Universität zu Lübeck, l.schulz@student.uni-luebeck.de
[2] Brütsch Elektronik AG, heinz.macher@brel.ch
[3] KARL STORZ GMBH & Co. KG, denis.braun@karlstorz.com
[4] Institute for Electrical Engineering in Medicine, Universität zu Lübeck, philipp.rostalski@uni-luebeck.de

Abstract

Endoscopic suction is used in the treatment of different symptoms within endoscopic surgery to remove fluids and gas. Therefore a system creating a vacuum pressure is needed. Depending on which operation is executed, there are different requirements concerning flow and pressure and various additional specifications. The aim of the project described in this article will be to evaluate and to compare several suction technologies and identify a suitable one for the given requirements. A schematic representation is derived, which exemplifies a possible design of an endoscopic suction device.

1 Introduction

Operative laparoscopy requires skills, facilities and equipment, as well as laparoscopic diagnostic and standardized sterilization procedures. Suction, irrigation, motor systems, high-flow insufflation, light source and instruments like angled scopes, special forceps and scissors are important tools for this advanced technique [1]. Furthermore, especially in the last 15 years, high frequency surgery and laser have increased influence in laparoscopy.

The impact and the importance of equipment and instruments are much different in endoscopic surgery than in open surgery. This is because visualization and tactile exploration of the operation field is always achieved indirectly through optical systems and instruments [2]. Suction, as well as irrigation, is a highly important topic, especially for clear visual field. Depending on the pump the suction is measured in l/min or bar. For some pumps the user can set the flow or pressure level by himself, others just have preset levels. If it is needed to remove a large amount of blood or other substances, for instance small spliters, the surgeon will select a high flow level. If this function is used in combination with a shaver system, the suction prevents uncontrolled movements of resected tissue. Many manufactures offer a combination of suction and irrigation in a single dual-control instrument. This is useful in peritoneal toileting, in case of peritonitis due to appendicular perforation or duodenal ulcer perforation. But it is also possible to create a specialized pump for each function. Instead of finger dissection in conventional surgery, the suction tip is highly useful for intermittent suction as a blunt dissecting instrument [2].

2 Material and Methods

Requirements are based on customers' needs. The development and the design of the devices follow these requirements as well as norms, regulations and the current state of the art.

2.1 Requirements

Initially, the intended use of endoscopic suction has been considered within the requirements. For some manufactures, the only indication is the use within an endoscopic procedure. But it is possible to divide the indications into two different groups: the application area and the different kinds of fluid and tissue. Suction pumps, especially vacuum-assisted systems, can be used for lymph node, blood via epicardia ablation, secretion, mucus, fat and also for nasopharyngal suction. Endoscopy is one, but not the only application area for suction. It is also a topic of surgery (especially neurosurgery and plastic aesthetic surgery) as well as anaesthesia, nursing, emergency and dental health. Concrete procedures where vacuum is the main component are: vacuum assisted caesarean operation, liposuction and negative pressure wound therapy.

Due to suction of relatively strong tissue like lymph nodes or relatively huge amounts of fluids, which often occurs in case of liposuction, there are corresponding performance requirements. The first topic of performance is the flow, which should be adjustable and controllable between 30 and 50 l/min. As well as the maximum flow, there is an infinitely variable and controllable maximum vacuum pressure between 620 and 680 mmHg. Besides, the acoustic atmosphere in the operating room should be as quite as possible. Manufacturers advertise a noise level comparable to somebody whispering or talking in a low voice, that is a noise between 45 to 50 dB in a distance of one meter away from the device. That limit is the actual state of the art, but there is no upper limit given in the medical device directive [4].

To provide the minimum and maximum pressure, as well as the actual value of vacuum pressure and the actual flow, it is useful to utilize a display, for example a simple manometer or a digital display. This is a design requirement. In addition, the dimension of the device is very variable from one manufacturer to another manufacturer. Some prefer to design a very small pump, so it is possible to use it in an equipment cart while others prefer a large device, so it can be sold as a stand-alone system with a large fluid collection container.

If the suction device works within a system, it has to fulfill specific requirements, especially communication with other devices. For example, a motor system is used for driving a hand piece, which is the direct connection to the patient. The hand piece is a specific tool for cutting different kinds of tissues and suck these tissues together with the irrigation fluid. According to the number of revolutions of the hand piece, the flow and the maximum vacuum pressure should be tunable. As well as the number of revolutions, it is also possible to handle the suction via an angle or a position of the hand piece, which has to be defined. The first option is a mechanical resolution and does not need any further components. The second option prescribes a certain intelligence of the system. Consequently, this is not the first choice in case of simple and save handling. Insufflation is also an important aspect of suction. Via laparoscopy there are difficulties to maintain a constant cavity while suctioning. It seems to be very useful to have a suction flow tunable according to the insufflation flow. The flow and the maximum vacuum pressure should be tunable according to the intracavitary pressure. Concerning the topics of high frequency, lithotripsy and laser, the flow and maximum depression has to be given at activation.

Furthermore, requirements concerning the working conditions of the devices are given. For some pumps it is useful to offer supply voltage between 100 and 240 V, so it is possible to use them all over the world. Power frequency is 50 to 60 Hz. The pump has to work under operating conditions of 15 to 80 % air humidity (relative humidity (RH), non-condensing), temperatures between $10°C$ and $40°C$ and a maximum operating height of 3000 m. The storage conditions are: air humidity from 15 to 95%, temperatures between $10°C$ and $60°C$ and an atmospheric pressure of a minimum of 500 hPa and maximum of 1080 hPa [3].

Some classifications are derived from the medical device directive, especially *DIN EN ISO 10079-1*. The type of protection against electrical shocks is protection class

one. Related to the degree of protection against electric shocks, there is applied part of type BF acoording *IEC 60601-1* [3]. A filter is a requirement of safety for the pump and it should prevent against moisture. In addition, the drip-water protection of IPX1 is needed. Sometimes it is useful to examine if a higher protection class is required [4].

2.2 Fail-safe operation

The requirements should be related to the potential difficulties and errors caused by the users, especially by nurses and physicians. What happens for example when the container is filled to capacity? For suction pumps it is a necessary requirement to insert the filter before starting the device. Thereby manufacturers prevent that the substance penetrate into the pump. As soon as the filter gets wet, the pump turns off automatically. Another method of protection would be a safety floater in the secretion bottle. The overflow protection is given upwards by the liquid which is suctioned until it closes the valve.

Another problem is that sometimes there is a poor or no suction. The reason for this is that the plug valve on the shaft of the endoscope should be open to suction. Otherwise the depression in the tube will be generated and the normal suction function cannot be performed. The pumps will do this job for the surgeon and will regulate the pressure and the flow. To avoid this problem, there should be a corresponding information in the user manual.

Sometimes the pressure does not build up. If the pressure at the treated area differs from the pressure the device displays, the reason could be the position of the pump. If the pump is placed lower than the patient, the pressure will be higher than the mmHg patient reach due to hydrostatic pressure. In contrast, if the pump is installed higher, the pressure will be lower. Due to that the surgeon needs to make sure that the device is placed on the same level as the patient. The resulting requirement is, that the device displays the hydrostatic pressure or calculates the optimal position.

Furthermore, the pump could be very noisy. So, a vacuum pressure could be the reason for dull noises. Often the cause for this is the hose clip. If the hose clip is tightened too much, there is no air the pump can suck in.

2.3 Evaluation and Comparison of Suction Technologies

Several different options excists for appropriate drive technologies of suction. The following task is, to get an overview of some possible technologies and identify the most suitable option. The named requirements especially flow and vacuum pressure are the crucial features, which have to be fulfilled. A comparison between pump technolo-

Table 1: Pump technologies - performance, based on [5]

Technology	max flow[l/min]	max pressure [bar]	max vacuum pressure [mmHg]
articulated piston	152.9	12.1	685
diaphragm/membrane	91	3	736
linear	375	0.7	609
peristaltic	3	29	749
rotary vane	283	1	751

Table 2: Comparison between articulated piston and membrane pump, based on [5]

Feature	BLDC	DC motor 2-fold
type of the pump	piston	membrane
power supply	24V DC	14V DC (24V DC possible)
current consumption	6A	5A (24V ca 2.5A)
thermic protection	yes	external
suction	50l/min	36l/min
pressure	-0.836 bar 627 mmHg	-0.92 bar 690 mmHg
hours of operation	10000 h	3000-5000 h
noise level	58dB	@ 14V DC = 62dB @ 12V DC = 58dB @ 10V DC = 54dB
operation temperature	10°C - 40°C	10°C - 40°C
air humidity (RH, non-condensing)	15%...80%	15%...80%
weight	1.26 kg	1.25 kg
size	105x77x118	172x73x174

gies is given in Table 1. An articulated piston is especially made for strong conditions, that means working during long times under load. A long lifetime and a high pressure are positive aspects of piston pumps, but on the other hand they are often noisy. The membrane or diaphragm pump is very suitable if a relatively low pressure and a moderate vaccum is needed. "Linear" technology leads to moderate flow and noise, but it had a low pressure. Therefore the pump runs by reduction of a volume by direct linear action of a shuttle on a membrane. Linear pumps have a low noise-level. A peristaltic pump works with high pressure and high vaccum according to the size and weight. This technology is even more efficient than the systems using a membrane. Rotary vanes are only suited to provide a high flow, therefore they can not be used to create a high pressure [5].

A piston pump generally consists of a piston which moves within a cylinder. The cylinder has an inflow and an outflow. First, the piston moves up, the inflow valve opens and the liquid or the gas insufflates into the cylinder. The outflow valve is closed. In the second step the piston moves down. The inflow valve closes and through the pressure the outflow valve opens. The liquid or the gas streams out into the tube. The membrane pump is a modification of the standard piston pump where the conveyed liquid is separated from the gear. Due to that, the gear is protected from dirt and contaminated liquid/gas. Similar to the piston pump, the membrane pump has two valves, an inflow and an outflow valve. For suction, the gear lifts the membrane and a vacuum condition occurs, the inflow valve opens and the liquid or the gas is aspirated [6]. A comparison between these technologies is shown in Table 2. The information given in the tables is extracted from the data sheets of the manufacturer Gardner Denver Medical.

The most suitable technologies for endoscopic suction seems to be articulated piston and membrane pumps. Gardner Denver Medical offers these two systems, *007BDC19 DC (Direct current)* membrane pump and *260Z WOB-L BLDC (Brushless Direct current)* piston pump, which have been evaluated for this project [5].

Especially the features flow, pressure and temperature have to be proved over at least 6 h with and without load. The temperature of the pump was also controlled during this time. Two tests were performed to the given membrane and piston pump. The first test includes a load of maximum flow every 30 min for duration of 2 min. The second test was performed under continuous load and maximum flow. Normal water was used for suction and there was a room temperature of 20 °C.

3   Results and Discussion

Almost all features given in Table 2 could be proved during the described test. The piston pump had a maximal flow between 47 and 50 l/min and a pressure range of plus

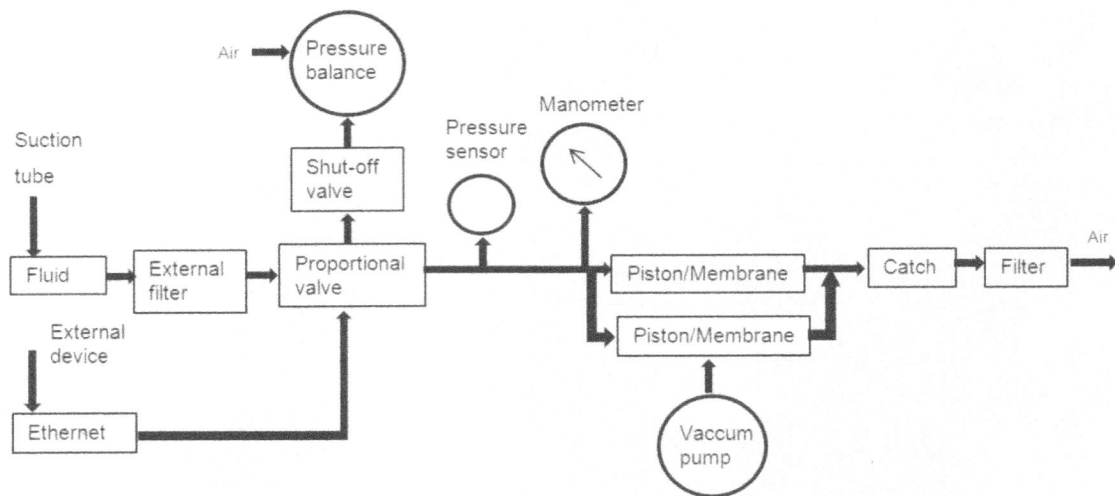

Figure 1: Possible composition of a pump system. The fluid is sucked because of negative pressure which is created by one or two pistons or membranes. Filters are a protection against overflow and contamination of the pump. The two valves regulate hydrostatic pressure of different patient levels. The Ethernet interface is useful to communicate with insufflation and drive systems of hand pieces.

minus 10 mmHg. The second test shows that the pump warmed-up to 39 °C. In comparison with the membrane pump, which only warmed-up to 35 °C, it could be disadvantageous in daily clinical use. The membrane pump works within a maximal flow of 35 up to 36 l/min and a pressure range of plus minus 10 mmHg. According to this test, the membrane pump seems to be more suitable for suction pumps, although the difference between these technologies is marginal. Therefore it could not be finally stated that one technologie will be the best overall. It is useful to take also into account several different aspects, as for example economic questions. In reality, some pumps need a software for excitation. Those are topics which have to be respected within the development of suction pumps.

A possible composition of an adjustable system is shown in Fig.2. If the membrane technologie is used, two pumps have to be integrated in parallel to get more flow. But often the space of the housing is the limit for using more than one drive. The design shown in Fig. 2 is a composition of a suction device. It represents that both technologies could be part of a device, which fulfills the named requirements. Especially safety is given via the shut-off valve and filters. Furthermore, there is an interface for communication within a system and a manometer for pressure display.

4 Conclusion

Based on the intended use, a list of different requirements of endoscopic suction devices was created. Specific technologies were evaluated and piston and membrane pumps were compared in detail. The features flow, pressure and temperature were proven in two test under different load. Both

technologies are suitable for medical use, but the membrane technology needs two pumps placed in parallel to get desired flow, which might be disadvantageous for economical reasons.

Acknowledgement

The work has been carried out at KARL STORZ GMBH & Co. KG at Tuttlingen relating to Brütsch Elektronik AG and supervised by the Institute of Electrical Engineering in Medicine, Universität zu Lübeck.

5 References

[1] M. E. Arregui, R. J. Fitzgibbons, Katkhouda Namir, McKernan, J.Barry, Reich, Harry *Principles of Laparoscopic Surgery - Basic and Advanced Techniques.* Berlin/Heidelberg: Springer-Verlag, 2008, p. 21.

[2] C. Palanivelu, *Atlas of Laparoscopic Surgery.* New Delhi: Jitendar P Vij, 2000, p.3-5

[3] *KV-6 Endoscopic Suction Pump.* Available: https://www.olympus.co.uk/medical/en/medicalsystems/ productsservices/productdetails/productdetails111872.jsp [last accessed on 2017-02-06].

[4] *MDD.* § 7-9, 5th Edition, p.28/29

[5] *Verdrängerpumpen.* Available: http://www.herold-gefrees.de/de/kompendium/verdraengerpumpen [last accessed on 2017-01-14].

[6] *Pumpen-Technologien.* Available: https://www.gd-thomas.com/de/technologien/kolbenpumpen [last accessed on 2017-02-04].

Development of a surface electromyography-based mechanical ventilation procedure

Jonas Kühne [1] and Eike Petersen [2]
[1]Medizinische Ingenieurwissenschaft, Universität zu Lübeck, jonas.kuehne@student.uni-luebeck.de
[2]Institute for Electrical Engineering in Medicine, Universität zu Lübeck, eike.petersen@uni-luebeck.de

Abstract

In modern medicine, one of the most important procedures is mechanically applied ventilation to substitute or support a patient's own breathing effort. Modern ventilation modes offer mechanically delivered breaths, which are proportional to the patient's breathing effort. This increases the patient's comfort and decreases long-term damage such as diaphragmatic atrophy. For the purpose of proportional ventilation it is required to estimate or invasively measure the airway resistance, the lung compliance and the pressure P_{mus}, generated by the respiratory muscles. The aim of this project was to modify a mechanical ventilator to support the patient's respiratory effort proportionally via surface electromyography triggered mechanical ventilation. By measuring the electrical activity of the muscles the respiratory effort can be observed. For this purpose the control board of the ventilator has been replaced with a rapid prototyping system and coupled with an EMG sensor.

1 Introduction

Since the invention of mechanical ventilation, many different ventilation modes have been developed, capable of supporting patients without any breathing effort as well as spontaneously breathing patients. The modes which assist respiratory activities most often notice the patient's own breathing effort and support it with a preset pressure value (e.g. Pressure Support), while newer modes vary the pressure value and tidal volume in regard of the patient's work of breathing (e.g. Proportional Assist Ventilation). This is a step to minimize cases of diaphragmatic atrophy [1] and shorten the period of weaning [2].

There are two main problems in mechanical ventilation. Firstly, the pressure generated by the respiratory muscles (P_{mus}) and the respiratory parameters R (resistance of the airways) and C (lung compliance) have to be estimated for use in mechanically ventilation. These estimations are difficult. Secondly, it is important to synchronize the respiratory effort of the patient and the ventilator breaths to optimize trigger efficiency and comfort. Lung damage and discomfort could be the result [3], as well as an prolonged mechanical ventilation duration, shorter ventilator-free survival, and increased length of stay [4].

One method to ensure the synchronization of ventilator and patient is to measure the muscular activity (Edi) of the respiratory muscles and especially of the diaphragm. This mode is used in the recently developed ventilation mode NAVA (Neurally Adjusted Ventilation Assist) [2]. The measurement of Edi triggers the ventilation, while the amplitude and duration of the signal determines the pressure of the assisting ventilation. Therefore the ventilation is synchronous and proportional to the patient's effort. However, one important disadvantage of this method concerns the measurement method itself: Edi is measured by an electrode array, attached to a nasogastric tube, which is positioned in the esophagus at the level of the diaphragm [5]. These electrodes obtain reliable and valid EMG signals from the respiratory muscles, but may be unpleasant for the participant and carry risks of regurgitation, aspiration, and vagally mediated bradycardia.

The aim of project reported here was to realize a ventilation mode, with the synchrony of NAVA, but without the unpleasantry or risks. It was tried to implement an alternative method to measure the muscular activity and control a mechanical ventilator with the derived respiratory parameters. Therefore the respiratory surface electromyogram (sEMG) was measured to determine Edi. In a first step it was planned to use the sEMG as a trigger for the ventilation. In a second step the activity generated by the respiratory muscles should be derived from the sEMG to regulate the pressure and volume of the mechanically delivered breath. Within the scope of this first months of this project, the modification of the ventilator's controller hardware and the connection to a rapid prototyping system could be realized, as well as communications between an EMG amplifier and a wireless-to-serial converter, which will be used later on as a connection between ventilator and EMG amplifier. The ventilator is able to provide mandatory ventilation and proportional assist ventilation. The processing of the sEMG signal on the rapid prototyping system and following triggering could not be realised and will be the subject of a subsequent master's thesis as well as sEMG proportional pressure support.

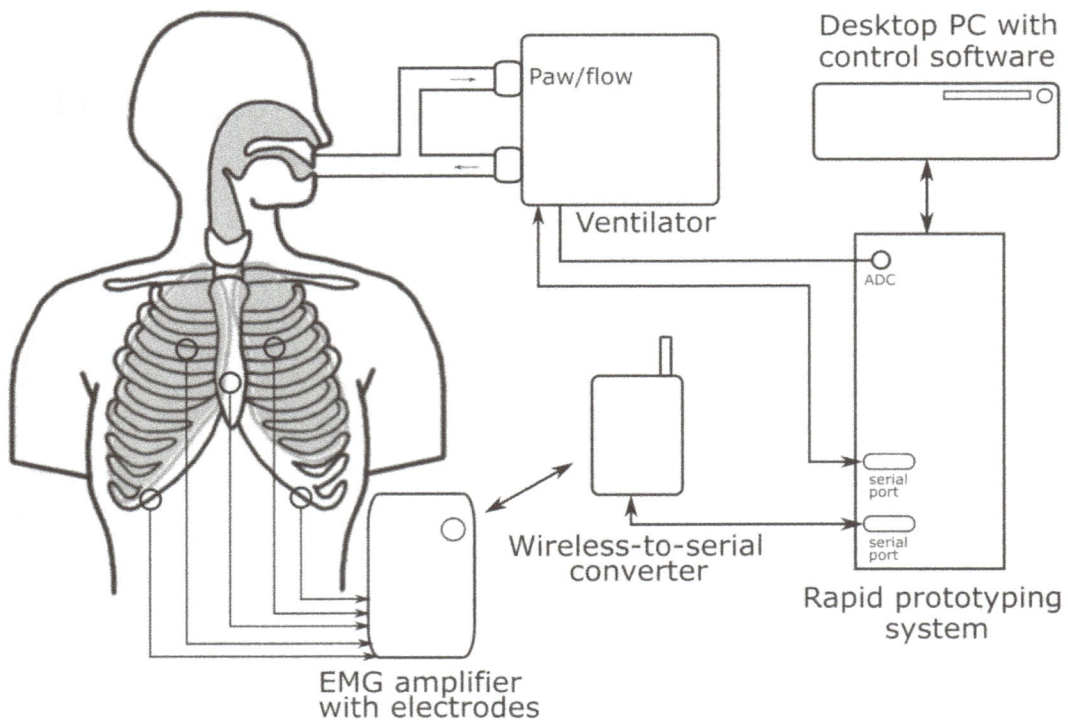

Figure 1: The planned experimental setup: The rapid prototyping system is the center of this concept. It is connected to the mechanical ventilator, the wireless-to-serial adapter and the Desktop PC. The control software of the rapid prototyping system is installed on the PC. It controls the ventilator and processes the EMG data, which is transferred wirelessly from the EMG amplifier to the wireless-to-serial adapter. The sEMG measuring electrodes are placed on the costal margin, the intercostal space and the sternum. The control software then controls all parameters and variables of the wireless-to-serial adapter and therefore the ventilator. The ventilator supports the patient and measures P_{aw} and $\dot{V}$ for regulation purposes.

2 Material and Methods

The envisioned experimental setup can be seen in Fig. 1. It is able to provide mechanical ventilation initiated by the machine or by the patient, as well as several hybrid modes. In this setup, patient-initiated ventilation is used, more precisely proportional pressure support (PPS), combined with an sEMG trigger for synchronisation purposes. This ventilation mode supports spontaneously breathing patients proportionally to their own breathing effort. To enable the communication between ventilator and EMG amplifier, the controller board of the mechanical ventilator had to be exchanged with a custom-made connector board. It is connected to the I/O panel of a rapid prototyping system capable of real-time processing and communicating with the ventilators hardware, sensors and actuators. This system controls the ventilator, which also measures the P_{aw} and the flow $\dot{V}$ and sends the data back to the rapid prototyping system. All variables and parameters of the ventilator control algorithm are set in the control software of the rapid prototyping system on a desktop PC.

In order to measure the EMG signal, an EMG amplifier is used. This is a wearable wireless sensor capable of measuring two EMG channels. The measuring electrodes are placed on the costal margin, the intercostal space and the sternum. Because of the wireless design, the EMG amplifier is decoupled of the mains hum. The sensor is paired

with a wireless-to-serial adapter on the rapid prototyping systems I/O panel, thus the data could be processed within the system.

2.1 Respiratory mechanics

Respiratory mechanics are mainly described through various pressures P, lung volume V and air flow $\dot{V}$. These quantities are dependent on the airway resistance and the lung compliance, as well as pathological changes of these parameters. A simple way to model the respiratory mechanics is the linear single compartment model shown in Fig. 2. This pneumatic model consists of cylindrical tube and an elastic bag. It can be transformed into an electrical analogue [6] as seen in Fig. 3, where the pressure P corresponds to electrical voltage and the air flow $\dot{V}$ corresponds to electrical current. The resistance R of the airways can be seen as an electrical resistance, while C can be seen as the capacity of a capacitor. Increased resistance leads to reduced flow and reduced compliance leads to reduced tital volume. From this model, the equation

$$P_{aw} + P_{mus} = R \cdot \dot{V} + 1/C \cdot V, \qquad (1)$$

can be derived, where P_{mus} corresponds to the breathing effort of the patient against the airways resistance and the compliance of the lung. To relief the patient and compen-

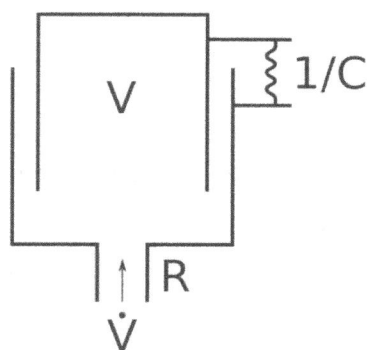

Figure 2: Shown is the linear single compartment model of the lung. The flow $\dot{V}$ fills up the volume, which leads to a higher resilience caused by a linear spring with spring constant 1/C. Additionally the resistance R works against the flow.

sate a higher breathing effort, he is mechanically supported by the mechanical ventilator, which provides the external pressure support $P_{aw} > 0$.

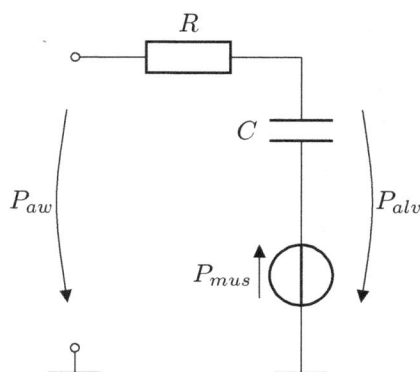

Figure 3: The lung and its parameters can be described analogously to an electrical RC circuit [6]. Equation (1) can be derived from this analogue.

2.2 Edi and P$_{mus}$

The breathing effort of the patient and therefore the pressure generated by the respiratory muscles (P_{mus}) is tightly related to the muscular activity Edi [7]. With measurable sEMG of the respiratory muscles it is possible to estimate P_{mus} from the muscular activity as well as the pneumatic signals [8] and therefore the system should have the advantage of showing less patient-ventilator asynchronies than regular Proportional Assist Ventilation and while being non-invasive unlike NAVA.

It should be noted, that the ratio

$$PEI = P_{mus}/Edi \tag{2}$$

is quite stable within a patient [8]. The index in (2) (P_{mus}/Edi Index) represents the neuromechanical efficiency and has dimensions of mbar/µV. It indicates the pressure

that is generated by the respiratory muscles of the patient per µV of sEMG activity. However, deviations of the index within a patient are possible, since the EMG-force relation depends on a variety of geometrical factors. These can be taken into account by means of a model-based analysis such as that proposed by Ratnovsky et al. [10].

Additionally the sEMG measurements have to be cleaned of the electrocardiogram (ECG) and other artefacts, before they can be used as a viable trigger. This contamination, especially the QRS-complex of the ECG, affects the accuracy of data derived from the signal and has to be substracted from the sEMG [11]. The filtered data can be used as a trigger as well as a measurement of the patient's breathing effort for later applied proportional pressure support.

3 Results

The resulting setup can be seen in Fig. 4. Firstly, the controller board of the mechanical ventilator was exchanged with a custom connector board. It enables communication with the rapid prototyping system. Therefore ventilation modes and parameters (e.g. pressure, flow) could be modified on-line in the ventilators specific control environment. Additionally the controller boards of the flow- and pressure sensors were connected to two ADC ports of the rapid prototyping system, so that the data could be read for regulation purposes. This setup enables the use of the ventilator to provide mandatory ventilation as well as assistance to spontaneous breathing patients. These standard functions have been verified in use with a breathing bag as a simple mechanical patient simulator.

Secondly, communications between the EMG amplifier and wireless-to-serial adapter has been implemented and verified by connecting the serial side of the adapter to a desktop PC. To this end, a script has been written in Python, that specifies and opens the necessary serial port for a wireless connection, configures the two EMG channels of the sensor and starts the measurement of the sEMG data with a sampling rate of 1024 Hz to ensure sufficient measurement resolution. Measured data is then streamed to the PC, printed in the command window and plotted against time.

4 Conclusion & Outlook

Within the scope of this project both the hardware and software of a mechanical ventilator were modified to allow controlling the ventilator by means of a rapid prototyping system. Additionally an EMG amplifier was connected wirelessly to a wireless-to-serial converter.

Firstly, the EMG amplifier has to be connected to the rapid prototyping system by means of the wireless-to-serial converter. Therefore the control software has to be adapted.

Secondly, the sEMG trigger has to be implemented. However, before it can be used, the data has to be rid of ECG and cross-talk from other adjacent muscle fibres. Additionally the data has to be rectified and filtered with a low-pass filter, to obtain the linear envelopes [12], that can be used as

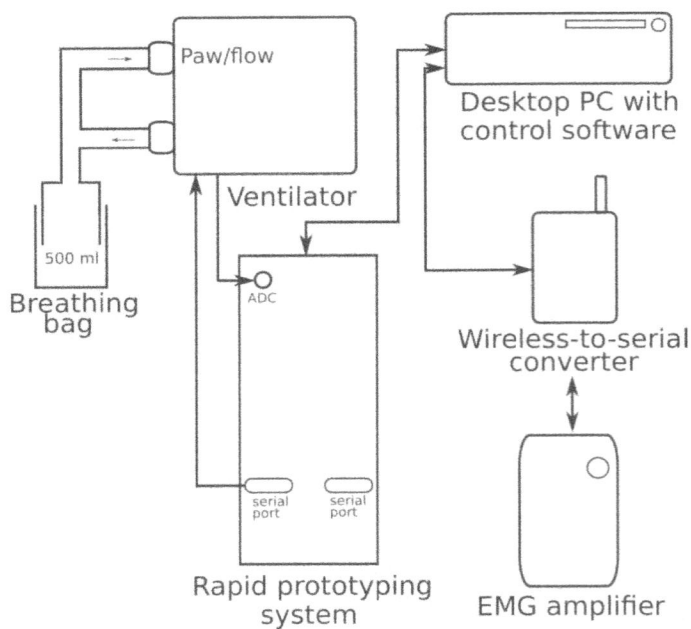

Figure 4: The resulting experimental setup: The most parts are as described as in Fig. 1. However, the rapid prototyping system and the wireless-to-serial adapter are not yet communicating. Therefore the serial adapter is connected to the Desktop PC. Additionally, a patient's breathing is simulated with a 500 ml breathing bag.

a trigger.
At last the proportional assist ventilation has to be implemented. Therefore the amount of muscular activity has to be measured, to derive the necessary extent of pressure to assist the patient's breathing.

Acknowledgement

The work has been carried out at Institute for Electrical Engineering in Medicine, University of Lübeck.

5 References

[1] Levine S. et al., *Rapid disuse atrophy of diaphragm fibers in mechanically ventilated humans.* New England Journal of Medicine 2008; 358: 1327–1335.

[2] Brander L., *NAVA - oder wenn der Patient das Beatmungsgeraet steuert.* Timisoara Medical Journal 2008; 19-22.

[3] Tobin M., Jubran A., Laghi F., *Patient-ventilator interaction.* American Journal of Respiratory and Critical Care Medicine 2001; 163: 1059–1063.

[4] de Wit M. et al., *Ineffective triggering predicts increased duration of mechanical ventilation.* Critical Care Medicine 2009 Oct; 37(10): 2740-2745.

[5] Sinderby C. et al., *Neural control of mechanical ventilation in respiratory failure.* Nature Medicine 1999; 5: 1433–1436.

[6] Leonhardt S., Mersmann S., *Automatisierungstechnik fuer die kuenstliche Beatmung - eine Standortbestimmung.* Automatisierungstechnik 2007; 244-254.

[7] Beck J., Gottfried S.B., Navalesi P. et al., *Electrical activity of the diaphragm during pressure support ventilation in acute respiratory failure.* American Journal of Respiratory and Critical Care Medicine 2001; 164: 419–424.

[8] Bellani G. et al., *Estimation of Patient's Inspiratory Effort From the Electrical Activity of the Diaphragm.* Critical Care Medicine 2013 Jun; 41(6): 1483-1491.

[9] Gandevia S.C. and McKenzie D.K., *Human diaphragmatic EMG: changes with lung volume and posture during supramaximal phrenic nerve stimulation.* Journal of Applied Physiology 1986; 60: 1420–1428.

[10] Ratnovsky A. et al., *Integrated approach for in vivo evaluation of respiratory muscles mechanics.* Journal of Biomechanics 2003 Dec; 36(12): 1771-84.

[11] Bartolo A. et al., *Analysis of diaphragm EMG signals: comparison of gating vs. subtraction for removal of ecg contamination.* Journal of Applied Physiology 1996 June; 80(6): 1898-1902.

[12] Winter D.A., *Kinesiologogical electromyography. In: Biomechanics and Motor Control of Human Movement.* Wiley 1990, New York, 191–212.

Development of a stopped-flow apparatus for rapid mixing of small fluid volumes in reaction kinetic studies

J. Kappel[1], C. Hübner[2],
[1] Medical Engineering Science, Universität zu Lübeck, janosch.kappel@student.uni-luebeck.de
[2] Institute of Physics, Universität zu Lübeck, huebner@physik.uni-luebeck.de

Abstract

The observation of dynamic interactions of biomolecules can reveal profound insights into molecular details of as, for example, enzyme function, protein interactions and molecular transport. Investigation of these kind of processes is of great interest for biophysics and the pharmaceutical industry. One method to study the mentioned processes is the stopped-flow method. In this work the design and development of a stopped-flow apparatus for rapid mixing of small fluid volumes in reaction kinetic studies using confocal fluorescence microscopy is presented. Certain functional requirements that are set by a microfluidic helical mixer have been identified and taken into account during the design process. A vertical design approach and the technical configuration of the device have been created using CAD software. Furthermore, appropriate components for the technical realisation have been selected.

1 Introduction

The observation of dynamic interactions of biomolecules can reveal profound insights into molecular details of as, for example, enzyme function, protein interactions and molecular transport [1]. Investigation of these kind of processes is of great interest for biophysics and medicine, as well as for the pharmaceutical industry.

To determine kinetic parameters like reaction rates between different states, conformational changes in molecular structure and intermediate states, the reaction can be observed in real time by optical (e.g. absorption and fluorescence measurements) or electro-chemical (e.g. conductivity measurements) methods [2]. As described in [1], the simplest reaction frequently encountered in kinetic experiments is the first-order reaction:

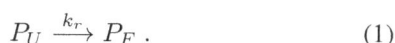

$$P_U \xrightarrow{k_r} P_F \ .$$ (1)

Various reactions, such as several protein folding reactions, that do not involve binding of ligands, are described as shown in (1). In this example the transition from the unfolded state (U) to the folded state (F) of the protein has to be achieved and initiated by mixing two reactants. Integration of the rate law

$$\frac{d[P_U]}{dt} = k_i[P_U]$$ (2)

yields the time dependence of a first order reaction:

$$A_t = A_0 e^{-k_r t} \ ,$$ (3)

where A_t describes the change in signal at time t, A_0 is the total amplitude expected for the reaction, and k_r stands for

the observed rate constant in units of s^{-1}.

There are several methods that are used in kinetic experiments to study the kinetics of fast reactions, such as continuous-flow (CF), quenched-flow (QF) and stopped-flow (SF). Depending on the type of the reaction, each of these methods has its advantages and disadvantages. One of the advantages of the SF method is low sample consumption, which is of high importance if sample preparation is expensive and time-consuming.

In this work the design and development of a stopped-flow apparatus for rapid mixing of small fluid volumes in reaction kinetic studies is presented. The device is designed to meet the requirements set by a certain type of microfluidic mixer with the aim of lower sample consumption and a shorter dead-time compared to commercially available systems. Furthermore, extendibility and modifiability of hardware and software of the system have been considered within the design process.

2 Material and Methods

In the following, the basic principles of the stopped-flow method and the combination of that technique with confocal fluorescence microscopy are described. Furthermore, the design of the microfluidic helical mixer will be explained.

2.1 The stopped-flow technique

Stopped-flow rapid mixing is a commonly used method for the investigation of fast biochemical reactions with high temporal resolution such as, for example, protein interactions, conformational changes in protein structure or enzyme reactions [3]. The stopped-flow method was first introduced in 1937 by Chance [4], and has been improved and modified over the past decades.

In this technique two or more reactants are mixed together at high velocities directly into an observation chamber. After a very short period of time, the flow is stopped rapidly when the observation chamber is filled [5]. The reaction between the mixed solutions then can be monitored as the change in signal over time by a data acquisition system with high temporal resolution. To do so, the reaction can be observed by optical absorbance, fluorescence or circular dichroism (CD), for example [1].

The time between first contact of the reactants and the mixed solution getting rapidly stopped in the observation chamber and, therefore, the initiation of data acquisition, is called dead-time. It mainly depends on the time of mixing, the flow rate and mixer geometry. The dead-time can be determined from signals of fluorescence and absorption measurements in first-order reactions [6]. Another important characteristic of the SF method is the low consumption of reactants. Once the flow stopped the mixed solution can be observed without further sample consumption. Therefore only an isolated area needs to be focused by the detection system and no scanning is necessary.

2.2 Microfluidic mixer

Concerning chemically induced reactions, for example the adding of a denaturant to a protein solution in order to induce unfolding, both reactants have to enter into contact and need to reach the area where the change of physical properties can be measured. Mixing by diffusion is a slow process, in particular with respect to the fast reactions mentioned before and, therefore, too slow for the observation of these kinds of reactions.

In [7] the design of a microfluidic helical mixer for submillisecond mixing of small fluid volumes for SF applications is described, suitable for fabrication in quartz glass using in-volume selective laser etching (ISLE). Mixing and observation chamber are incorporated in a single optically accessible element. This holds the advantage of a short distance between the point of mixing and the point of observation and data acquisition, and shorter transport distances result in a reduction of the dead-time. The mixing performance of this mixing-observation-block has been evaluated by using computational fluid dynamics (CFD).

The block is of dimensions of $11.87\,\mathrm{mm}$ of length, $10\,\mathrm{mm}$ of width and $6\,\mathrm{mm}$ of height. The inlet and outlet channels have a diameter of $0.5\,\mathrm{mm}$ and the combined internal volume of mixing and observation compartment is designed to be $0.0136\,\mathrm{\mu l}$.

2.3 Confocal fluorescence microscopy

As mentioned before, the SF method can be combined with fluorescence measurement techniques, such as confocal fluorescence microscopy (CFM) when working with small sample volumes and low sample concentrations.

The light source excites the fluorescent solution, which is pushed through the mixing chamber into the observation compartment by means of the syringe drives. The emitted fluorescence is separated from the excitation light by a dichroic mirror, and focused on the detection system, which can be a charge-coupled device (CCD) or an avalanche photodiode (APD), as seen in Fig. 1.

To give an example of how this setup can be used, the measurement of unfolding green fluorescent protein (GFP) using sodium dodecyl sulfate (SDS) is shortly described hereafter. As described in [8], GFP is a highly stable, autofluorescent protein, that is widely used for cell and molecular biology applications. GFP is vulnerable to denaturation by SDS at a pH of 6.5 and looses its fluorescence within a minute. This leads to a single, exponentially decaying timecourse where A_t decays from a maximum value A_0.

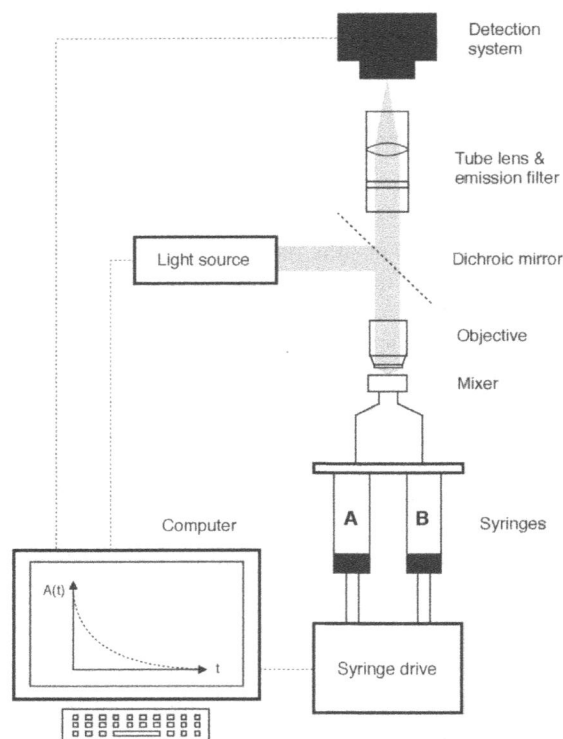

Figure 1: Planned setup of the confocal fluorescence microscope and the stopped-flow apparatus. The light source excites the fluorescent solution, which is pushed from syringes A and B through the mixing chamber into the observation compartment by means of the syringe drives. The emitted fluorescence is separated from the excitation light by a dichroic mirror, and focused on the detection system (CCD/APD).

3 Results and Discussion

In the following, the identified functional requirements concerning the configuration of the SF system are explained. In addition the main components of the design and the criteria for their selection, as well as the resulting design of the device will be presented.

3.1 Functional requirements

Experiments and computer simulations show that various proteins begin folding in the nano- and microsecond time range. It is assumed that the hydrophobic collapse occurs in less than $100\,\mu s$ [9]. Thus, a volumetric flow rate of $116\,\mu l\,s^{-1}$ ($58\,\mu l\,s^{-1}$ per inlet) is required to fill the mixing compartment of the micromixer. The volumetric flow rate of $58\,\mu l\,s^{-1}$ per inlet corresponds to a mean time of $117\,\mu s$ for the mixed solution to reach the end of the observation chamber. According to [7], the CFD simulations have shown that an inlet pressure of almost 50 bar is required to achieve a flow rate of $58\,\mu l\,s^{-1}$. The mechanical force that is required to push the piston into the syringe in order to cause an outlet pressure of 50 bar can be calculated using the Bernoulli equation:

$$p_1 + \frac{1}{2}\varrho v_1^2 = p_2 + \frac{1}{2}\varrho v_2^2 \qquad (4)$$

where p_1 denotes the pressure inside the syringe and p_2 the pressure inside the tubing that connects syringe and mixer. Correspondingly, the velocity of the fluid inside the syringe and inside the tubing are indicated by v_1 and v_2, and ϱ describes the density of the fluid.
The piston force then can be calculated with $p_1 = \frac{F}{A_1}$ and $v_2 = \frac{A_1 v_1}{A_2}$ as follows:

$$F_p = (p_2 + \frac{1}{2}\varrho \left(\frac{A_1 v_1}{A_2}\right)^2 - \frac{1}{2}\varrho v_2^2)A_1 . \qquad (5)$$

Calculations are done by choosing $\varrho = 997.99\,\mathrm{kg\,m^{-3}}$ for water at a temperature of $21\,°C$, since most proteins are dissolved in aqueous solution. A_1 and A_2 denote the cross-sectional area of piston and tubing. With $A_1 = 4.1 \times 10^{-3}\,\mathrm{mm}^2$ for a $250\,\mu l$ syringe with a piston diameter of usually $2.3\,\mathrm{mm}$, and $A_2 = 2.8 \times 10^{-4}\,\mathrm{mm}^2$ for standard PEEK tubing with a diameter of $0.5\,\mathrm{mm}$, (5) yields a force of $20.78\,\mathrm{N}$.
Calculations with different combinations of these parameters and variation of syringe and tubing size as well as velocity of fluid have been performed using MATLAB® (MathWorks, Inc., Natick, MA).

3.2 Choice of components

3.2.1 Syringes

Concerning the low sample consumption of the SF method and the relatively high outlet pressure of 50 bar, two $250\,\mu l$ precision glass syringes (cetoni, Korbußen/Gera, Germany) that withstand a maximum pressure of 77 bar, were selected.

In addition a $500\,\mu l$ syringe is required to counteract the inertia of the fluid in the channel system allowing for a sharper stop of the flow. The selected screw-in syringe holders (cetoni, Korbußen/Gera, Germany) are made of stainless steel, and facilitate the connection of syringe and micromixer via $^1/_4$"-28 UNF fittings and PEEK tubing. Unfortunately there was no information about friction between plunger and tube of the syringes and, therefore, this has to be determined experimentally.

3.2.2 Stepper motors and lead screws

In order to move the plunger of the syringe reliable and accurate linear actuators are required. Piezoelectric actuators have been taken into consideration, as they come with high precision positioning, however, most of them are either insufficient in linear load and travel range or very expensive. On the other hand, motorized lead screws offer a good compromise between costs and performance. Hence, a combination of hybrid stepper motors and precision lead screws (Thomson Industries, Wood Dale, IL) was selected, using the performance plot generator on the manufacturer's website. According to that plot the selected combination of stepper motor (rotating screw, NEMA 23, single stack, bipolar, 3 A per phase) and lead screw (length: $100\,\mathrm{mm}$, diameter: $10\,\mathrm{mm}$, lead: $10\,\mathrm{mm}$) still handles a load of around $90\,\mathrm{N}$ at linear velocities of $500\,\mathrm{mm\,s^{-1}}$, enabling a wide dynamic range.
Generally a safety factor of 2 is recommended when sizing an application. Even when applying a safety factor of 4 to the result of (5), taking into account friction inside the syringe and resistance at elements of the fluid system, the generated dynamic range is still within the limits. Bearing in mind a safety factor of 4, the pressure drop at tubing connecting syringe and mixer can be neglected.
For the sake of completeness the maximum buckling load as well as the maximum pressure and tensile load have been calculated for this setup, considering a fixed-free mounting method for the lead screw.

3.3 Design

The syringe drive actuates by having the stepper motor rotate the lead screw and translate the load to the lead nut. To avoid wedging and buckling, it is important that the attached load exerts at the centre line of the lead screw. As seen in Fig. 2, a vertical design approach of the device has been taken into consideration. The stepping motors are located at the bottom of the system, providing a firm state of the construction by their weight.
The syringes are located at the top, in order to simplify removing air bubbles from the system. They are mounted on the system by means of the screw-in syringe holders. The plungers are connected to the motorized drive via M3 threads. This makes the application more flexible, and allows for easy exchange of syringes (i.e. different syringe volumes). To provide a better view of the important components, some elements of the frame of the system are not presented in Fig. 2.

Separation of syringe drive and the microfluidic mixer facilitates the possibility of mechanical decoupling of the driving part and the mixing part of the system, which is important in order to minimise vibration during data acquisition. The mechanical design of the presented SF apparatus has been developed using Autodesk® Inventor® 2017 (Autodesk, Inc., San Rafael, US).

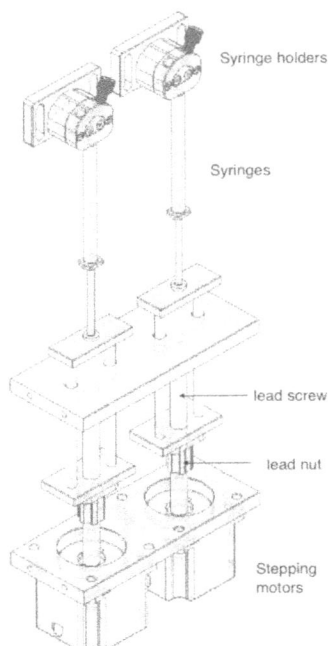

Figure 2: Technical drawing of the vertical design of the stopped-flow apparatus. The hybrid stepping motors are located at the bottom of the system, thus stabilising the construction by their weight. The syringes are located at the top, in order to simplify removing air bubbles from the system. To provide a better view of the important components, some parts of the frame of the system are omitted in this drawing.

4   Conclusion and Outlook

In this work the design of a stopped-flow apparatus for mixing of small fluid volumes in reaction kinetic studies has been developed. The SF device has to meet special functional requirements caused by application of the microfluidic mixer. These requirements have been identified after thorough research and taken into account while developing the technical configuration of the SF device. The aim is to achieve lower sample consumption and a shorter dead-time compared to commercially available systems.

Further work should consider the completion and optimisation of the mechanical and electronic configuration. A solution for simple venting and flushing of the fluid system has to be developed, and high pressure 3-way valves should be selected to allow for simple sample loading.

Moreover, motor controllers have to be selected and pro-grammed to control and synchronise the syringe drives. In order to enable accurate positioning motion control encoders or external triggers have to be implemented. Synchronisation of the SF device with the detection system has to be implemented, as well as the possibility to control the setup using an application software such as LabVIEW. Finally, the performance of the whole system in terms of hardware and software control and data acquisition needs to be evaluated by means of appropriate experiments.

Acknowledgement

The work has been carried out and supervised by the Institute of Physics, Universität zu Lübeck.

5   References

[1] S. M. Hargrove, *Ligand binding with stopped-flow rapid mixing*. In: Protein-Ligand Interactions: Methods and Applications, Humana Press, Totowa, New Jersey, pp. 323-341, 2005.

[2] S. M. H. Khorhassani, A. Ebrahimi, M. T. Maghsoodlou, M. Shahraki, D. Price, *Establishing a new conductance stopped-flow apparatus to investigate the initial fast step of reaction between 1, 1, 1-trichloro-3-methyl-3-phospholene and methanol under a dry inert atmosphere*. Analyst, 136(8), pp. 1713-1721, 2011.

[3] R. Bleul et al., *Compact, cost-efficient microfluidics-based stopped-flow device*. In: Analytical and bioanalytical chemistry, 399(3), pp. 1117-1125, 2011.

[4] B. Chance, *The stopped-flow method and chemical intermediates in enzyme reactions–a personal essay*. Photosynthesis Research, 80, pp. 387-400, 2004.

[5] R. A. Harvey, *A simple stopped-flow photometer*. Analytical biochemistry, 29(1), pp. 58-67, 1968.

[6] B. I. Tonomura, H. Nakatani, M. Ohnishi, J. Yamaguchi-Ito, K. Hiromi, *Test reactions for a stopped-flow apparatus: Reduction of 2, 6-dichlorophenolindophenol and potassium ferricyanide by L-ascorbic acid*. Analytical biochemistry, 84(2), pp. 370-383, 1978.

[7] H. Mueller, *Design and simulation of a passive micromixer for rapid mixing of small fluid volumes in reaction kinetic studies*. Master Thesis, University of Lübeck, 2016.

[8] K. M. Alkaabi, A. Yafea, S. S. Ashraf, *Effect of pH on thermal- and chemical-induced dnaturation of GFP*. Applied biochemistry and biotechnology, 126(2), pp. 149-156, 2005.

[9] S. V. Kathuria et al., *Advances in turbulent mixing techniques to study microsecond protein folding reactions*. Biopolymers, 99(11), pp. 888-896, 2013.

Nav-CARS-EVAR: Comparison of patient individual vessel anatomies with an abdominal aortic aneurysm to their corresponding rapid prototyping printed 3D-models

J. Tesche[1], S. Matthiensen[2], M. Horn[3], E. Stahlberg[4], J. P. Goltz[4], M. Schenk[5], D. Wendt[6] and M. Kleemann[3]

[1] Medizinische Ingenieurwissenschaft, Universität zu Lübeck, jula.tesche@student.uni-luebeck.de
[2] Humanmedizin, Universität zu Lübeck, sarah.matthiensen@student.uni-luebeck.de
[3] Bereich Gefäß- und endovaskuläre Chirurgie, UKSH Lübeck, marco.horn@uksh.de, markus.kleemann@uksh.de
[4] Klinink für Radiologie und Nuklearmedizin, UKSH Lübeck, erik.stahlberg@uksh.de, janpeter.goltz@uksh.de
[5] Fraunhofer MEVIS, Lübeck, mark.schenk@mevis.fraunhofer.de
[6] Fraunhofer EMB, Lübeck, dennis.wendt@emb.fraunhofer.de

Abstract

Over the past twenty years, the usage of minimally invasive procedures (EVAR) in patients with aneurysms increased. Disadvantages, like radiation dosage and renal toxicity of contrast agent, shall be limited by the Nav-CARS-EVAR project (Navigated- Contrast Agent and Radiation Sparing- Endovascular Aortic Repair). To evaluate an experimental environment, 3D-models have to be developed. CT-angiographic datasets of patients with aortic aneurysms serve as the base for the models. So far, one phantom was produced by rapid prototyping. Measuring points were defined, measured and compared to each other for patient and model. The results showed the model was very similar to the patients aorta. As an experimental environment, models printed by rapid-prototyping seems to be a valid opportunity for training, simulation and testing. Further testing of different materials and remaining models are necessary. In future, the models will help to develop the navigation system, which will achieve the aim of radiation- and contrast agent reduction.

1 Introduction

An aneurysm is a bulge of a blood vessel, where its dilatation is at least one and a half times bigger than the normal diameter. Most aneurysms (60 to 65%) are localized in the infrarenal segment of the aorta (fig. 1), whose normal diameter is about 2 cm. Often aneurysms volumes are filled with thrombotic material, so they consist of a bood supplyed part (lumen) and a non-perfused part [1] - [4]. With enlarging diameter of the aneurysm, the risk of rupture increases due to the higher vasculare wall tension. In case of rupture, the mortality is about 80 to 90 %, since the ruptur is connected to a big blood loss. That is why detected aneurysm with high rupture chances will be surgical attended [1] - [3], [5]. There are two methods for treating an aortic aneurysm to prevent ruptures and further growing: The open surgical replacement (OS) and the endovascular aortic repair (EVAR). The first method describes a replacement of the aneurysm effected aortic segment by a synthetic implant. The second one is a minimally invasive method, where a covered stent is inserted by crossing the groin to bypass the aneurysms bulge [1] - [3]. During the past 20 years, the number of EVAR surgeries increased. Reasons are a few advantages of EVAR over OS, for example a lower perioperative mortality and less "aneurysm related deaths" [3]. EVAR is less straining than OS, so even patients "deemed unfit" [3] for OS can have a surgery [2], [3], [6]. One problem of EVAR is the high radiation- and contrast agent exposure, which is needed to place the stent correctly while using digital subtraction angiography (DSA). By using DSA, two xray-images are generated, one before and one after the patient takes contrast agent. The images get substracted from each other, only the contrasted vessels are seen on the screen. [7]. Contrast agent can cause a deterioration of the renal function, which is already reduced in many patients (40%) with aortic aneurysm. The radiation affects the health, not only of the patient, but also of physicians and assistants. Furthermore, endoleaks can appear after EVAR because of stent "missmatch and missplacement" [6]. Endoleaks are characterized by a blood flow into the aneurysm bulge outside the endoprosthesis, if the stent is placed incorrectly [4]. The two-dimensional view from DSA in a three-dimensional (3D) vascular system is not very precise, a 3D-navigation could simplify the procedure and lead to a perfect fit of the stent, following less endoleaks [4], [6], [8] - [10].

The Nav-CARS-EVAR project (Navigated- Contrast Agent and Radiation Sparing- Endovascular Aortic Repair) tries to develop such a 3D-navigation system by using a glass fiber and a connected light source. Fibre system and stent are attached to each other during positioning. Bending of the

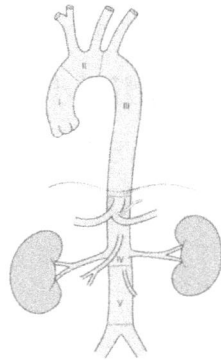

Figure 1: Sections of the aorta: Section IV and V are part of the abdominal aorta, as well as the following iliac vessels (not shown in the picture) [11]

fibre while moving through the vessels effects different reflections of the light inside the fibre, caused by Bragg grating. This data will be used to estimate the position of the stent inside the body. The future aim is to display the previously created CT-angiographic image (CTA) of the patient on a monitor and overlap the position data durig the surgery. This way allows to follow the stent position without using further contrast agent or radiation [6].

To test the navigation system, a realistic test environment has to be provided. Therefore, three-dimensional aorta models of real patients will be created by applying rapid prototying. Those models will be used to replace the patients while analyzing the navigation systems precision. Since the corresponding CTA-images will be used while testing as well as during the real surgery, the aorta models have to be as alike as the real patients vessels [6]. The aim of this project is to validate, how accurate the models match to the original patients aorta.

2 Material and Methods

Selecting patients and analyzing their CTA-images

For testing the navigation system, 25 models with different peculiarities of their aneurysm will be generated. To find proper ones, the hospital information system Orbis (Agfa HealthCare GmbH, Bonn, Germany) from the UKSH (Universitätsklinikum Schleswig-Holstein) was used. In this information system all data of the patients are saved, likewise the CTA-images. This way, 341 patients with aortic aneurysm, who have had a CTA-examination during the last 3 years, were examined. Finally, 25 patients were selected, to have a selection of different aneurysmal peculiarities. Afterwards, their CTA-images were examined by using the radiological software Syngo.via (Siemens AG, München, Germany). 23 significant measuring points were specified, which were measured for each patient, as shown in table 1. The measuring points include distances between aortic bifurcations (number a-h, fig. 2a), as well as diameters of the aortic lumen at certain positions on the aorta (number 1-14,

fig. 2b). Depending on the aneurysmal peculiarity, further individual measuring points were added.

Table 1: Measuring points for all patients

Number	Measuring point
a	celiac trunk to aortic bifurcation
b	neck (renal arteries to aneurysms beginning)
c	renal arteries to aortic bifurcation
d	aneurysms length
e, f	aortic bifurcation to iliac bifurcation right, left
g, h	iliac bifurcation to femoral bifurcation right, left
1	neck
2.1	aneurysms maximum diameter
2.2	lumen at 2.1
3	celiac trunk
4	superior mesenteric artery
5, 6	common iliac artery proximal right, left
7, 8	common iliac artery distal right, left
9, 10	internal iliac artery proximal right, left
11, 12	external iliac artery proximal right, left
13, 14	femoral artery distal right, left

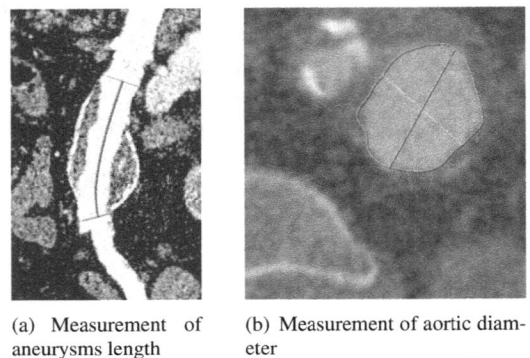

(a) Measurement of aneurysms length

(b) Measurement of aortic diameter

Figure 2: Examples of measurements

Segmentation

In order to create the models, the CTA-images were segmented. Therefore, the software Mimics Research 19.0 (Materialise NV, Leuven, Belgium) was used. A printable STL-file (standard triangulation language [13]), which describes the surface geometry of the object as a net of triangular areas [12], is created in several steps: First of all, the aorta was seperated from the other structures inside the body. In this case, the contrast agent within the perfused vessels is helpful and easily distinguishable from the surrounding tissue. All the corresponding voxels of the contrast agent are identified inside every image slice and get connected to a three-dimensional volume, by selecting the voxels with the same color (fig. 3a). For now, this volume forms the lumen of the aorta. In the second step, the same procedure is performed creating a volume, which represents the non-perfused part of the aneurysm. The trombothic material has a darker gray value than the contrast agent. Using a second software (3-matic Research 11.0, also from Materialise NV), both three-dimensional objects can be edited. Actually the lumen of the aortic model has to be hollow, while the vessel wall will be printed. That is why a two millimeter thick wall is created around the former segmented lumen, which is deleted afterwards. Finally, both volumes are joined together (fig. 3b) and vessel endings are cut open, to allow the inserting of the navigation system in later experiments. In a final step, the created model can be saved as an STL-file, which can be red and performed by the 3D-printer. In this way the 25 CTA-images were segmented and corresponding STL-files were created.

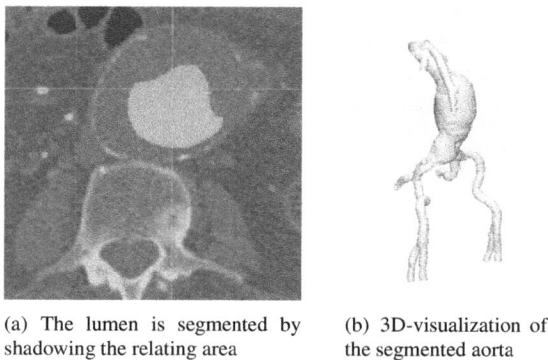

(a) The lumen is segmented by shadowing the relating area

(b) 3D-visualization of the segmented aorta

Figure 3: Segmentation of the aorta

Generating the aortic models

For printing the three-dimensional models, the 3D-printer Connex500 from Stratasys, Eden Prairie, Minnesota, USA will be used. It can read the created STL-files and prints each model slice by slice by using polyjet technology, which is a form of rapid prototyping: The material is applied by nozzles and hardened by ultraviolet radiation [12]. There are two different materials available: VeroClear, a very stable material, and TangoPlus, which is elastic with different possible stages of flexibility. The printer has a high resolution: about 40 micrometer for the x- and y-axis and circa 32 micrometer in z-range, which is also the slice thickness. During the printing process a supporting material is used, exemplarily for printing hollow structures. After printing, the supporting material is removed from the model.

Since the printing takes a long time and is very cost-intensive, only one model was printed so far (fig. 4a). Using TangoPlus with the least state of flexibility, which has a similar consistence to a real vessel, this model was the first try to investigate if the material is suitable for the following experiments. After printing, a CT-image of the model was made. Using this image, the same measuring points were analyzed, which were specified for the patients CTA-image in the beginning (fig. 4b). Now the corresponding measurments from model and patient were comperend to each other.

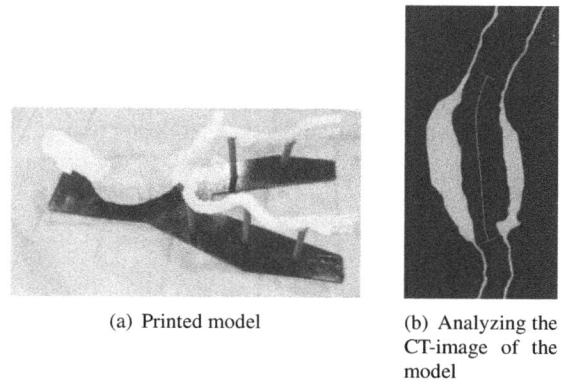

(a) Printed model

(b) Analyzing the CT-image of the model

Figure 4: Analyzing the printed model

3　Results and Discussion

Table 2 shows exemplary a selection of results which were made by analyzing both, the patients CTA-image and the models CT-image.

Table 2: Selected measurement results

Measuring points	Patient in mm	Model in mm	Difference in mm
c	128,9	130,9	2,0
d	113,4	115,7	2,3
e	53,5	53,3	0,2
f	65,1	67,7	2,6
8	12,7	15,1	2,4
9	8,0	6,5	1,5
10	7,4	9,1	1,7
11	9,6	9,2	0,4

In total, two thirds of the measuring points show very similar values for both, the patients CTA- and the models CT-image. Those measuring points did not differ from the original about more than 2 mm, which would be an acceptable variation. When stents are implanted, a maximum of 2 millimeter deviation would be passable. But still, in all other cases there were differences higher than 2 mm, for three measuring points even more than 6 mm. Difficulties appeared in areas with low image quality, basing on the CTA of the patient. These difficulties were already foreseeable while measuring the patients CTA-images: Blurred images were tough to analyze, especially if protheses were inside the patients body. Vessel shapes were not clearly definable from the surrounding tissue. Furthermore, some measuring points were not exactly locatable, for example beginning and ending of some aneurysms, which had very slightly increasing bulge diameters. If a patients CTA-image was difficult to analyze, the segmentation was not easy to perform as well. Another problem was the missing contrast agent after vascular stenosis. Vessels without contrast agent are not identified by the software, so helping tools cannot be used, accordingly analyzes and segmentation last much longer. Additionally, the handling with the software needs a lot of practice to perform neat and reproducable measurments and segmentations. All these difficulties lead to deviations of the model from the original patients aorta. But still, most measuring points attuned or were very close to an acceptable value. This shows the tendency that creating aortic models, which are similar enough to the original, is possible.

4 Conclusion

So far, the first step of the Nav-CARS-EVAR project is made. The remaining models still need to be printed, as well as different materials still need to be tested. In addition, more experiments are planned to test the developed navigation system by using the models as test environment. Since most measurements were adequate by the first try, despite a lot of difficulies, the creating of the models definitely can be improved to achieve good quality models. In future, models printed by rapid prototyping will be used as an experimental environment and enable the avoiding of animal experiments. Additionally, those models can be used for training and teaching in university environment. As soon as the navigation system functions properly, it can be used during EVAR surgery and reduces the use of radiation and contrast agent.

Acknowledgement

The work has been carried out and supervised by the Division of Vascular- and Endovascular Surgery, Universitätsklinik Schleswig-Holstein, Lübeck.

5 References

[1] M. P. Harward, *Medical secrets*. In: Secrets series, Elsevier / Mosby, Philadelphia, 2012.

[2] M. M. W. Jakubaß, *Perioperative und Langzeit-Morbidität und -Mortalität nach elektiver Therapie infrarenaler Aortenaneurysmen – retrospektive Analyse eines universitären Zentrums*. Dissertation, Technische Universität München - Fakultät für Medizin, München, 2016.

[3] J. L. Eliason and G. R. Upchurch, *Endovascular abdominal aortic aneurysm repair*. American Heart Association, Dallas, 2008.

[4] A. Greiner, J. Grommes and M. J. Jacobs, *The place of endovascular treatment in abdominal aortic aneurysm*. Deutsches Ärzteblatt International, Aachen, 2013.

[5] J. P. Barral and A. Croibier, *Visceral Vascular Manipulations, translated by A. Mackenzie (Manipulations vasculaires viscérales)*. Churchill Livingstone / Elsevier, Edinburgh, 2011.

[6] M. Horn, M. Kleemann et al., *Ein Prototyp für die navigierte Implantation von Aortenstentprothesen zur Reduzierung der Kontrastmittel- und Strahlenbelastung: Das Nav-CARS-EVAR-Konzept (Navigated-Contrast-Agent and Radiation Sparing Endovascular Aortic Repair)*. Georg Thieme Verlag KG, Stuttgart, 2015.

[7] D. Raschke, *Halsgefäßstenosen - Computertomographische Angiographie (CTA) versus Digitale Subtraktionsangiographie (DSA). Eine Validitätsstudie zur Stenoseermittlung in der Arteria carotis und Arteria vertebralis*. Dissertation, Georg-August-Universität zu Göttingen - Medizinische Fakultät, Göttingen, 2013.

[8] E. R. Ketteler and K. R. Brown, *Radiation exposure in endovascular procedures*. In: Journal of Vascular Surgery, Elsevier, 2011.

[9] S. R. Walsh, T. Y. Tang and J. R. Boyle, *Renal consequences of endovascular abdominal aortic aneurysm repair*. J Endovasc Ther, 2008.

[10] V. Tacher et al., *Image Guidance for Endovascular Repair of Complex Aortic Aneurysms: Comparison of Two-dimensional and Three-dimensional Angiography and Image Fusion*. In: Journal of Vascular and Interventional Radiology, Elsevier, 2013.

[11] B. Luther, *Techniken der offenen Gefäßchirurgie. Standards, Taktiken, Tricks*. Springer, Berlin, 2014.

[12] P. Fastermann, *3D-Druck/Rapid Prototyping. Eine Zukunftstechnologie - kompakt erklärt*. Springer, Berlin / Heidelberg, 2012.

[13] R. Lachmayer, R. B. Lippert and T. Fahlbusch, *3D-Druck beleuchtet. Additive Manufacturing auf dem Weg in die Anwendung*. Springer, Berlin / Heidelberg, 2016.

Construction of an Elastic-Motion Phantom for MRI

T. Hinz[1], A. Moeller[2], T. Parbs[2], and A. Mertins[2]

[1] Medizinische Ingenieurwissenschaft, Universität zu Lübeck, torben.hinz@student.uni-luebeck.de
[2] Institute for Signal Processing, Universität zu Lübeck, {moeller, parbs, mertins}@isip.uni-luebeck.de

Abstract

A major problem of present MRI techniques is movement of the patient during the scanning time. It causes artifacts that lower the quality of the resulting images. To compensate emerging motion artifacts, several reconstruction algorithms are under investigation. These algorithms need to be validated by a phantom providing ground truth. To achieve this, an elastic-motion phantom was built based on a pneumatic control system which inflates and deflates an elastic tube with the respiration rate of a human. This tube is placed between other two pre-filled elastic tubes containing detectable fluids. The induced deformation is the reproduction of a patient's respiration during an MR scan. The results lead to an enhanced model of the phantom that is presented. The phantom presented in this work is suitable for the purpose of validating reconstruction algorithms on MR data acquired with gradient echo sequences.

1 Introduction

Magnetic resonance imaging (MRI) is one of the most important non invasive medical imaging modalities of the present time. It provides high soft tissue contrast by the use of magnetization characterization of spins. Furthermore, it is safe by the avoidance of ionization radiation [1].

Despite the advantages of MRI technology, movement of the patient or even a single organ during the scanning time is still a major problem. It yields to different unwanted motion artifacts e.g. ghosting and blurring. To reduce or even delete the appearance of those artifacts, several approaches are already used in clinical routine [2]. E.g. a heart magnetic resonance (MR) scan is performed with an additional electrocardiography sensor [3]. This leads to the knowledge of the hearts respiration behavior during the scanning time. Specialized gating algorithms are able to use this information to create an optimized resulting image by the use of the best acquired data or determining selected acquisition times [4]. In the case of an abdominal MR scan, the patient is often restricted to holding breath over the entire scanning time which is most likely not viable for most patients.

Attempts to avoid this restriction are made by using position tracking or applying navigators [5]. Additional information are recorded to calculate the motion of the objects of interest. A disadvantage of this technique is that the scanning progress has to be interrupted several times to set the navigator. This results in a longer overall progress.

Several attempts are under investigation to compensate the motion without prolonging the overall progress and restrictions for the patient [6], [7]. Blind reconstruction algorithms are using only the standard acquired MR data as the basis for the compensation. Thus, they are a promising method for optimizing the procedure. It results in a shorter scanning progress and removes the restriction of holding breath during that time in contrast to the current techniques. Before the algorithms can be used in a clinical context, it is necessary to validate the quality of those. To achieve this, complete knowledge of the performed motion is essential. This paper introduces an elastic-motion phantom for MRI that is capable of providing reproducible data sets with complete knowledge of the performed motion. This can be used to validate the quality of any given reconstruction algorithm.

2 Material and Methods

This section describes the setup of the elastic-motion phantom. It is constructed with two supply components and one signal emitting component. A compressed air system that provides the air flow for the phantom, a control system that regulates the compressed air system and the phantom geometry itself that is used to generate MR data. A precise overview of the components structure and functionality is given below. The two supply components are combined upon an aluminum panel. This reduces the number of single components, hence it provides a high movability. The entire phantom setup is operated by a single scientist.

2.1 Compressed Air System

To induce any sort of motion, it is necessary to select a source medium that is capable of expanding or displacing other objects. In this scenario, compressed air was used to achieve this goal. It has the advantage of being completely safe as well for the MR scanner itself as for the practicing scientists. Compressed air is also insusceptible to minor system parameter changes. Hence, it is good to handle for

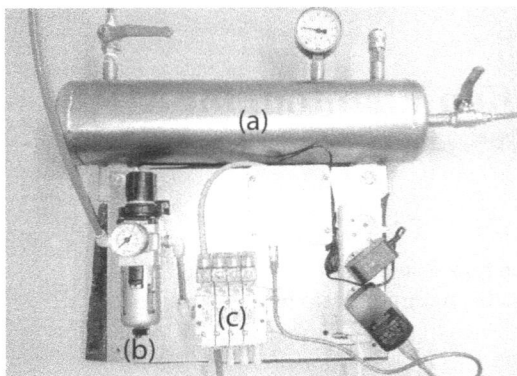

Figure 1: (a) Compressed air is stored in a pressure tank. (b) The tank is connected to a pressure regulator leading to (c) a switchable air valve. It regulates the air flow for the phantom.

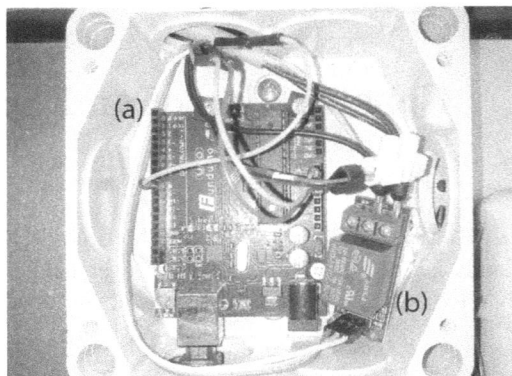

Figure 2: The control system of the phantom is realized with (a) a *Funduino Uno R3* and (b) a related relay to toggle the switchable pressure valve.

an experimental setup that has to be constantly optimized towards its changing aims.

A 4.5 l pressure tank contains the compressed air with up to 8 bar. The tank is refillable by any source of compressed air due to different adapters. It is even possible to refill the pressure tank while the actual data acquisition is in progress. A *SMC AW-30 F02H* pressure regulator is directly connected to the pressure tank. It lowers the incoming air pressure to 0.57 bar. It is adjustable from 0.5 bar up to 8.5 bar. The lowered pressure is forwarded to an electrically switchable 5/2-way solenoid valve *SMC SY-5140-5YOE-Q* that provides the compressed air for the actual motion geometry. Fig. 1 shows the complete setup of the air system.

2.2   Control System

To utilize and control the supplied compressed air system, it is necessary to implement a control system that works precisely and constantly over time. This system also has to be small due to weight and capacity in order to be transportable including the entire setup.

This is accomplished with a *Funduino Uno R3* as the operating microcontroller. The *Funduino* controller is based on the *Arduino* development environment hence it is compatible with most common microcontroller components.

It is wired to the valve that provides the compressed air for the phantom system. The state of that switchable valve is either *open* or *closed* and can be toggled by an electrical input of 0 V and 12 V. In this setup, the high voltage sets the state to *closed* and the low voltage sets the state to *open*. Whereas, the microcontroller itself only delivers a voltage of 5 V, it is necessary to add a relay to the system that is capable of toggling the valve with higher voltages. The implemented setup is shown in Fig. 2. It is inserted in an acrylic box for protection purposes.

To create a behavior that is similar to the respiration phase of a patient, the inspiration phase is preset to 5.5 s and the expiration phase to 4.5 s. These values can be adjusted to receive different respiration phases. This provides the sim-

ulation of different patient types, e.g. children, adults and patients with diseases that influence the respiration.

Additionally, the microcontroller generates a transistor-transistor logic (TTL) signal that is connected to a Bayonet Neill Concelman (BNC) interface of the MR scanner. The signal is used to synchronize the induced movement of the phantom with the MR data of the scanner. With each toggle point, which is the switch from 0 V to 5 V, the signal triggers the next measurement step. The toggle point is set to the beginning of the inspiration phase.

2.3   Phantom Geometry

The control system and the compressed air system are used to supply the actual elastic-motion phantom geometry. The phantom geometry is the only component that is placed inside the MR scanner.

A wooden structure is used to keep two balloons and an expandable tube at fixed spots. It is distinguishable in three planes. The bottom and top plane are used to ensure the stability of the setup. The middle plane holds the two balloons in place by the use of brackets. All planes are circular with an outer diameter of 11 cm and inner diameter of 6.5 cm. They are connected with four wooden rods with the length of 13 cm, resulting in a overall height of 15 cm. The phantom geometry is shown in Fig. 3.

The two balloons are filled with detectable and deformable materials. They are placed parallel to each other with a distance of 6 cm at the brackets. An inflatable butyl tube is placed perpendicular between the balloons. This centered tube is connected to the solenoid valve. With the incoming air flow it yields the deformation of the signal generating balloons. To obtain a steady and slow deformation, an air flow regulator is added to the connection.

This construction ensures the reliability of the phantom to execute the exact same deformation at every step of the experiment. The entire structure is placed within a waterproof container to protect the scanner from any damage caused by the phantom e.g. a balloon filled with a fluid breaks. The butyl tube is fixed at the top of the container.

(a)

(b)

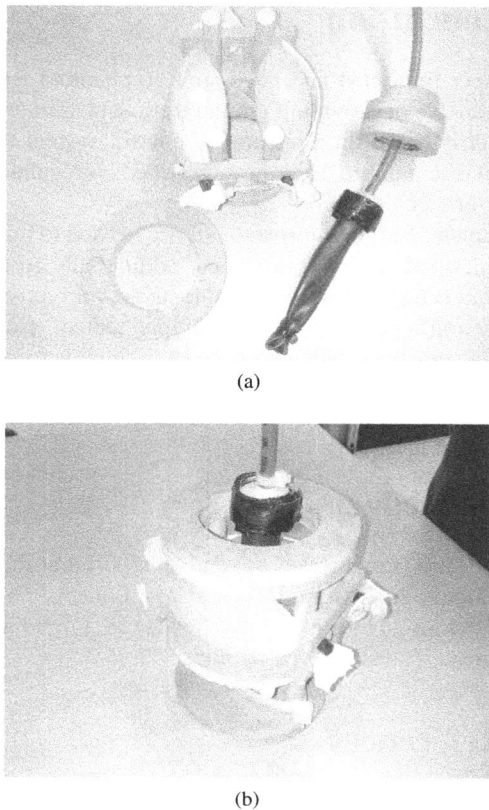

Figure 3: (a) The fundamental parts of the phantom geometry are shown. (b) Two balloons filled with water are placed at opposite sites. An inflatable butyl tube is placed within the center of the balloons to execute elastic motion. The setup is kept together by a wooden structure. The combined setup is shown without the waterproof container.

3 Results and Discussion

Three MR sequences were performed to identify an adequate sequence that results in good MR data for the blind reconstruction algorithms. Additionally, the results lead to a second model for the phantom geometry.

3.1 Results

To validate the functionality and quality of the constructed elastic-motion phantom towards its intended purpose, it was tested with an *Ingenia 3.0T Philips* MR system.

The results of the performed sequences show a high resolution without unintended artifacts. In this case, unintended artifacts are the occurrence of those without any direct relation to the phantom's motion. Each sequence was performed with three different respiration phases. The characteristics of the phases are shown in Tab. 1. Resulting images of an exemplary inspiration phase is shown in Fig. 4. The expiration phase looks the same but in reverse order.

The main characteristics of the performed sequences are displayed in Tab. 2. All three are spoiled gradient echo sequences with a slice thickness (ST) of 6 mm and a flip angle of $10°$. The only difference between these are the repetition

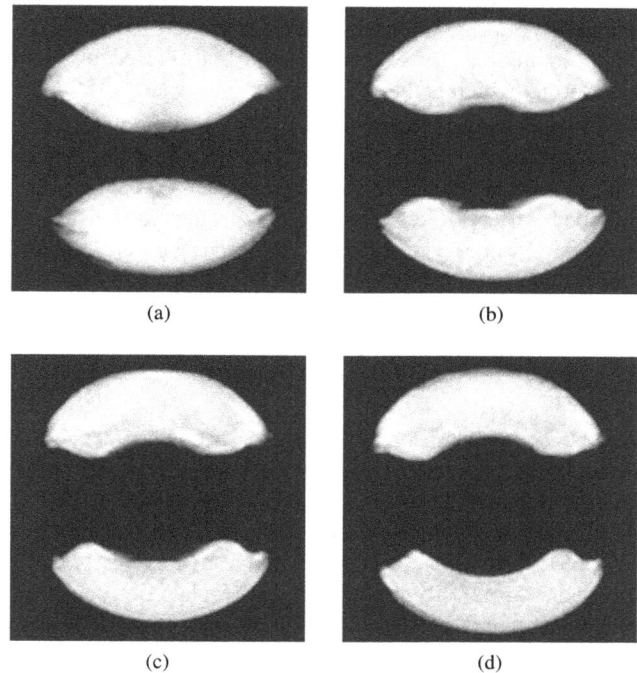

Figure 4: Inspiration phase of the phantom with the phase no. 1. (a) The phantom without any movement at $0\,\text{s}$. (b) The inspiration phase at $1\,\text{s}$. (c) The inspiration phase at $3\,\text{s}$. (d) The inspiration has reached its maximum at $4.5\,\text{s}$.

Table 1: Three different phases that are used to validate a sequence. The inspiration time and expiration time of the phases are listed.

Phase no.	Inspiration	Expiration
1	$5.5\,\text{s}$	$4.5\,\text{s}$
2	$3.5\,\text{s}$	$3\,\text{s}$
3	$7.5\,\text{s}$	$6\,\text{s}$

Table 2: Main characteristics of high resolution sequences with the parameters slice thickness (ST), repetition time (TR) and echo time (TE). The sequences (SEQ) are numbered chronologically.

SEQN	ST	TR	TE
SEQ 1	6.00	3.09	1.47
SEQ 2	6.00	4.13	2.00
SEQ 3	6.00	2.87	1.38

time (TR) and echo time (TE). The slight changes in TR and TE were used to provide different data for the blind reconstruction. SEQ 1 shows no unintended motion artifacts at high contrast and resolution.

3.2 Discussion

The results show that the constructed phantom is capable of providing controlled and reproducible data. It successfully accomplished its intended purpose of providing ground truth for the validation of reconstruction algorithms.

At the current time, it is suitable for gradient echo sequences. Further improvements are needed to make the phantom suitable for every sequence that is used in clinical routine, which is the overall goal to look at.

At this point, the experiment was only executed with water as the detectable fluid inside the balloons. Other fluids with a higher viscosity, e.g. ultrasound gel, will cause less unintended artifacts due to minor inner turbulences during the deformation phase. Hardening materials, e.g. an agar solution which is a gold standard for creating phantoms [8], were not viable due to their susceptibility to breaking at higher grades of deformation. Thus, fluids with a high viscosity are promising to increase the number of suitable sequences.

Another possible reason for the occurrence of artifacts is the tissue interface between the signal emitting material water and the non signal emitting material compressed air. High differences in magnetic susceptibility of connected materials can lead to signal voids and image distortions [4]. To lower the strong differences, the entire phantom construct can be submerged within a signal emitting fluid. The susceptibility of the fluid has to be different to the balloons susceptibility. Thereby, the straight interface can be eliminated.

The elastic tube in the center still causes a straight interface in that scenario. This can be eliminated by adding another elastic tube filled with a layer of emitting material. The original tube is placed inside the emitting tube. Thus, the interface of the object that causes the deformation and the object of interest is adjusted. In the meaning of reconstruction, this can be a problem because the interface is just shifted to the two tubes but still is in the field of view. To entirely eliminate that problem, it is necessary to either set the compressed air tube outside the field of view or to fill that tube completely with an emitting material that is capable of expanding and also contracting it.

3.3 Outlook

In awareness of the aspects given above, an improved phantom is under investigation and will be tested within the next time. The idea is to set the elastic tube outside the field of view, hence avoiding the straight interface. Additionally, the contributing objects will be submerged into an emitting fluid. In terms of consistency, the carrying structure will be entirely built with acrylic polymer. This guarantees a high stability against damage caused by the fluid over time, e.g. porosity or mold.

The waterproof container will be replaced by an acrylic box that is permanently connected to the phantom geometry. The elastic motion deformation is caused by a connection between the tube and the balloon. In this setup, the number of emitting balloons is reduced to one. While the tube expands, the center of the balloon will be dragged out of its position. The connection can be either realized with a strand or another emitting balloon, leading to more usable signal emitting objects to work with in the reconstruction.

4   Conclusion

This paper presented an elastic-motion phantom for MRI that is capable of providing ground truth. The implemented construct is based on a pneumatic control system that induced elastic motion. This was performed by regulating the air flow of an elastic tube.

Nevertheless, further improvements are needed to make the phantom suitable for more setups, additionally to the investigated setup. These improvements are currently under investigation and are about to be implemented. However, the presented elastic-motion phantom is suitable to validate blind reconstruction algorithms on data that is acquired with gradient echo sequences.

Acknowledgement

The work has been carried out and supervised at the Institute of Signal Processing, Universität zu Lübeck.
Special thanks go to Alex Frydrychowicz for providing time and his expertise at the MR scanner.

5   References

[1] O. Dössel and T. M. Buzug, *Medizinische Bildgebung.* De Gruyter, Berlin, 2014.

[2] Y. Hirokawa et al., *MRI Artifact Reduction and Quality Improvement in the Upper Abdomen with PROPELLER and Prospective Acquisition Correction (PACE) Technique.* American Journal of Roentgengenology, vol. 191, pp. 1154-1158, 2008.

[3] G. Liney, *MRI in Clinical Practice.* Springer, London, 2006.

[4] D. Weishaupt, V. D. Köchli and B. Marincek, *How does MRI work? An introduction to the physics and function of magnetic resonance imaging.* Springer, Berlin/Heidelberg, 2006.

[5] T. F. Uckermann, *Einsatz eines Navigator Echo Biofeedbacksystems zur Verbesserung der Navigatoreffektivität bei der Koronarangiographie im Magnetresonanztomographen.* Doctoral dissertation, Universität Ulm, 2012.

[6] A. Möller, M. Maaß and A. Mertins, *Blind Sparse Motion MRI with Linear Subpixel Interpolation.* Springer, Berlin/Heidelberg, 2015.

[7] D. Atkinson et al., *Automatic compensation of motion artifacts in MRI.* Magnetic Resonance in Medicine, vol. 41, pp. 163–170, 1999.

[8] A. Hellerbach, V. Schuster, A. Jansen and J. Sommer, *MRI Phantoms - Are There Alternatives to Agar?.* PLoS One. vol. 8, no. 8, pp. 1–8, 2013.

A design to adapt a bleaching laser to a research microscope – An easy hardware extension to perform FRAP/FRED/photoswitching –

T. Kutscher[1], R. Duden[2] and I. Majoul[3]

[1] Medizinische Ingenieurwissenschaft, Universität zu Lübeck, tonio.kutscher@student.uni-luebeck.de
[2] Institut für Biologie, Universität zu Lübeck, duden@uni-luebeck.de
[3] Institut für Biologie, Universität zu Lübeck, irina.majoul@bio.uni-luebeck.de

Abstract

The aim of this project was to design a supplement laser-bleaching slider that can be fitted into an existing Nikon Ti-E research fluorescence microscope utilizing its empty side port. This will be an original development designed to enable additional types of experiments like photo-activation/inactivation or photobleaching of different fluorescent proteins. Photobleaching of fluorophores is an important part of techniques like FRET or FRAP [1] . Further, laser induced color switching of fluorescent proteins makes it possible to follow the fate of sub-populations of fluorescent proteins. The hardware adapter design had to include the whole equipment to initialize the laserbeam with fitting parameters like energy exposure and laser spot radius. We developed a construction which allows adjusting the focal diameter of the illuminatinated spot on the biological sample as well as controlling the laser energy emitted by adjusting the exposure times as well as the laser intensity with neutral grey filters as required for the experiment.

1   Introduction

Current advances in biomedical science are based in large part on data based on the observation of fluorescent compounds in living cells. Some organic molecules are able to absorb light of a specific wavelength and then emit radiation with lower energy at a longer wavelength, a phenomen that is called fluorescence [2].

Förster resonance energy transfer (FRET) is a mechanism involving the energy transfer between two chromophores [1] in the case of FRET, the so-called donor chromophore transfers energy to the acceptor chromophore, by way of a nonradiative dipole–dipole coupling [3] FRET is extremely sensitive to small changes of distance between donor and acceptor since the efficiency of this transfer is inversely proportional to the sixth power of the distance [4]. FRET-Measurements can thus be used to determine the distance of two fluorophores from each other. These kinds of measurement are broadly used for research in the fields of medical biology and biophysics [5].

If the radius of interaction is smaller than the wavelength of the emitted light FRET is comparable to near-field communication [6]. In the near-field area, the donor emits a photon which is absorbed by the acceptor instantly. Since FRET is known as a radiationless mechanism, the photons which are transferred between the chromophores are not detectable [6].

Another way to display the FRET effects is by the photobleaching rates of the donor in the presence or absence of an acceptor [7] This approach can be implemented on most modern fluorescence microscopes.

For such an experiment, the sample is exposed to a narrow band of excitation light such that only the acceptor fluorophore gets excited. This procedure is performed on samples with and without the acceptor fluorophore and monitors only the donor fluorescence without the fluorescence of the acceptor by using a bandpass filter. The time scale for the decay is seconds to minutes. The photobleaching decay depends on the presence or absence of the acceptor. Due to the fact that most scientists need to analyze dynamic processes like aggregation or complex formation of fluorescently labeled molecules, it is necessary to have an option to switch or destroy the fluorophores in living cells without inflicting harm to the cell. Dynamic bleaching is a complex technical procedure prone to drift-induced artifacts.

One of the main methods in this context is Fluorescence recovery after photobleaching (FRAP). It is a technique to measure the kinetics of diffusion through cell membranes. FRAP is able to measure the lateral diffusion of molecules within a narrow area in fluorescently labeled cells. It is often used in biological studies of cell membrane diffusion or protein binding.

The laboratory of Drs. Irina Majoul and Rainer Duden uses a Nikon inverted Ti-E fluorescence microscope[9]. The microscope is set up with a Yokogawa Spinning disk unit and to enable single-molecule detection sensitivity, an Andor iXON EM-CCD camera is used, to provide an essential tool for examining a wide variety of live cell dynamics, biological structures and processes.

The idea was to connect a 10mw Laser to an unused side-port of this microscope to enable photobleaching and photoswitching with this setup. This would be a good extension to the possible research. It will be used for pilot experiments, FRAP, photobleaching and photoactivation techniques - in analyses of the formation and dynamics of Connexin-43 and Drebrin –containing protein complexes underneath the plasma membrane of living cells.

2 Material and Methods

Initially starting from an older design which had been originally custom-built for a Zeiss 200M inverted microscope, we needed to develop a functional progression and adaptation for the Nikon microscope mentioned above.

Further, the illumination spot on the microscope sample slide needed to be variable to adapt the illuminated area to the analyzed cell parts. Depending on the analyzed cell parts the illuminated spot size thus should correspond to an area in cells from 5μm to 30μm. The regulation of the energy transferred onto the sample will be achieved through a combination of adjusting shutter timing and variable insertion of different neutral density filters.

2.1 Construction

From the begining on it was our aim to find a solution which is preferably easy to use. We began to assemble the components made built into a casing made of aluminum square pipe. The casing was based on 2 mm thick aluminum foil. Three sides of the casing are fixed and the fourth is attached with screws. This achieves that the lenses cannot get exposed to dust and grime. It also ensures that no reflected light from the laser endangers the eyes of persons operating the equipment.

During the research we came to the conclusion that one of the main difficulties would be the precise adjustment of the shutter time. The easiest way would have been a spinning disk with a small hole. However, due to the fact that this solution would be quite expensive, we decided to use instead the shutter of an old single-lens reflex camera (Minolta 7000 AF). Therefore, we disassembled the camera and took out the parts required and reassembled them in our model.

The lens system had to be adjustable to get a fitting spot size on the biological sample. The lenses and the prism had to be placed with an accuracy of 0.1mm to get a proper focus.

3 Results and Discussion

The resulting construction appears as a 30cm long square tube made of aluminum painted with black paint to prevent any kind of back reflection of laser light to its inside. The mounting allows the application of different lasers to choose the appropriate wavelength required for the intended experiment. For bleaching of the acceptor of the CFP/YFP

fluorophore FRET pair, a 532 nm laser is required. Other possibilities would be the use of a 560 nm laser for a FRET pair based on the donor Cyanine Dyes3(Cy3) or with the donor Alexa Fluor 647 or Cyanine Dyes 5 (Cy5) and a 647 nm laser. The laser light intensity hitting the sample can be adjusted through the use of neutral gray density filters. To achieve a logarithmic extinction, we built in two filter slots. When it is necessary to get just 1% of the irradiated laser-energy onto the sample a combination of two 90% grey filters is an easy way to meet this requirement. Through this, it is possible to combine several filters to achive nearly every decrease of laser intensity desired. To ensure that the foldable mirror (Fig. 1(3)) which is covered with black paint does not heat up due to the incoming laser-energy it is necessary to place one slot in the upper appliance, in front of the mirror.

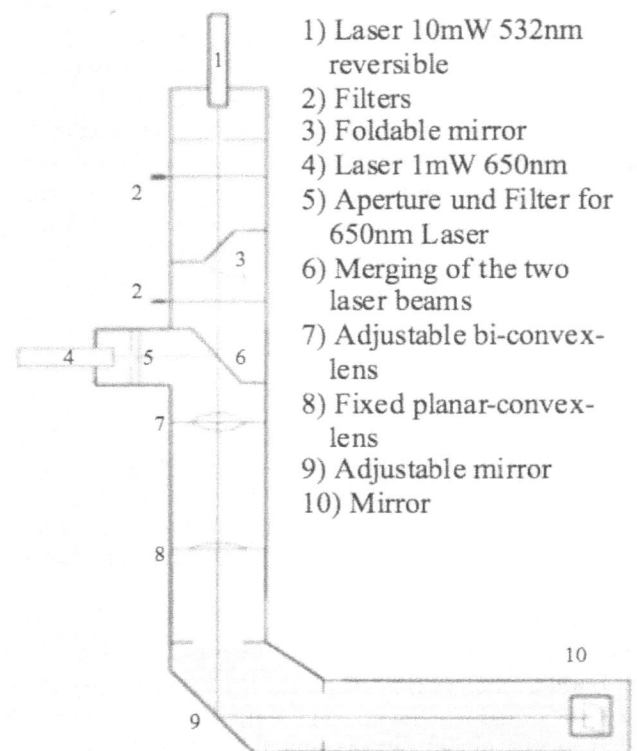

1) Laser 10mW 532nm reversible
2) Filters
3) Foldable mirror
4) Laser 1mW 650nm
5) Aperture und Filter for 650nm Laser
6) Merging of the two laser beams
7) Adjustable bi-convex-lens
8) Fixed planar-convex-lens
9) Adjustable mirror
10) Mirror

Figure 1: Sketch of the focal tuning mechanism.

Due to the usability, the other slot had to be placed below the mirror. Otherwise, it would have been very difficult to change one of them. To ensure an easy and fast replacement of the filters, those were glued onto an aluminum frame which fits the slot in the casing. The construction of the foldable mirror initially presented various difficulties. After I removed the retainer of the Minolta 7000 AF camera, I masked it off with black foil, to prevent back-reflection of incoming laserlight.

The foldable mirror is not as originally used to lead the image into the viewfinder but to seal the laser ray entrance. Because we ascertained that taking a precise aim would be rather complicated, we came up with the idea to build a default beam. First, we used the foldable mirror to lead the beam parallel, past to the shutter (Fig. 2).

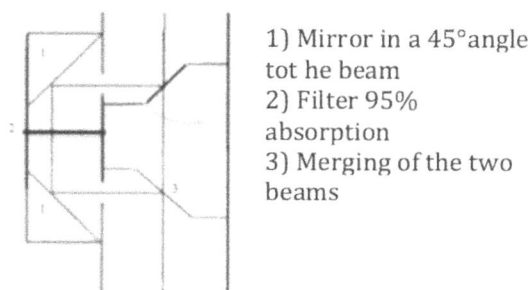

1) Mirror in a 45°angle tot he beam
2) Filter 95% absorption
3) Merging of the two beams

Figure 2: Sketch of a possible design for the target beam

Since cells react to the slightest light intensities nevertheless, when the light reaches the respective excitation wavelength, we decided to use a second laser pointer with 1mW power and a wavelength of 650 nm (Fig. 1). In case the wavelength of 650nm is too close to the excitation wavelength it is also possible to change this laser. With this implemented default beam it is essentially easier to illuminate the intended part of a cell. Since the red targeting laser proceeds the lens system for focusing as well, the emerging red spot has just the same diameter as the spot of the bleaching laser.

For a precision tuning of the energy, which is supposed to shine in, the mirror's controller is utilized. The mirror of the Minolta 7000 AF can realize exposure times as short as of 1/2000 sec. Thus it is possible to regulate the energy exposure of laser onto the cell sample. For working with cells only a very small area neededs to be illuminated. Because of the construction of side port's at the Nikon- microscope, the precise positioning of the focus to the object slide has to be adjustable in the adapter.

Consequently, the best way to achieve a focused spot of the laser beam appears to focus the laser first and to convert it with a plan-convex-lens to a parallel light bundle. This bundle needs to have a smaller diameter than the original laser beam, but simultaneously the same energy. The knowledge can be illustrated by formula (1).

$$E(Target) = \frac{E(Laser)}{(FilterI * FilterII)} * Belichtung[t] \quad (1)$$

As shown it is necessary to be able to adjust the shutter time therefore we had to use the foldable mirror. If we worked only with exchangable filters and a fixed shutter time it would have caused lots of drawbacks for the usage. Because of the fact that the light spot diameter has to be adjustable in the diameter size range of 5 μm to 30 μm apparent spot size on the biological sample, the lens system needed to be adjustable too. To modulate the distance between the biconvex lens and the plan-convex-lens we used a threaded rod (Fig.3). The lens is attached to the rod by

an aluminum frame. Diagonally towards this, another rod is located, though without a thread. Its primary function is to serve as the guideway for the lens frame.

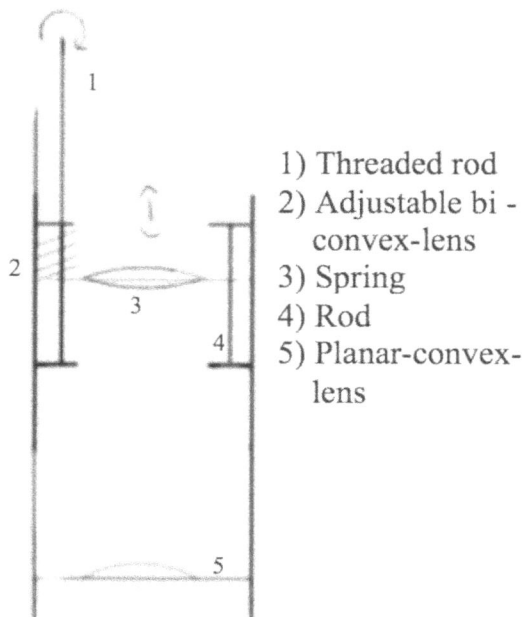

1) Threaded rod
2) Adjustable bi - convex-lens
3) Spring
4) Rod
5) Planar-convex- lens

Figure 3: Sketch of the focal tuning mechanism.

When the rod is twisted, the lens moves upwards or downwards. To prevent wiggling of the lens a spring was placed at the top of the rod (Fig3(2)). The emerging beam is redirected by the adjustable mirror in the corner piece (Fig1(9)) to the slide-in. This mirror is the final possibility to align the spot to the center of the field of view in the microscope. The last mirror (Fig1(10)) redirects the beam into the microscope twards the biological sample.

4   Conclusion

The main challenge in this project was to find a way to focus the laser precisely to the biological sample. The usage of a laser with a higher beam quality would make it possible to improve the definition of the focal spot.

A useful addition would be a completely automatized regulation of the target energy. It would allow a quick and precise adjustment of the components. For this, custom-built software and hardware components would need to control the filter settings and the flipping shutter. The only thing that would need to be be changed for this are the filter slots. A combination of two filter disks would be easier to handle for an automized system. The hardware what would be required is actually conceptually quite simple. For changing the filters two small electric engines would be required. An ATMega should fulfill the needs of a computing unit.

The original idea of the project to construct an adapter for a bleaching laser designed to be as a versatile applicable device for use with the Nikon Ti-E research fluorescence microscope. It opens lots of possibilities to the microscope user like FRAP or laser light induced color switching of flourophores or ON/OFF switching of fluorescent proteins (like mEOS or Dendra), making it possible to follow the fates of individual sub-populations of fluorescent proteins expressed in living cells after this experimental manipulation.

Acknowledgement

The work has been carried out and supervised at the Institute of Biology, Universität zu Lübeck.

5   References

[1] P.-C. Cheng, *The Contrast Formation in Optical Microscopy*, In: Pawley, J. Handbook Of Biological Confocal Microscopy (3rd ed.). Springer, New York, pp. 162–206, 2006.

[2] I. Majoul, Y. Jia, R. Duden, *Practical Fluorescence Resonance Energy Transfer (FRET) or molecular nanobioscopy of living cells.* In: Handbook of Biological Confocal Microscopy, pp: 788-808, Springer US, 2006.

[3] V. Helms, *Fluorescence Resonance Energy Transfer* In: Principles of Computational Cell Biology, Wiley-VCH, Weinheim, pp. 202, 2008.

[4] D. Harris, *Applications of Spectrophotometry* In: Quantitative Chemical Analysis (8th ed.), W. H. Freeman and Co., New York, pp. 419–444, 2010.

[5] H. Plattner, J. Hentschel, *Zellbiologie.* Georg Thieme Verlag, 2006.

[6] D. Andrews, *A unified theory of radiative and radiationless molecular energy transfer.* Chemical Physics, 135(2), pp. 195–201, 1989.

[7] R. Clegg, *Förster resonance energy transfer—FRET: what is it, why do it, and how it's done.* In: Gadella, Theodorus W. J. FRET and FLIM Techniques. Laboratory Techniques in Biochemistry and Molecular Biology, Volume 33, Elsevier, pp. 1–57, 2009.

[8] J. Szöllősi, D. Alexander, *The Application of Fluorescence Resonance Energy Transfer to the Investigation of Phosphatases.* In: Klumpp, S; Krieglstein, J. Protein Phosphatases. Methods in Enzymology, Volume 366. Elsevier, Amsterdam, pp. 203–24, 2007.

[9] Nikon *Handbook Nikon Inverted Microscope Eclipse Ti-e.* Nikon.

Development of a variable gradiometer coil
to determine the thermal properties of magnetic nanoparticles

M. Sasse[1], A.Behrends[2] and T. M. Buzug[2]

[1] Medizinische Ingenieurwissenschaft, University of Luebeck, miriam.sasse@student.uni-luebeck.de

[2] Institute of Medical Engineering, University of Luebeck, {behrends, buzug}@imt.uni-luebeck.de

Abstract

A useful application of magnetic nanoparticle is the magnetic hyperthermia for the cancer treatment. Thus, it is necessary to know the thermal properties and their dependence. To determine the thermal properties of magnetic nanoparticles, the non-linear magnetization response of the nanoparticles is measured with a spectrometer which is based on the principle of Magnetic Particle Imaging. Previous constructions use a band-stop filter to suppress signals at excitation [4]. The aim is to design a gradiometer coil which can attenuate the excitation frequency. To achieve this, the components of the coil are designed in *Solidworks* in accordance with theoretical results obtained in *Matlab*. The result is a gradiometer coil with a variable adjustment to attenuate the excitation signal. However, a complete cancellation was impossible, because the electrical characteristics of the coils will never be identical to those obtained theoretically. This should be developed in furthers studies to optimise the coil setup.

1 Introduction

Magnetic nanoparticles are widely used in diagnostic imaging and therapy due to their ability to be magnetised [1]. One example is magnetic hyperthermia, a method for cancer treatment. Here, the nanoparticles can be excited by an external magnetic field. Therefore, they act as a heat source to destroy surrounding tumor cells [3]. In order to allow a treatment, as precise as possible, it is necessary to know the thermal properties and their dependence. In addition to the size, shape and viscosity of the environment, the behaviour of the nanoparticles is largely influenced by the parameters of the magnetic field especially the frequency [3].

Therefore a spectrometer, that covers a large frequency range in the excitation field is necessary in order to investigate the properties by means of the spectrum of the nanoparticles. With a Magnetic Particle Spectrometer (MPS), which can be interpreted as a zero dimensional Magnetic Particle Imaging scanner (MPI) [1], the thermal properties should be determined more precisely. In the case of MPS, the magnetic nanoparticles are excited by a magnetic field. The magnetization response of the nanoparticles is recorded by an induction coil. Due to the non-linear magnetization of the nanoparticles, a spectrum is formed which, in addition to the excitation frequency, contains integer multiples of this frequency (harmonics) [1]. On the basis of the higher harmonics, nanoparticles properties can be derived. In contrast to the excitation frequency the particle signal is much smaller.

In this paper a set-up is presented which uses a gradiometer coil as a receive coil in order to suppress the excitation frequency instead of a regular analog band-stop filter [1].

The combination of the filtering method and a gradiometer coil achieves a high attenuation for suppressing the excitation signal for MPI [5]. In this work, the gradiometer coil should not attenuate the excitation signal in one only frequency but also be adjustable to the respective frequency.

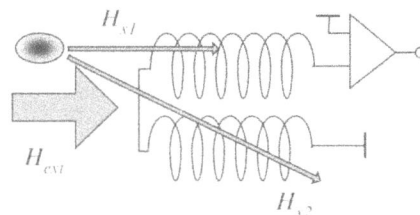

Figure 1: The gradiometer coil of first order consists of two oppositely wound coils connected to each other at a small distance. If both coils are in a strong homogeneous magnetic field H_{ext}, the source of which is at a greater distance, the signal is canceled out due to the opposite windings and becomes zero. [2]

The principle of the gradiometer coil is that two connected oppositely, wound coils are exposed to a homogeneous magnetic field H_{ext} as can be seen in Fig. 1. A smaller source produces a magnetic field which is closer to one coil H_{x1} than to the other coil H_{x2}. This field gradient can then be measured. The stronger field induces a voltage that is canceled due to the different polarity between the coils, while the magnetic field of the smaller source induces voltages of different magnitudes in the coil sensors. A gradiometer coil is called gradiometer of a first order when it consists of two connected oppositely, wound coil. A gradiometer coil is called gradiometer of a second order when it consists of two gradiometers of first order [2]. By changing the frequency of the excitation coil, the magnetic flux and thus the induced voltage changes in the receive coil. The aim is to design a variable gradiometer of a second order, which can attenuated the excitation signal in an independent manner during the measurement.

2   Material and Methods

The developed gradiometer coil consists of a center coil and two oppositely wound outer coils. The center coil and an outer coil are on one part of the gradiometer coil. The other outer coil is wound on a second part. The second part of the gradiometer coil should move upwards or downwards. As a result, the excitation signal should be attenuated efficiently in a frequency-independent manner.

The development of the adjustment for the coil is described in chapter 2.1. From the concept to the final model, several versions and printouts have been developed in order to optimise the design step by step. The models were designed with the computer-aided design software *Solidworks2015* (EDU 2015-2016) and the components were made by 3D printing technology. The printing material is a photopolymer. In addition, a program was written in *Matlab*, which is described in chapter 2.2. The program calculates the theoretical attenuation by using various parameters. On this basis the coil was further developed. To test the set-up it is used an excitation coil consists of a copper tube. The internal diameter of the excitation coil is 30 mm. It has 10 turns whose pipe diameter is 6 mm. The excitation frequency is set to 557 kHz and the magnetic field has 30 mT. The induced voltage of the gradiometer coil is determined on an oscilloscope.

2.1   Development of threads for adjustment

Figure 2: The design consists of a threaded suspension a), a bracket b) and a wheel thread c). The thread suspension is intended to move upwards or downwards via the thread wheel and thus also to be able to move a coil.

The final concept is a further development from the existing design. It is shown in Fig. 2. The design consists of a threaded suspension a), a threaded wheel c) and a bracket b). The threaded suspension moves upwards or downwards by turning the threaded wheel, but is prevented from rotating by the cross braces of the bracket.

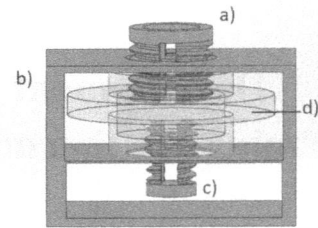

Figure 3:   The first construction for adjustment (Fig. 2) was extended by an opposite thread c). Due to the rotation of the wheel d) in one direction the threads a) and c) run toward each other and in the other direction, the threads turn apart.

The coil, which is on the thread suspension, can thus be calibrated via the rotational adjustment. When the lower part of the coil is moved downwards, the distance to the center coil increases and an additional wire is required. As the lower coil moves upward toward the upper coil, there is an excess of wire which should remain internally tensioned. In addition a second thread should keep the wire tensioned inside. The previous design of the thread has been extended by an opposite winding threadd, which can be seen in Fig. 3 c). It pulls the wire down during calibration and therefore is tense. Fig. 3 shows the first model that was printed. The focus laid on the thread construction and detection of problems therefore the coil is absent.

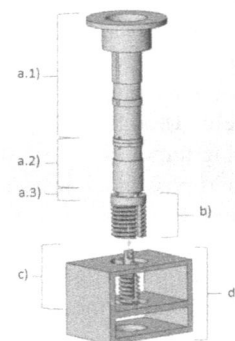

Figure 4: The first construction of the variable gradiometer coil, which was used for the first measurement. It consists of the upper part a.1) for the upper and middle coil, the bottom part a.2), a large thread b) for the calibration of the part a.2), the small thread c) with guide for the wire and a holder d).

In the reworked model the bracket and the threads have been revised, as can be seen in Fig. 4. The large thread suspension b) together with the lower coil a.2) component has now a bayonet mount a.3) to connect them together. The smaller thread has a guide for the wire, which is formed the coil. The model was used for the first measurement to check whether there is any change in the voltage at all. For this purpose, the coil components were wrapped with a wire having a diameter of 0.7 mm. In the practical design, the oppositely wound outer coils have 12 and 13 windings. The middle coil has 28 windings. As a result of a missing fixation the coil does not remain in the center of the excitation coil. This test did not give any accurate measurement results and served only to try out the mechanism. The final construction for the variable gradiometer coil is shown in Fig. 5. The coil can now be placed in the excitation coil and the holder is attached subsequently. The holder a) is located on a plexiglass frame h) and is connected through a bayonet mount with the upper coil component b). Moreover the thread holder g) is changed for the plexiglass frame. In the head of the small thread f) is now a recess for a small plastic wheel (d = 7.5 mm).

2.2   Development of the coil

The gradiometer coil can be divided into two parts, which can be seen in Fig. 4. On the first coil part a.1) are the middle and the oppositely wound upper part of the coil. The lower half a.2) of the oppositely wound coil is located on the second adjustable coil former. In order to determine the optimum number of turns, the distances between the coils and the diameter of the wire, a Matlab program was written.

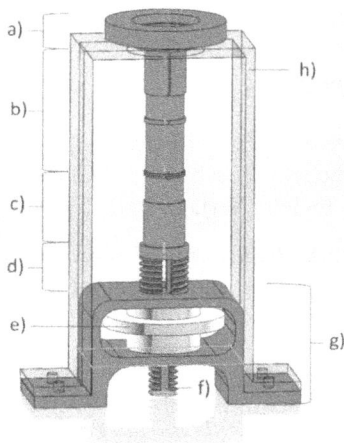

Figure 5: The final model of the gradiometer coil, which was printed and used for the measurement. It has a reworked holder g) and the small thread f) has been modified. The coil is wound outside the excitation coil and assembled. The holder a) is located on a plexiglass frame h) and is connected through a bayonet mount with the upper coil component b). The lower coil component c) is connected with the thread d). Due to the rotation of the wheel e) in one direction the threads run toward each other.

By means of the program, the maximum induced voltage of the coil and the resulting attenuation were calculated. The induced voltage was calculated based on the given field profile of the excitation coil. The magnetic fi eld from the excitation coil has 30 mT, but it is not homogeneous. As a result, a voltage is induced in the two oppositely wound outer coils which is different from the voltage induced in the center. The maximum induced voltage is calculated according to the formula

$$U_{\mathrm{max}} = n\, A\, t B_0\, 2\, \pi\, f \qquad (1)$$

with n windings, the area of cross-section of the coil A, the magnetic flux density of the excitation field B_0, and the frequency f [6]. For the value B_0, the average value is calculated over the width of cross-section of the magnetic field in which the gradiometer coil is located. The frequency f is 557 kHz.

For each partial coil, the maximum induced voltage is calculated according to its position in the magnetic field. The center coil is in the middle of the magnetic field. This is followed by the calculation of the voltage difference $U_{\mathrm{diff}} = U_c - (U_b + U_t)$ between the center coil U_c and the sum of the outer coils $(U_b + U_t)$. From this, the attenuation of the entire coil can now be calculated [7]:

$$v_u = 20\, log \frac{U_{\mathrm{diff}}}{U_{\mathrm{c}}} \qquad (2)$$

The voltage and attenuation was calculated for various combinations of winding numbers and wire diameter. On the one hand the winding number and the wire diameter determine the length of the coil and thus the position in the magnetic field. On the other hand the voltage U_{max} is depends on the windings and the area of cross-section of the coil (1).

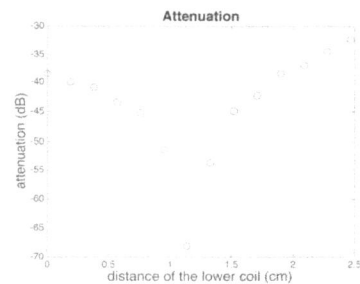

Figure 6: The attenuation (dB) is calculated as a function of the distance (cm) between the lower coil and the center coil. The theoretic coil has a wire diameter of 0.4 mm, 46 windings of the center coil and 23.5 windings of the outer coils. The attenuation increases from -39 dB with increasing distance of the lower coil to -68 dB. After the maximum at 1.1 cm the attenuation decreases again strongly.

The diagram in Fig. 6 shows the attenuation in decibel (dB) as a function of the distance (cm) between the lower coil and the center coil. The real minimal distance amounts to 2 mm and is zero in the diagram. If the distance increases, the attenuation increases to a certain point. In this case at a distance of approximately 1.1 cm, the attenuation has a maximum of -68 dB. As the distance increases further, the attenuation drops again. The curve also shows that around the maximum the attenuation reacts very sensitively to the change of the distance.

A clear tendency of the attenuation is not apparent as a function of the number of windings and the wire diameter. Therefore, no clear solution can be found. In the final model there are used a winding number of 46 is used for the center coil and 23.5 is used for the outer coils. The copper wire has a diameter of 0.4 mm.

3 Results and Discussion

The first model of the coil (Fig.4) was used to test whether the mechanism of rotation adjustment works and the change of the distance has an effect on the induced voltage. The measurement worked, but was very inaccurate due to the missing fixation. However, an effect of the voltage change of several volts could be perceived.

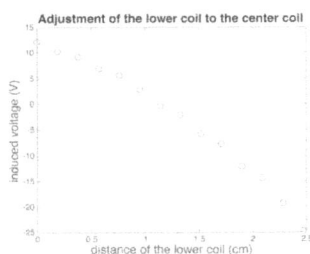

Figure 7: The voltage (dB) is calculated as a function of the distance (cm) between the lower coil and the center coil. The parameter of the coil are a wire diameter of 0.4 mm, 46 windings of center coil and 23.5 windings of outer coils. At the minimum distance there is a induced voltage of 12 volt. With increasing distance the voltage decreases approximately linearly. The voltage is zero at the distance of 1.1 cm.

The final version with the bracket was less susceptible to interference, because the coil was held more stable in the center of the excitation coil during the adjustment. The minimum voltage that was first measured was about 15 volt. In order to further decrease the minimum voltage, the wire was guided internally close the coil. By the reduction of the measuring circuit the maximum measured voltage was then only two volt. A further attenuation could not be achieved at the time. However, it is known from the theoretical results (Fig. 6) that the damping is very sensitive to the distance. At the point, where the attenuation has its maximum, the gradiometer coil reaches an induced voltage of zero volt (Fig. 7). Nevertheless it can not be ascertained at which actual distance between the two coils this exactly occurs, and therefore not be compared with the theoretical value. But the expected voltage for the center coil without the improved gradiometer approach, amounts to -1022 volt, which was precalculated in Matlab. In order to get exactly this result corresponding to the zero volt, the electrical characteristics like the inductance and resistance of the coils have to be identical to those obtained theoretically. So, the achieved attenuation to 2 volt is therefore a good result.

4 Conclusion

To determine the thermal properties of magnetic nanoparticles a spectrometer is needed, which covers a large fre-

quency range in the excitation field. The aim is a receive coil for MPS, which can attenuate the excitation signal in a frequency-independent manner. The focus lays on the development of a mechanism for a variable gradiometer coil. The measurements should show whether the excitation signal is attenuated. In theory, the attenuation increases as the distance between the lower coil and the middle coil increases. At a certain distance, the maximum of the attenuation is reached. In Fig. 6, the distance is approximately one centimetre. When the distance increases the attenuation decreases again. In the experiments, an effect of attenuation was observed, but not as strong as in theory. One weakness is the induced voltage by the measuring circuit and by the loop of the wire in the interior of the coil. Therefore a perfect cancellation is impossible. Accordingly, solutions still have to be developed to compensate this voltage. However, the practical implementation could be shown that the voltage be attenuated and varied by the rotational adjustment.

Acknowledgement

The work has been carried out at Institute of Medical Engineering, Universität zu Lübeck and supervised by André Behrends.

5 References

[1] S. Biederer *Magnet-Partikel-Spektrometer*,Springer Fachmedien Wiesbaden GmbH, Wiesbaden, Vieweg+Teubner Verlag , 2012

[2] S. Tumanski *Induction coil sensors - a review*, Measurement Science and Technology, Vol. 18, No. 3 , 2007.

[3] E. Garaio, J. M. Collantes, F. Plazaola, J. A. Garcia and I. Castellanos-Rubio *A multifrequency eletromagnetic applicator with an integrated AC magnetometer for magnetic hyperthermia experiments*. J Measurement Science and Technology, Vol. 25, No. 11 , 2014

[4] S. Biederer, T. Sattel, T. Knopp, K. Lüdtke-Buzug, B. Gleich, J. Weizenecker, J. Borgert, T. M. Buzug *A Spectrometer for Magnetic Particle Imaging*, Springer Berlin Heidelberg , 4th European Conference of the International Federation for Medical and Biological Engineering, 2009

[5] M. Graeser, T. Knopp, M. Grüttner, T. F. Sattel, T. M. Buzug *Analog receive signal processing for magnetic particle imaging*. Medical Physics, Vol. 40, No. 4, April 2013

[6] P. A. Tipler, G. Mosca *Physik für Wissenschaftler und Ingenieure*, Heidelberg: Springer Spektrum, 2014

[7] http://www.elektronik-kompendium. de/sites/grd/0303311.htm, last accessed on 2017-01-05

Automated cavity-length optimization of a passively Q-switched microchip-laser by a piezo actuator control

K. Schmidtke[1], R. von Elm[2], and F. Reinholz[3]

[1] Medizinische Ingenieurwissenschaft, Universität zu Lübeck, katja.schmidtke@student.uni-luebeck.de
[2] Coherent LaserSystems GmbH Lübeck, Ruediger.vonElm@coherent.com
[3] Institute of Biomedical Optics, Universität zu Lübeck, reinholz@bmo.uni-luebeck.de

Abstract

Passively Q-switched microchip-lasers offer the possibility to generate short pulses less than 100 ps with high repetition rates (up to 2 MHz). The microchip-laser in this project is based on a solid-state laser with Neodymium doped yttrium vanadate (Nd:YVO$_4$) as the active laser medium, which emits at 1064 nm. This is very interesting for ophthalmologic applications.
Because of the very small cavity-length of 140 μm, the laser is very instable and sensitive to outside influences. Smallest changes of the cavity-length cause changes of the laser parameters, which is very critical. The goal of this project was to design and implement an automated stabilizing circuit with fast response time. Based on the measurement of the Fabry-Pérot-Etalon, a circuit board was designed and could be implemented. The first measurements show encouraging results and the completion of the whole stabilizing device can be implemented in near future.

1 Introduction

Especially in medical use it is very important to ensure a reliable laser with exact properties (wavelength, pulse length and so on). The laser emits at 1064 nm, which is a very interesting wavelength i.a. for ophthalmologic applications. In Q-switch mode, it is used especially in the front region of the eye (iris, lens capsule) [2]. Because the laser is designed as a microchip-laser, the active medium Nd:YVO$_4$ is formed as a thin wafer of 120 μm. This is convenient, because the medium has a low thermal conductivity [2]. So the active medium is bonded on YVO$_4$, which is very important for the heat removal. Additionally it is making the thin wafer more stable.
Fig. 1 shows the setup of the cavity. The active medium Nd:YVO$_4$ is bonded on YVO$_4$. Between the bonding, there is a mirror coating with a reflectivity of about 80%, which builds the outcoupling mirror. Between the medium and the Saturable Absorber Mirror (SAM) is free space of about 10 μm. The piezo actuator and the SAM are glued together. The SAM builds the endmirror. The whole cavity-length, including the SAM (only the first few μm are involved in the laser activity, the rest is substrate), amounts to approx. 140 μm. Thereof 130 μm are accounting for the gain medium with the SAM. In other words, just 10 μm are free space, in which the precise alignment takes place.
Because the laser operates in pulse mode, the small cavity-length is a big advantage. The smaller the cavity-length, the smaller is the build-up-time for the pulse (speed of light) and until the SAM is saturated. And these factors affect the pulse duration [3]. The problem behind that, is the sensivity

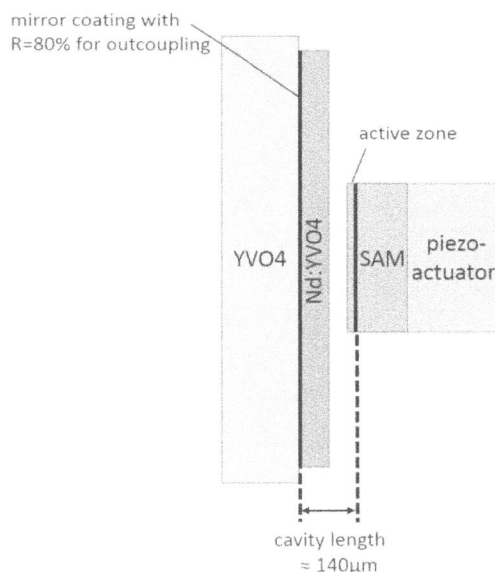

Figure 1: Extract from the cavity-setup of the microchip-laser. The outcoupling mirror is in between the active medium and the bonded material as a coating with a reflectivity of about 80%. There is about 10 μm free space between the active medium and the Saturable Absorber Mirror (SAM). The endmirror is built by the SAM. The active zone describes the laser active zone of the SAM.

of the setup. Smallest changes of the cavity-length, just less than 0,5 μm, cause changes in wavelength, output power, pulse energy and so on. This is very critical in medical applications. Minor deviations from the prescribed parameters can be very dangerous for human health.

Until now, the cavity-length is regulated manually with the piezo actuator. When a voltage is applied to the piezo actuator, it deforms mechanically (expand or contract). The manual regulation is not very effective, because it is way too slow as well as it is uncertain in which direction the voltage has to be regulated. Now a way has to be found to stabilize the cavity-length and so the laser.

2   Material and Methods

To overcome the instability of the laser, an electronic circuit should be designed. The circuit should facilitate an automated regulation of the cavity-length using a piezo actuator. At first the single pulses have to be visualized and brought into a shape, that they can be controlled. The idea is to measure the pulse energy. Additionally, these parameters have to be controlled, weather they are in a specified range. In case of deviation from set values, the piezo actuator will adjust them. To understand the background, the main components will be described in the following.

2.1   Q-switch with SAM

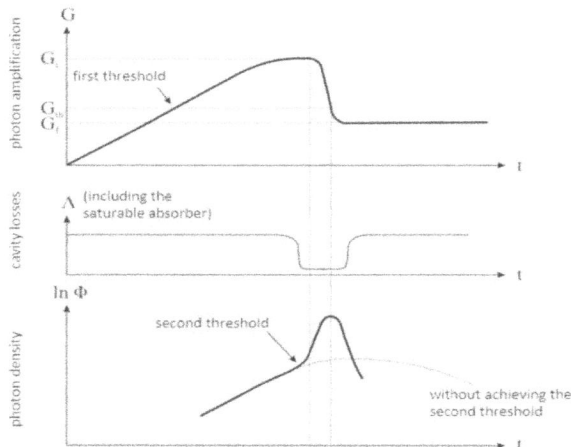

Figure 2: Timing of the pulse formation under the view of the photon amplification, the cavity losses with the SAM and the photon density. The high losses created by the SAM cause an enrichment of the pump energy. When the SAM is satured (second threshold), the photon density increases immediately. The maximal photon density is achieved, when the inversion is decreased to the first threshold (laser threshold) [4].

The passive Q-switch is realized by a SAM, which is the end mirror at the same time. The SAM is the simplest method for realizing a Q-Switch. By all Q-switches in comparison, the SAM has the shortest switching time ($\approx$ 1ns). As shown below, this is very important for the shortness and intensitivity of the pulse [4].

Fig. 2 shows the origination process of the pulse. The laser oscillation cannot start, because the SAM increases the intra-cavity losses. So the laser threshold is increased and the population inversion in the laser medium rises. The higher the light intensity, the more transmissive is the SAM. When a certain level is reached, the SAM is full transmissive, and from one second to another there is a massive light amplification, that exceeds the losses. This is the time where the pulse is generated. After a relaxation time of about 1 μs to 1 ps, all molecules returned in the ground state, so the process can start again. It is very important, that the SAM is saturated before the maximal population inversion is achieved. Otherwise the second threshold cannot be achieved and no pulse can be built [5].

2.2   Fabry-Pérot-Etalon

One of the main tools in this setup is the Fabry-Pérot-Etalon (henceforth referred to as "Etalon"). An etalon consists of a small, plane parallel plate, e.g. quartz glass, which is half-mirrored on both surfaces. It has frequency-dependent, periodic properties when light passes through it. As in Fig. 3 shown, it has a thickness d and the refraction index n_2. Because of the refractive index difference to the surrounding medium, reflections occur at the interface. The light is splitted in several partial-beams [6].

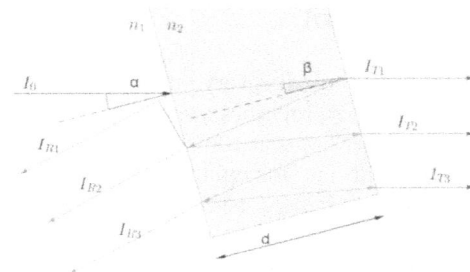

Figure 3: Beam path in the etalon. Transmission- and reflection beams are generated by a single incident beam. The different transmitted beams get in a constructive or destructive interference with each other, so a sinusoidal signal is built as shown in Fig. 5 [3].

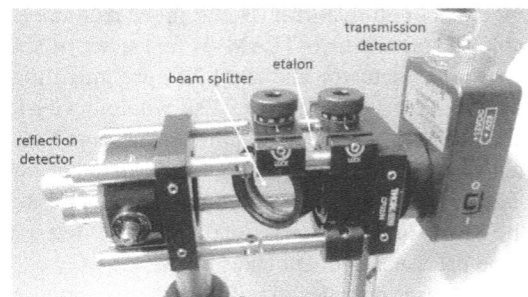

Figure 4: Whole etalon setup. The beam splitter led the incoming beam into the etalon. The transmitted and reflected beams are registered from the respective detector.

The reflection- and transmission-intensities change as a function of the wavelength. When the tilt angle of the etalon to the incident beam is changed, the resonance frequencies change accordingly (Fig. 5). So an Etalon can be used as an adjustable optical filter, to tune the wavelength of the laser. This is the important property, on which this project is based [7].

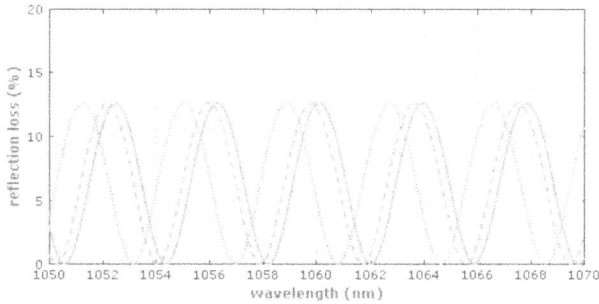

Figure 5: Transmission spectra of an etalon for normal incidence (solid curve), and for tilt angles of 2° (dashed curve) and 4° (dotted curve) [7].

2.3 The Idea

As mentioned in section 2.2, the etalon can be used for tuning the wavelength. The microchip-laser has to operate at a wavelength of 1064 nm. So the middle of the slope between resonance (high transmission) and anti-resonance (high reflection) has to be at 1064nm (Fig. 6).

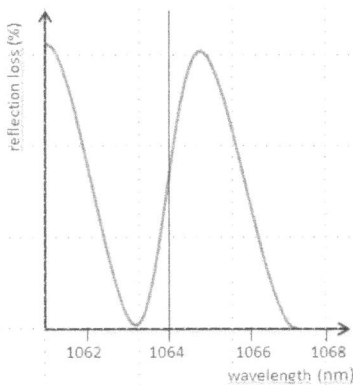

Figure 6: Tuning of the laser wavelength with the etalon. The wavelength is located on the rising slope of the etalon between resonance and antiresonance.

This is important for the measurement because on the one hand it is known where the signal is located. When there are fluctuations in the wavelength, e.g. caused by cavity-length change, the signal moves up or down the slope. On the other hand, the signal is intensity-independent, which is realised with a divider positioned after the circuit board. These characteristics are used for the cavity-length stabilization. The adjustment of the etalon can be realized by manual adjustment.

So, two known signals were generated by the etalon, which are the input for the electric circuit (Fig. 7). Afterwards, the signal is led into a regulator, which compares the actual existing signal (U_{act}) with the one, which is set by the etalon. The difference of both is amplified and led to the powerstage. There the signal is amplified again, so that it can be handled from the piezo-actuator, which needs higher currents.

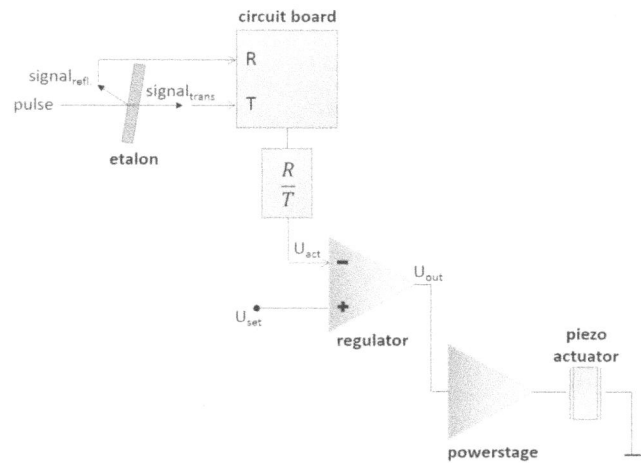

Figure 7: Schematic view of the experimental setup. The signals from the etalon are the input for the circuit board. Its purpose is the pulse energy measurement. So the pulses are shown on the oscilloscope as a rectangular function. Then R is divided by T, for intensity-independency. The regulator compares U_{set} with U_{act} and amplifies the difference between them. Because the piezo needs high currents to operate, the powerstage has to amplify the signal from the regulator. U_{act} shows where the signal is located on the slope. U_{set} is the ideal adjustment of the laser with the signal on the middle of the slope.

3   Results and Discussion

The electronic circuit could be implemented successfully.

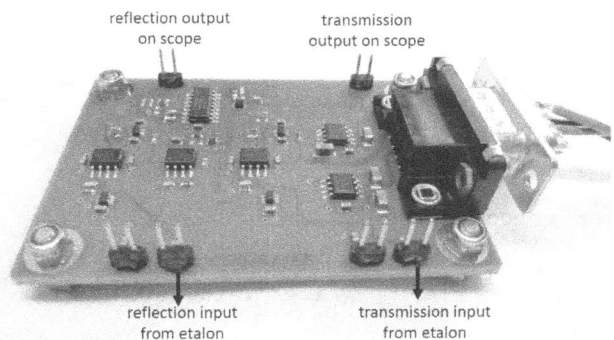

Figure 8: Here the finished circuit board with the different connections is shown.

Fig. 8 shows the circuit board with the in- and outputs. The circuit is based on a sample and hold principle. Practically this is shown as a rectangular function on the scope. The pulses are transformed into a rectangular shape, so that the circuit has enough time to evaluate them. The height of a single rectangle displays the pulse energy.

Fig. 9 and 10 show the output signal of the circuit on the scope. The reflected signal in Fig. 9 contains an error signal which causes the wave form. Ideally it should be noisefree like the transmitted signal.

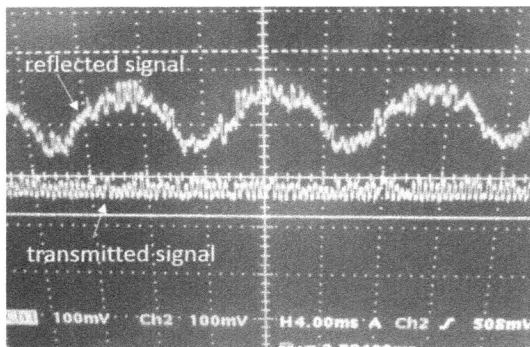

Figure 9: Output of the implemented electronic circuit on the scope. The transmitted signal (below) shows a noisefree function. The reflected signal is disturbed by the ambient light. Thats the cause of the wave shape.

Fig. 10 shows the same signal but zoomed in. Here the, from the circuit generated, rectangular function is shown. So both input signals are converted as they should. Whereby also the influence of the error signal on the upper, reflected signal is visible.

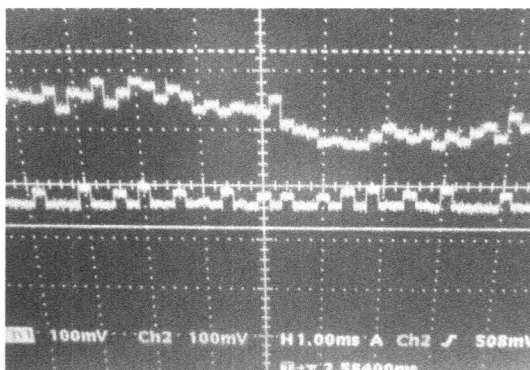

Figure 10: This is the same signal as in Fig. 9. Here it is zoomed in to show the desired rectangular function in detail.

This is a good example how sensitive the setup is for outside influences, i.e. ambient light and touch-sensitive. To correct the reflected signal, a black box has to be build around the etalon to exclude ambient light.
If this is stable, the etalon can be adjusted as shown above. The next step is to vary the piezo current manually to see the effects. If they are as imagined, an analog regulator can be implemented, which recognizes the deviations from the

ideal position. The piezo actuator get the correction information and adjusts.
It is planned to use a regulator, because the implementation is very fast and it shows less noise performance. Instead of the regulator a microcontroller can be used, because it is much more flexible and offers other possibilities. This should be later tested in the setup.

4 Conclusion

The results show that there is a good chance of success for realizing the idea. The etalon shows in combination with the electric circuit very good results. It is necessary to improve the shielding of the single components to avoid error signals. In the near future, the remaining components, such as the regulator and the piezo actuator, have to be implemented.

Acknowledgement

The work has been carried out at Coherent LaserSystems GmbH & Co. KG, Lübeck and supervised by the Institute of Biomedical Optics, Universität zu Lübeck. Especially I want to thank my supervisors Rüdiger von Elm and Fred Reinholz for their great support.

5 References

[1] Available: http://www.batop.com/information/microchip-laser.html [last accessed on 2017-01-25].

[2] M. Kaschke, K. Donnerhacke, M. Rill, *Optical Devices in Ophthalmology and Optometry: Technology, Design Principles and Clinical Applications*. Wiley-VCH, 2014.

[3] P. Steyer, *Optimierung eines passiv gütegeschalteten Nd:YVO4 Mikrochiplasers mit sättigbarem Absorberspiegel*. Master thesis, Beuth Hochschule für Technik, Berlin, 2016.

[4] M. Eichhorn, *Laserphysik*. Springer, Berlin / Heidelberg, 2013.

[5] F. K. Kneubühl and M. W. Sigrist, *Laser*. Vieweg & Teubner, Wiesbaden, 2008.

[6] H. Bauer, *Lasertechnik: Grundlagen und Anwendungen*. Vogel, Würzburg, 1991.

[7] Available: https://www.rp-photonics.com/etalons.html [last accessed on 2017-01-25].

Developing a Calibrated Setup to Investigate markerless Multi-View Reconstruction of Freely Moving Rodents

P. Strenge[1], A. Kyme[2], and H. Handels[3]

[1] Medizinische Ingenieurwissenschaft, Universität zu Lübeck, Germany, paul.strenge@student.uni-luebeck.de
[2] Brain & Mind Research Centre and School of AMME, Faculty of Engineering, University of Sydney, Australia, andre.kyme@sydney.edu.au
[3] Institute of Medical Informatics, Universität zu Lübeck, Germany, handels@imi.uni-luebeck.de

Abstract

Motion-compensated positron emission tomography (PET) enables brain imaging of animals, while they are free to move and respond to environmental or drug stimuli. A key requirement for motion-compensated PET is accurate pose estimates of the head and body of animal. One way to estimate the motion of an animal is by using multiple cameras, which locate the position of the animal in space. This paper describes the calibration process of a multi-camera setup. The acquired image data of the multi-camera setup is used to reconstruct a 3D model of a rodent from multiple camera views. The resulting 3D point cloud is the base for developing and testing new methods of markerless motion tracking of rodents.

1   Introduction

In todays brain research small rodents, like mice or rats, are used to see the effects of certain drug or environmental stimuli on the brain function of the animal [1]. For the evaluation of studies and experiments, positron emission tomography (PET) is used to visualize the brain function. Critical for the quality of the images and, therefore, for the outcome of the study is the movement of the rodent during the image acquisition. The movement creates motion artifacts resulting in blurring and loss of quantification of the image (Fig. 1).

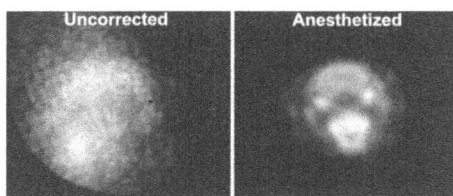

Figure 1: A reconstructed PET image of a rat head with motion-artifacts (left). Reconstructed PET image of the rat being anesthetized, showing no motion-artifacts [1] (right).

There are two main ways to reduce motion artifacts in the images:

1.   Anesthetize the rodent: This method eliminates the movement of the rodent and avoids motion blurring in the resulting images (Fig. 1), but it also has severe disadvantages [1]. Firstly, if the rodent is asleep, it is not possible to study its behavioral responses to various stimuli. Secondly, it is common for the anesthetic to impact the process of interest and, therefore, change the outcome of the study.

2.   Motion-compensated imaging: The movement of the rodent is tracked during the image acquisition and compensated for the image reconstruction. The motion data is needed to eliminate the motion blur [1]. There are different methods to perform the motion-tracking, which are either markerbased or markerless.

Kyme et. al. introduced a rat tracking method in 2012, where a marker is glued on the forehead of a rat, which is placed inside a tube within the field of view of the PET scanner (Fig. 2) [2]. The position and orientation of the head marker is tracked using a binocular optical motion tracking system. This method relies on the marker position remaining fixed with respect to the head, which is difficult to ensure without invasive attachment. If the marker is displaced (e.g. marker becomes loose) during the data acquisition, the motion tracking and, therefore, the compensation of the motion-artifacts worsens.

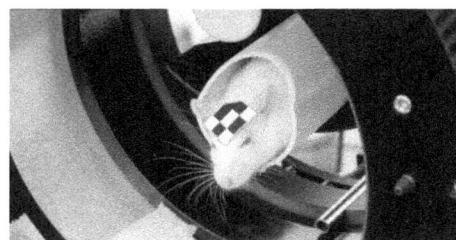

Figure 2: Rat with marker on the forehead inside a small animal PET scanner, as used in [2].

A markerless solution would get rid of that dependency. One possible solution to achieve a markerless motion tracking is by locating the position of the rodent in space using multiple cameras. This paper describes the process creating a calibrated multi-view setup for the reconstruction of 3D point cloud of a rat in real-world space from 2D silhouettes. The 3D point cloud is the base to determine the motion of the rodent.

2   Material and Methods

The multi-camera setup consisted of four grayscale, USB 3.0 cameras with an image resolution of 1280 x 1024 pixel. All four cameras were evenly arranged around a volume of interest (VOI), 250 x 250 x 130 mm (Fig. 3). The VOI is the space, where the rodent is freely moving while the cameras collecting the required data for the reconstruction. To ensure full coverage of the VOI, all cameras are additionally equipped with low-distortion lenses, which have a wide viewing angle of around 82 degrees.

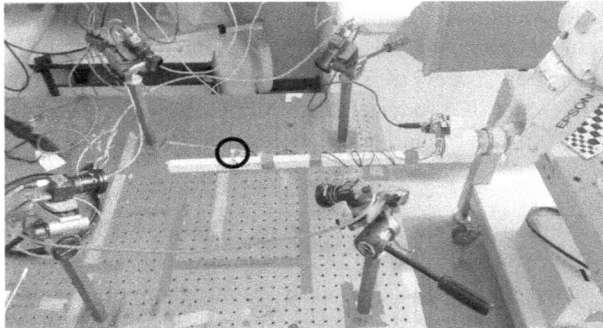

Figure 3: Setup of the four cameras with the wide angle lenses around the VOI. The modified LED mounted to the *EPSON RC+ 5.0* robot arm via a custom holder is circled.

To allow a precise reconstruction from multiple views the cameras had to acquire images at the same time. We modified the hardware and developed a data acquisition software in conjunction with using hardware triggering to handle synchronized frame capture from the four cameras at a frame-rate of 45 Hz [3][4]. Our software provided two operations: free-running and manual data acquisition. In free-running mode the cameras acquire images constantly at up to 45 Hz. In manual mode an image is taken, if the user triggers the cameras manually. Both modes are used in the camera calibration. The calibration can be divided into two parts, the intrinsic and the extrinsic calibration [5][6]. The resulting matrices of both calibrations are necessary to map 3D space coordinates into 2D image coordinates. For a given camera the relation to the 3D point can be expressed by (1) [6]:

$$\lambda \tilde{\mathbf{x}} = \begin{bmatrix} \mathbf{\Lambda} & \mathbf{0} \end{bmatrix} \begin{bmatrix} \mathbf{\Omega} & \tau \\ \mathbf{0}^T & 1 \end{bmatrix} \tilde{\mathbf{w}} \qquad (1)$$

λ is a scaling factor. The 2D-point $\tilde{\mathbf{x}} \in \mathbb{R}^{3x1}$ and the 3D-point $\tilde{\mathbf{w}} \in \mathbb{R}^{4x1}$ are represented in homogeneous coordinates [6]. The first matrix includes the intrinsic matrix $\mathbf{\Lambda} \in \mathbb{R}^{3x3}$, which is the result of the intrinsic calibration of one camera. The intrinsic matrix is composed of intrinsic

camera parameters such as the focal-length and principal point [6]. The second matrix includes the extrinsic parameters, the rotation $\mathbf{\Omega} \in \mathbb{R}^{3x3}$ and the translation $\tau \in \mathbb{R}^{3x1}$ [6]. Together, they describe the camera position in respect to an arbitrary coordinate system. The following sections will explain how the extrinsic and intrinsic parameters for our setup were acquired.

2.1   Intrinsic Calibration

The intrinsic calibration characterizes internal parameters of a certain camera. Since our setup consisted of four cameras, we needed to find four intrinsic matrices. For the intrinsic calibration we use the *MATLAB®*- calibration toolbox [7], which is based on the method introduced by *Zhang et. al.* in 1998 [8]. As input it requires images of a checkerboard pattern in different positions from a single camera [7]. For the image acquisition each camera is placed in front of the *EPSON RC+ 5.0* robot arm as seen in Fig. 4. A checkerboard pattern with squares of side length 14 mm was attached to the *EPSON RC+ 5.0* robot arm and moved in 30 different positions with different poses, while we used the manual acquisition mode to sample an image each time the position changed. The first step of the intrinsic calibration processing is the detection of saddle points of the checkerboard pattern in each of the 30 images. For a single camera the intrinsic parameters $\mathbf{\Lambda}$ are then calculated through a maximum likelihood estimation from a certain number of corresponding 2D-points $\mathbf{x}_i$, as shown in (2) [6]:

$$\mathbf{\Lambda} = \underset{\mathbf{\Lambda}}{argmax} \left[\sum_{i=1}^{I} log \left[P\left(\mathbf{x}_i | \mathbf{\Lambda}, \mathbf{\Omega}, \tau, \mathbf{w}_i \right) \right] \right] \qquad (2)$$

The term $P\left(\mathbf{x}_i | \mathbf{\Lambda}, \mathbf{\Omega}, \tau, \mathbf{w}_i \right)$ is the likelihood, the 2D-point $\mathbf{x}_i$ is the projection of $\mathbf{w}_i$ for the given parameters $\mathbf{\Lambda}$, $\mathbf{\Omega}$ and τ. The extrinsic parameters $\mathbf{\Omega}$ and τ are fixed, while $\mathbf{\Lambda}$ is estimated.

Figure 4: Image acquisition for the intrinsic calibration using the robotically controlled checkerboard pattern.

2.2   Extrinsic Calibration

For the extrinsic calibration the cameras were arranged as shown in Fig. 3. It matches the final setup since the extrinsic calibration determines the spatial relationships of the four cameras to an arbitrary coordinate system [5]. For the extrinsic calibration we used a *MATLAB®*- based multi-camera self-calibration toolbox [9]. The extrinsic parameters Ω and τ are calculated, similarly to the intrinsic parameters, through a maximum likelihood estimation (3) [6]:

$$\Omega, \tau = \underset{\Omega, \tau}{argmax} \left[\sum_{i=1}^{I} \sum_{j=1}^{J} log \left[P\left(\mathbf{x}_{ij}|\mathbf{\Lambda}_j, \mathbf{\Omega}_j, \tau_j, \mathbf{w}_{ij}\right) \right] \right] \quad (3)$$

The term $P\left(\mathbf{x}_{ij}|\mathbf{\Lambda}_j, \mathbf{\Omega}_j, \tau_j, \mathbf{w}_{ij}\right)$ is the likelihood, that 2D-point $\mathbf{x}_{ij}$, which is visible in multiple cameras j, is the projection of $\mathbf{w}_i$ for the given parameters $\mathbf{\Lambda}_j, \mathbf{\Omega}_j$ and τ_j. The extrinsic parameters $\mathbf{\Omega}_j$ and τ_j are estimated by maximizing the likelihood. To improve the convergence of the algorithm we supplied the intrinsic matrices $\mathbf{\Lambda}_j$ from the intrinsic calibration.

The input for the extrinsic calibration is a series of images from a calibration target by all four cameras. We used a small light source, a modified LED, which was sanded to produce a very small light spot. The LED was then placed on a custom holder, which consisted of a 50 cm long piece of balsa wood and a mount for the *EPSON RC+ 5.0* robot arm (Fig. 3). The custom holder allowed the *EPSON RC+ 5.0* robot arm to raster the LED safely through the VOI for a dense sampling. To have a enough data for a good estimation of the extrinsic parameters, each camera acquired 5000 images of the LED, while operating in free-running mode. After collecting the images, the data was loaded into the toolbox [9], which automatically determined the extrinsic parameters for each camera. The camera configuration can not be changed after the extrinsic calibration because the calibration would become obsolete and needed to be repeated.

2.3   Transfer to a World Coordinate System

The result of the extrinsic calibration is four matrices (one per camera) describing the position and orientation of the cameras with respect to an arbitrary coordinate system, which differs from real-world coordinates. Therefore, we need to transfer the coordinates of the arbitrary coordinate system (ACS) into a world coordinate system (WCS), which represents real-life proportions [1]. 1987 *Arun et. al.* introduced a method to find the transformation between two sets of corresponding 3D-points: $\mathbf{w}_1$ and $\mathbf{w}_2$ [10]. The transformation consists of a rotation matrix $\mathbf{R} \in \mathbb{R}^{3x3}$, a scaling factor λ and a translation vector $\mathbf{t} \in \mathbb{R}^{3x1}$. (4) shows how these three parameters can be used to transform one 3D-point $\mathbf{w}_1 \in \mathbb{R}^{3x1}$ into the 3D-point $\mathbf{w}_2 \in \mathbb{R}^{3x1}$ [10].

$$\mathbf{w}_2 = \mathbf{R}\lambda\mathbf{w}_1 + \mathbf{t} \quad (4)$$

To obtain two 3D-point sets $\mathbf{w}_{ACS}$ and $\mathbf{w}_{WCS}$ we use the *EPSON RC+ 5.0* robot arm. The robot was programmed to navigate a circular target, which is placed on the custom holder (Fig. 5) to 15 different positions inside the VOI. The four cameras were operated in manual mode to acquire an image, each time the circular target reached one of the 15 predefined destinations. After the data acquisition, the center of the circle target was determined in every image

for each camera. This provided a set of 15 2D-points $\mathbf{x}_{ij}$ for each camera. Corresponding 2D-points $\mathbf{x}_{ij}$ between the four cameras and the respective camera calibration matrices were then used to calculate the corresponding 3D-point $\mathbf{w}_i$ by triangulation [5]. The resulting 15 3D-points $\mathbf{w}_i$ represent $\mathbf{w}_{ACS}$. Since the *EPSON RC+ 5.0* robot arm used a coordinate system with real-life proportions to navigate the circular target through space, we can use these 3D-points as our $\mathbf{w}_{WCS}$. We used the method of *Arun et. al.* to determine the transformation between the two 3D-point sets [10]. Finally, the transformation was applied to the extrinsic matrices so that all 3D point reconstructions would now be in the new world coordinate system (5) [1]. The vector $\mathbf{0} \in \mathbb{R}^{3x1}$ contains only zeros.

$$\begin{bmatrix} \mathbf{\Omega}_2 & \tau_2 \\ \mathbf{0}^T & 1 \end{bmatrix} = \begin{bmatrix} \mathbf{\Omega}_1 & \tau_1 \\ \mathbf{0}^T & 1 \end{bmatrix} \begin{bmatrix} \mathbf{R}\lambda & \mathbf{t} \\ \mathbf{0}^T & 1 \end{bmatrix}^{-1} \quad (5)$$

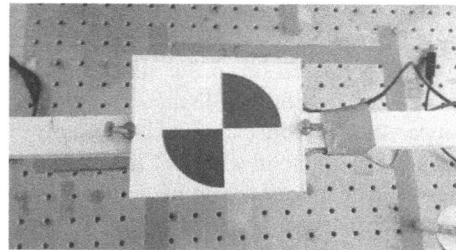

Figure 5: Circular target used in the robot controlled data acquisition to determine the transformation parameters between the arbitrary coordinates system and the world coordinate system.

3   Results and Discussion

We assessed the accuracy of the setup based on the errors of each calibration step. Since the four cameras and lenses are nominally identical, we expect, that the result for the intrinsic calibration are roughly the same. This is confirmed by the results shown in Table 1. The results for each camera are in good correspondence and show only small calibration errors.

Table 1: Intrinsic Calibration Parameters

Camera	focal-length		principal point	
	x (mm)	y (mm)	x (pixel)	y (pixel)
1	762.055	763.951	645.298	516.763
	± 1.393	± 1.381	± 1.014	± 0.985
2	757.072	759.152	647.463	510.766
	± 1.250	± 1.244	± 0.960	± 0.932
3	761.040	762.643	634.780	531.541
	± 1.199	± 1.199	± 0.904	± 0.887
4	759.218	760.721	658.414	508.264
	± 1.202	± 1.198	± 0.898	± 0.880

The results for the extrinsic calibration, shown in Table 2, behave similarly to the intrinsic results. The errors were

Table 2: Error for the extrinsic Calibration

Camera	1	2	3	4
Error (Pixel)	0.178 ± 0.098	0.161 ± 0.090	0.181 ± 0.114	0.169 ± 0.105

roughly the same for the four cameras. The small errors show, that our calibration has subpixel accuracy, which should be sufficient enough for our purpose. The small errors obtained for these wide-angle views are likely in part due to use of the robot and the large number of images with dense and even sampling of the VOI. The large amount of data also helps the self-calibration algorithm [9] to solve (3) with low errors (Table 2).

Fig. 6 shows a 3D point cloud of a taxi-dermal rat model, which was reconstructed using the space carving algorithm and our calibrated multi-view setup [11]. The algorithm uses (1) to project a number of 3D points w_i on to four images, which showed segmented silhouettes of the taxi-dermal model from the four different camera views. If a 3D point w_i is projected on to the silhouettes in all four images, it becomes part of the 3D model. It shows, that the calibration parameters were calculated accurately enough to reconstruct a 3D point cloud with real-life proportions from four images. The quality of the point cloud is limited by the number of camera views.

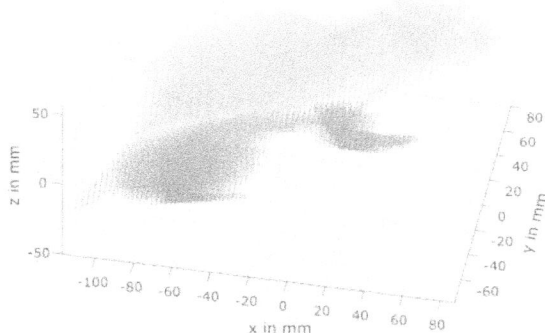

Figure 6: 3D point cloud of a taxi-dermal rat using space carving algorithm and four camera views.

4 Conclusion

The calibrated multi-camera setup provides an accurate reconstruction of 3D-points w_i in real-world coordinates from corresponding 2D-points x_{ij} (Fig. 6). There are still some small issues in terms of reconstruction accuracy, which can be improved by recalibrating the cameras. Due to the use of the *EPSON RC+ 5.0* robot arm, all calibrations steps are highly reproducible and easily to repeat. In future experiments, we will use this setup to investigate the feasi-

bility of accurately tracking the motion of a freely moving rodent.

Acknowledgment

The work was carried out at the Brain and Mind Centre, University of Sydney, in cooperation with the Universität zu Lübeck.

5 References

[1] A. Kyme, *Optimised Motion Tracking in Small Animal Positron Emission Tomography*. Thesis (Ph. D.)– University of Sydney, 2012.

[2] A. Kyme, S. Meikle, C. Baldock and C. Fulton, *Tracking and characterizing the head motion of unanaesthetized rats in positron emission tomography*. Journal of the Royal Society Interface, 9(76), 3094-3107, 2012.

[3] Point Grey, *ConFiguring Synchronized Capture with Multiple Cameras*. Point Grey, 2016.

[4] Point Grey, *Using Linux with USB 3.0*. Point Grey, 2016.

[5] R. Hartley and A. Zisserman, *Multiple View Geometry in Computer Vision*.2nd edition, Cambridge University Press, 2004.

[6] S. J. D. Prince, *Computer vision: models, learning and inference*. Cambridge University Press, 2012.

[7] J. Y. Bougouet, *Camera Calibration Toolbox for MATLAB®*. Available: `http://www.vision.caltech.edu/bouguetj/calib_doc/`, [last accessed on 2016-01-26].

[8] Z. Zhang, *A Flexible New Technique for Camera Calibration*. IEEE Transactions on Pattern Analysis and Machine Intelligence, 22(11):1330–1334, 2000.

[9] T. Svoboda, *Multi-Camera Self-Calibration*. Available: `http://cmp.felk.cvut.cz/~svoboda/SelfCal/`, [last accessed on 2016-01-26].

[10] K. S. Arun, T. S. Huang and S. D. Blostein, *Least-Squares Fitting of Two 3-D Point Sets*. IEEE Transactions on Pattern Analysis and Machine Intelligence PAMI-9(5):698 - 700, 1987.

[11] K. N. Kutulakos and S. M. Seitz, *A Theory of Shape by Space Carving*. International Journal of Computer Vision 38(3), 199–218, 2000

5

E-Health

Standardisation of Surgical Terms
for the Extension of the ISO/IEEE 11073-1010X Nomenclature

S. Baumhof[1], B. Andersen[2], and J. Ingenerf[2]

[1] Medizinische Informatik, Universität zu Lübeck, simon.baumhof@student.uni-luebeck.de
[2] Institute of Medical Informatics, Universität zu Lübeck, {andersen, ingenerf}@imi.uni-luebeck.de

Abstract

To overcome the lack of medical device interoperability, a service-oriented architecture for systems of networked medical devices has been developed within the OR.NET project as a part of the ISO/IEEE 11073 family of standards. Whereas this architecture enables the automatic exchange of information between devices, the interoperability remains dependant on semantic descriptors for the exchanged information. The preferred vocabulary of those within the ISO/IEEE 11073, the nomenclature, was initially designed for only a small set of devices and hence does not cover the descriptors needed within the operating room. Therefore, the nomenclature needs to be extended for the modelling of surgical devices. For this purpose, a set of new terms is proposed in this paper based on literature research and the aid of device manufacturers.

1 Introduction

Today's medical treatments within the operating room (OR) include a variety of procedures that usually involve and rely on many different electronic devices. With the rising number of devices, these procedures become increasingly complex. One reason for this is the limited interoperability between devices, i. e. the ability of two or more devices, including software, to automatically exchange information and interpret the information that has been exchanged.

There are already surgical devices available that are capable of communicating with each other. In this case, proprietary protocols and vocabularies are used that are rarely able to interact with devices from other manufacturers. Therefore, the manufacturer-independent interoperability of medical devices is becoming a key enabler to handle the increasing amount of different devices within the OR. To achieve this, technical, structural, and semantic standards need to be established and used: Technical standards to define the physical requirements needed for communication; structural or syntactical standards to define how the data should be represented; and semantic standards to precisely define the content.

The ISO/IEEE 11073 family of standards on medical device communication already defines components of a system making it possible to exchange vital data between different medical devices, to evaluate them and to remotely control these devices. Within the OR.NET-project, three new standard proposals have been developed as a part of the ISO/IEEE 11073 family of standards in order to fulfil the specific needs within the OR [1]. While these newly developed proposals only involve technical and structural definitions, they remain dependant on semantic descriptors of exchanged informations.

The nomenclature, defined in the ISO/IEEE 11073-1010X ('X' denotes a placeholder for different last digits), is the preferred vocabulary of medical device communication within the scope of the ISO/IEEE 11073 family of standards. It contains semantic descriptors, referred to as *terms*, needed for the description of device components and the semantic of exchanged informations. However, the nomenclature itself is currently limited to devices of specific domains. In particular, it lacks terms for the domain of surgical point-of-care (PoC) devices needed in the OR. Therefore, extending the nomenclature for the modelling of surgical devices is one key factor to achieve manufacturer-independent interoperability.

This paper presents the efforts of standardising terms for the domain of surgical devices. It briefly summarises the development of the ISO/IEEE 11073 nomenclature in Sec. 2.1. The recommended tool for working with terms is presented in Sec. 2.2, followed by the description of the terms' representation in Sec. 2.3. The proposed terms are described in Sec. 3, grouped by their device type.

2 Material and Methods

2.1 ISO/IEEE 11073-1010X Nomenclature

The ISO/IEEE 11073-10101 was released in 2004, representing the core nomenclature standard within the ISO/IEEE 11073 family of standards. It was initially developed with focus on capturing patient vital signs information. Therefore, the initial nomenclature contains terms for domains such as heamodynamics, electrocardiography (ECG), respiration, blood gas, and more [2]. Beside these domain-specific terms, the core nomenclature includes terms concerning units of measurement, device related events and alarms, and body sites.

In the recent years, the nomenclature was extended several times. For the first time, in 2012 the ISO/IEEE 11073-10102 added terms for the annotated ECG. Further terms concerning implantable cardiac devices were added with the ISO/IEEE 11073-10103 in 2013. The latest extension of the nomenclature is the ISO/IEEE 11073-10101a on ventilator-, respiratory-, and anaesthesia-related terms.

There are currently two major extensions under development. The ISO/IEEE 11073 part 10101b will include terms related to several domains including neuromuscular transmission, infusion pumps, ventilator modes, ECG-waveforms, real-time location services, device management control, generalised signal quality index, dialysis as well as additional observation identifiers and settings. Another extension, part 10101c will cover physiologic alerts as well as essential technical events and alerts.

2.2 Rosetta Terminology Mapping Management System

The *Rosetta Terminology Mapping Management System (RTMMS)* is a web application, provided by the U.S.-american *National Institute of Standards and Technology* (NIST). It was developed to support the *Rosetta Terminology Mapping (RTM)* working group of the *Integrating the Healthcare Enterprise's (IHE) Patient Care Device (PCD)* domain. The RTM working group's task is to specify uniform terms and codes for clinical and technical observations from devices, to reduce risk of mistaken or lost measurement identity, and also to declare permissible valid units of measure for observations [3].

To facilitate this task, the RTMMS allows manufacturers to map the proprietary semantics communicated by their medical devices to a standard representation using ISO/IEEE 11073 semantics. Fig. 1 illustrates this process: Manufacturers can create and edit vendor terms which are stored in the *Rosetta Table (RTM)*. Each vendor term, or each row respectively, represents the mapping of one proprietary term to a term existing in the ISO/IEEE 11073 nomenclature. Vendor terms describing the same semantic are then grouped into one *harmonised* Term. The *Harmonised Rosetta table (hRTM)* contains these harmonised terms agreed upon during the harmonisation process. These terms can then be used for further purposes like conformance testing. If there is no appropriate term available in the ISO/IEEE 11073 nomenclature to match a specific proprietary semantic content, manufacturers are able to suggest a completely new term. In this way, manufacturers can additionally aid the harmonisation process by identifying missing terms. These missing terms can later be added to the nomenclature.

Additionally, each term in the Rosetta tables can be enriched with parameter co-constraints, a set of permissible units of measurement, body sites, and enumerated values that may be associated with a given parameter, thus enabling even more rigorous validation of exchanged medical device semantic content [4].

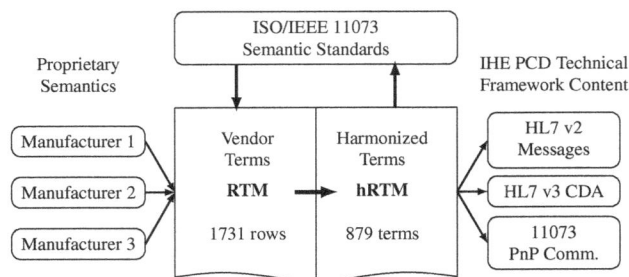

Figure 1: Schematic overview of the mapping process. Manufactors can map their proprietary semantics (vendor terms) to the existing ISO/IEEE 11073 nomenclature.

2.3 Data Collection and Representation

In the ISO/IEEE 11073 nomenclature, each term consists of a set of attributes. The human-readable and -understandable part of this set defines its semantic content. It includes a short description or definition respectively, and optionally, a commonly used name or an acronym. A unique numeric code and a reference identifier (REFID) string form the machine-readable and -processable part of a term [2].

Vendor terms within the RTMMS also consist of a set of mandatory and optional attributes. Since vendor terms are intended to map proprietary vendor semantics to nomenclature terms, the most important attribute in this set is the ISO/IEEE 11073 nomenclature's REFID that the manufacturer believes is an appropriate match to their existing terminology. Other mandatory attributes are the description, the *display name*, i. e. a common term, an appropriate unit of measurement, and the *Virtual Medical Device (VMD)* to which this term is assigned [4].

The VMD is part of the modelling paradigm of medical devices within the ISO/IEEE 11073. Medical devices are represented in a hierarchical model, called *Device Containment Tree*, that is made up of four core components: *Medical Device System (MDS)*, VMD, *Channel*, and *Metric* [1]. Fig. 2 illustrates an example of this. The MDS is the root element that represents the device. It usually contains several VMDs, that are used to represent a subsystem of the device. Metrics are used to model settings or measurements. Related Metrics are grouped as a Channel that belongs to a certain VMD.

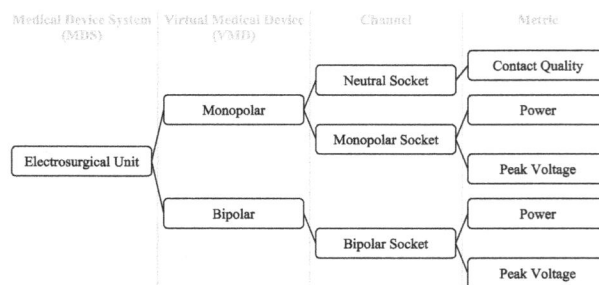

Figure 2: Simplified schematic overview of a *Device Containment Tree* on the example of a HF-Device.

The proposed terms for the modelling of surgical devices should be enriched with attributes which fulfil the requirements of both the nomenclature and the RTMMS. Therefore, it would be beneficial to propose a Device Containment Tree as well.

In this work, for each device type a Device Containment Tree is proposed, an example of this is illustrated in Fig. 2. Each Channel in a Device Containment Tree is then represented as a table, see Tab. 1 for an example. The proposed terms are represented as rows in these tables. To satisfy the term requirements of the nomenclature and the RTMMS, each term is represented by the following attributes: (proposed) *REFID* as identifier, *Name* as a common used term or display name, *Symbol* as an abbreviation or acronym, *UoM* as permitted unit of measurement, and *Definition* for a short textual description of the content.

The proposed terms and Device Containment Trees are mainly based on literature research, including standard documents and the devices' operating manuals. Additionally, manufacturers have been asked to provide lists of relevant parameters that, in their opinion, should be standardised. Previous research results from the OR.NET project have been used as well to extract relevant parameters.

3   Results

Terms for four surgical device types are proposed in this work. In this this section, the devices as well as their important parameters are described. These parameters are used as the proposed term.

3.1   Electrosurgical Instruments

High frequency surgical cutting devices (HF-devices) are essential instruments in the field of electrosurgery. The term electrosurgery (also called radiofrequency surgery) refers to the process of applying high-frequency electrical current to the tissue in order to achieve a specific surgical effect, such as cutting or coagulation [5].

The core component of HF-devices is the *electrosurgical unit (ESU)* that is used to generate the high frequency electrical current. It is herein used as the MDS in the Device Containment Tree. In order to direct the current to a patient's tissue, one or more handpieces (electrosurgical instruments) can be connected to the ESU. Since each handpiece can be operated individually, the proposed Device Containment Tree separates the ESU into its connection sockets. Each socket should contain Metrics concerning the provided current, see Fig. 2. Handpieces differ in their form and their connection socket, based on the procedure. There are two classes of procedures: Monopolar, where the handpiece has one active electrode that requires a neutral electrode to disperse the electrical current, and bipolar, where the handpiece has two electrodes between which the current flows. This difference is modelled as separate VMDs. The applied current is decisive for the desired effect [5]. ESUs therefore have many different operating modes, that provide differently modulated electric current. Because

these modes differ from manufacturer to manufacturer, the provided current should be described in as much detail as possible. Fig. 3 shows the general differences in the waveform of the provided electric currents. Common attributes to describe these waveforms have been worked out to standardise the different operating modes, see Tab. 1. This includes the *fundamental frequency*, the *average* and the *peak voltage*, and the *current's power*. Additionally the *Duty-Cycle* is used to describe the percentage of the total time that there is actually electrical current being produced by the generator. In this way, the coagulation and blend modes, illustrated in Fig. 3, can be separated from each other. Often times, the *Crest-Factor* is needed as well. It defines the peak-to-average ratio of the voltage.

These attributes should satisfy both the mono- and the bipolar operations. The monopolar operations additionally need descriptors for the neutral disperse electrode: The *type*, because there are three different forms of neutral electrodes, and the *contact quality*. Bipolar operations need attributes that indicate if *AutoStart* or *AutoStop* are enabled. They describe functions to automatically start the current flow if a tissue is detected between the electrodes or to automatically stop at a certain resistance between the electrodes.

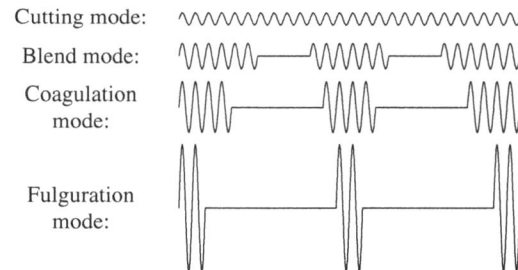

Figure 3: The differences in the current's waveform based on the different operating modes [5].

Table 1: Example of a Channel-table on the example of the HF-device's *Monopolar-Socket*. The Column *REFID* is left out here.

Name	Symbol	UoM	Description
Power Setting	-	W	The power of the applied current
Peak Voltage	V_p	V	The peak voltage of the applied current
Average Voltage	V_a	V	The average (effective) voltage of the applied current
Frequency		kHz	The fundamental frequency of the applied current
Duty-Cycle	DC	%	The percentage of the total time that there is actually an electrical current produced by the generator
Crest-Factor	CF	-	The peak-to-average ratio of the voltage

3.2   Moveable C-Arms

A mobile or moveable C-arm is a medical imaging device which works with x-ray radiation. The name is based on the c-shaped arc connecting the x-ray source to the x-ray detector. This arc is mounted on a mobile base so that it can be used flexibly at several locations. Due to its flexibility in the movement of the x-ray apparatus, the mobile C-arm is used for intraoperative imaging. The device provides high-resolution x-ray images in real-time, so that the treatment success can be checked at any time of the procedure and any corrections can be applied immediately.

The proposed model separates the motion of the c-arm from the parameters used for the image acquisition. Concerning the motion parameters, it is common that C-arms are capable of up to five individual motions. Each of these motions are represented as Channels in the C-arm's Device Containment Tree, containing the *minimal*, *maximal*, and the *current state* of each motion. The exact positioning therefore can be described by five parameters.

The image acquisition parameters can be separated into those related to the x-ray dose and the general parameters of the C-arm. The x-ray dose is influenced by the *electrical current* that is used to heat the x-ray tube's cathode and by the *voltage* that is applied between the cathode and the anode. To capture the radiation exposure, parameters like the *exposure time* and the *Dose Area Product* (DAP), which describes the dosimetric quantity, need to be standardised as well.

3.3   Endoscopic devices

Today's endoscopic device systems include several components. In general, endoscopes are used in the field of minimally invasive surgery to acquire images from body cavities. Its core components are light sources, used to illuminate the desired area, and the camera system. Beside the core components, numerous peripheral devices can be attached, depending on the procedure. For example, the gas flow produced by an insufflator can be used to stabilise cavities.

Light sources are describable by their used *light source type*, i. e. Xenon, Halogen or LED, and their *illumination intensity*. Additionally, the *operating hours* of the light source is an important parameter.

Insufflators need terms describing the pump activity. This includes the *targeted pump flow* and *pressure* as well as the *actual* values of both. The *total volume of gas delivered* is also significant. Additionally, insufflators may include a heating unit or a nebuliser to change the gas' *temperature* and its *humidity*. These additional functionalities can be modelled as separate VMDs.

4   Discussion and Conclusion

In this paper, we present a harmonisation proposal including several new nomenclature terms. This proposal is mainly based on the similarities between lists of relevant parameters provided by manufacturers and parameters found during literature research. The terms may not be sufficient to fully model each device type, but they constitute a starting point for the development of further terms. The proposed terms are focused on the specific and characteristic observable metrics for each device type. Components needed for the device communication, like terms for device specific alarms or events, are not part of the herein proposed terms. However, they need to be standardised as well to ensure full semantic interoperability. While working with the RTMMS, it was noticeable that the RTMMS may be not well suited for the development of terms for an entirely new domain, especially if there are no device models for comparison. The use of an additional tool for device modelling could facilitate the process of standardising terms. A fitting tool for modelling devices according to the ISO/IEEE 11073's domain information model is currently being developed. It could be extended to fit the specifications of the new standards from the OR.NET including the surgical device domain.

Acknowledgement

This work has been carried out at and supervised by the Institute of Medical Informatics, Universität zu Lübeck.

5   References

[1] B. Andersen, M. Kasparick, F. Golatowski, and J. Ingenerf, *Extending the IEEE 11073-1010x nomenclature for the modelling of surgical devices.* In: 2016 IEEE-EMBS International Conference on Biomedical and Health Informatics (BHI), pp. 244--247, 2016.

[2] *ISO/IEEE Health Informatics - Point-Of-Care Medical Device Communication - Part 10101: Nomenclature.* ISO/IEEE 11073-10101:2004(E), 2004.

[3] J. G. Rhoads, T. Cooper, K. Fuchs, P. Schluter, and R. P. Zambuto, *Medical device interoperability and the Integrating the Healthcare Enterprise (IHE) initiative.* In: Biomedical instrumentation & technology, vol. Suppl, pp. 21--27, 2010.

[4] J. Garguilo, S. I. Martinez, and N. Crouzier, *RTMMS user's guide – Rosetta Terminology Mapping Management System*, National Institute of Standards and Technology. Available: `https://rtmms.nist.gov/rtmms/index.htm#!help` [last accessed on 2017-02-09].

[5] A. Taheri, P. Mansoori, L. F. Sandoval, S. R. Feldman, D. Pearce, and P. M. Williford, *Electrosurgery: part I. basics and principles.* In: Journal of the American Academy of Dermatology, Vol. 70, no. 4, pp. 591.e1—591.e14, 2014.

A framework for testing the compatibility of medical devices with new communication standards of the ISO/IEEE 11073 family

D. Labitzke[1,2]
[1] Medizinische Informatik, Universität zu Lübeck, labitzke@student.uni-luebeck.de
[2] Institute of Telematics, Universität zu Lübeck, labitzke@itm.uni-luebeck.de

Abstract

This paper presents a developed solution for automatic black box testing the standard conformity of medical devices with new ISO/IEEE 11073 standards and soon to be standards. Given that nearly every medical device is an encapsulated system with manufacturers not revealing a device's blueprint and program code, the ability to evaluate a device's capability of using a standardized communication is necessary. To develop a tool for testing the standard's requirements, test cases were designed to make differentiated statements about standard conformity. The developed framework delivers a report about a connected device to communicate with the new group of standards.

1   Introduction

This paper describes the development of a framework to perform automatic standard conformity tests on medical devices used in modern operating rooms of hospitals and similar facilities. The framework – called *Testsuite* – performs black box tests with a device under test (DUT) by using exclusively the information provided by the tested device.

The following sections motivate the importance of standardized communication in context of medical devices.

1.1   Related Work

Several approaches and research projects have been carried out in order to establish a standardized interoperability of medical devices. Great progress was achieved by Goldman et al. in 2009 with publishing the ASTM F2761-1:200 Integrated Clinical Environment (ICE) standard [1] in context of the Medical Device „Plug-and-Play" Interoperability Program (MDPnP). Here, functional elements for point-of-care related IT systems are defined, focussing on patient data communication and equipment controlling.

A well known german approach is the OR.NET project [2] in which many german research groups were involved. OR.NET combined and consolidated the concepts of several predecessor projects in order to propose a new standard – the Open Surgical Communication Protocol (OSCP).

1.2   The necessity of standardized communication for medical devices

The amount of modern medical devices with embedded computer systems in today's operating room (OR) has greatly increased over the last years. While there are many manufacturers providing complete solutions to equip an OR with (see for example [3] and [4]), these are often rather

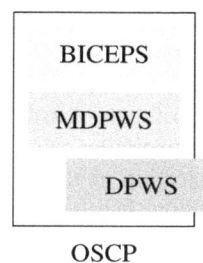

Figure 1: The standard family 11073-SDC as a combination of DPWS, MDPWS and BICEPS in a layer architecture

costly and thus not suitable for smaller facilities with a budget more humble. This issue could be addressed with obtaining different devices from multiple vendors. Being much more cost efficient, such assembled ORs lack a proper option for the individual devices to communicate with each other. To solve this problem, standards are needed, preferably open ones with support for a plug-and-play functionality [5]. With many approaches and standards released, the 11073 standard family [6] released at the Institute of Electrical and Electronics Engineers (IEEE) propably contains the most famous ones. IEEE 11073 comprises device communication standards to enable communication between medical, health care and wellness devices as well as communication with external computer systems.

1.3   New standards of the IEEE 11073 family

Article [7] introduces three new standards to be part of the IEEE 11073 family. These describe the possibility of a service-oriented device communication for distributed medical systems. The proposed standards cover the Domain Information & Service Model (IEEE 11073-10207) following the SOMDA architecture paradigm (service-oriented medical device architecture) – also called Basic Integrated Clinical Environment Protocol (BICEPS) – the communication technology for data transmission (IEEE 11073-20702)– named Medical Devices Profile for Web Services (MDPWS) – and the binding between the first two standards (IEEE 11073-20701). They form a standardized method for communication of medical systems and devices in an operation room, thus being called Open Surgical Comunication Protocol (OSCP). MDPWS is an extension of the Devices Profile for Web Services (DPWS) [8], adjusting it to work with medical devices. The conceptual design of the OSCP communication stack as described in [9] is shown in Fig 1: BICEPS, MDPWS and DPWS form a layer architecture. While the first two layers (BICEPS and MDPWS) are exclusive parts of the OSCP stack, DPWS is the only third party layer from which parts are used in OSCP.

with the first two layers are exclusive parts of the OSCP stack, while only parts of DPWS are used.

1.4 The need for a test framework

The main scope of this paper is to describe the development of a framework that tests medical devices for the capability of using the communication standards mentioned in section 1.3. As described in section 1.2, standardized communication techniques grant a lot more possibilities for healthcare facilities to equip their operating rooms. Since the implementation of communication methods is the responsibilty of the different vendors, one has to ensure the standard compatibility and to which extend the standard is supported before the purchase. Due to the complexity of modern OR's and medical devices themselves [9] and the large quantity of requirements in the standardization documents, a manual check of standard conformity would be time consuming. A more convenient method is the use of a testing framework. The developed framework gets connected to a device and tests the specified requirements automatically. The result is a statement for each requirement, whether the tested device fulfills it or not.

2   Material and Methods

In the following section the concept of the developed test framework is described as well as the specific tools used to clarify the capability of a device to communicate with the new 11073 standards mentioned in section 1.3

2.1   The process of testing a device

To test the standard conformity of a device, it is connected to the testsuite, either directly via cable or indirectly with simply being in the same local network. Figure 2 schematically shows the procedure of testing a device.

At first the testsuite searches for compatible devices in the network. Once a device is found, the implemented testclient subscribes to all offered information and begins to send messages invalid in respect of the BICEPS message model and xml specifications. Due to the subscribtions, the

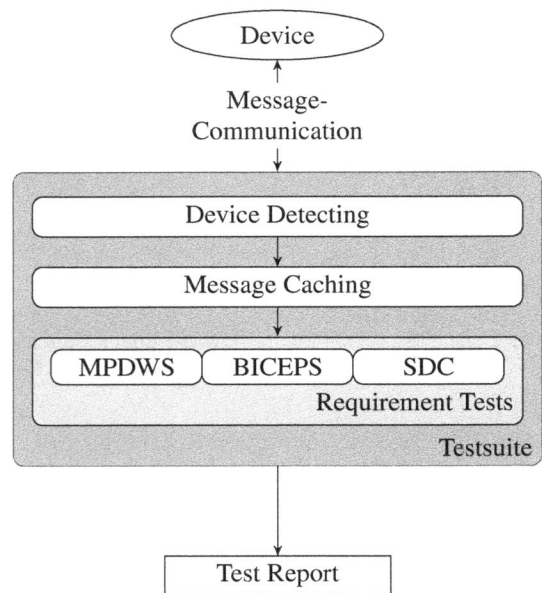

Figure 2: The concept of the test framework

testframework receives several messages from the device – e.g. reports about changed metric values or operation invocation states – and sends messages as well – e.g. a request messages to get the recent snapshot of the devices MDIB (Medical Device Information Base). All these messages – sent and received – are saved in a message cache for further processing. After some time passed, the framework goes on to the stage of processing the saved messages. Since the testsuite only gathers information through the message communication, the information contained in the messages sent is the only one available for the framework. Therefore only black box tests are implemented and processing the messages saved in the cache. The specified requirements in the three standardization documents are then tested on the information gathered from the messages – each with one seperate test case.

2.1.1   The MDIB history

The MDIB is a core component of a medical device modeling in IEEE 11073 and therefore of great interest during the testing process. It contains all the information about a device, as there are component descriptions and their respective states.

Due to the MDIB's functional range, it is a central point of interest for testing a device. A cache called MDIB history was implemented saving all received snapshots of the DUT's MDIB.

To fill the cache with snapshots of the MDIB, everytime a report about a change of the MDIB is received, the testsuite sends a *GetMdib* message to the concerning device in order to receive a snapshot of a device's current MDIB which is added to the cache. This results in a chronologically ordered collection of MDIB snapshots.

As there is a number of requirements concerning the temporal behaviour of a device's components located in it's

MDIB, this cache is used to easily iterate over all MDIB snapshots to examine and evaluate the over time changes of contained information.

2.2 openSDC

As presented in [9], the three OSCP (soon to be) standards are released as the openSDC protocol stack [10]. It uses a modified version of the WS4D JMEDS stack [11], which grants the functionality of DPWS (for details see section 1.3. openSDC serves as a reference implementation written in the Java programming language and is not intended to be used in clinical studies, clinical trials or in clinical routine.

2.3 The key tools used

The main task of the developed framework is to cover each of the requirements specified in the three standards described in section 1.3 with one test case. To make the programm code easier to read and more comprehensible, Hamcrest is used as a tool to write individual test assertions. All specified requirements follow the same pattern: they all contain the subject of the respective requirement, a requirement level and the actual demand. The requirement level indicates the importance of fullfilling the recent requirements, examples are *shall*, *should*, *must* and *may*. A requirement subject for example can be a service provider or a service consumer. Hamcrest enables the phrasing of linguistic tests. A fictional requirement starting with „A service consumer shall do this" then maps to the program code

```
it (SHALL) . beThat ( serviceProvider ,
    doesThis ).
```

Next to Hamcrest, aspect oriented programming as described in [12] is used with AspectJ. Several methods from the openSDC implementation mentioned in section 2.2 are marked as aspects. E.g. the method of receiving a message in openSDC is marked as an aspect. Every invocation of the receive method results in a call of the frameworks functionality to save a received message to the message cache (see section 2.1).

The framework is developed using the Java programming language.

2.4 The development process

The testsuite is developed in several phases, corresponding to the three new standards to test (see section 1.3) due to some of them still facing the standardization process of the IEEE. Several deadlines exist for the Testsuite to cover the requirements from the MDPWS-, the BICEPS- and the SDC standard. The development in each phase is organized using the scrum model, with each sprint lasting three weeks. Each sprint ends with a review meeting in which new requirements are chosen to develop new test cases for. For evaluation purposes, the testsuite is connected to medical devices, which are able to communicate using the openSDC implementation presented in section 2.2, at each review meeting. Additionally, weekly telephone conferences are

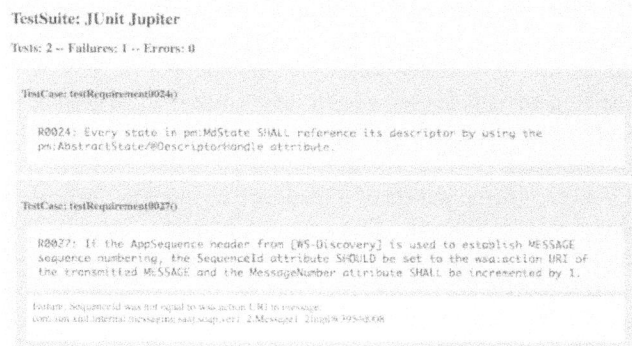

Figure 3: Visual representation of the test report generated while testing the minimal test devicewith a reduced number of tests. Two tests were performed, one of them evaluating positively, the other one resultung in a failure, to be recogniced from the second box containing the failure message.

held to discuss and review syntactic and semantic issues found in the standardization documents.

Besides the routinely connection to real medical devices, the functionality of every test case is evaluated by running it against a software-simulated medical device. This device delivers all the basic functionalities a real device could have. This way, possible mistakes in the implementation of a test case can be eradicated long before running it against the real device.

3 Results and Discussion

The evaluation by using several different devices as well as software simulations of medical devices testified the ability of the testing framework to make a statement about the conformity of a device with the new group of standards IEEE 11073-SDC. With the detailed test report as a result it is possible to testify which specific functionalities of the SDC standards a device supports and which it doesn't.

3.1 Evaluation using a minimal test device

To evaluate the developed framework and to present some valid results – the other tested devices are not meant to have their information published – a minimal software simulation of a minimal device was implemented and the number of tests to be performed was reduced to two. The device was implemented for one of the tests to fail, while the other one should succeed. Figure 3 shows the resulting xml test report visualized using a stylesheet.

4 Conclusion and Future Work

In this paper, a framework was presented for the conformity of medical devices with newly formulated communication standards. Given that black box tests are performed, no additional information about a device is necessary.

In future work semantic testing of medical devices will be enabled. More complex tests could be done with help of device specialisations, following the example of [13]. The reaction of devices to specific messages will be tested as well as the semantic validity of device generated data and information.

Finally implementation of a feedback channel to the device under test is planned. This way, the complete information flow can be validated. The reaction of devices to specific situations can be tested, e.g. the active regulation of the devices parameters.

Acknowledgement

The work has been carried out at and supervised by the Institute of Telematics, Universität zu Lübeck.

Special thanks goes to Prof. Dr. Stefan Fischer, director of the Institute of Telematics, for granting me the possibility to work at his institute, independently and with all the responsibility. Additional thanks goes to Daniel Burmeister, research assistant at the Institute of Telematics, from whom I took over the project, for giving me such a good break into the complex project structure. Last but not least, I'd like to thank Johannes Thorn, Institute for Software Engineering and Programming Languages, my partner of development for the project.

5   References

[1] ASTM International, *ASTM F2761 – 09 Medical Devices and Medical Systems – Essential safety requirements for equipment comprising the patient-centric integrated clinical environment (ICE) – Part 1: General requirements and conceptual model*, 2009.

[2] OR.NET, "OR.NET project website." `http://www.ornet.org`. Accessed: 2017-01-16.

[3] K. Storz, "OR1." `https://www.karlstorz.com/de/de/karl-storz-or1.htm`. Accessed: 2017-01-13.

[4] Olympus, "ENDOALPHA." `https://www.olympus-europa.com/medical/en/medical_systems/products_services/systems_integration/productselector_service_solutions_8.jsp`. Accessed: 2017-01-13.

[5] H. U. Lemke and M. W. Vannier, "The operating room and the need for an IT infrastructure and standards," *International Journal of Computer Assisted Radiology and Surgery*, vol. 1, pp. 117–121, November 2006.

[6] ISO/IEEE, "ISO/IEEE 11073: Health informatics – Point-of-care medical device communication." `11073.org`, June 2004. Accessed: 2017-01-11.

[7] M. Kasparick, S. Schlichting, F. Golatowski, and D. Timmermann, "New ieee 11073 standards for interoperable, networked point-of-care medical devices," in *2015 37th Annual International Conference of the IEEE Engineering in Medicine and Biology Society (EMBC)*, pp. 1721–1724, August 2015.

[8] D. Driscoll and A. Mensch, editors, "Oasis standard: devices protocol for web services version 1.1," tech. rep., OASIS, July 2009.

[9] D. Gregorczyk, S. Fischer, T. Busshaus, S. Schlichting, and S. Pöhlsen, "An Approach to Integrate Distributed Systems of Medical Devices in High Acuity Environments," in *5th Workshop on Medical Cyber-Physical Systems* (V. Turau, M. Kwiatkowska, R. Mangharam, and C. Weyer, eds.), vol. 36 of *OpenAccess Series in Informatics (OASIcs)*, (Dagstuhl, Germany), pp. 15–27, Schloss Dagstuhl–Leibniz-Zentrum fuer Informatik, 2014.

[10] "openSDC Website." `https://sourceforge.net/projects/opensdc/`. Accessed: 2017-01-16.

[11] Web Services for Devices (WS4D), "JMEDS (Java Multi Edition DPWS Stack)." `https://sourceforge.net/projects/ws4d-javame/`. Accessed: 2017-01-16.

[12] G. Kiczales, J. Lamping, A. Mendhekar, C. Maeda, C. Lopes, J.-M. Loingtier, and J. Irwin, *Aspect-oriented programming*, pp. 220–242. Berlin, Heidelberg: Springer Berlin Heidelberg, 1997.

[13] "IEEE 11073-10404: Device specialization – Pulse oximeter." `https://standards.ieee.org/findstds/standard/11073-10404-2010.html`. Accessed: 2017-01-17.

Development of data interfaces for opthalmologic devices

G. C. Teeuwsen[1], G. Buehrle[2] and S. Fischer[3]

[1] Medizinische Ingenieurwissenschaft, Universität zu Lübeck, grischa.teeuwsen@student.uni-luebeck.de
[2] Carl Zeiss Meditech AG, Oberkochen, guenter.buehrle@zeiss.com
[3] Institute of Telematics, Universität zu Lübeck, fischer@itm.uni-luebeck.de

Abstract

The aim of this paper work is to create modern flexibility data interfaces according to risen customer requirements from doctors and hospitals. Following different data interfaces are described and the structure of data communication from given devices will be investigated. It is important to separate the post-processing and the output-interfaces to allow the best possible flexibility. Post-processes convert the proprietary data into a readable output-format and the output-interfaces transmit the output-formats to external database-interfaces. The result of these scripts enables a flexible conversion into different data-formats, which can be distributed to different output-interfaces. Moreover, there is the opportunity which support the output-interfaces, so that the device and the electronic health record system can communicate among each other. Due to these approaches, the doctor is able to use modern interfaces with a complete flexibility. In addition, the ongoing digitalization progress facilitate the simplification of the workflow from doctors.

1 Introduction

The digitalization of healthcare is moving forward at an unstoppable rate affecting outpatient and clinical care, patients and physicians. This progress is an excellent opportunity for the healthcare sector allowing software-supported visualization techniques for life-saving brain surgery, digital logistics for intraocular lenses, the networking of the ophthalmic practice or the pooling of information on 50,000 cataract operations in a comprehensive clinical database that helps doctors to improve treatment outcomes [1]. More and more the digitalization becomes the key factors of medical devices.

Many devices used in practice and hospitals all over the world are still based on obsolete data interface technologies. Moreover only proprietary measurement data which can be read from the internal Practice Management System (PMS) of the manufacturers only are transmitted. This means a restricted advantage for the user and therefore don't correspond to the modern age of the digitalization any more. The possibilities which originate from the digitization generates an increased claim from the user concerning compatibility and performance of the devices. The aim of this work is to implement these new user requirements in form of adaptable software in combination with modern data interfaces. This includes essentially the bipartite data production to post-processing and output-interfaces. Moreover, the communication between the equipment and how doctors can use their workflow more efficiently with the new interfaces, will be clarified.

2 Material and Methods

In this section an overview is given about the materials and methods subcategories. Section 2.1 describes the post-processing. In this process, measuring data is converted into different formats, such as different readable XML (Extensible Markup Language) and portable document formats (PDF), which are introduced there. In section 2.2 the output interfaces which are able to transmit the new generated formats to the respective final system about the intended output interfaces, are described. The separation allows adaptable software integration and enables the manufacturer to support other output-formats and/or other external PMS. At least section 2.3 explains how the implemented server can be used for a configuration interface.

2.1 Post-processing

The post-processing describes the conversion of the attained proprietary measuring data to readable data formats. The post-processing represents the basis for the following output protocols. Every PMS/EHR (Electronic Health Record System) can only read a certain number of different data formats. The user has the possibility to select from different formats to provide the most important interfaces. Thus the conversion concentrates on four basic file conversions, as shown in fig. 1: *xml2pdf, xml2txt, xml2xml* und *bmp2jpg*.

Xml2txt's transformation represents the most simple form of conversion. The script changes the proprietary XML format into a text file. The opposite interface in the PMS is able to decode the text file and store the measurement data in its database. The picture data attained during the measurement, can be stored in reference to its measuring data. Besides, the system transmit the picture data as a bitmap which requires a conversion in *.jpg* file. The direct linkages at the pictures, for example lighting parameters, are noted in references to the xml measurement. The most complicated form of transformation is described by *xml2xml*, where every manufacturer has his own *.xml* or *.svg* (Scalable Vector Graphics) format, which can just be read by his own PMS/EHR. Due to this fact, the ophthalmology manufactures has concerns the standard of the JOIA-XML (Japan Ophthalmic Instruments Association) which allows that the measuring data, and therefore the device can communicate among each other and with not house-internal PMS [4].

An exception is the transformation of *xml2pdf*. Here, it is about an optical processing of the measuring data for the preparation of a printing process. In this case the patient does not get one small slip of paper from the device internal printer where his measuring values are quoted. He gets a DIN A4 (German Institute for Standardization Format) PDF format with his measurement values which can be provided with own office logos and transmit the print to the practice network printer interface.

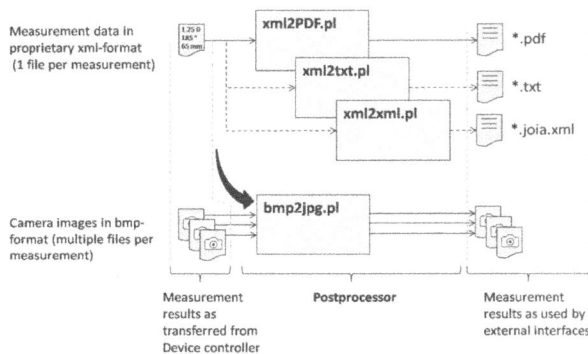

Figure 1: Structure of internal data processing. On the left side the measured data can be seen. These data are processed in the post-processing and converted into several formats (.pdf, .txt, .jpg, .xml). The measurement results can be used by external interfaces.

2.2 Output protocols

The output protocols enables the transmitting of the processed measurements from the post-processing to the intended interfaces of the respective PMS/EHR. Every interface shows its own output protocol only. Fig. 2 shows an example data transfer.

Therefore, the postscripts can be simply dispatched by different output protocols. The connection to a PMS/EHR is carried out by different interfaces SOAP (Simple Object Access Protocol) or REST (Representational State Transfer). The majority of the equipment uses the SOAP method guarantees a safe data transmission. The SOAP protects always requires writing or using a provided server program (to serve data) and a client program (to request data) [2].

Every SOAP-protocol expects a request and returns an acknowledge on successful receipt to confirm perfect or default data transmission. The SOAP and the REST interfaces are declared for all systems identically, nevertheless, every device needs a different implementation. The flexible design of the interfaces allows the system to create more output protocols for further connected systems to achieve the greatest benefit for the customers.

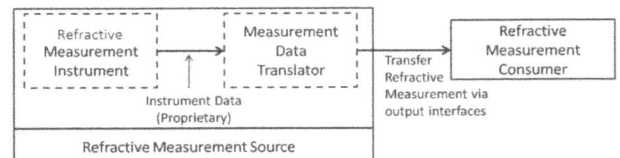

Figure 2: Schematic example for a refractive measurement data transfer. The measurement source translate the proprietary data and transmit it via given output interfaces to the electronic health record system.

2.3 Configuration interfaces

A server application is implemented on the devices, therefore, every device acts as its own server which can be approached by his IP address. The server communication is based on the REST interface. In general, the requirement has to correspond to all characteristics of the client server architecture. Therefore, the server provides a service which can be required on a request by the client [3]. The REST interface is convinced by the uniformity of the interface and its easy use. As well as it is visible in the succeeding application, the use of REST is often preferred compared to the more difficult SOAP styles because REST does not need as much bandwidth, which makes it more suitable for using the internet. Fig. 3 shows the communication interfaces, enhanced before.

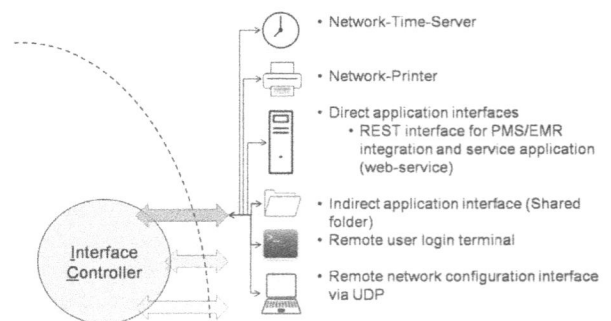

Figure 3: Given output interfaces from device Interface Controller. Connection via LAN interface, for example to server applications or electronic health record systems.

3 Results and Discussion

3.1 Web interface

Different service applications could be recalled by the PMS/EHR via the web interface. The customer can get specific requests to the web interface by an installed server on device. With these interfaces many new possibilities are available for the customer or the service technician. In this way doctors can see the last measuring data on the device. They can distribute it by a new post-process and transfer them if necessary to a network folder. Besides, it enables doctors to carry out settings at the device without leaving the workstation. Settings like user-defined profiles for measurement parameters can be done. Also of interest, particularly in regard to research purposes, is the function of *.csv* generation, where the customer can get the measurement data. An example therefor is the analysis of doctors workflow or the scientifically analytic. Even the setup of a printer-network is now available for the doctor to print measures values at the network printer to get a high-quality output of the measuring data for the patient.

On the other side the web interface makes the maintenance of the device easier for the service technician. Due to the network interface, updates can be transmitted automatically on the device by an update function, for example a new firmware or new output interfaces. Now also *.log* files are visible and are sorted for the service technician. Based on these list he can carry out device settings by the interface.

Both the doctor and the technician are able to see which other devices are in its surroundings and, without changing the workstation, switch to another device and communicate automatically with the new server to get the measurement data and setting parameters. In addition to this, in the following will be described how broadcast questions are used.

3.2 Remote Network

In this section an overview will be discussed the remote network subsections. Section 3.2.1 explains how devices can communicate via broadcast among each other. And at least section 3.2.2 describes an example workflow and explains how the JOIA protocol is a part of it.

3.2.1 Broadcast

The respective device is coupled with an interface LAN to the internal network. The device can interrogate its environment by a broadcast in this network. Here it acts on the one hand in the remote function only as an responder, on the other hand however as a server. Therefore a direct communication among the devices is given and the user is able to switch from one device to another by the interface. The broadcast doesn't have its value only in the web interface,

it also offers the possibility of connecting the equipment intelligently. The following example describes a consultation: A patient is having several measurements with different devices one after another, accompanied by the user (doctor or nurse). The measuring starts at the first device which received the patient-related master data sent from a SOAP interface before, as shown in Fig. 4. At the next device a measurement is also carried out. However, there were no measurement announced previously by the PMS. Now, due to the intelligent connecting of the broadcast, the device can request its direct environment, whether a measurement with master data was carried out during the last minutes. So these master data can be transmitted among each other. Now the user gets all patient's data indicated which have been measured in its environment, during an ascertain passed time and select the concerning patients. The device matches the patient meta-data directly with the measuring values, which learn a qualitative revaluation in this way and will be assigned directly in the PMS to the patient.

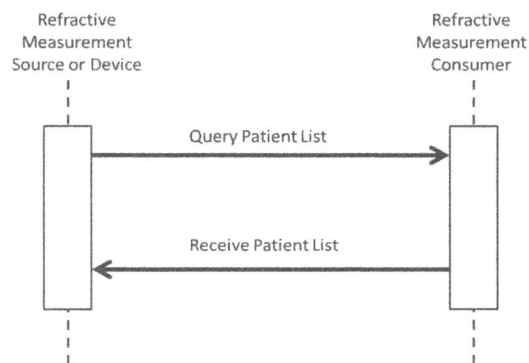

Figure 4: Communication between Device and PMS/EHR. The measurement source transmit a patient list query. Due to, the electronic health record system receive the patient list and the source can matches the measurement and patient data together.

3.2.2 JOIA and Workflow

Refractive measurement devices typically last for many years. Therefore devices may not have the ability to conform to the JOIA XML specification. They are only capable transmitting measurement data in a proprietary format. This requires that the instrument's measurement data are transmitted to an intermediary, a software application that captures the instrument's measurement data (in a proprietary format) and translates it into a standardized message format before transferring it to the target.

Since neither the instrument nor the capture/translation program have access to patient information, the capture/translation program will set the *PatientIDSource* field in the JOIA XML data stream to "unreliable" in order to notify the Refractive Measurement Consumer (target system or instrument) of that fact. The Refractive Measurement Consumer is required to associate the received measurement data with a specific patient session (an appointment or examination). The Refractive Measurement Consumer can only accept this measurement data if a valid patient session context exists for the data.

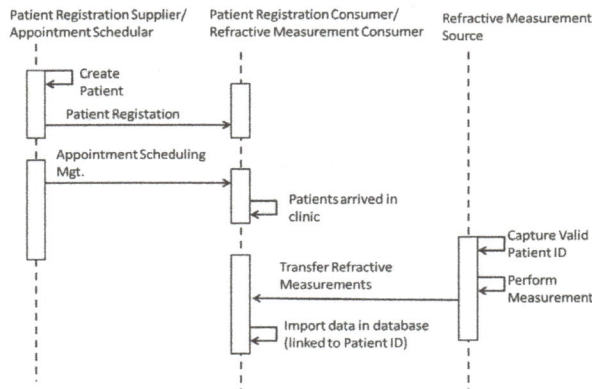

Figure 5: Refractive Measurement Model (with valid Patient ID) - Workflow Diagram to Administrative Process.

The model at Fig. 5 specifies the transactions and actors which are required for the scenario where organizations are integrating refractive measurement devices that are able to incorporate a valid Patient ID when providing refractive measurements to their EHR system. The Refractive Measurement Source accomplish this directly or by a Refractive Measurement Source Importer. Therefore, the Refractive Measurement Consumer (typically an EHR) is required to use the Patient ID provided by the Refractive Measurement Source or a Refractive Measurement Source Importer to establish the context of the patient. It uses the context to provide the correct patient information when importing the measurement(s) into its database. Refractive measurements are conveyed using the JOIA XML based specification [4].

4   Conclusion

The result of this work is that the implemented scripts and its uses give a clear contribution to the digitization of the data transmission. All interfaces post-processors or output interfaces allow a maximum flexibility of utilization to the user as well as the manufacturer. In particular, the external processes like broadcasting and the servers application ensure in practice a relief of the workflow and connects the device at the same time. This integration allows an increase in the data quality and supports the standardized JOIA-XML format. In that way the user can utilize devices with different practice management systems. Besides, the integration of an interface LAN enables the access to the practice-internal network, for example the printing of patient's data in DINA4 over the network printer. This gives the patient an increased feeling of valency. Applications like the web interface or shared folder can support the doctor at his daily routine. Moreover, it offers also a better support by the company's service or technician.

The task of converting the increased customer requirements in the area of the digitalization is a progressive process. Further requirements that join new possibilities will find approval. The integration of the data interfaces and its split in two processes enables the implementation of post-processors or interfaces in future and can be installed by a software update via the network-interface on the device. Thus the demanded flexibility is fulfilled and represents the basic for further expansions of the different interfaces.

Because of strict regularization and new laws, the medical technology is at the very beginning of digitalization. How politics and companies handle the problem with the data security will be one of the biggest questions within the next years. Without basic regulations concerning patient data, this process will not progress faster. The customer will have an enormous advantage by digitalization progressing especially regarding better diagnosis and better personalized treatment [5].

5   Acknowledgements

The work has been carried out at Carl Zeiss Meditech AG, Oberkochen and supervised by the Institute of Telematics, Universität zu Lübeck.

6   References

[1] R. Noffsinger, S. Chin *Improving the Delivery of Care and Reducing Healthcare Cost with the Digitization of Information*. Jornal of healthcare information management, Vol. 14, No. 2, 2000.

[2] F. Curbera, M. Duftler, R. Khalaf, W. Nagy, S. Weerawarana *Unraveling the Web services web: an introduction to SOAP, WSDL, and UDDI*. IEEE Internet Computing, Vol. 6, Issue: 2, 2002.

[3] R. T. Fielding *Architectural Styles and the Design of Network-based Software Architecture*. University of California, Irvine, 2000.

[4] IHE Eye Care Technical Committee *Unified Eye Care Workflow Refractive Measurements (U-EYECARE Refractive) Based upon JOIA 1.5 Realease*. IHE International, Inc., 2016.

[5] P. J. Gorman, A. H. Meier, C. Rawn, T. M. Krummel *The future of medical education is no longer blood and guts, it is bits and bytes* The American Journal of Surgery, Vol. 180, Issue: 5, 2000.

MEDEAS – Development of an Medical Device Development Assistance System

J. Abeler[1], A. Mildner[2] and M. Leucker[3]

[1] Medizinische Informatik, Universität zu Lübeck, julia.abeler@student.uni-luebeck.de
[2] UniTransferKlinik Lübeck GmbH, a.mildner@unitransferklinik.de
[3] Institute for Software Engineering and Programming Languages, Universität zu Lübeck, leucker@isp.uni-luebeck.de

Abstract

Medical devices are an essential part of today's medical care. The increasing complexity of these systems require the consideration of various specifications and regulations. Speed and success of the development can be increased significantly by the use of methods that monitor the documentation process. Up to this date there is no comprehensive assistance system that combines regulatory information with an engineering model. Here we describe the development of a web-based medical device development system called MEDEAS. It connects a method-based engineering model with the respective regulatory information. In addition it includes currently developed regulations and standards for networkable medical systems. MEDEAS is a versatile tool for the creation of a compact documentation and leads to a more targeted development of medical devices.

1 Introduction

Today, medical devices are an integral part of the everyday life of registered doctors, surgeons and operating-room-staff. In particular, the environment for surgical procedures has become increasingly complex in the past decades. New surgical techniques are being developed and improved constantly. This increasing complexity of medical devices results in a challenging development, which is accompanied by a regulatory process. Directives in Europe obligate manufactures to adhere to regulatory requirements regarding performance and safety [1]. The manufacturer may use harmonized standards, which reflect the current state of the technology, to demonstrate the complicance. The variety of the complex regulations hinders the manufacturer to develop and commercialize a product confirming to the regulatory requirements. In addition several publications [2], [3] demonstrated that the usage of a methodical product development often results in better and more competitive products. Till today this approach does not find approval in the development of medical devices. Although there is software which supports the development of medical devices with methods like e1ns from PLATO GmbH [4] and CARAD from SurgiTAIX AG [5] but they do not support the whole engineering process and provide no information about the regulatory affairs.

For this reasons a web-based medical device development assistance system called MEDEAS was developed, which represents the regulatory affairs in a structured manner and gives an example of an engineering model that integrates development- and quality-management-methods in the product development process.

2 Material and Methods

The regulatory and normative requirements have to be considered from the engineering process and the launch right up to the decommissioning of medical devices. The european Medical Device Directive (MDD) which imposes requirements on the development is transposed into German national law by the Medical Devices Act (short MPG from German Medizinproduktegesetz) and supplemented by several regulations [6]. It ensures the safety for patients, users and third parties through a minimum technical level and, from a regulatory perspective, provides steps for the development of a medical device. One of this steps is the observance of the essential requirements on safety and performance. The essential requirements demand the consideration of four main disciplines: risk management, quality management, software life-cycle processes and usability. Manufactures may use harmonized standards to monitor the compliance in these disciplines. Fig. 1 represents the relationship between the law, disciplines and standards.

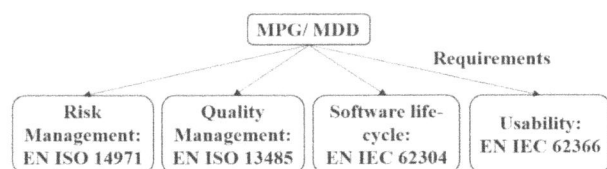

Figure 1: The European directive MDD is converted into German law by the MPG. They requires the compliance of four disciplines. To demonstrate the compliance manufacturer may use harmonized standards.

2.1 Networkable Medical Systems

The development of networkable medical systems becomes increasingly important. Medical devices produce a growing amount of medical data. Connection of this data will accelerate the diagnosis and therapy. To ensure this, medical devices should be integrated into the hospital information system and connected to each other. In the past, this was only feasible by using isolated proprietary solutions from a single manufacturer, because there were no open standards for the interconnection of devices [7]. Currently an open IEEE-standardfamily 11073 is developed, which should simplify the connection between devices of different manufactures. It defines a standard communication protocol and a data format whereby the devices not only communicate but also may interpret the received data [8].

In addition the networking imposes new requirements on the development process and generates new hazards during the operation. In a recent work a new concept for the accreditation and risk management process was presented [7]. They defined a standardized device- and service-profile which includes all necessary information for a safe and reliable interconnection. The profile can be used on the one hand by manufacturers of medical devices during the development process in order to make their devices compatible with intended network partners and on the other hand by operators to verify the safety and functionality of specific device pairs. The content of this profile is briefly described in the following.

The medical intended use of the original medical device to be interconnected is usually not affected, but the intended use of the interconnection itself must be documented. This is considered in an extension of the original intended use of the device. The manufactures have to determine whether services for the connected device can be visualized, used or controlled and in addition the type of data exchange. For every medical device a risk analysis regarding the patient or user harm has to be performed. By the interconnection the risk analysis changes only marginally. Additionally the manufacturer has to perform the so called technical risk analysis regarding hazards concerning changed parts, components and software modules of a medical device and hazards concerning the interconnection and communication over an IT-Network. Before the medical devices can be interconnected a comparison of the specification of the components used to implement the function and the protocol specification has to be performed. Only when this specification requirements are met, devices may be connected. For this reason manufactures have to define a technical interface description based on the IEEE 11073.

2.2 Engineering- and Quality-Methods

To support the development of medical devices and to increase the success of the product, the manufacture may use engineering- and quality-methods. These methods determine, in the scope of quality management, how steps and engineering phases have to be performed and how results have to be documented [2]. Therefore they define how specific inputs may be prepared into outputs.

The regulatory affairs do not provide methods for each development phase but provide isolated proposals, like the Failure Mode and Effective Analysis for the risk management process [9]. These methods were supplemented in MEDEAS with the following methods [10]:

- Quality Function Deployment (QFD): transfers customer wishes into technical requirements

- Preliminary Hazard Analysis (PHA): facilitates a preliminary risk analysis without knowing the product and his construction; usually checklists are used

- Fault Tree Analysis (FTA): analyses relationships between hazards and their causes; represented in a tree structure

- Failure Mode and Effects Analysis (FMEA): analyses the possibility of hazards, the causes and the consequence during the engineering process

- System Modelling Language (SysML): expanded UML for the representation of electronic systems

- Matrix Analysis: connects the ideas and expectations of the customer with the functions and elements of the system

- Design Verification Plan & Report (DVP&R): supports the verification of the specifications

- Design Review Based on Failure Mode (DRBFM): integrates change management into the engineering process

2.3 Implementationtool of MEDEAS

To make MEDEA available to as many manufacturers as possible it was implemented as a web-application. To keep the development short the open source content development system (CMS) Contao was used. Contao is a lean but powerful and easy-to-expand CMS that is comparatively easy to learn. Interactive functions such as navigationbars, picture galleries are already installed or easily integrable and do not have to be programmed individually.

3 Results and Discussion

The main aim of the assistance system is the structured overview of the regulatory requirements in the development of medical devices. In order to ensure a plain and precise overview the user can focus either the respective regulatory framework, networkable medical systems or an engineering model. This separation in combination with the hierarchical organisation ensures quick access to the required information.

The navigation through MEDEAS is divided into the main navigation, with only two sub points, and a breadcrumb, to make it more clear. The different navigations and their structures are shown in Fig. 2.

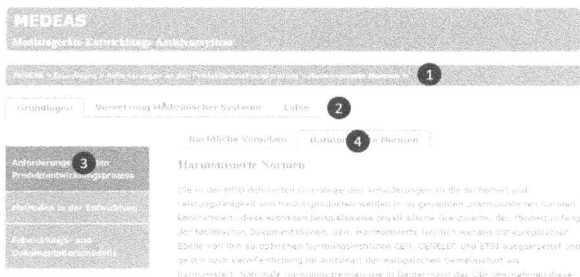

Figure 2: Navigation through MEDEAS. 1: breadcrumb 2: main navigation 3: second navigation level 4: third navigation level

The main navigation is divided into three topics and are called: *Basics*, *Networkable Medical Devices* and *Guide*. The navigation point *Basics* provides information on the regulatory requirements for the product development process of medical devices like it is briefly described in Sec. 2. This allows the manufacturer to get an overview of the legislation in Europe and obtain summaries of the essential requirements of the MDD and the harmonized standards. Furthermore different engineering methods are explained (see Sec. 2.2). The described methods are not a complete list of engineering methods, but only the one, which are used during the presented engineering model. The description in MEDEAS contains the benefit of the method, the aim, which should achieved through the usage, the process of the method and the pro and contra.

The trend in the development of medical devices is to integrate them into the hospital information system or to interlink them. For this reason, the state of research in the development of networkable medical devices is presented under *Networkable Medical Systems* (see Sec. 2.1). The main focus is the difference of the development of proprietary networked and cross-linkable medical devices.

3.1 Engineering Model – Guide

The main aspect of MEDEAS is the representation of requirement methods in an engineering model [9]. It shows the manufacturers how an integrated software and hardware engineering process can be applied to the regulatory requirements of medical devices with the aim of a slimmer and more targeted development documentation. The model integrates and fulfills the four disciplines of the MDD, which may be applied generically to medical devices. As studies in the product development of mechatronic systems have shown, better and more competitive products are developed if the development process is supported by methods [9]. For this reason the model applies developing methods, such as FMEA and DVP&R, on the development process of medical devices.

The individual requirements of the MDD and the relevant standards were divided into phases of product development: pre-analysis, requirements analysis, functional specification, system-modelling, realization and verification, integration and system verification, change management and

problem solving. Each phase was stored with a possible method. The respective output of each phase documentation always forms the input of the following phase with connection to the regulatory requirements. Fig. 3 displays a schematic of the structure of the engineering model with the documentation requirements.

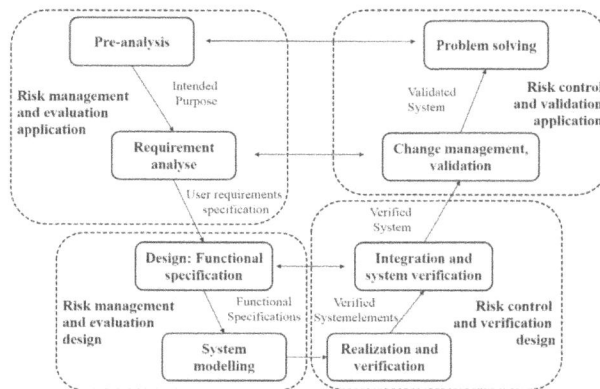

Figure 3: Structure sketch of the engineering model. The V-model represents the phases of the product development with the documentation requirements as the in- and output. The dotted parts define the risk management activities.

This illustrated engineering model is represented in MEDEAS under the navigational point *Guide*. The navigation through the single steps of the model occurs in the same structure as the engineering model and is demonstrated in Fig. 4 number 1. This structure directly shows the manufacturer his progress in the development process, including preceding and following phases.

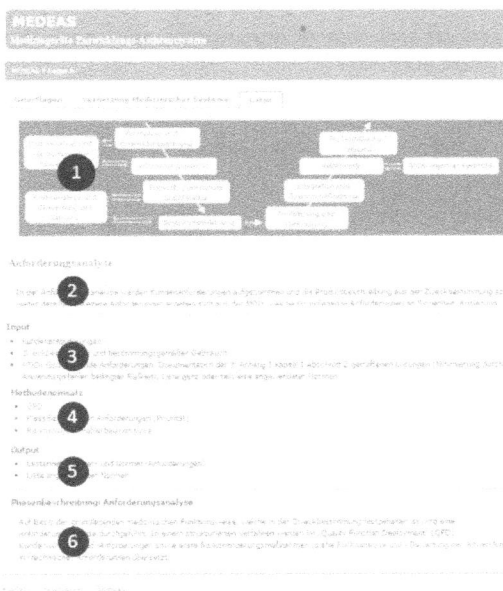

Figure 4: Page Layout of the guide. 1: Navigation 2: Summary of the phase 3: Input 4: Methods 5: Output 6: Phase description

The composition of the particular phases is always based on the same principle. First of all there is a short summary of the phase, so the manufacturer gets an overview of what the phase focuses on (Fig. 4 number 2). In the next segment three lists specify the inputs, the methods, which will be used, and the outputs of the phase (Fig. 4 number 3-5). The inputs are composed of regulatory requirements and the documentation output of the previous phase. The output always portrays the generated documentation e.g. the intended purpose or the requirements specification. The last segment gives a comprehensive step-by-step instruction of the phase with examples, possible documentation and the benefit of the methods (Fig. 4 number 6).

3.2 Discussion

As mentioned previously the support of the development process of medical devices through engineering methods is not a new innovation. Software solutions of several manufacturers exists, which provide the use of methods. But these software solutions do not facilitate the representation of the whole life-cycle of medical devices by methods. They provide only isolated methods, especially for risk management. Furthermore, no solution is linked to the regulatory requirements. Additionally, there are many individual standards which together form the regulatory framework for the development of medical devices, but do not employ the usage of engineering methods, which may accelerate the development process.

For these reasons, the implemented web-based development assistance system MEDEAS is a pioneer in the linking of regulatory requirements in medical technology with development methods. Through its use the engineering process might be shortened and allows a more streamlined and targeted documentation.

4 Conclusion

A web-based assistance system MEDEAS was developed. It outlines the development process from the regulatory perspective. The requirements of the MDD and the harmonised standards are summarized and expanded with the current state of research in the development of networkable medical devices. Futhermore the requirements of regulatory affairs are connected to an method-based engineering model [9]. It describes a methodical and purposeful, standard and regulation-compliant systematic of an integrated development documentation.

Using this connection MEDEAS demonstrates and supports the preparation of a compact, more targeted development documentation of medical devices and accelerates the development process.

Acknowledgement

The work has been carried out at UniTransferKlinik Lübeck GmbH and supervised by the Institute for Software Engineering and Programming Languages, Universität zu Lübeck.

5 References

[1] A. Gärtner *Medizinproduktesicherheit Band 1: Medizinproduktegesetzgebung und Regelwerk*. TÜV Media GmbH, Köln, 2008.

[2] K. Ehrlenspiel, H. Meerkamm *Intergierte Produktentwicklung - Denkabläufe, Methodeneinsatz, Zusammenarbeit*. Carl Hanser Verlag, München, 2013.

[3] K. Ehrlenspiel *Integrierte Produktentwicklung - Methoden für Prozessorganisation, Produkterstellung und Konstruktion*. Carl Hanser Verlag, München, 1995.

[4] Plato GmbH *e1ns*. Available: `http://www.plato.de/plato-e1ns.html`, [last accessed on 2017-07-02].

[5] SurgiTAIX AG *Riskmanagementsoftware CARAD*. Available: `http://www.surgitaix.com/cms/index.php/Risikomanagementsoftware_CARAD` [last accessed on 2017-02-07].

[6] C. Johner, M. Hölzer-Küpfel, S. Wittorf *Basiswissen Medizinische Software*. dpunkt.verlag, Heidelberg, 2011.

[7] A. Mildner, et al. *Development of Device-and Service-Profiles for a Safe and Secure Interconnection of Medical Devices in the Integrated Open OR*. In: Risk Assessment and Risk-Driven Testing - Third International Workshop, RISK 2015, Berlin, pp. 65-74, 2015.

[8] M. Kasparick, S. Schlichting, F. Golatowski, T. Timmermann *Medical DPWS: New IEEE 11073 standard for safe an interoperable medical device communication*. In: IEEE Conference on Standards for Communications and Networking, CSCN 2015, Tokyo, pp. 212-217, 2015.

[9] A. Mildner *Konzept für die integrierte und methodenbasierte Entwicklung interoperabler Medizingeräte*. Lübeck, 2016.

[10] G. Kamiske *Handbuch QM-Methoden*. Carl Hanser Verlag, München, 2013.

Creating Reports with BIRT to support the Development Process of Medical Software

M. Baake[1], A. Henke[2], E. Kessner[2], and J. Ingenerf[3]

[1] Medizinische Informatik, Universität zu Lübeck, merle.baake@student.uni-luebeck.de
[2] MeVis BreastCare Gmbh & Co. KG, {anna.henke,ekaterina.kessner}@mevis.de
[3] Institut für Medizinische Informatik, Universität zu Lübeck, ingenerf@imi.uni-luebeck.de

Abstract

The development process of software is accompanied by a huge amount of data. Especially companies which develop medical software have to be particulary detailed in the documentation to make sure errors in the software do not lead to patient risk. Generating reports helps to comprehend bugs and get an image of the current state in the development process. They have to be built on the basis of distributed data within the company's complex data infrastructure. But not only collecting data is important, furthermore it has to be displayed in a user friendly way. In this paper, reports for MeVis BreastCare GmbH & Co. KG (MBC) with the help of Business Intelligence and Reporting Tools (BIRT) are discussed. BIRT helps to gather, organise and present heterogenous information. Contrary to standardized reports, the user can create reports that visualise the data in tables or different kind of charts and personalize the reports with the help of parameters.

1   Introduction

MBC is a joint-venture between MeVis Medical Solutions AG (MMS) and Siemens Healthcare GmbH [1]. It develops software for the diagnosis and therapy of diseases based on different kind of images coming from Magnetic Resonance (MR), Computed Tomography (CT), as well as Mammography (MG) and Tomosynthesis. Most of the created applications are integrated into Siemens' imaging software syngo.via[2]. The respective MBC team extends syngo so that it depicts the radiologists workflow depending on the imaging technique.

Implementing medical software, MBC has to comply different a lot of requirements. *ISO 13485* is one of many standards for the medical device industry that sets the tone for the quality management system. Depending on the country where the software is distributed, specific regulations have to be fulfilled. The U.S. Food and Drug Administration (FDA) for example determines the restrictions for medical devices in the USA [12]. MBC has to ensure the integrity of the documentation to make sure requirements are met and the risk of undetected bugs is reduced. For that, reports are generated. In general, they show the productivity, potential and perspectives of a company in the process of software development [4]. They also help to get to the sources of errors by finding and displaying bugged objects before they interfere with the patient data.

The thorough documentation of the development process means a huge amount of data descending from customer requirements, functional specifications, test results and bugs. This data needs to be gathered, processed and connected. There are different kinds of both commercial and free tools helping to collect the mostly heterogenous and distributed data and visualise it in the form of charts, tables and lists [5].

2   Material and Methods

To get a hold of the amounts of data, MBC uses Jama. The Jama web application serves the purpose of helping organisations in the management of their projects [3]. Some built-in reports can be generated with the web application that are based on common reporting requirements. Those standardized reports often lead to very comprehensive results [6]. Reports not being customized to specific use cases harbor the danger of loosing important key facts in the mass of information. For custom reporting, Jama supports the reporting tool BIRT and the template engine Velocity. The latter is used as well to generate reports for MBC. Contrary to BIRT, with Velocity the data can not be visualised in the form of charts and graphs [7]. Moreover, rich HTML text that is used in the description of Jama items can not be rendered. Therefore, BIRT was chosen to generate more flexible and user-friendly reports. This open-source software is integrated in Eclipse so apart from the development environment, no further software is needed [8]. The reports created with BIRT are dynamic. With every run, the information of the source is pulled again so that changes in the data source have direct impact on the report's results.

As shown in Fig. 1, BIRT connects to the Jama database and retrieves information through SQL queries to generate the reports. Apart from databases, BIRT can utilize different kinds of data sources, for example XML files. After

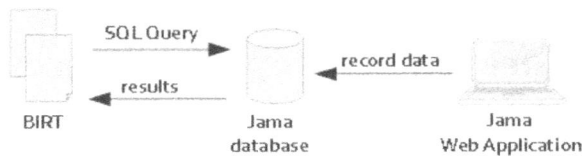

Figure 1: Data is retrieved from the Jama database which holds the information from the web application through SQL queries. The results are visualised in the reports.

defining a source, data sets are built by executing custom SQL queries that extract the relevant fraction. In the query, report parameters can be considered to adapt the outcome in favor of the use case. The following recursive SQL query shall exemplify the limitless complexity in the retrieval of information provided by BIRT. It builds the base for the report described in section 3.3.

```
WITH temp (passed_key, derived_key, derived_id) as
(SELECT passed.documentKey as passed_key,
    derived.documentKey as derived_key, derived.id as
    derived_id
FROM document as passed, document as derived,
    document_document as link
WHERE link.toDocument=passed.id AND
    link.fromDocument=derived.id AND
    link.relationshipType=8
AND passed.documentKey like ?
UNION ALL
SELECT all_derived.id, all_derived.documentKey,
    temp.from_Key
FROM document as all_derived, document_document as
    all_link, temp
WHERE link_all.toDocument=temp.derived_id AND
    link.fromDocument=all_derived.id
AND link.relationshipType=8)
SELECT DISTINCT * from temp
ORDER BY derived_id DESC
```

With the help of a temporary created table and the operator *UNION ALL*, it recursevly searches the Jama database for an item's predecessors. The table *document_document* links two items based on different relationship types. In this case the *toDocument* ist *derived from* the *fromDocument*. The item in question is passed as a report parameter, indicated by the question mark. Instances of the same table are named to address them seperatly.

The actual report then consists of different elements like tables, charts or text fields, which bind the datasets' data. To position the elements on the working space, they can be structured into a grid. For better readability, the elements as well as their content can be further formatted individually.

Every element triggers a sequence of events while being generated. With a handler in Java Script, the elements can be changed or used dynamically in these phases. In the following code snippet, the content of a table row is dynamically put into a global variable. With that, it can be accessed from anywhere in the report, for example as a parameter for another data set.

```
my_var=
    .getRowData().getColumnValue("     ");

reportContext.
    setGlobalVariable("         ",my_var);
```

Furthermore, a reference to another report can be depositted in an element and data can be passed to this subreport as its report parameter.

Even if only a web browser is needed to look at the reports, the report can be exported to other formats like PDF, XLS or CVS. The web viewer can be deployed in a servlet engine to make the reports accessible through a URL.

3 Results

The following descriptions exemplify the practicability of reports in general and also the reporting tool BIRT. The reports are based on general use cases or requests by other employees after bumping into bugs or inconveniences in the web application.

3.1 Test Runs on outdated specifications

Per work instructions valid for MBC, test cases should be executed in the most recent version to make sure that all changes of the software are considered while testing. It was found out that quite a lot of test runs were executed on outdated specifications as Jama does not support to filter on this use case. The changes in the test case after the last test run may lead to different results. A BIRT report was created to list the affected test cases. The underlying SQL query queries the *documentKeys* of test cases that were tested at least once. However, the entry in the row *Test Case Version Number* of the related test runs does not match the most recent version number of the test case. The output was put into a simple list element in BIRT and the keys were linked to the Jama application. Fig. 2 shows the resulting report. Furthermore, a parameter was set that makes the user choose a specific project. With that, irrelevant test cases are not presented to the user. The report lists test cases that need to be examined further to decide if the changes would not affect the result of another test execution or if it should be tested again to make sure no errors were overlooked.

Figure 2: The report lists the keys and additional information of tests that were only executed in outdated versions.

3.2 Test Cases with wrong status

Another report does not evaluate the omission of test management but rather a bug in the web application. Some test cases in the web application have the status *In Testing \Scheduled for testing*, but no test runs are associated with them. If a test case is planned to be executed, test runs are created and the *Test Case Status* is changed from *Not Scheduled* to *In Testing \Scheduled for testing*. After the execution, the status is changed to *Passed* or *Failed*, depending on the result. Looking into the database, it showed that the test cases with the incorrect status all had test runs that were inactive. Inactive items in the database are items, that have been deleted in the web application. The buggy test cases seemed to be planned once, but the test runs were deleted. The *Test Case Status* is automatically changed by Jama, but this does not work in case test runs are deleted prior to their execution. A SQL query was written that first pulls all the *documentKeys* of the test cases with the status *In Testing \Scheduled for testing*, and then narrows them to the ones only having inactive test runs. As seen in Fig. 3 the output is listed in a simple list element and a parameter was set that restricts the output to test cases in a specific project. The report should make sure that no test executions are missed and therefore no bugs stay undetected. By highlighting the faulty test cases conclusions can be drawn concerning the source of the error.

Figure 3: The report lists the keys of test cases with a false status.

3.3   Complete version history of an item

After the release of a software version, the project including all items are copied and further changes are based on those copies. In Jama's version history, only the changes on the copy of the current project are displayed. The former released version of an item and its changes can be accessed by looking at the relationships. A BIRT report was created to bypass this indirect way. As described in section 2, with a recursive SQL query all the items of which a passed item key is derived from can be retrieved. The output was put into the first row of a table with two rows. The second row was filled with another table that contains the entries of the *version* table, which holds all changes of an item, for each item from the first dataset. For that, the output of the previous dataset is stored in a global variable and passed as a parameter to the second dataset. The resulting report in Fig. 4 displays a complete version history of an item that not only shows the changes of the input item, but also the changes made within former releases.

3.4   License usage in Jama

Currently, MBC shares the Jama web application and database with MMS. In the future, the resources will be

Figure 4: The report displays the change history of an item as well as the changes of it's predecessors.

seperated. The quantity of concurrent user accessing the application is restricted by licenses. Alongside licenses that allow certain users to access the application at all times, some users have *floating licenses*. MMS and MBC share five *floating licenses*, which means that only five people with those licenses can use the application at the same time. The licenses now have to be distributed. For that, a report was created, showing a bar chart in which the employee's activities per day are counted in a specific time frame. The bars are devided into the affiliation of the users to MBC or MMS. With that, the report shown in Fig. 5 shall help in the decision of which company needs how many licenses.

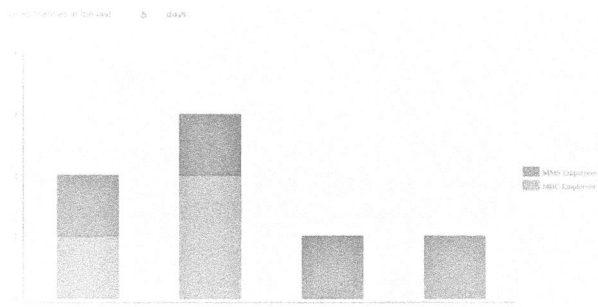

Figure 5: In a bar chart, the activities from users with a *floating license* are counted over a specific amount of days. Moreover, the affiliation of each user to the company highlighted.

Clicking on a bar of the bar chart, the user is redirected to a subreport listing the individual names of the employees responsible for the counted avtivities. The bar chart report passes a list of user ids to the subreport as the report parameter. The subreport thereupon uses this list to query the database for the employees' names.

3.5   Discussion

The use of the individual reports described in this paper is apparent from the specific use cases that led to their generation. A detailed evaluation of BIRT can be done in the future when it is clear if the reports are benifitial in the company's everyday life.

The biggest obstacle while creating the reports was to understand the structure of the Jama database. Not every connection between the tables is obvious and different ways exist to express certain attributes of an item. For example, different *ids* are used to express the same test case status. As seen in section 2, complex queries were needed for some reports including multiple joins or a recursive structure.

The report designer was quite easy to understand due to the existence of many tutorials. BIRT leaves out aspects like viewing rights and scheduling updates [5]. The latter would definetly be an improvement as it would enable the reports to run incidentally to keep the company up-to-date at all times without specifically looking for new information. As of now, the created reports can be accessed through a HTML page, shown in Fig. 6. By providing the software

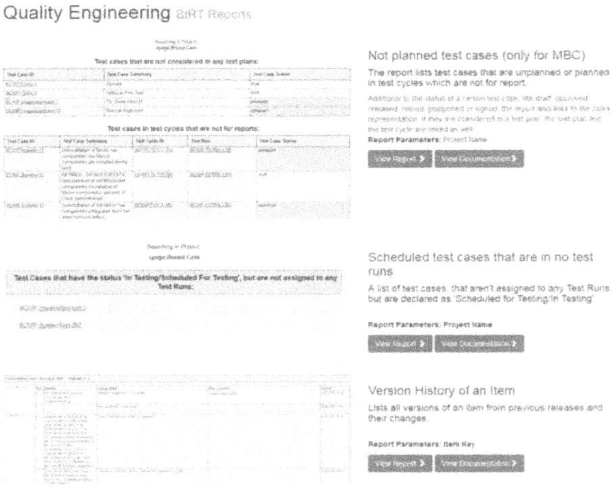

Figure 6: All reports relevant for the MBC employees can be accessed from a common HTML page that not only links the reports, but also holds detailed information concerning their usage.

Information Hub (iHub), the BIRT sponsor Actuate makes a more dynamic dashboard possible [10]. With that, the created BIRT reports can be run and displayed next to each other. Furthermore, properties can be set to make all reports refer to the same paramater, which can be changed individually by the user, and reports are updated automatically after a specific time. The iHub dashboard was evaluated but not employed yet.

4 Conclusion

With BIRT reports, business data can be provided purposeful, user-friendly and flexible to users. Being able to visualise the company's data in charts and tables and render rich HTML text, it has significant advantages over other free reporting tools. Coming from an informatics background, it was quite easy to understand the tool's workflow. Nevertheless, the creation of SQL queries to get the needed information can be tricky depending on the used database.

Next to their adaptability, the main advantage of reports is their up-to-dateness. They do not create a unique image of the data, but pull the information from the data source with every run. Changes of the underlying database have a direct effect on the report's outcome. A self-updating dashboard displaying all reports in one place would make even more use of this advantage.

Acknowledgement

The work has been carried out at MeVis BreastCare Gmbh & Co. KG and supervised by the Institute of Medical Informatics, Universität zu Lübeck.

5 References

[1] K. Holtmann, *Ertragssprung durch Akquisition: MeVis Medical Solutions übernimmt substantiellen Teil des gemeinsam mit Siemens betriebenen Joint Ventures*, Corporate News, Bremen, 2008.

[2] syngo.via, https://www.healthcare.siemens.de/medical-imaging-it/clinical-imaging-applications/syngovia [last accessed on 2017-01-15].

[3] Jama Software, https://www.jamasoftware.com [last accessed on 2017-01-15].

[4] D. Keil and M. Gasper, *Berichte erstellen mit BIRT*. eBusiness-Lotse, Oberschwaben-Ulm, 2015.

[5] C. Brell and T. Kieninger, *Freie Reporting-Tools im Vergleich*. JAVASPEKTRUM, 2007.

[6] W. Mende, J. Robert and N. Ladas, *BIRT - Das Berichterstellungsprogramm für (fast) jede Anwendung*. Institute for Transfusion Medicine, Hannover Medical School, Hannover, Germany.

[7] Jama User Guide 8.10-OP. Jama Software, Inc, 2016.

[8] BIRT - About. http://www.eclipse.org/birt/about [last accessed on 2017-01-15].

[9] Eclipse documentation - Current Release. http://help.eclipse.org/neon/index.jsp, [last accessed on 2017-01-15].

[10] OpenText Information Hub (iHub). http://www.opentext.com/what-we-do/products/analytics/opentext-information-hub, [last accessed on 2017-01-15].

[11] ISO 13485:2016 http://www.iso.org/iso/catalogue_detail?csnumber=59752, [last accessed on 2017-01-23]

[12] Medical Devices - U.S. Food and Drug Administration http://www.fda.gov/MedicalDevices/, [last accessed on 2017-01-23]

Replicating and Synchronising a Medical Database in an Object Relational Mapping Context

J-H. Mathes[1], O. Meincke[2], H. Ulrich[3], and J. Ingenerf[4]

[1] Medizinische Informatik, Universität zu Lübeck, mathes@mi.uni-luebeck.de
[2] Epikur Software GmbH, Berlin, Oliver.Meincke@epikur.de
[3] IT for Clinical Research, Universität zu Lübeck, hannes.ulrich@itcr.uni-luebeck.de
[4] Institut für Medizinische Informatik, Universität zu Lübeck, ingenerf@imi.uni-luebeck.de

Abstract

Connecting distributed database systems is an omnipresent topic in the database community. It varies strongly for different database systems and their environment. This paper describes a way to synchronise and replicate a database system, which uses object relational mapping to store data. Considering the ease that object relational mapping offers to save complex data models in the database, a generic solution is of great use. To develop a solution, a test environment in a setting of a medical practise was selected. Within this model the doctor has to work offline, if no connection to the database can be established. In this setting possible problems where identified. After collecting and comparing different methods to solve these problems, a solution was implemented, tested and evaluated. The evaluation showed that the used methods are not sufficient to solve all occurring problems and that there is no possibility to implement a fully generic solution.

1 Introduction

Database replication is a basic mechanism to increase availability and performance of distributed databases. Although there are many different protocols for relational databases, these solutions do not solve the problem within an object relational mapping (ORM) context.

In addition the ability of the systems to work in a portable, partially offline, environment becomes more and more important. This changes the requirements for the strategies used to synchronise distributed databases, since concurrent transactions can no longer be prohibited. If they were prohibited the usability of the software would suffer.

The utilization of ORM is reasonable for numerous use cases for example: multimedia systems, modelling and analysing statistical and scientific systems, knowledge based systems and models relying on numerous relations between objects, like medical databases. To overcome the differences between the relational model and the object oriented model without ORM, the ease that object oriented programming provides is partly lost. [1]

This paper describes the the process of development of a synchronisation and replication protocol for a software using ORM. As a test environment for the design of the protocol the requirements of a medical practise with a depending medical practise and a doctor who attends house calls were chosen. The data needs to be as up-to-date as possible, since a second depending practise may need the newest information about the patient. If one of the practises is offline or the doctor is on a house call, the software should allow the doctor to work without any connection the whole system.

2 Material and Methods

The base for this work was set by the existing database management system, for which the solution should be build. The methods described in the following are the foundation to develop a synchronisation and replication strategy for the existing system.

2.1 Database Setting

The existing database uses object relational mapping to store data in the database. The mapping is done by an extra layer between the application (objects) and the database (relations). This step is used to overcome the differences between application and database. This discrepancy is called *paradigm mismatch*. The described problem arises from different concepts, for which there is no logical equivalent in the other. These are, for example, transient objects and their counterpart persistent elements. The characteristics of an object are stored as attributes and properties, whereupon characteristics of elements are stored in columns. ORM now provides a mapping, which allows automated transfer of the transient objects into a persistent state and contrariwise [2]. Performing the communication between the systems by its own, the mapper reduces the amount of code needed to transform the objects into a persistent state, within the database. The translation is done by an object relational mapper, in this case *Hibernate*. It utilizes predefined mapping descriptions, which have to exists for every object, characterizing the way the objects are stored in the database. The constraints, type definitions and relations of every used object are defined within their class definition. Hibernate provides methods to perform searches, inserts and updates without using any SQL instructions. But these instructions are concealed and cannot be accessed. Because these instructions are vital for the synchronisation of the data, one of the requirements is to find a way to work around this circumstance. [3]

To develop a specific strategy to synchronise and replicate the existing database a smaller database was created. This database consists of diagnoses, patients and billing details. These elements are sufficient to illustrate the key problems, shown in section 2.4. Figure 1 shows the relations between these classes/tables.

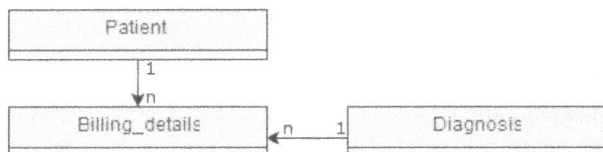

Figure 1: This figure shows the relations between the used objects in the test environment.

2.2 Replication and Synchronisation

Replication and synchronisation are highly linked to one another. Both have slightly different meanings in distributed databases and distributed systems. In distributed systems the term synchronisation describes the organisation of processes in a right order between the systems. Replication in the context of distributed systems describes that different copies of one system share information. In the context of distributed databases replication describes the exchange of data between databases. Synchronisation describes the act of restoring or ensuring the consistency of the databases.

Using the same database in different locations leads to better performance for single databases (**a**vailability) and increases the stability of the whole system (**p**artition tolerance). But working on different databases leads to the question how to keep the data equal and correct across the whole system (**c**onsistency). Since, as presented within the CAP theorem by Brewer, not all of the three named properties are feasible at the same time. There is a trade-off between these characteristics, which leads to different replication strategies. These strategies are distinguished by three attributes. The first is the correctness of data which differentiates between strict and weak consistency. Weak consistency allows a read on data which may not be up-to-date. Strict consistency only allows reads on data which is up-to-date. The second property to classify the replication strategy is the moment of update, which could either be eager/synchronous, leading to an update on every transaction,

or lazy/asynchronous, having another specified update time. The third attribute are the rights to read or write data on a specific replica. If all replicated database systems are equal regarding their rights to change data, the strategy is called symmetric. An asymmetric strategy would consist of at least one master/primary replica receiving all changes and at least one slave/secondary where data could only be read. This classification allows to find the needed replication method.

Most partial offline scenarios offer the possibility to work parallel on the same data and not all replicas are available at all times to observe changes. This setting leads to the problem of possible inconsistency. To overcome the disadvantages of the trade-off, different replication strategies were developed.

These strategies define in which way the changes on replica were propagated among one another, otherwise, the data could become inconsistent. The different methods to keep track of inconsistencies are discriminated by their way to handle problems. The pessimistic strategies avoid possible conflicts by restrictions for example by locking the data or define one server to schedule. The optimistic strategies resolve conflicts after they happened, presuming conflicts are rare. [4]

If optimistic strategies were applied, the transferred data has to be checked for possible inconsistencies originating in changes within the data or incorrect order of transactions. This procedure is centralized by the term synchronisation. Synchronisation methods are divided in two groups: data-oriented and transaction-oriented [5]. Strategies which are data-centric detect and resolve conflicts by comparing the different states of the replicas. Transaction-centric methods detect dependencies between transactions and resolve them. Both groups need exact specifications for the resolution of the conflicts.

2.3 Time in distributed systems

To identify dependencies and possible conflicts it is required to have a definition of time, which is well-defined and exact in every system. Since the local system-time can differ between replicas, distributed systems use logical clocks to assign a time value to actions.

Logical clocks run detached from the physical time and are defined by actions within the system. One specification of a logical clock is the *Lamport* clock. By definition this clock is incremented every time an action is performed. In a distributed system, where different parts could work parallel, a single clock does not solve the problem, because actions which are performed parallel are indistinguishable. Since not every system performs the same number of actions, every system needs its own clock.[6]

To solve the problem *vector clocks* are needed. Each system has to hold a vector with the size of the number of systems. Each system now uses a Lamport clock to assign timestamps to their actions. Each message between the systems inherits the current time-vector of the sending system. The receiving system updates its counter with the values from the systems, if they are bigger then their own. With this setting it is possible to determine whether an action depends on an earlier action or not. Action x depends on an action y if the elements in x are less than or equal to the corresponding elements in y and at least one of these elements is strictly smaller.[6] As shown in Fig. 2 this solution may produce a number of dependencies, which do not exist. Since not all objects related.

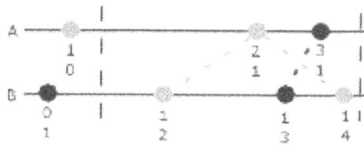

Figure 2: This diagram shows a sequence of different actions on the databases and their vector clocks. The colors show two datasets which could produce conflicts. The dashed lines mark the moment of synchronising. The dotted lines connect possible conflict actions.

2.4 Test Setting

To specify a test setting, first the possible problems have to be defined. If an update is overwritten by another update from the replica, it is called *lost update*. Another problem occurs, if a replica reads its own changes before they were committed and commits only the newest version of the data. When uncommitted data is read by another transaction, the process is called *dirty read*. Executing the same read instructions consecutively and getting different results, due to a concurrent update is called *unrepeatable read*. The problem of *phantom reads* occurs, if the same read instruction produces different outcome due to a concurrent update.

To test these scenarios different *JUnit* tests were set up. JUnit is a test framework for the Java programming language. For every test the databases were set to their origin and filled with the data needed for the test. Every test consists of specific transactions creating one of the conflicts mentioned before. Since this protocol is still in development, no benchmark tests were implemented.

For example the test whether lost updates can be detected is set up as follows. A patient is created on database A, then the databases A and B are synchronised. After the data is integrated both A and B change a specific value of the patient and synchronise again. After the synchronisation the data should be equal and correct. The tests only cover tests on the logical level, semantic problems will be tested after further developments of the replication strategy.

3 Results and Discussion

As described before strategies to replicate data differ, due to the demand on availability, partition tolerance and consistency. For the chosen setting the system should work offline, allow changes on multiple servers and provide the best possible consistency. Since the software ought to be used in medical context like house calls or for the connection of distant medical practises different conditions have to be fulfilled. This restricts the number of possible solutions of the problem.

3.1 Replication Strategy

The first option is called *snapshot replication*, every replica only stores the needed data and works with it. This leads to the question how the data which is needed for the snapshot is selected. This was the reason for the decision against this strategy. It could not be decided which data should be replicated for one snapshot, if one element is missing the replica could be useless. It also needs one server which is online all the time and keeps track of all snapshots.

The second option, *push/pull replication*, offers the user the option to decide, when to replicate and synchronise the data, leading to the possibility that the user forgets to update the database. Since consistency and up-to-dateness is vital to a medical database, especially when the doctor is on house calls, this option did not fit our demands.

The third option is one specific master server which coordinates all updates and writes. This option is not possible, because the system does not work, if no connection to the master can be established.

The fourth strategy keeps track of all changes that are made on the replica and exchanges them, when online. Since all replicas are as up-to-date as possible and have a log of all committed changes, every copy is able to replace the master replication. Presuming they have the needed hardware capabilities. All our demands where met, so we decided to implement this method for an ORM context.

3.2 Identifier problem

If replicates work concurrent it could happen that the same identifier is assigned to different elements of the same type. Pawel et al.[7] described two solutions for this problem. One solutions distributes a pool of possible IDs to every representation of the database. For a medical database this idea is not suitable, because it is not possible to define a proper amount of IDs needed by the user in its offline time. Additionally the number of billing details have to be consecutively. The decision was made to use the other option and define a range for each replica. Every newly generated ID is now computed with information which is unique for the server and set during the set up of the server.

3.3 Protocol

The main problem, during the development of the protocol, was the lack of information about the transactions. The protocol defines four steps to replicate the data. The first step is to save all transactions which happen on the database. If the system is online for the next time the transaction log is send to the master. The master integrates the received log elements into the existing log and checks if there are any possible conflicts. If all conflicts are resolved the complete log of the master is send back to the replica. To reduce the work, the master saves the information about the last synchronisation.

The used object relational mapper Hibernate does not allow to access the SQL instructions, which are performed. This is why single transactions of Hibernate had to be saved in a log element. These elements inherit information about the type of transaction, which was executed, the object itself and a time vector. During the synchronisation, the replica does not log the transaction events, since they are already represented in the log.

The first step on the server is to insert the received elements into the existing log element list, by comparing their time vectors. After integrating, the list is examined for possible conflicts. Succeeding an offline time all log elements depend on the last synchronised element in the log list, due to the definition of the vector clock. But not all elements really have a connection to the changed object and depend on it. This is why the list is checked for dependencies regarding specific objects, done by comparing the type of the altered object and the ID. If both properties are equal the conflict has to be resolved, otherwise there is no conflict. This corresponds to a log list which is kept for each element in the database.

3.4   Evaluation of the protocol

This strategy detects all problems described in section 2.4, but the resolution of consistency problems is not trivial. Lost updates can be detected, but they currently can only be resolved by the user, since all strategies to resolve the problem for an object are very specific and not yet integrated. If uncommitted data is read by the same transaction, the possible misinformation is integrated in all commits from this server.

Dirty reads cannot be performed since only committed/logged data is available for other transactions. Unrepeatable reads are not possible, due to the circumstance that the integration with logical time produces a sequential procedure of commits. Inconsistencies can only appear if an element of the integrated log list depends on previously existing elements. Phantom reads cannot occur, due to the sequential order within the log.

The strategy works for the described test setting, but it develops its flaws if more than two databases have to be synchronised. It is not clear where the log element list has to be added. This problem emerges in the usage of the logical clock. If the physical time would be used, the commits of different replicas are mixed.

It is still to be tested, whether this leads to inconsistencies in a larger scenario or not. If updates occur simultaneously, it is not clear which update is the correct one, this has to be decided individually for every case.

The shortcomings of the strategy arise from the small scenario which was selected to test its capabilities. This may be changed within the next steps of development. Also the lack of tests for semantic inconsistency, due to a very complex relations in the real world will be rectified.

4   Conclusion

As stated there is currently no solution to the problem of ORM replication and synchronisation. This paper directs a possible way to replicate and synchronise data which is stored by an object relation mapper. The main problem is the difference between the normalized data within the database and the object relational mapping. This makes it harder to identify possible problems on the semantic level. The complexity of the medical context also plays a major role in the problem.

The solution of problems on the semantic level is only solved for one specific database. Every time a new object is introduced into the database, new rules have to be developed.

Solutions for replication and synchronisation of data in an enterprise environment are very specific if it comes to synchronising data and resolving semantic conflicts. The problem of logical conflicts should be resolved generic.

Acknowledgement

The work has been carried out at Epikur Software GmbH , Berlin and supervised by the Institut für Medizinsiche Informatik, Universität zu Lübeck.

5   References

[1] D. Bonjour, "The need for an object relational model and its use," in *International Conference on Extending Database Technology*, pp. 135–139, Springer, 1996.

[2] M. Keith and M. Schincariol, *Pro EJB 3: Java Persistence API (Pro)*. Berkely, CA, USA: Apress, 2006.

[3] C. Bauer and G. King, *Java Persistence with Hibernate*. Greenwich, CT, USA: Manning Publications Co., 2006.

[4] Y. Saito and M. Shapiro, "Replication: Optimistic approaches," tech. rep., HP Laboratories Palo Alot, February 2002.

[5] M. Liebisch, "Synchronisationskonflikte beim mobilen Datenbankzugriff: Vermeidung, Erkennung und Behandlung," Diplomarbeit, Friedrich-Schiller-Universität Jena, 2003.

[6] G. Bengel, *Grundkurs Verteilte Systeme: Grundlagen und Praxis des Client-Server und Distributed Computing*. Wiesbaden: Springer, 2014.

[7] P. Gruszczynski, S. Osinski, and A. Swedrzynski, "Offline business objects: Enabling data persistence for distributed desktop applications," in *On the Move to Meaningful Internet Systems: CoopIS, DOA, and ODBASE*, vol. 3760, 2005.

Linked Data Applications through Ontology Based Data Access in Clinical Research

C. Kamann[1,3], A. Kock[3], H. Ulrich[2,3], P. Duhm-Harbeck[3] and J. Ingenerf[2,3]

[1] Medizinische Informatik, Universität zu Lübeck, Germany, christian.kamann@student.uni-luebeck.de
[2] Institut für Medizinische Informatik, Universität zu Lübeck, Germany
[3] IT for Clinical Research, Lübeck (ITCR-L), Universität zu Lübeck, Germany

Abstract

Clinical care and research data are currently widespread in isolated systems with heterogeneous data models. Biomedicine predominantly makes use of connected datasets based on the *Semantic Web paradigm*. *Electronic Healthcare Records* are generated and processed in diverse clinical subsystems within hospital information systems utilizing relational database systems with proprietary schemata. Semantic integration and access to the data is hardly possible. This paper describes ways of using the *Ontology Based Data Access (OBDA)* approach for bridging the semantic gap between existing raw data and a user-oriented view supported by ontology-based queries. By employing mappings between ontologies and relational data the information can be made available as *Resource Description Format (RDF) triples* for subsequent querying and processing. The performed experiments apply test data from a demo system for biobanking and study management. This work shows that a professional platform for *Linked Data Applications* is recommended due to inherent complexity as well as scalability and performance.

1 Introduction

There is a demand to semantically process and integrate clinical care and biomedical research data from different heterogeneous resources especially in the context of translational research. Biomedicine as a knowledge based discipline predominantly make use of connected datasets based on the *Semantic Web paradigm*. Initiatives like *Bio2RDF* created *RDF*-versions of genomic, proteomic or metabolomic resources, enabling sophisticated *Linked Data Applications* [1]. Instead of schema matching approaches (e.g. for relational databases) that have been tried without great success for decades, the *Semantic Web paradigm* favors the separation of the *Resource Description Format (RDF)*. The flexible representation of facts by sets of interconnected triples on the one hand and semantic information describing the corresponding instantiated entities and their relationships by ontologies on the other hand are major advantages of this approach. All three formats or description languages are standardized by W3C specifications. *SPARQL Protocol and RDF Query Language (SPARQL)* is sensitive to the semantics of the *Resource Description Framework Schema (RDFS)*- or the *Web Ontology Language (OWL)*-ontologies as far as a reasoner is activated to deduce implicit knowledge. Efficient frameworks for distributed queries across multiple *RDF* data sources are used intensively in many application areas; amongst others in biomedicine and less often in healthcare [1], [2]. Especially in healthcare context many data sources are not provided as *RDF triplestores*. There are good reasons for using *Relational Database Management Systems (RDBMS)* in operational contexts like clinical care, e.g. transactional application systems are supported for a very long time by mature tools of database software vendors like Oracle, Microsoft SQL or MySQL. The *OBDA* approach has the potential make data stored in relational database systems accessible by *Semantic Web technologies* and has the potential to facilitate the necessary translation process. This approach allows bridging the semantic gaps between existing patient data in clinical information systems and the ways which medical researchers interpret those records, e.g. trying to identify patients with diseases that are located at specific body parts. This intention is only possible when the existing relational data is processed by linking it with this exact information represented by *concepts* from the *Systematized Nomenclature of Medicine (SNOMED)*. The linking with additional knowledge source is described and visualized in section 3.2.

2 Materials and Methods

OBDA is a new paradigm for accessing and integrating data, which key idea is to resort to a three-level architecture, constituted by an ontology, data sources, and the mapping between them [3]. The ontology defines a high-level global schema of (already existing) data sources and provides a vocabulary in terms of concepts, roles, i.e., binary relations between them, and attributes for user queries. The mapping layer explicitly specifies the relationships between the domain concepts and the data sources. Then an *OBDA* system rewrites such queries into the vocabulary of the data sources and delegates the actual query evaluation to a suitable query answering system such as the *Structured Query Language (SQL)* for *RDBMS*. The ontology in combination with the mappings expose a virtual *RDF graph*, which can be queried using the standard query language in the *Semantic Web* community *SPARQL*. This virtual *RDF graph* can be materialized by generating *RDF triples* or alternatively it can be kept virtual and queried only during query execution. In the following several ways to adopt the *OBDA* approach by primarily accessing data from the *RDBMS*-based *CentraXX* system for biobank and study management [4] are presented. In this work the provided *CentraXX* system is used with "realistic" demo data.

2.1 Ontop used as a plug-in within Protégé

Ontop is one of the most popular *OBDA* systems [5].This open source software is available amongst others as a plug-in for the ontology editor *Protégé*. First, a domain ontology with relevant concepts like patients, encounters and diagnoses and their relationships are defined. After activating the *Ontop* plug-in data sources, e.g. *Java Database Connectivity (JDBC)*-access to a relational database, and mappings can be managed. The mapping define in this context how instances of classes and relationships are mapped to the demo database entries by *SQL* statements.

Figure 1: Realizing *OBDA* with the *Ontop* plug-in within *Protégé*: *R2RML*-mappings of relational database content

Finally *SPARQL* queries can be created that are executed by a query answering engine with *OWL 2 QL/RDFS* entailment called *Quest*. Fig. 1 shows an example of this process. The mappings expressed in W3C standard *RDB to RDF Mapping Language (R2RML)* can be constructed semi-automatically by domain experts or by using a bootstrapper that creates an ontology and corresponding mappings automatically by analyzing the relational database schema. Furthermore *Ontop* works together with *Teiid* as an

open source *Java* software for data virtualization. In principle, this combination can be used for federating different heterogeneous *RDBMS* behind one *JDBC* interface.

2.2 Optique platform with OptiqueVQS as a Visual Query System

Optique (Scalable End-user Access to Big Data) is a EU-funded project where novel solutions based on the *OBDA* idea have been developed [6]. The *Optique Platform* has been made available as an *app (application with plug-in technology)* that can be installed and deployed within the *Information Workbench (IWB)* that is introduced in section 2.3. Regarding the aspect of usability the most interesting feature is the *Visual Query System (VQS)* where query dialogues are rendered based on the ontology. Fig. 2 shows a screenshot of the *OptiqueVQS* interface to query the data of the demo system.

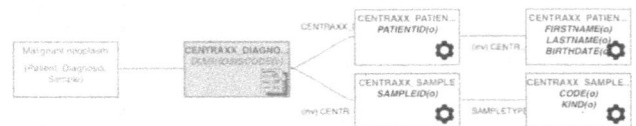

Figure 2: Exemplary query formulated with the *OptiqueVQS*: Patients and corresponding tissue samples for diagnoses of malignant neoplasms

2.3 Information Workbench (IWB) - A platform for Linked Data Applications

The *Information Workbench* by Fluid Operations [7] provides a generic front-end for customizable user interfaces based on *Semantic Wiki* technologies, enriched with a large set of widgets for data access, navigation, visualization, analytics and data mashups with external data sources. It can be further customized and extended for domain specific applications through a proprietary *Software Development Kit (SDK)*. Techniques for *OBDA* from section 2.1 and 2.2 like *Ontop* are included in the platform. For federated queries, i.e. sending decomposed *SPARQL* sub-queries to various data services and integrating the results virtually, the *FedX* module is available. The *Information Workbench* is available as a Community Edition under an Open Source license as well as an Enterprise Edition with a commercial license. The *IWB* was used to create a *demonstrator* that provides an ontology based query front-end for accessing *RDBMS* data of the demo system enriched by *Linked Data*. This approach is described in the following section 3.

3 Results and Discussion

The use of *Ontop* as a plug-in in *Protégé* at the beginning of this work was very helpful and offered various valuable insights. However, for mainly two reasons the decision was made to look for alternatives. First, the end users of the envisioned *OBDA*-based query system should not be forced to enter *SPARQL* queries. Second, there is a limitation when trying to follow up with *Linked Data Applications* based on the resulting *RDF triples*. The *OptiqueVQS* offers an impressive potential but for this approach the flexibility concerning the desired query front-end emerged as insufficient. The customization can only be realized by some cumbersome modifications of the applied ontology. For flexibility reasons it exposed to be even more promising to work directly with the *IWB*. It became apparent that the modular structure of the *IWB* greatly facilitates the creation of a *demonstrator* to access the data provided in the demo system and semantically enrich this data by linking it with additional sources. In the following the architecture of the *demonstrator* is described, the linked data is explained and the *demonstrators* front-end is presented.

3.1 Architecture

The architecture of the demonstrator using the *IWB* is shown in Fig. 3. The *IWB* enables the generation of an ontology with concepts, relationships and attributes of interest and relevant *R2RML*-mappings to the corresponding data entries within the relational demo database. The *RDBMS* of the demo system serves as the data source. The core content of the system e.g. patient, sample and diagnostic data is queried via *SQL* and mapped with corresponding *R2RML* mappings to generate *RDF triples*.

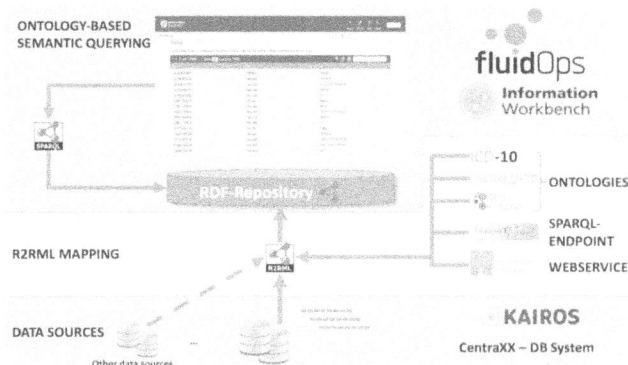

Figure 3: Architecture of the *Information Workbench* for *Linked Data Applications*

The materialized *RDF triples* are stored in the internal *RDF repository* of the *IWB* and then processed by linking them with data from ontologies, *SPARQL-Endpoints* and *Web services*. The semantical enrichment with *Linked Data* from different sources is described in the following subsection 3.2. The *RDF repository* is accessible via customized *Wikipages* of the *IWB* that allow to query the processed triples with *SPARQL*. Examples of the *demonstrators* front-end are explained in subsection 3.3.

3.2 Linked data

The major advantage of this approach becomes apparent in the linking of supplementary data to the triples in the *RDF*

repository. For the demonstrator the *International Statistical Classification of Diseases and Related Health Problems 10th revision (ICD-10)-* and the *Systematized Nomenclature of Medicine Clinical Terms (SNOMED CT)*-ontologie as well as the *Disease Ontologie* [8] are used. Moreover the *Web service* of the *Unified Medical Language System (UMLS)* [9] and the *SPARQL-endpoint* of the *DisGeNet* [10] are accessed to *Linked Data* of interest. Fig. 4 shows an example of a diagnosis from the demo system enriched with additional knowledge through ontological mappings. By extracting finding sites from mapped disorder concepts in *SNOMED CT* it now becomes possible to query for example all data from patients or samples that are located at the digestive tract which is shown in Fig. 5. The initial *ICD-10* codes from the demo system do not offer such specific information due to the lack of localisation information of the *ICD-10* terminology. Additionally, *Disease Ontology* and *UMLS* codes are annotated. These codes enable the possibility to retrieve again additional information about e.g. disease-associated genes from the *DisGeNet*.

Figure 4: Technical view of a diagnosis resource with *Linked Data*

3.3 Demonstrator

The *demonstrator* utilizes customizable *Wikipages* of the *IWB* to provide access to the *RDF triples* of the *repository*. The customization is done in a completely declarative way, resorting to a rich pool of widgets and creating template pages in *Wiki markup*, which are associated with elements of domain ontologies. It has shown in preliminary experiments that this significantly simplifies and speeds up the application development.

Figure 5: Query form of the IWB front-end

Fig. 5 shows a query page of the *demonstrators* front-end that provides a form to specify parameters to search for. In this case a user is searching for tissue samples of over 30 year old patients who are diagnosed with either an *ICD-10* code from the code group C00-C14 (Malignant neoplasms of lip, oral cavity and pharynx) or with the *SNMOED CT* finding site of the diagnose (Upper digestive tract structure).

User frontend: result

100014565355	492523	MALE	62	C01	Malignant neoplasm of base of tongue	DOID_0008649	[illegible]
100021903921	295389	FEMALE	75	C02_1	Border of tongue	DOID_0008649	[illegible]
100027331757	505463	FEMALE	58	C02_1	Border of tongue	DOID_0008649	[illegible]
100052276632	591424	FEMALE	66	C01	Malignant neoplasm of base of tongue	DOID_0008649	[illegible]
100078456721	983764	MALE	69	C00_1	External lower lip	DOID_0008863	[illegible]
100080931347	619542	MALE	33	C07	Malignant neoplasm of parotid gland	DOID_0009036	[illegible]
100084633646	918017	FEMALE	81	C02_1	Border of tongue	DOID_0008649	[illegible]
100085109535	528407	MALE	33	C02_1	Border of tongue	DOID_0008649	[illegible]

Figure 6: Result of an executed query showing patient demographics and diagnosis from the relational data source as well as linked data from the *Disease Ontology* and *DisGeNet*

The search result of the query is shown in Fig. 6 in the representation of a table with test data from the demo system and additional information from the *Linked Data* like the related *Disease Ontology* concept and the corresponding genes associated with the diagnose from *DisGeNet*.

4 Conclusion

The *OBDA* paradigm is a promising approach within clinical research informatics since the majority of existing applications are based on relational database systems. The application of the *Information Workbench* as a solution to generate *RDF data* from the *CentraXX* demo system with subsequent processing and querying has shown promising potential. Although this work was done mainly by a master student in short time it was possible to develop a small *Linked Data Application* with core features that are of interest for semantic data querying and integration. There are much more *Linked Data* of interest that can be included similarly, e.g. accessing literature from *Medical Literature Analysis and Retrieval System Online (MEDLINE)* etc. The modular architecture of the *IWB* allows furthermore to integrate additional data sources like *FHIR* resources, *Extensible Markup Language (XML)*- or *Excel*-files. In the long run it is planned to develop an *app* for the *Information Workbench* which will be able to integrate different kinds of clinical research data and link it with supplementary knowledge. In addition to that the *app* in combination with the *IWB* could also serve as a data provider for analytical applications. This would extend this approach by further addressing the issue of interoperability regarding heterogeneous data sources.

Acknowledgement

This work has been carried out at the Institut für Medizinische Informatik, Universität zu Lübeck in cooperation with the IT for Clinical Research, Lübeck (ITCR-L), Universität zu Lübeck. The authors would like to thank fluid Operations AG and KAIROS GmbH for their cooperation and support.

5 References

[1] M. del Carmen Legaz-García, J. A. Miñarro-Giménez, M. Menárguez-Tortosa, and J. T. Fernández-Breis, "Generation of open biomedical datasets through ontology-driven transformation and integration processes," *Journal of biomedical semantics*, vol. 7, no. 1, p. 1, 2016.

[2] H. Sun, K. Depraetere, J. De Roo, G. Mels, B. De Vloed, M. Twagirumukiza, and D. Colaert, "Semantic processing of EHR data for clinical research," *Journal of biomedical informatics*, vol. 58, pp. 247–259, 2015.

[3] S. T. Liaw, J. Taggart, H. Yu, S. de Lusignan, C. Kuziemsky, and A. Hayen, "Integrating electronic health record information to support integrated care: Practical application of ontologies to improve the accuracy of diabetes disease registers," *Journal of Biomedical Informatics*, vol. 52, pp. 364–372, 2014.

[4] Kairos, "Centraxx." http://www.kairos.de/centraxx/. [Last accessed: 2017-01-26].

[5] D. Calvanese, B. Cogrel, S. Komla-ebri, R. Kontchakov, and D. Lanti, "Ontop : Answering SPARQL Queries over Relational Databases," *Semantic Web journal*, 2015.

[6] A. Soylu, M. Giese, E. Jimenez-Ruiz, G. Vega-Gorgojo, and I. Horrocks, "Experiencing OptiqueVQS: a multi-paradigm and ontology-based visual query system for end users," *Universal Access in the Information Society*, vol. 15, no. 1, pp. 129–152, 2016.

[7] A. Gossen, P. Haase, C. Hütter, M. Meier, and A. Nikolov, "The Information Workbench – A Platform for Linked Data Applications," *Semantic Web Journal*, pp. 1–8, 2013.

[8] L. M. Schriml, C. Arze, S. Nadendla, Y. W. W. Chang, M. Mazaitis, V. Felix, G. Feng, and W. A. Kibbe, "Disease ontology: A backbone for disease semantic integration," *Nucleic Acids Research*, vol. 40, no. D1, pp. 940–946, 2012.

[9] O. Bodenreider, "Biomedical ontologies in action: role in knowledge management, data integration and decision support.," *Yearbook of medical informatics*, vol. 3841, pp. 67–79, 2008.

[10] N. Queralt-Rosinach, J. Piñero, À. Bravo, F. Sanz, and L. I. Furlong, "DisGeNET-RDF: Harnessing the innovative power of the Semantic Web to explore the genetic basis of diseases," *Bioinformatics*, vol. 32, no. 14, pp. 2236–2238, 2016.

Pipeline Insertion Sort on an FPGA
- Special hardware for special solutions -

R. Süs[1] and S. Groppe[2]

[1] Medizinische Informatik, Universität zu Lübeck, raphael.sues@student.uni-luebeck.de
[2] Institute of Information Systems, Universität zu Lübeck, groppe@ifis.uni-luebeck.de

Abstract

This paper describes the method and the implementation of the pipeline insertion-sort algorithm on an FPGA. Furthermore, this sorting method for semantic data in the triple form was realized. A further additional point is the use of medical data, which are present in this triple form. The amount of data collected and processed continues to grow. This applies to all areas of mankind, special attention is paid to the medical field, because in this area you will have a huge enlargement of data in the near future. In this case it possible to find a way to sort a set of data very fast. An FPGA is a possible solution for this task, because it can be solve such tasks by its special hardware, also for the special triple datatype, which is used by the Semantic Web and RDF context.

1 Introduction

The Internet offers the possibility to browse many sources and data. Futhermore, this amount of data to be sorted and searched is continiously increasing. This growth is shown in the figure 1. As it can be seen in this figure, the volume of

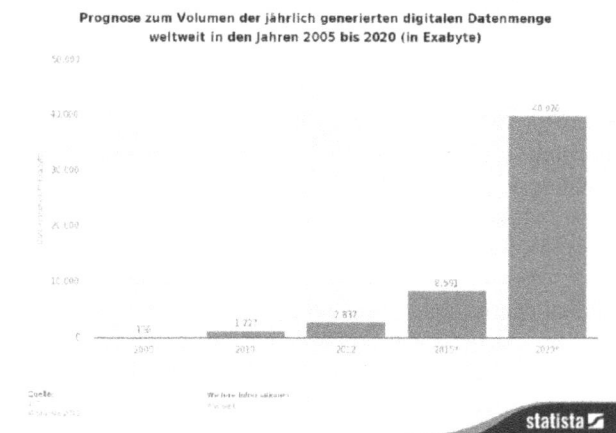

Figure 1: Datavolume Internet [1]

data estimate to grow non-linearly, but exponentially. The growth of this data is thus faster than the development in the hardware. This development is evident in all areas, but in particular the focus is set on the medical field. In our day and age, more and more data are collected and made available across different areas. This applies to all areas of society. Thus, in all areas of society, it is necessary to search this data volume for specific results, at a specific time. In the medical field, this applies both to the publication of new medical findings and the networking of operating rooms,

e.g. the OR.NET project[10] of the universtiy of Lübeck tries to implement an interface which connects all machines in an operating room with each other. Thus, the amount of data in this area increases particularly rapidly.

2 Material and Methods

For programming the sorting algorithm, the "VHDL" language was used and the algorithm was implemented on an Field-programmable gate array (FPGA). The mapping on this particular hardware minimizes runtime. The FPGA represents the possibility to map the algorithm in hardware and logic blocks. The advantage of this solution is that the runtime is optimized, by the use of massive parallelism. Furthermore, triple data of the "Semantic Web" context is used. These data are fundamentally different from ordinary data from the IT environment, since these data always consist of three information units, as described in the section 5. Hence algorithms can be designed for triples, such that they are optimized, based on the simple structure of triples.

3 The idea of the pipeline insertion sort algorithm

The idea of the pipeline insertion sort algorithm is that data arriving sequentially is sorted into a pipeline. This pipeline is sorted automatically after all data arrive, and the smallest or largest element depends on the sort order, at a specific location in the pipeline. Thus the data can be read out sequentially from the pipeline and are sorted correctly. In the figure 2 the general structure is clarified. The squares represent the individual memory locations. The grey squares represent the memory-part of the storage pipeline and the squares that are not filled represent the storage locations of the sorter pipeline. The arrows represent the connections between the memory locations.

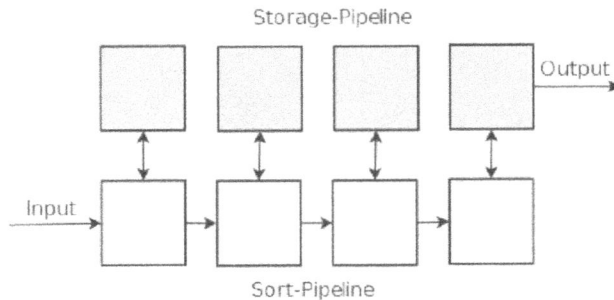

Figure 2: Pipeline-Insertion-Sort

The exact operation of the algorithm is described in the following:
The algorithm consists of two pipelines. A pipeline is referred to as a sorting pipeline, the other as a storage pipeline. The unsorted data is loaded into the sorting pipeline. Each location of the sort pipeline is connected to a particular location of the storage pipeline according to the following pattern: The first location of the sorter pipeline is connected to the first location of the storage pipeline, etc. Thus, each location of both pipelines can be compared to a particular location of the opposing pipeline. It follows that both pipelines must have the same number of storage locations. The operation of the algorithm is described in Algorithm 1, on page 2 for a descending list, but only the first step. This algorithm is only partial because each point of both pipelines is linked to its counterpart, so it is theoretically possible that all elements are compared and exchanged with their respective counterparts in just one clock cycle. This is only possible theoretically on classical hardware (from Neumann architecture), since here the data can only be loaded sequentially from the main memory into the processor register. Also, a processor can always process only one command at a time. Thus, the pipeline insertion sort algorithm would produce a very large overhead to management, and the run time would thus extend. This algorithm has a $O(n) = n^2$ which means that the runtime of this algorithm increases quadratically as the list length is increased. This results in an expenditure which can also be achieved by other sorting methods, which, however, require less memory space than this algorithm. The disk space requirement for this method is $2 \cdot n + c$ where n is the number of elements and c is the general overhead. It can thus be seen that this sorting algorithm is not suitable for a von-Neumann computer.

4   FPGA

An FPGA[5] is a special hardware that can be used for special tasks. This hardware differs fundamentally from the von-Neumann architecture. The FPGA does not have rigid hardware, but consists of a set of logic devices that are con-

Data: Unsorted data from Host
Result: A pipeline of sorted data stored in Storagepipeline
Create Sorterpipeline;
Create Storagepipeline;
$number\_of\_elements\_in\_pipeline = 0$;
Sorterpipeline[0] = Data from Host;
$number\_of\_elements\_in\_pipeline + +$;
while *Sorterpipeline is not empty* **do**
$\quad i = 0$;
$\quad element\_sort = 0$;
$\quad element\_storage = 0$;
$\quad$**while** $i < number\_of\_elements\_in\_pipeline$ **do**
$\quad\quad a = $ Sortpipline[$element\_sort$];
$\quad\quad b = $ Storagepipeline[$element\_storage$];
$\quad\quad$**if** $a > b$ **then**
$\quad\quad\quad$ Storagepipeline[$element\_storage$] $= a$;
$\quad\quad\quad$ Sorterpipeline[$element\_sort + 1$] $= b$;
$\quad\quad$**else**
$\quad\quad\quad Sorterpipeline[element\_sort] = a$;
$\quad\quad$**end**
$\quad\quad element\_sort + +$;
$\quad\quad element\_storage + +$;
$\quad\quad i + +$;
$\quad$**end**
$\quad$Sorterpipeline[0] = Data from Host;
end
Algorithm 1: The Pipeline-Insertion-Sort algorithm

nected to each other by a written program. Thus, certain algorithms and tasks can be imaged in hardware and processed very quickly. This hardware is particularly suitable for sequential data which is to be processed with a very highly parallelized method. All the logic modules on the FPGA work in parallel and thus a very fast processing of the data can be made possible. However, not every algorithm is suitable for the FPGA and not every application can be executed by an FPGA. The FPGA should be seen as a support to an existing processor that provides the FPGA with data and the FPGA is used for special tasks to relieve the actual processor.

A further disadvantage of the FPGA compared to a von-Neumann computer lies in the fact that all logic blocks, which are installed on the FPGA, can not be changed at runtime, besides more complicated processes namily dynamic partial reconfiguration. All parts of the algorithm, such as the maximum number of elements that can be stored, must be known at compile time.

The pipeline-insertion-sort algorithm gets a linear runtime on an FPGA, in difference of a von-Neumann architecture. The data sorting part should be much faster on an FPGA, because the linear runtime grows slower than an exponential runtime.

5 Semantic Web

The Semantic Web[6] is part of the Web. This part tries to create a machine-understandable web, which means that information that is stored in the internet can be found and interpreted by a machine. At the moment the internet is optimized for the communication between humans, not between machines (machine-to-machine communication). For a better communication between machines, it is needed to store more information, the so called meta-information, for a set of data. These information, the meta-information and the set of data, is stored in a set of triples. The form of the triple is given by the "Resource Description-Framework" (RDF)[7]. This framework is specified by the W3C. The defined triple contains three parts of information, the subject, the predicate and the object. These three topics are used to link a set of data with an other set of data. These links and the information, which is stored in the data, can be interpret by a machine, so the information is stored in a machine-understandable way.

5.1 Benefit of the Semantic Web

The benefit of a semantic web is that the data can be linked together. Thus, an image and text for a machine can not be linked together. If the metadata for the text is that the main theme of the text is Tim Berner's Lee and the metadata for the picture is that there is a profile of Tim Berner's Lee, the machine can create a link by name . So you would get both files if you were looking for Tim Berners Lee on this machine.

This results in the advantage that a search query can be made by a human being who resembles his natural language and the machine completes the missing parts independently. Thus, the search result would meet the expectations of the searcher without having to submit the search query a second time. The classification of the data in this context also allows the machines to communicate with machines. Because of the classification data get a semantic background and can thus be used better. This is necessary for the "Internet of Things" (IOT). This is an extension to the actual Internet, which is mainly used for communication between man and machine or between people. This extension is also intended to simplify communication between machines. This simplification would be very helpful in many areas, for example in the medical sector, which is described and explained in more detail in the section 6.

6 Medical Data

The amount of data in medicine is very large and continues to grow. Also, people are accustomed to the fact that the search speed is still very high even with larger amounts of data. Thus, computer science faces the problem that increasingly large amounts of data have to be searched in a very short period of time. The computer is also to assist a doctor with the diagnosis and to protect the doctor from possible errors. In order to ensure this, the program, thus also the computer, has to be able to analyze and sort very large data very quickly. It must be possible in the future to search this growing data very quickly, for diagnoses, examinations, publications, etc. This will become more and more necessary in the future since this data volume can no longer be searched by a human being, but must be searched by a machine become. Also, these data are often sorted into a medical Semantic Web. This increases the amount of data, but the amount can be searched automatically. A further advantage is that the data volume can be interpreted in terms of content and is therefore available to the user in a processed manner. Furthermore, it is possible that these data are automatically searched for possible errors as well as complete incomplete registered data and thus keep the Semantic Web complete. A further advantage of this arrangement is the fact that the communication between different medical devices is thus simplified. Thus, it would then be possible for all medical devices to communicate via a communication interface, and thus operations can be monitored very well. Furthermore, it is also an advantage that facilitates communication and billing with the health insurance companies, since a report on the treatment and the drugs administered can be submitted in such a way as to be machine-usable. Thus it would also be possible to settle international treatments very well.

This international standardization has been tried for a very long time and the loinc coding system [2] is called a representative. This assigns a special LOINC code to each diagnosis and treatment, which is unique. Thus it is possible to allow an international treatment of the disease, also avoid mistakes and problems with the medication and it is easier for the health insurance funds to settle and book the individual treatments.

7 Results and Discussion

This implementation can only sort 256 elements in one Sortrun because there is a hardware limitation, which is given by the used hardware [1]. Where each element represents a triple, each triple contains three integer numbers, each with 32 bit digits, so the required memory space of a triple is 96 bits. The implementation of the algorithm was adapted to the idea that not only simple data types can be sorted, but also the triple data from the "Semantic Web" environment. Thus, not only the memory requirement was adapted to this data structure, but also the comparison process was adapted in such a way that the different triples can be compared with each other. In addition, care was taken that the program would be as extensible as possible. For this reason, all parts were divided into separate modules. Furthermore, care was taken that a jamming of the individual modules is not possible, thus ensuring the fastest possible processing of the data. Furthermore, care was taken that no data is lost and all data is correctly displayed. Also, an additional information unit, which is not prescribed by the algorithm, has been used. This part of the software,

---

[1]Xilinx Virtex V6

the so-called "Metavector", is used to process the information about the sort order and the number of records. It is also possible to tell the sorting algorithm that a duplicate elimination should take place. This has not yet been implemented, but the possibility has already been considered. In addition, the "Metavector" also takes over the task of controlling the sorting process and, above all, also ensures that the sorting module is emptied after the filling process and the data is written to the BRAM. All the data, during one sorter-run, is not stored in the BRAM, they are stored in the logical part of the FPGA. So the number of triples is limited by the number of logical blocks, which can be used on the FPGA. This is not optimal but these kind of storage is the fastest, but also the smallest, on an FPGA. You are able to see that the algorithm is not yet optimally implemented, either with regard to the runtime or the set of triples that can be sorted in a run. One of the so-called "bottlenecks" by the usage of a FPGA is the transmission of the data, between host and FPGA. This is done using a PCI Express interface. This interface can only transfer 32 bits at once. All the data that the FPGA uses is transferred through this interface, so reading the data takes at least $3 \cdot$ #triples cycles. This number of cycles is also required to return the data from the FPGA to the host. This results in the transmission in both directions, i.e. from the host to the FPGA and from the FPGA to the host. For technical reasons, this number of cycles can not be exceeded. At least three cycles are also required for sorting a triple. It is thus recognizable that as large a sort-runs as possible have a more optimal processing time. For these small sorting-runs it would be faster to sort them at the host, with a software-based algorithm, but if the number of triples will be increased to a huge number, then you are going to see that the sorting at an FPGA will become faster than a software-based sorting algorithm. All of these results are only theoretical, because the implementation of the algorithm is just running in a simulation, not even on the hardware.

8   Conclusion

The need of a very fast sorting algorithm is estimated, because the data volume of the World-Wide-Web increasing in future. Hence the search algorithms will be forced to work very fast to find the data. But there is a physically border, which is impossible to cross. This border defines the space between two conductors, that they won't influence each other. So it is not possible to add more flip-flops on a chip, without increasing its space on the motherboard. For the future it is necessary to look for other ways to search and sort the data. One way to do that job is to use a speacial hardware, like an FPGA or an GPU. Both ways are promising and there are a lot of good results, like [8], [9]. An other view to the future provides the idea of a connected architecture of a von-Neumann architekture and an FPGA architecture. This means a fusion between host and FPGA, with a shared memory[11]. Hence it is possible to transfer the data between host and FPGA very fast. The FPGA speed will raise in future. So that you will have a speed-up

during worktime.
In the future work the implementation of this algorithm should be optimized. On the one hand it is possible to sort more triples in one run and on the other hand it could be a good idea to sort two runs in one pipeline. That means that you can write the data of the second run when the first run is going to be stored in the BRAM.

Acknowledgement

This has been carried out and supervised out by the Institut of Information Systems, Universität zu Lübeck.

9   References

[1] Statista.com, *Datenaufkommen im Internet*, Available: https://de.statista.com/statistik/daten/studie/267974/umfrage/prognose-zum-weltweit-generierten-datenvolumen/

[2] McDonald et al, *LOINC, a universal standard for identifying laboratory observations: a 5-year update*, Clinical chemistry, 2003.

[3] World Wide Web Consortium, *Main internation standards organisation for the World Wide Web*, Available: https://www.w3.org/

[4] World Wide Web Consortium, *Web Ontology Language (OWL)*, Available: https://www.w3.org/OWL/

[5] I. Kuon, R. Tessier and J. Rose, *FPGA Architecture: Survey and Challenges*, In: Found. Trends Electron. Des. Autom., Now Publishers Inc., Hanover, MA, USA, pp. 135-253, 2008

[6] S. Groppe, P. Rupino da Cunha, *Semantic and Web: The Semantic Part*, In: Open Journal of Semantic Web (OJSW), RonPub, pp. 1-3, 2015

[7] Eric Miller, *An Introduction to the Resource Description Framework*, Available: http://www.dlib.org/dlib/may98/miller/05miller.html

[8] S. Werner, D. Herinrich, M. Stelzner, V. Linnemann, T. Pionteck and S. Groppe, *Accelerated join evaluation in Semantic Web databases by using FPGAs*, In: Concurrency and Computation: Practice and Experience, pp. 2031-2051, 2015

[9] S. Werner, D. Heinrich, S. Groppe, C. Blochwitz, T. Pointeck, *Runtime Adaptive Hybrid Query Engine based on FPGAs*, In: Open Journal of Databases (OJDB), Vol.3, (1), pp.21-41, 2016

[10] Or.Net, Available: http://www.ornet.org/

[11] D. Sheffield, *IvyTown Xeon + FPGA: The HARP Programm*, Available: https://cpufpga.files.wordpress.com/2016/04/harp_isca_2016_final.pdf

System Testing on a Medical Software
– A conception for the procedure –

L. Coskun [1] and M. Paul[2]

[1] Medizinische Ingenieurwissenschaft, Universität zu Lübeck, leyla.coskun@student.uni-luebeck.de
[2] EUROIMMUN Medizinische Labordiagnostika AG, Lübeck, m.paul@euroimmun.de

Abstract

This paper aims to describe how to complete one of the most important step of software development by realizing the workflow of a system test. Standard guidelines such as the German Institute of Standardization, the DIN (Deutsches Institut für Normung), the European Standard, EN (Europäische Norm) or the Food and Drug Administration (FDA) define conditions, which have to be accomplished. The basis of this work is a medical software for the calculation of the CSF-Serum-Quotient (CSQ). By this software and these regulations, a concept for a software system test was developed. As a method of functional testing, the black box technique was applied. The results demonstrate that the procedure of software system test has to be integrated into the software development process and is a continuous and ongoing process. Therefore, the concept of the software system test has to be completed during the development of the software.

1 Introduction

The CSF-Serum-Quotient (CSQ) is significant for the diagnosis of acute and chronic inflammatory processes of the central nervous system. Usually this kind of evaluation of test results is used in laboratories. The worst case of this evaluation is to display a false negative result with the consequence that the patient will not be treated. Testing of medical software is necessary to prevent failure and at worst, life-threatening effects for the patient. Most often, software failures with risky consequences are caused in the development process due to time pressure [1]. A system test has to be seen as profitable and is usually not assumed to be time-consuming. Time and money are mostly the reasons why project managers keep the process of sofware testing terminable, not reflecting that exactly these two reasons cause iteratively an enormous amount of time and costs. Based on the V-Model as recommended in DIN EN 62304, the goal of this work is to reduce the amount of time and risk required to show the process of software testing in the overall process of the development of medical software. Fig. 1 shows one of the procedure models for the software lifecycle processes, where the integration of the system test process in the development of software can be seen. Test activities already begin in the development phase where the software requirements specification (SRS) are defined. At this point it is necessary to also define test methods and test cases in parallel. In the following the possible test methods and test activities will be discussed.

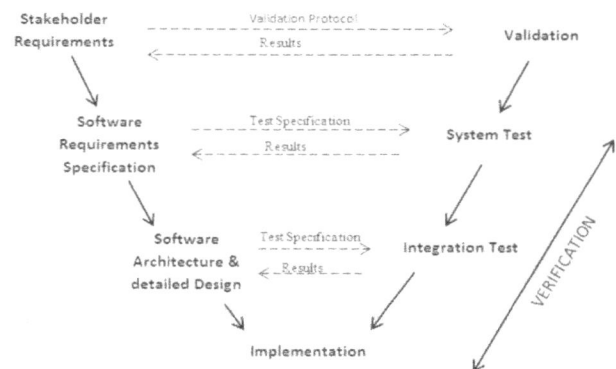

Figure 1: V-Model, an iterative procedure model of software development lifecycle [2].

2 Material and Methods

In this section the materials and methods for testing the medical software are presented. The regulations and guidelines will be mentioned and afterwards the chosen test methods and their basic operation will be introduced.

2.1 Medical Software

This work is about testing the medical software for the calculation of the CSQ. It is about a complex stand alone software, which is used in laboratories for the evaluation of patient samples to give a result on central nervous system diseases. The software is, as usual, connected with a **L**aboratory **I**nformation **M**anagement **S**ystem (LIMS) and imports patients requests. Simply described, as shown in the following Fig. 2, the medical software passes the patient requests to the measurement device and imports the single value results to calculate and evaluate them and export them as the CSQ result back to the LIMS. Otherwise, it is possible to enter patient requests and values manually and the software calculates, evaluates and exports them as a CSQ result to the LIMS.

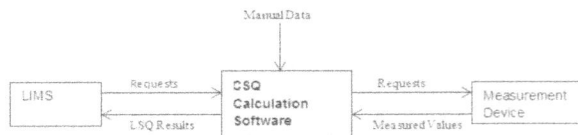

Figure 2: The connection between the medical software and LIMS.

2.2 Application Lifecycle Management

The development of the software was documented in an Application Lifecycle Management (ALM) Tool where the Stakeholder Requirements and the SRS were listed and traced to Use Cases. Including the risk analysis reports, which were documented in the ALM-Tool as well, the test basis was complete. All following workflow data were entered into this tool too.

2.3 Legal Demands on the Software Development

Medical software development has to be conformed to a lot of legal demands, especially the development level of software system testing. As mentioned in EN DIN 62304 and FDA, a software system test has to be executed and documented. In DIN EN ISO 14971 "Medical devices - Application of risk management to medical devices", it is regulated to define risk analysis reports which have to be considered and tested as described in EN DIN 62304 and FDA [3], [2].

2.4 Black box technique

Because of the complexity of the medical software, black box testing was used to execute the system testing. The black box technique describes a specification-based test design method. To start with the black box testing definition of test cases for use cases and SRS is needed. In order to define test cases, it is required to give all test cases an ID, for example test conditions, input values, execution preconditions, expected results and execution postconditions and the traceability to a use case or an SRS [2], [4]. In table 1 an example template is presented.

Table 1: Example for a Test Case Template [6]

Test Case	
ID:	TC-0001
Title:	Export result to LIMS
assigned:	name of tester
execution precondition:	result is calculated
input value:	Press button to export results
expected result:	result is exported to LIMS
execution postcondition:	result is available in LIMS
software version:	Build-39
traceability:	UC-0023, SRS-0102
status:	fail/pass

2.5 Test execution

The current software version was installed and a LIMS simulator was connected to the medical software. With these preconditions the test execution started. If test cases failed, the logfile or a screenshot proved the procedure and was added to the attachments of the test case. While installing new software versions, regression testing was used to make sure that the already tested features were still available. The software developer has to understand which conditions and steps were used to execute failed test cases to be able to fix the bug. It will cost a lot of time to ask the tester for his practice.

2.6 Bug report or change request

After defining test cases, the execution starts. It is even impossible implementing software without failures, so expectedly bugs will be detected. Every discovered bug or change request has to be documented in a bug report or change request as shown in table 2.

Table 2: Example: bug report or change request [6]

Bug report or change request	
ID:	Bug-0001, CR-0052
Title:	missing search function
creator:	name
Date:	02.02.2016
affected software version:	build24
evidence:	add to attachment
risk:	high risk

3 Results and Discussion

Medical software will never be in the entire state of development. Even if no bugs are reported, there will be change requests, for example to increase the performance. Anyway, medical software has to come to the point where the SRS and the software risks measures are implemented. For a new release, the software anomalies were documented and no risky test cases were left open. But if the remaining of the software anomalies had a risky impact, the software release was to be reasoned by the project manager. With increased regulations, fewer and fewer test cases failed. If test cases which deal with interfaces failed, regression testing was executed to make sure no risky failure could influence the LSQ results. At this point it must be said that the upstream operations had a huge impact on the system testing. If a test case fails, it is not possible to decide if and when regression testing is needed and in which dimensions it needs to be executed. It could be enough to iterate test cases which are traced to the same component, but it might need a completely new start of running all the test cases. The main focus in testing medical software is to document the procedure with all test conditions and execution, beginning with the test case design and development, up to the tester who executes the test case and the steps of reproducability. The tester has to write down all the information to be available for both, other testers and developers. It will save time, effort and obscurities.

3.1 Non-functional Requirement

However, it has to be mentioned that it was not enough to get to know the SRS. The quality features listed in the following table 3 had to be considered at the beginning of the software system testing too.

Table 3: Additional requirements [6]

quality features
Functionality
Reliability
Efficiency
Maintainability
Portability
Runtime performance
Capabilities

During testing the functionality, the quality might not exactly be evident, and at that point of realizing, it is usually associated with expense to fix it.

3.2 Basic procedure for software system testing

By testing Medical Software, a basic procedure was developed as shown in Fig. 3. As illustrated in Fig. 1, the test case draft should begin as early as possible, right after the definition of the SRS. The more explicit the SRS is defined, the more detailed the test cases and use cases could be generated and the bugs cleared. Also, the procedure for the software developer to fix the bugs will not take too much time and spending. As shown in the V-Model, the development of medical software is an iterative process. The more detailed the beginning of the development is scheduled and considered, because of the strong dependence, the more the following steps of the software lifecycle development are implemented. Furthermore, Fig. 3 shows that change re-

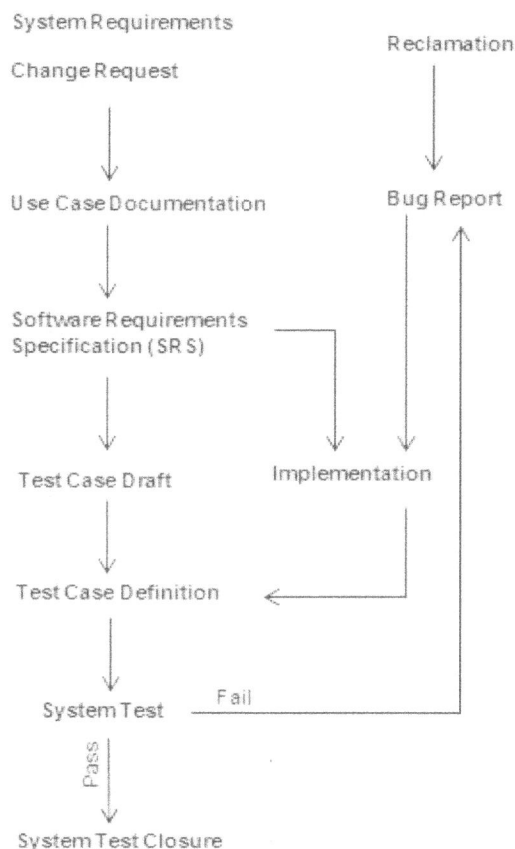

Figure 3: A basic software system testing procedure for medical software, including bug reports and change requests.

quests had been newly seen as system requirements, therefore test cases had to be newly developed.

Testing of medical software is dependent on writing down the lessons learned and the best practices. It is not possible writing down one test concept for every medical software. But it is about the know-how to do it better in the next development of medical software than before.

Each software has to be well-conceived at the beginning of the software development. The more software testing is included in the development lifecycle the more the software is on the right track as required. A test concept and test plan will support the tester, the developer and the project manager. Especially the project manager has to calculate about the same time for the development of the software as for the testing of the software. In doing so, the project manager must bear in mind, that the procedure of testing software is not a standing alone process. The iterative arrows in Fig. 1 underline the dependence of all development steps in software development. The sooner possible system failures are detected, the sooner the correct implementation will start.

Fig. 4 represents the dependence between all development phases and shows the costs of each phase. As shown, ev-

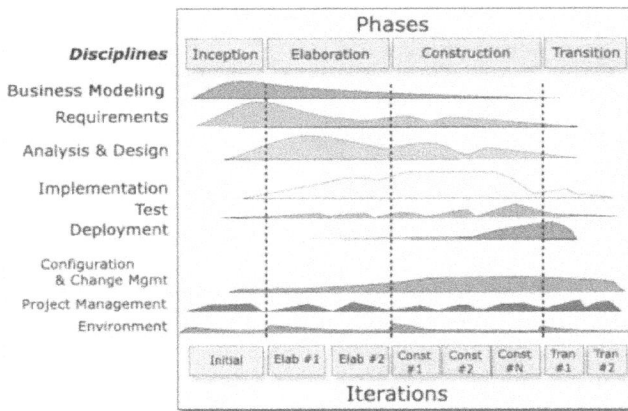

Figure 4: The Rational Unified Process (RUP), a procedure model for the development of software. Representation of the development phases with possible iterations and the work processes of the RUP with the schematic distribution of the effort [8].

ery phase is more or less included in the entire process of software developement. The costs of each phase are sometimes higher and sometimes lower, but they exist during the whole development.

4   Conclusion

Summarizing, this paper shows that software system testing is a development phase, that is as important as the implementation of software. The sooner the software system testing is included in the entire development process, the sooner failures can be predicted and prevented. Software system testing is not about finding failures as effectively as possible but about preventing failures as effectively as possible [7]. For the utility of this conception for the software system test procedure, it should be applied to one or more medical software. The conception should be completed in order to cover more development and test aspects, which might not be considered.

Acknowledgement

This work has been carried out at EUROIMMUN Medizinische Labordiagnostika AG in Lübeck and supervised by the Institute of Physics, Universität zu Lübeck. I would like to thank my supervisor for giving me the possibility to carry out this project and Hauke Paulsen for his assistance.

5   References

[1] G. E. Thaller, *Verifikation und Validation - Software-Test für Studenten und Praktiker.* Vieweg, Braunschweig/Wiesbaden, 1994.

[2] DIN EN 62304 (VDE 0750-101):2007-03, *Medical device software - Software life-cycle processes.* Beuth, Berlin, 2007.

[3] DIN EN ISO 14971, *Medical devices - Application of risk management to medical devices (ISO 14971:2007, Corrected Version 2007-10-01).* Beuth, Berlin, 2009.

[4] ISTQB®, *Certified Tester Foundation Level Syllabus*, Released Version 2011, Available: http://www.istqb.org/downloads/syllabi/foundation-level-syllabus.html [last accessed on 21.01.2017]

[5] ISTQB®, *Certified Tester Advanced Level Syllabus Test Analyst*, Released Version 2012, Available: http://www.istqb.org/downloads/syllabi/advanced-level-syllabus.html [last accessed on 21.01.2017]

[6] Johner Institut, Available: https://www.johner-institut.de/blog/iec-62304-medizinische-software/software-systemtest/ [last accessed on 18.01.2017]

[7] U.S. Department Of Health and Human Services Food and Drug Administration Center for Devices and Radiological Health Center for Biologics Evaluation and Research,*General Principles of Software Validation; Final Guidance for Industry and FDA Staff*, Version 1.1, 1997.

[8] INFFORUM SIMON, Know-how-Transfer in der Software-Entwicklung Available: http://infforum.de/ *Rational Unified Process (RUP)* [last accessed on 23.01.2017]

Development of a verification technique for Automatic Intra-operative SpineMask™ Registration (AIM) Fallback workflow

V. Semrau[1], A. Hettich[2], R. Spahlinger[2] and R. Brinkmann[3]

[1] Medizinische Ingenieurwissenschaft, Universität zu Lübeck, vanessa.semrau@student.uni-luebeck.de
[2] Stryker® Leibinger GmbH und Co. KG in Freiburg, {alexander.hettich, roland.spahlinger}@stryker.com
[3] Institute of Biomedical Optics, Universität zu Lübeck, brinkmann@bmo.uni-luebeck.de

Abstract

In navigated spinal surgeries the surgeons are guided by navigation technology. It helps the surgeon to simplify procedures, set screws with improved accuracy and achieve improved patient outcome with minimal invasive incisions. Navigation systems track the instruments relative to the patient's anatomy. With the usage of the adhesive SpineMask™ tracker, it is now possible to avoid extra incision for patient trackers, which are fixed to the patient's spinous processes. However, in some cases it was observed that the automatic detection of the SpineMask™ tracker is not possible when used with certain imaging devices. To avoid this, a new fallback mode was implemented in the software to give the user the possibility to register the tracker again. This paper shows a verification technique for the new intra-operative SpineMask™ Registration Fallback workflow.

1 Introduction

The navigation system is intended as a planning and intra-operative guidance system to enable computer assisted surgery. The system is designed for any medical condition in which the use of computer assisted planning and surgery may be appropriate. Procedural simplification, reliability and accuracy are the main goals of navigation systems in supporting surgeons [1]. Due to the possibility to plan surgeries based on the patient's CT data, it is, for example, possible to preplan and set virtual screws. During spinal surgery the software gives the surgeon guidance by displaying an instrument's position on patient's image series. The system tracks the instrumentation relative to the patient's anatomy and less imaging is necessary. Therefore, navigation systems optimize the accuracy during spinal surgeries and decrease radiation exposure [2]. Stryker® estimated that approximately 30.000 spine surgeries per year are done with navigation [3].

1.1 Navigation technology

The gathering of the area and the accurate positioning of the navigated surgical instrument are the challenges of navigation technology. Similar to Global Positioning System (GPS), there are several different attempts to navigate instrument positions [4]. The system can be used for intra-operative guidance, where a reference to a rigid anatomical structure can be identified. Patient's rigid anatomical structures are captured by medical imaging systems (CT,

MRI) [1]. Most of the Stryker® navigation systems are based on active optical tracking technology.

1.1.1 Possibilities of optical tracking

Intra-operative navigation technology requires information exchange between the camera-computer system and the surgical instruments. Sensors for optical tracking can be active or passive [4]. Compared to other tracking systems, the line of sight is the biggest drawback for optical tracking [5]. Surgeons, who stand in the line of sight and disturb the communication between camera und instrument, can cause an interruption of navigation or inaccuracies.

Passive tracking
Communication in passive tracking technology occurs by reflecting light signals on the instruments. These pulsed light signals are exposed by the camera. The advantage of passive tracking is the ease of use and that the instruments do not need a power supply. The instruments have a limited lifetime compared to active tracking instruments because of the reflector's sensitive surface [5]. Furthermore, the instruments leads to confusion on account of identically geometrical instruments [5].

Active tracking
In active tracking technology the instruments emit light signals to the camera. Stryker® instruments provide Infrared (IR) Light Emitting Diodes (LED) for active tracking [5]. In contrast to passive tracking, active tracking needs needs batteries for power supply [5]. As a consequence of strong LED signals, the accuracy is higher compared to passive

tracking. Furthermore, active tracking instruments have a longer lifetime than passive tracking instruments [5].

1.1.2 Navigation at Stryker®

Navigation at Stryker® is based on active optical tracking. The localization of the position of the IR light emitting instruments in the working area takes place by the camera (Fig. 1). A patient tracker mounted on a rigid anatomical structure with a SpineClamp is the reference. In relation to this tracker, positions of other instruments can be detected by the camera. Afterwards, the computer system calculates the positions of the instruments on the patient's images. In image guided surgery, the surgeon controls the instruments.

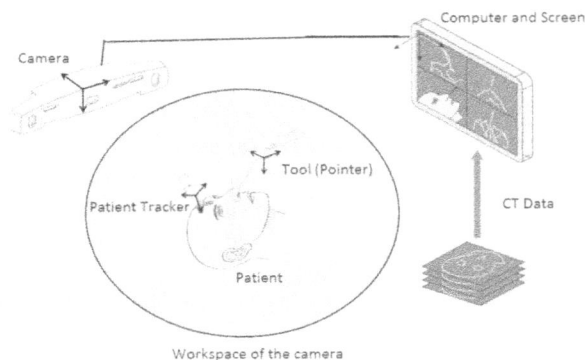

Figure 1: Stryker® navigation system with its components. The camera tracks the signal from tools and patient tracker. The computer analyzes the patient's images and signals from the camera and provides intra-operative guidance for the surgeon. Modified from [1].

1.1.3 Requirements for navigation

Requirements for navigation technology are separated in three parts.

Effectiveness and efficiency
The system shall simplify complex surgeries and provide ease of use and intuitive handling. To help support the surgeon, it is important that the solution includes a personalized workflow for optimized surgical experience [1].

Accuracy
Because of the high accuracy requirements, the system provides a good reproducible success rate also in complicated surgeries [1].

Reliability
Furthermore, the system shall provide the best possible outcome for the individual patient and surgery [1].

The listed points are well implemented by Stryker® navigation systems. A major progress is the SpineMask's™ development as patient tracking without the need of an extra incision for invasive patient trackers (Fig. 2). The SpineMask™

works as a patient tracker. Many LEDs are located on the mask which generate nearly a plane in the working area. These LEDs are detected and analyzed by the camera. It is expected that the usage of the SpineMask™ tracker for registration and tracking will be frequently used, since it shows benefits compared to the previous procedure. Nevertheless, in some cases the Automatic Intra-operative SpineMask™ Tracker registration (AIM) fails, due to bad image quality of the used image technologies. To cover these cases, a new mode was implemented in the navigation software called AIM Fallback mode. This mode enables the user to delete false positive detected LEDs on the mask or to add new LEDs, which were not detected.

To verify the AIM Fallback mode a system test was created and executed to show that the new mode was correctly implemented.

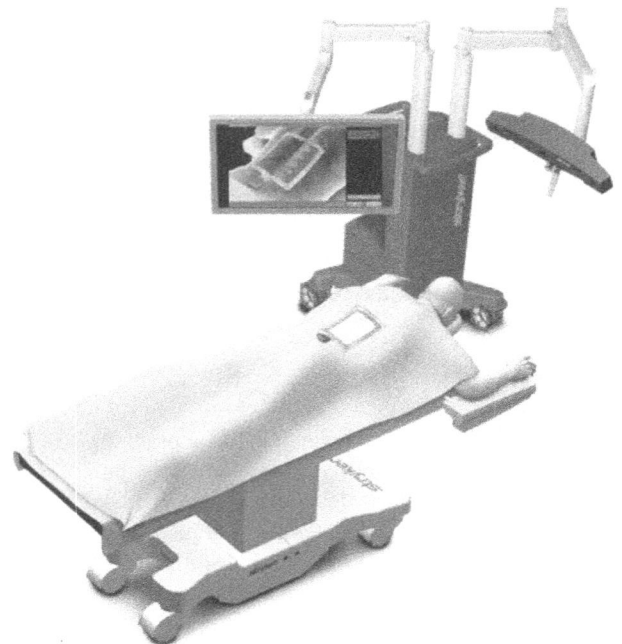

Figure 2: Stryker® Navigation System. The SpineMask™ is attached to a patient. The camera detects the LEDs and the software matches the patient's image data with the detected LEDs [6].

2 Material and Methods

In this section the developed test method is presented. For the test cases' development the AIM Fallback mode workflow is illustrated in detail and the requirements are analyzed. Afterwards the newly established test method is described.

2.1 AIM Fallback workflow

This use scenario concentrates on the new functionality for AIM Fallback, further, it focuses on the case of not having enough LEDs detected, rather than on detecting too many (false positives) LED positions. It should be possible to perform AIM registration, although image quality of intra-operative acquired image scans prevent algorithm from detecting all SpineMask™ LED positions.

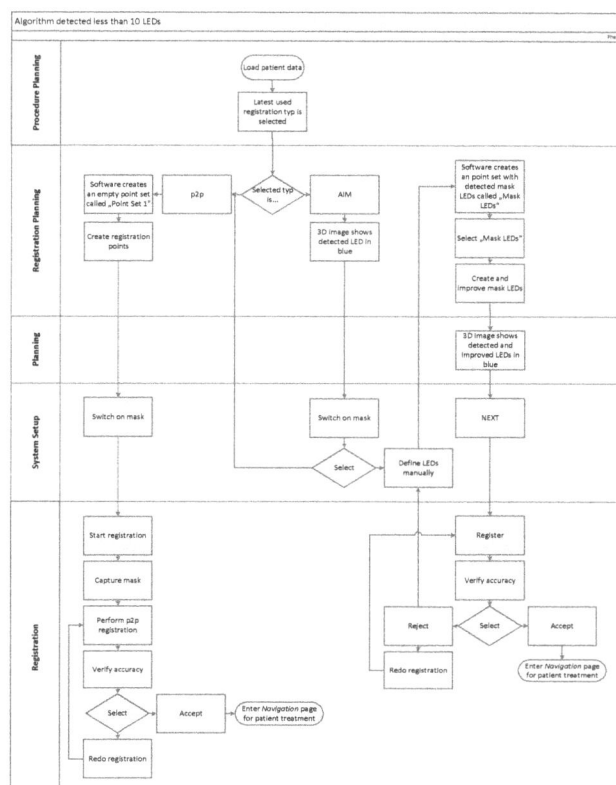

Figure 3: Flowchart of AIM Fallback workflow.

For the development of the test method it is important to give an overview of the workflow for AIM registration providing AIM Fallback workflow (Fig. 3). The first step is that the surgeon imports intra-operative acquired CT image data from the patient. The patient was scanned with a SpineMask™ attached to the patient's back. The software algorithm detects less than 10 LEDs and automatically sets the registration type to point-to-point on the *Procedure Planning* page. Now the surgeon selects Mask Registration in *Procedure Planning* page and goes to *System Setup*. The surgeon has the possibility to activate all instruments and to turn on the SpineMask™ for navigation. Because of not having enough LEDs detected, Mask registration will fail. The software automatically switches to *Registration Planning* page and creates a new point set called "Mask LEDs Set" with all detected mask LED positions. At this point the surgeon has the possibility to delete false positive detections and add missing LED detections. If the user deletes true positive LED positions or adds inaccurate LED positions, the registration will not be accurate. Afterwards the surgeon goes to *Registration* page and the software captures the mask and performs AIM registration again. Before entering the *Navigation* page, the surgeon has to perform a

landmark check to confirm accuracy. If the landmark check is accepted, the surgeon has the possibility to enter the *Navigation* page for patient treatment.

2.2 Test method

Thanks to sufficient testing methods, bugs can be detected, nonetheless it is not possible to show that there are no bugs in the software [7]. Only efficient testing can help to improve software and product quality [8], [9]. Due to this fact, software tests should be carried out as early as possible [9], [10]. System tests are executed after all component tests from the parts where successfully concluded [8], [9]. Therefore, a whole functional navigation system, with all its components and instruments for spine surgery, is necessary for the test. To conclude a suffcient test plan, the dependencies of a function are systematically varied. This can be repeated for all combinations of functions, or at least for all pairs to save effort. Errors can be provoked by disconnecting a camera cable or by intentionally use functions in the wrong way. The main goal of this system test is to check whether the AIM Fallback mode meets the requirements or not. This means, test cases are generated upon the system requirements for the fallback mode. The system requirement specification is the test base for the verification and all generated test cases in AIM Fallback mode.

Requirements which have to be fulfilled:

Planning Page View
SR4510: Adjusted LED position visualization on *Planning* page
The software application shall display the adjusted LED positions as blue spheres in 3D and as blue crosses in 2D on *Planning* page if the software is in AIM Fallback workflow [SR6000]. Software allows an assessment of the adjusted LED positions and the LEDs will be visualized.

Plan Registration
SR2953: Detected LED visualization on *Registration Planning* page
The software application shall display the detected LED positions as blue spheres in 3D and as blue crosses in 2D on *Registration Planning* page if the mask registration feature was selected and the software is not in AIM Fallback workflow [SR6000]. To allow an early assessment of the scan quality in respect to the detected LEDs, the LEDs will be visualized before System Setup.

Registration
SR6000: Provide Mask LED point set
On entering the *Registration Planning* page, the software shall create a point set with all detected Mask LEDs if the following two conditions apply:
• *Mask Registration* is selected manually as registration method or was selected automatically because more than 9 LEDs were detected.
• Less than 10 LEDs were detected or the mask registration

failed or was rejected at least once.
So the software gives the user the possibility to change the position of detected LEDs.

SR6010: Use adjusted Mask LEDs
On performing mask registration, the software shall use the adjusted mask LEDs point set (see [SR6000]) for the next mask registration. This enables mask registration although not all LEDs could be detected in CT image set.

SR6020: Clear Mask LEDs button
On clicking the button *Clear Mask LEDs* in *Registration Planning* page, the software shall delete all detected mask LEDs. This provides a convenience method if the point set contains a lot false positive detections.
Before the test plan was executed, it was reviewed. The reviewers were the test leader and a software developer.

3   Results and Discussion

To verify the functionality of AIM Fallback mode a test plan was created. Test functions in the context of the use cases were created. This test cases are based on the system requirements SR4510, SR2953, SR6000, SR6010 and SR6020. Because the requirements are not sufficient to generate a satisfying test plan, a flow chart from AIM Fallback mode from chapter 2.1 was generated. This flowchart gives an overview about all possibilities for the user in the software. However, the generated test plan is based on both points and was completed by adequate testing practice from experienced testers. Experienced testers tipically know critical parts in the software where bugs often occur. All in all, 53 test cases are carried out where the actual result meets the expected result.

All in all, the exact meaning, interpretation and implementation of the given test method are not fixed. They are only an orientation for testing and testers need to be creative to generate adaquate tests. The characteristics are self-determinable. This test plan is only based on the requirements, the workflow in AIM Fallback and the best practice from experienced testers. The test methods are not described in the requirements and there is the possibility of interpretation and the possibility of interpretation in the context of one's own environment. This means it is not possible to discover all bugs in the software, albeit, to provide a good quality of the software and to provide a good user experience with the new implemented feature.

4   Conclusion

In conclusion, the verification of the new AIM Fallback mode was successful. To provide good quality of the software for the user and to show that all requirements are fulfilled, a test plan was executed. Only one test ID was not accurate. The execution showed one issue regarding the change from AIM registration to point-to-point registration

before the user set or delete LEDs in AIM Fallback mode. This issue was documented in an error management software and will be fixed in a new version from the spine software. The system test will be repeated as soon as the issue is corrected. No other issues were found.

Acknowledgement

The work has been carried out at Stryker® Leibinger GmbH und Co. KG in Freiburg and supervised by the Institute of Biomedical Optics, Universität zu Lübeck.

5   References

[1] D. Mucha, B. Kosmecki and T. Krueger, *Das Konzept der Navigation*. In: Computerassistierte Chirurgie. Hrsg. by P. M. Schlag und J. Adermann, Munich: Elsevier, Urban and Fischer, Kap. 7, 2011.

[2] V. Kosmopoulos and C. Schizas, *Pedicle screw placement accuracy: a metaanalysis*. In: Spine 32.3, 2007.

[3] J. Kaltenbrunn, *Post Market Surveillance Spine/ OrthoMap 3D 2012*. Stryker® internes Dokument, 2013.

[4] M. Kleemann, et al., *Navigation und Medizin*. In: Focus Mul 23 Jahrgang, Paper 4, 2006.

[5] W. Birkfellner, *Lokalisierungssysteme*. In: Computerassistierte Chirurgie, Hrsg. by P. M. Schlag and J. Adermann, Munich: Elsevier, Urban and Fischer, Kap. 8, 2011.

[6] Stryker® Leibinger GmbH und Co. KG, *Training-US-Test-SpineMap3D3.0-Rev.2*. Stryker® internes Dokument, 2015.

[7] E. W. Dijkstra, *The Humble Programmer*. In: Communications of the ACM, Vol.15, Issue 10, 1972.

[8] E. H. P. Roitzsch, *Analytische Softwarequalitätssicherung in Theorie und Praxis*. Edition Octopus, 2005

[9] A. Spillner and T. Linz, *Basiswissen Softwaretest*. dpunkt Verlag, Heidelberg, 5. Edition, 2012.

[10] J. Schaeuffele and T. Zurawka, *Automotive Software Engineering*. Springer Vieweg, Wiesbaden, 5. Edition, 2013.

An Image Annotation Tool for the Preparation of Object Detection and Automatic Segmentation in Thermal Images

A. Essenwanger[1], J. Diesel[2], M.-F. Uth[2], and H. Handels[3]

[1] Medical Informatics, Universität zu Lübeck, Andrea.Essenwanger@student.uni-luebeck.de
[2] Drägerwerk AG & Co. KGaA, Jasper.Diesel@draeger.com, Marc-Florian.Uth@draeger.com
[3] Institute of Medical Informatics, Universität zu Lübeck, Handels@imi.uni-luebeck.de

Abstract

This paper presents an image annotation tool implemented with Matlab for the comfortable annotation of objects and their corresponding segmentation in Dräger's thermal imaging database. There were two main reasons to implement our own annotation tool: the lack of annotation tools to annotate sensitive clinical data and our pretty specific requirements, such as the tool's storage file format (`.mat` and `.json`) or the custom categories and attributes of the annotations. Moreover, the required annotation tool should also offer a fast method for a (semi-) automatic segmentation in gray-scale images of unconnected regions belonging to the same object, such as the human skin. This was implemented making use of Shai Bagon's MEX Graph-Cut functions wrapper [8], which was further adapted to interactively work with the user to correct the computed segmentations. With its segmentation method, the resulting tool helped to speed up the annotation process.

1   Introduction

One mayor and yet underrated problem of object detection studies is the previous annotation of an adequate dataset. Depending on the complexity of the desired task, the algorithms needed for detection may require at least a couple of thousands of annotated images. A manual generation of these annotations does not only take time, but also requires enough workforce. Some of the most known annotation databases such as ImageNet [1] make use of the fact that the internet is full of different images. To annotate these images, ImageNet uses Amazon Mechanical Turk, a crowdsourcing internet marketplace whose main objective is to get employers who would be willing to solve tasks computers are not able to do, so called Human Intelligence Tasks (HITs). This is a good proceeding when working with public images, but in the case of clinical studies where image data has to be protected, the story looks differently. This paper describes the implementation of an image annotation tool which should be used to generate a ground truth dataset with Dräger's thermal imaging database. The implemented annotation tool should not only enable the user to add custom attributes and annotation categories to the single bounding boxes, but it should also offer an easy solution for interactively segmenting single structures that can consist of multiple separated polygons, such as the skin of a human body.

2   Materials and Methods

This section with outline some of the materials and methods used to implement the annotation tool. These include a short outline of the dataset to be annotated, a brief description of the annotation file format `.json`, a short outline of the pre-processing methods used to improve the chance of the semi-automatic segmentation for directly getting the optimal segmentation and will finally focus on the tool's main feature, the semi-automatic segmentation with the Graph-Cut algorithm.

2.1   Dräger's Thermal Imaging Database

Dräger's internal thermal images database includes almost two thousand 288×384 gray-scale thermal images till present day.

2.2   Annotation File Format

For the annotation file format we used Microsoft's image recognition, segmentation and captioning dataset COCO (*Common Objects in Context*) [2] as an inspiration, which introduced the file type `.json` for their annotation files.

The advantage of a `.json` file is its structural compatibility with Matlab. Like a structure in Matlab, `.json` files can also be concatenated, being able to represent an image container, where every image can have one or more annotations, which on their side can contain one or more segmentations. In addition, every object has a unique ID, this enables a fast indexing and filtering of the single data with the help of the corresponding API.

Furthermore, in terms of storage, MS COCO seemed to have already found a very good solution. To effectively save the segmentation masks, MS COCO uses RLE (*Run-Length Encoding*), a lossless data compression method [9].

2.3 Pre-Processing

For improving the semi-automatic segmentation, the GUI offers a histogram stretching option to adjust the image's contrast. To overcome the smooth edges thermal images may have, the GUI also enables an unsharp masking with a Gaussian low-pass filter. The strength of the sharpening effect can be modified by increasing the contrast of the original image before laying an unsharp mask.

2.4 Interactive Graph-Cut Segmentation

The Graph-Cut (GC) algorithm should offer a semi-automatic option to segment areas of interest inside any bounding box. Therefore, the user should merely declare certain pixels as background or foreground pixels. This section will offer a description of the GC algorithm followed by its adaption for the implemented annotation tool.

2.4.1 Brief Description of the Graph-Cut Algorithm

The GC algorithm computes a set of possible segmentations that satisfy the hard and soft constraints best. To determine which of these segmentations could be the optimal segmentation, the image to be segmented is represented as an undirected graph, the minimal cut for this graph should be then computed.

The graph representation should be an undirected graph $\mathcal{G} = \langle \mathcal{V}, \mathcal{E} \rangle$ like the one in Fig. 1, with a set of nodes $\mathcal{V}$ and a set of edges $\mathcal{E}$. Each edge $e \in \mathcal{E}$ has a weight or capacity $w(e) \geq 0$, which determines how much "water" can flow through the edge. The nodes set additionally includes the terminal nodes S (source) and T (sink), $\mathcal{V} = P \cup \{S, T\}$, with P being an arbitrary set of pixels (non-terminal nodes). The set of edges has two kinds of edges, *n-links* (neighborhood edges) and *t-links* (terminal edges). The terminal nodes are each connected to all non-terminal nodes through t-links. While the actual image's neighboring system is represented by the connections of the non-terminal nodes with each other through n-links. The goal is to find a cut which separates the sink from the source with a minimal cost. The cost for removing an edge e is equal to its capacity $w(e)$. By cutting the undirected graph so that sink and source are separated, each non-terminal node gets exactly one of its t-link edges removed. In that way, the image is divided into two different clusters, object and background.

According to the theorem of Ford and Fulkerson [6], a graph can be divided into two disjoint subsets, with non-terminal nodes either belonging to the sink or source, by saturating its edges and thus finding the minimal cut. An edge is saturated when the flow value reaches the capacity value. The solution of the *Min-Cut Problem* is equivalent to the one of the *Max-Flow Problem*. This means, finding the maximum flow of a graph from source to sink, will at the same time reveal the value of the minimal cut.

Now, the question is how to solve this segmentation problem with the help of the solution of the *Min-Cut and Max-Flow* problem. Given a set of pixels or voxels P, a neighborhood system N of unordered pairs $\{p, q\}$ with $p, q \in P$

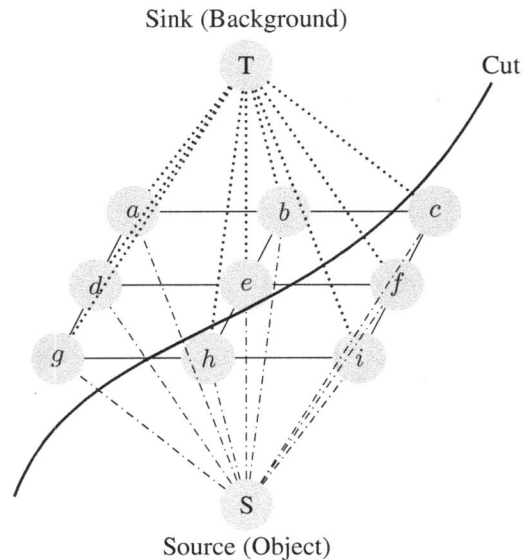

Figure 1: Image represented as an undirected graph. The image pixels are represented by the small nodes labeled from a to i. The undirected graph has two additional nodes, the terminal nodes T (sink) and S (source). It also has two different kinds of edges: t-links (dotted and dash-dotted lines) which connect both terminals with each image pixel node and n-links (straight lines) which connect all neighboring image pixel nodes with each other.

and an assignment vector A, which assigns the elements of P to object or background, the cost function, as described in [7], would be:

$$E(A) = \lambda \cdot R(A) + B(A) \qquad (1)$$

Where λ determines the importance of the region term $R(A)$ towards the boundary term $B(A)$.

$$R(A) = \sum_{p \in P} R_p(A) \qquad (2)$$

The region term $R(A)$ has elements of the assignment vector A as an input and can be interpreted as the penalty given for assigning a certain pixel p to the object or background, $R_p(obj)$ and $R_p(bkg)$ respectively.

$$B(A) = \sum_{\{p,q\} \in N} B_{\{p,q\}} \cdot \delta(A_p, A_q) \qquad (3)$$

$$\delta(A_p, A_q) = \begin{cases} 1, & \text{if } A_p \neq A_q \\ 0, & \text{otherwise} \end{cases} \qquad (4)$$

The boundary term on the other hand, describes the boundary properties of a segmentation by computing the similarities between neighboring pixels. The function $\delta(A_p, A_q)$ analyzes the equality of the neighboring pixels' assignments, being zero when both pixels were classified in the same class and one if they belong to different classes. The term $B_{\{p,q\}}$ can be called the discontinuity penalty and reflects the boundary properties. $B_{\{p,q\}}$ is large when both pixels p and q are similar and gets close to zero when they are very different from each other.

Putting it all together, the capacities of the edges reflect the region (Eq. 2) and boundary terms (Eq. 3). Now, to compute the total cost for each cut, sum up all the capacities of the cut. These capacities or weights can be computed as follows:

Edge	Weight	For
$\{p,q\}$	$B_{\{p,q\}}$	$\{p,q\} \in N$
	$\lambda \cdot R_p(bkg)$	$p \in P, \; p \notin \mathcal{O} \cup \mathcal{B}$
$\{p,S\}$	K	$p \in \mathcal{O}$
	0	$p \in \mathcal{B}$
	$\lambda \cdot R_p(obj)$	$p \in P, \; p \notin \mathcal{O} \cup \mathcal{B}$
$\{p,T\}$	0	$p \in \mathcal{O}$
	K	$p \in \mathcal{B}$

Table 1: Description of how the weights should be computed according to their corresponding nodes.

with $\mathcal{O}$ being the set of object pixels, $\mathcal{B}$ the set of background pixels and:

$$K = 1 + \max_{p \in P} \sum_{q:\{p,q\} \in N} B_{\{p,q\}} \qquad (5)$$

Having obtained a set $\mathcal{F}$ of all possible cuts $\mathcal{C}$, the next step would be to find the minimal cut $\hat{C}$. The cut C contains all severed edges to divide the graph G into two disjoint subsets. Thus, every cut can have a corresponding segmentation $A(C)$, as defined by [7]:

$$A(C) = \begin{cases} obj, & \{p,T\} \in C \\ bkg, & \{p,S\} \in C \end{cases} \qquad (6)$$

Remember that after the cut, every non-terminal node has exactly one t-link, either to the sink T or to the source S, classifying these as either object or background pixels. The resulting cost for each cut $C \in \mathcal{F}$, according to [7], would be:

$$|C| = \sum_{e \in C} w(e) \qquad (7)$$

3 Implementation and Results

The following sections will describe the segmentation's implementation and will briefly present how the resulting tool was used.

3.1 Graph-Cut Implementation

For the GUI we made use of Shai Bagon's MEX graph cut functions wrapper for Matlab [8], which implements the techniques described in the following publications [3, 4, 5]. This GC function uses the k-means algorithm as a starting point and therefore delivers k segmentation proposals, with $k \geq 2$ being the number of clusters. Since we did not only aim to differentiate between object and background,

this implementation seemed to fit. This section will describe how the GC function wrapper was applied to compute an interactive GC segmentation. The main function `GraphCut.m` has different modes depending on the task it should perform. For the realization of the GC algorithm in the GUI we applied the `'open'` mode for the generation of the graph object, `'expand'` to perform a multi-label energy minimization and `'close'` to close the graph's handle object. The `'open'` mode can have several input parameters, the ones used for the GUI include:

- `DataCost`: A three dimensional matrix, with the size of the input image as the two first dimensions, and an additional dimension to store the costs for assigning each label to each image pixel. E.g. `Dc(r,c,1)` represents the cost for assigning label 1 to pixel `(r,c)`. This matrix can be interpreted as the previously mentioned *region term*. This term was set by first computing default cluster centroids with Matlab's implementation of the k-means algorithm. Since the result of the k-means algorithm strongly depends of the initial cluster centroid positions, the clustering process was repeated ten times using different initial centroid positions. In the end, the function returns the centroids with the lowest sum of point to centroid distances within each cluster. The number of clusters k can be set by the user in the GUI. Next, to compute the `DataCost` matrix, compute the euclidean distances between each centroid and the image pixels, these represent the costs for each label. The graph's dimensions are determined by the `DataCost` matrix's dimensions.

- `SmoothnessCost`: Can be interpreted as the *boundary term*. It is a (number of labels)×(number of labels) matrix, where `Sc(l1,l2)` is the cost for assigning neighboring pixels with label 1 (`l1`) and label 2 (`l2`). With a known number of labels of four and a cost parameter c, the matrix would look as follows:

$$c \cdot \begin{bmatrix} 0 & 1 & 1 & 1 \\ 1 & 0 & 1 & 1 \\ 1 & 1 & 0 & 1 \\ 1 & 1 & 1 & 0 \end{bmatrix}$$

The matrix would have a zero diagonal because the costs for labels being correctly assigned should be minimal.

- `vC,hc`: Provide vertical and horizontal edge information respectively. This terms were determined by computing the directional gradient of the image with the Sobel gradient operator in both of the image's directions, x and y. Furthermore, the resulting gradients were normalized by their maximal values.

The GUI displays the image with the k computed regions and lets the user choose one of these regions as the desired segmentation mask. The decision made by the user will be then internally saved in the GUI's handles variable.

3.2 Interactive Graph-Cut Implementation

To additionally implement an interactive form of the common GC algorithm for the GUI, we let the user choose one or more regions that have been falsely classified by the initial segmentation process, and store these regions as *hard constraints* in two different binary masks; one for the regions the user would like to add to the resulting segmentation mask, and another one for the regions the user would like to exclude. These binary masks are kept in Matlab's function workspace and updated during a single correction phase.

First, the `DataCost` matrix should be computed as described in Sec. 3.1. Afterwards, this matrix should be modified. Since the applied GC function returns all k regions, the interactive GC function has to iterate through all labels in the `DataCost` matrix.

For the binary mask containing the regions that should be added to the segmentation mask, the function will prove whether the current label belongs to the region selected by the user. If so, the costs in the `DataCost` matrix at the positions determined by the binary mask should be set to its minimum. For all other labels, these should be set with a maximal cost.

For the binary mask containing the regions that should be removed from the segmentation mask, we need to increase the penalty of the extracted coordinates if they belong to the segmentation mask's label. All other elements of the data cost matrix can keep their original values.

3.3 Results

The result is a fully functioning annotation tool ready to be used to annotate the images in Dräger's thermal imaging database. To facilitate the usage of the annotation tool for any user, even for the ones with no Matlab licenses, the annotation tool was compiled with the Matlab compiler into a stand-alone application. The tool was then tested by six different users with different qualifications. During the annotation phase, approximately thousand images were annotated. All in all, the feedback was quite positive. Because of the help of the interactive GC the average annotation time per image varied from one minute to a maximum of three minutes. Specially when the size of the desired region was rather small and therefore hard to be manually selected by the user, the GC offered a better alternative than a manual polygon segmentation and/or region selection. Besides, the GC itself barely took ten seconds to deliver an output.

4 Conclusion and Future Steps

We successfully implemented an annotation tool in Matlab for the annotation of private images in a clinical study and distributed it among the project group members as a stand-alone application. This does not only offer a standard bounding box annotation with attributes for a later object detection but also offers an innovative way of easily segmenting the images with the help of an interactive GC algorithm. The generated ground truth dataset is already being applied to train and test different object detection algorithms and should later on be used to implement an automatic segmentation with the help of machine learning. Currently, the annotation tool can only work with gray-scale images, in the future it could be helpful to expand this tool to also work with RGB images.

Acknowledgement

The work has been carried out at Drägerwerk AG & Co. KGaA and supervised by the Institute of Medical Informatics, Universität zu Lübeck.

5 References

[1] O. Russakovsky, L. Fei-Fei, *Attribute learning in large-scale datasets*. European Conference of Computer Vision (ECCV), International Workshop on Parts and Attributes, 2010.

[2] T.-Y. Lin, M. Maire, S. J. Belongie, L. D. Bourdev, R. B. Girshick, J. Hays, P. Perona, D. Ramanan, P. Dollár, C. L. Zitnick, *Microsoft COCO: Common Objects in Context*. Available: http://dblp.uni-trier.de/rec/bib/journals/corr/LinMBHPRDZ14 [last accessed on 2017-01-07].

[3] Y. Boykov, O. Veksler, R. Zabih, *Fast approximate energy minimization via graph cuts*. IEEE Transactions on pattern analysis and machine intelligence, vol. 23, no. 11, pp. 1222–1239, 2001.

[4] V. Kolmogorov, R. Zabih, *What energy functions can be minimized via graph cuts?*. IEEE Transactions on pattern analysis and machine intelligence, vol. 26, no. 2, pp. 147–159, 2004.

[5] Y. Boykov, V. Kolmogorov, *An experimental comparison of min-cut/max-flow algorithms for energy minimization in vision*. IEEE Transactions on pattern analysis and machine intelligence, vol. 26, no. 9, pp. 1124–1137, 2004.

[6] L. R. Ford, D. R. Fulkerson, *Maximal flow through a network*. Canadian journal of Mathematics, vol. 8, no. 3, pp. 399–404, 1956.

[7] Y. Boykov, M.-P. Jolly, *Interactive graph cuts for optimal boundary & region segmentation of objects in ND images*. Computer Vision, 2001. ICCV 2001. Proceedings. Eighth IEEE International Conference, vol. 1, pp. 105–112, 2001.

[8] S. Bagon, *GCMex - Graph Cut Mex functions wrapper*. Available: https://github.com/shaibagon.GCMex [last accessed on 2016-10-12]

[9] D. Pountain, *Run-length encoding*. Byte, vol. 12, no. 6, pp. 317–319, 1987.

6

Image Processing

Joint registration and segmentation using statistical shape models

N. Broecker[1] and J. Ehrhardt[2]

[1] Medizinische Informatik, Universität zu Lübeck, Nora.Broecker@student.uni-luebeck.de

[2] Institute of Medical Informatics, Universität zu Lübeck, Ehrhardt@imi.uni-luebeck.de

Abstract

In order to fully utilize the possibilities of magnetic resonance imaging for diagnostic purposes, an accurate, automatic and fast segmentation procedure is required. The presented algorithm contributes to solve said problem by combining statistical shape models with a joint segmentation and registration approach. Firstly a statistical shape model was generated. The joint algorithm combines the advantages of registration and segmentation to produce more exact results. The main idea is to use the configurations of the statistical shape model to generate a fitting segmentation instead of only providing shape guidance during the segmentation process. We compare our results to an algorithm which uses shape models as segmentation guidance tool and also to sole registration and sole segmentation. Our joint approach achieves a higher accuracy in significantly less computation time. Evaluation of the segmentation accuracy is done by overlap coefficients, surface distances and noticeable visual differences.

1 Introduction

Cardiovascular diseases are the leading cause of death globally [1]. Therefore it is essential to discover the diseases in its early stages so therapeutic measures can be applied. Beside conventional cardiac examination methods, magnetic resonance imaging (MRI) is capable of giving a detailed impression of a patient's cardiac condition without imposing pain on the patient. Automatic segmentation of the left and right ventricles in cine MR images is required for correct diagnosis. Segmentation algorithms have to overcome certain difficulties such as poor image contrast, missing or diffuse boundaries and image noise. Regarding the characteristics of cardiac MR scans and the anatomical variety among patients, the problem of automatic segmentation remains unsolved. In this paper, we offer a solution to said problem by developing a new segmentation procedure which combines statistical shape models (SSM) with a joint registration and segmentation approach.

1.1 Relationship to Prior Work

Our work is based on commonly known and established procedures in medical image processing and combines their benefits into a new approach. In [5] Tsai et al. proposed the application of a parametric SSM for medical image segmentation. The presented approach combines shape and appearance information with level set functions while being robust to noise and topology changes. More results using SSMs are presented in [3]. In the article, a representation for deformable shapes is introduced with a defined probability distribution over the variances of a set of train-

ing shapes. The segmentation process starts with an initial curve as the zero level set. In each step of the surface evolution a maximum a posteriori position and shape of the object in the image are estimated based on the image data and prior shape information. The surface is then evolved based on the prior estimations, image gradients and curvature. The results are of the same robust nature similarly to [5]. However, the model representation and curve evolution technique are not coupled. Our attempt to combine registration with the SSM as segmentation tool is similar to the presented procedure in [2]. This paper proposes an algorithm for simultaneous non-linear registration and level set-based segmentation of multiple objects. Each object is described by a level set function and region competition is used as segmentation approach. The algorithm minimizes a variational functional that contains prior shape knowledge, prior intensity information and a diffeomorphic transformation model.

2 Material and Methods

2.1 Generating the Shape Model

The segmentation process can benefit from using a shape model which contains information about the object that is being segmented. Organs and other anatomical structures exist in a large variety. A statistical shape model can be used to express the similarity of a certain object and also contain its natural variations that are different from the mean shape. This knowledge about the shape of an object is acquired from training samples and their label images. An initial alignment of the label images is required by using affine registration. Thus the anatomical differences and pose variety of the organs do not interfere with the calculation of the SSM. This ensures that surface overlap is maximized as well as the relation between the datasets increased. The aligned images $I_1, ..., I_k$ are represented by corresponding level set functions $L_1, ..., L_k$. The level set functions contain the initial voxel values as distance values to the zero level set curves which are the contours of our given label images. The mean shape $\bar{\Phi}$ is calculated from the level set functions $L_1, ..., L_k$. The variance in shape is computed by using Principal Component Analysis (PCA), a complete description of the procedure can be found in [3]. After PCA, the variety of the shape is contained in the k principal components (PCs). By linear combination of these PCs with the mean shape, new shapes can be generated within the variety of the training data set. Some example shapes can be seen in Fig. 1. Our SSM Φ is defined as

$$\Phi(\bar{w}) = \bar{\Phi} + \sum_{j=1}^{k} w_j \Phi_j \qquad (1)$$

with $\bar{w} = (w_1, ..., w_k)$. The vector $\bar{w}$ defines the weights for PC images. For the exemplary shapes in Fig. 1 the weights w_1, w_2, w_3 were all set to 3.

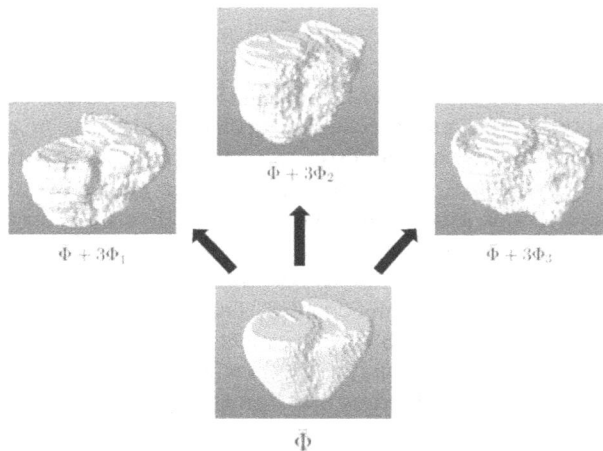

Figure 1: The mean model $\bar{\Phi}$ and the first, second and third PC images added.

2.2 Joint Registration and Segmentation

Our procedure combines registration and segmentation elements in a similar way to [2]. Instead of using level set segmentation, we use model guided segmentation. The entire method is described by:

$$\mathcal{J}^{joint}(\varphi, \bar{w}) =$$
$$\mathcal{D}[\mathcal{R}, \mathcal{T}, \varphi] + \mathcal{S}[\varphi] + \alpha E[\mathcal{T}, \Phi(\bar{w})] + \beta \mathcal{P}[\hat{\Phi} \circ \varphi, \Phi(\bar{w})].$$
$$(2)$$

$\mathcal{D}[\mathcal{R}, \mathcal{T}, \varphi]$ and $\mathcal{S}[\varphi]$ are distance measure and regularization of the non-linear registration. $\mathcal{T}$ describes the MR input image or template image whereas $\mathcal{R}$ stands for the reference image of the registration, the atlas image. The atlas and its label image are previously calculated using the images $I_1, ..., I_k$ and their respective label images. φ is the transformation calculated during registration. An iterative minimization is performed to obtain the final SSM configuration as segmentation for the MR input image.

The external energy $E[\mathcal{T}, \Phi(\bar{w})]$ which is the segmentation part of our algorithm can be formulated as

$$E[\mathcal{T}, \Phi(\bar{w})] =$$
$$\int_{\Omega} \mathcal{H}(\Phi(\bar{w})) \cdot \log(p_{out}(\mathcal{T})) + (1 - \mathcal{H}(\Phi(\bar{w}))) \cdot \log(p_{in}(\mathcal{T}))$$
$$(3)$$

where $p_{out}(\mathcal{T})$ and $p_{in}(\mathcal{T})$ describe the prior intensity probabilities outside and inside of the object. These probabilities are estimated prior to the procedure using the atlas label image. The Heaviside step function $\mathcal{H}$ is used to distinguish between background and object, for more information see [4]. The external energy term contributes the input image's intensity information to our algorithm while the model's weights $\bar{w}$ of the PC images get adapted in each iteration to fit the given intensity information.

The shape prior term $\mathcal{P}[\hat{\Phi} \circ \varphi, \Phi(\bar{w})]$ contains the squared distance between SSM $\Phi(\bar{w})$ and the atlas segmentation $\hat{\Phi}$ which is being registered during our procedure by transformation φ. The shape prior term $\mathcal{P}[\hat{\Phi} \circ \varphi, \Phi(\bar{w})]$ is defined as

$$\mathcal{P}[\hat{\Phi} \circ \varphi, \Phi(\bar{w})] = \int_{\Omega} (\Phi(\bar{w}) - \hat{\Phi} \circ \varphi)^2 dx. \qquad (4)$$

The shape prior term $\beta \mathcal{P}[\hat{\Phi} \circ \varphi, \Phi(\bar{w})]$ is the coupling term between registration and segmentation. The transformation φ is calculated by registering the atlas image $\mathcal{R}$ onto the original MR image $\mathcal{T}$. The given atlas label image will be registered onto $\mathcal{T}$ as well using the previously obtained transformation φ. Since this term considers the distance between the current shape model formation and the warped atlas segmentation $\hat{\Phi}$, it ensures that a large deviation from the warped atlas segmentation is penalized. As a result, our shape model $\Phi(\bar{w})$ finds a configuration of its weights $\bar{w}$ of the PC images which fits as closely as possible onto the warped reference segmentation $\hat{\Phi}$.

Regarding (2) again, we inserted the factor α to the external energy $E[\mathcal{T}, \Phi(\bar{w})]$ to weigh the influence of this term on the calculation. The same principle applies to the shape prior term $\beta \mathcal{P}[\hat{\Phi} \circ \varphi, \Phi(\bar{w})]$ with the factor β. Using these weighing factors α and β adjusts the influence of the respective parts in the segmentation process. This allows our procedure for example to compensate for poor MR image contrast by weighing the influence of the atlas segmentation higher. A visual explanation of our procedure can be seen in Fig. 2.

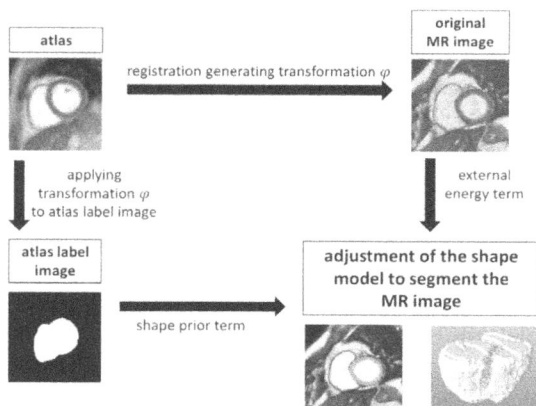

Figure 2: The presented procedure shown as graph. The configuration of the SSM is influenced by (2). Output is the calculated shape model which provides a segmentation for the MR input image.

2.3 Implementation

The presented method has been implemented as part of the image processing library at the Institute of Medical Informatics, Universität zu Lübeck.

In order to update the weights $\bar{w}$ of shape model Φ, we form the derivative of $E[\mathcal{T}, \Phi(\bar{w})]$:

$$\frac{\partial}{\partial w_j}E =$$
$$\int_\Omega \delta(\Phi(w_j)) \cdot (\log(p_{out}(\mathcal{T})) - \log(p_{in}(\mathcal{T}))) \cdot \frac{\partial}{\partial w_j}\Phi(j) \tag{5}$$

where δ stands for the Dirac function and is the derivative of the Heaviside function $\mathcal{H}$ (for more information, see [4]). The shape prior term $\mathcal{P}[\hat{\Phi} \circ \varphi, \Phi(\bar{w})]$ has to be derived twice, firstly with respect to the weights $\bar{w}$ and secondly with respect to the transformation φ:

$$\frac{\partial}{\partial \bar{w}}\mathcal{P} = \frac{1}{2}\int_\Omega (\bar{\Phi} + \sum_{j=1}^k w_j\Phi_j - \hat{\Phi} \circ \varphi)dx \cdot \frac{\partial}{\partial \bar{w}}\Phi(k) \tag{6}$$

$$\frac{\partial}{\partial \varphi}\mathcal{P} = \frac{1}{2}\int_\Omega (\bar{\Phi} + \sum_{j=1}^k w_j\Phi_j - \hat{\Phi} \circ \varphi)^2 dx \cdot \frac{\partial}{\partial \varphi}(-\hat{\Phi} \circ \varphi)) \tag{7}$$

The integral in (5), (6) and (7) forces us to iterate over the MR input image for every iteration step. Regarding the size of our data and the needed iterations, this is very costly. Instead, we implement our procedure using the sparse field method [6]. The outer contour of our shape model is represented by the indices in the sparse field layer. Instead of iterating over the whole image and checking our update criteria for every image voxel, the weight $\bar{w}$ updates will only be performed for voxels whose indices are in the sparse field layer. Based on the weight $\bar{w}$ change and thus the evolution

of the model, the layer is updated with the new indices until the algorithm reaches the maximum iterations or converges. For a more detailed description of the sparse field method, see [6].

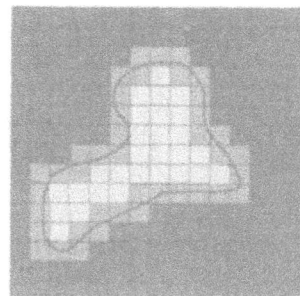

Figure 3: Visualization of the sparse field method [6]. The contour of our shape model is embedded in a sparse field layer which provides the indices of the points we need to consider to update our model accordingly.

3 Results and Discussion

Given 10 training images $I_1, ..., I_k$, our SSM is created using 8 of the 10 datasets. The remaining 2 are used as test images and their labels are used for validation of the results. We are comparing the results of our procedure to the results of the ITK implementation of geodesic active contours (GAC) presented in [3]. To show the benefit of a joint approach, we also evaluate the sole registration and sole segmentation part of our procedure.

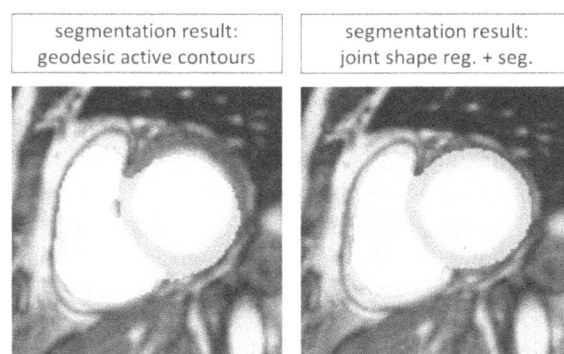

Figure 4: Visual comparison of the segmentation results for patient 02.

Comparing the segmentation results depicted in Fig. 4 and Fig. 5, it is noticeable that our procedure's segmentation fits the boundaries of the heart muscle better. Despite using the same SSM, our joint approach is able to extend the SSM further as well as to reach the edges of anatomically diverse structures. The difference regarding the myocardium segmentation is also visible - while the GAC segmentation does

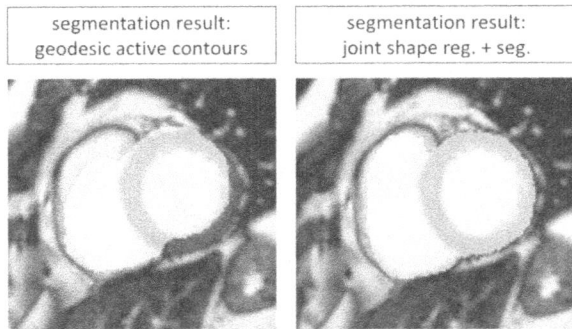

Figure 5: Visual comparison of the segmentation results for patient 04.

not cover the entire muscle structure, our joint approach adapted the SSM correctly. In the GAC segmentations, the level set curve did not evolve far enough because of the restricting effect of the curvature. Our algorithm does not rely on curvature since all possible and plausible forms of the segmentation are built into the SSM already. Adding the guidance of the registration part, our segmentation process was able to detect the form of cardiac muscle entirely and provide more accurate segmentation results. Another advantage is the runtime - our algorithm needs approximately 2 minutes per dataset, the GAC algorithm needs approximately 10 minutes per dataset.

Table 1: Validation of the segmentation results for patient 02 and 04 using GAC, shape segmentation, registration and our joint registration and segmentation procedure.

	Dice	Jaccard	Hausdorff
Patient 02 GAC	0.89	0.81	14.7
Patient 02 shape seg.	0.90	0.82	10.1
Patient 02 reg.	0.90	0.83	9.6
Patient 02 reg. + seg.	0.92	0.85	7.3
Patient 04 GAC	0.79	0.66	14.2
Patient 04 shape seg.	0.87	0.78	10.9
Patient 04 reg.	0.88	0.79	11.2
Patient 04 reg. + seg.	0.89	0.80	10.1

Table 1 shows that the presented joint approach outperformed the GAC algorithm. General similarity (Dice-coefficient) as well as segmentation overlap (Jaccard-coefficient) improved. The Hausdorff-distance which measures the longest distance between segmentation and label, also improved noticeably. Testing registration and segmentation separately produces better results than using GAC for our test images. Combining both parts to the joint shape registration and segmentation algorithm increases accuracy even further and lowers the Hausdorff distance.

4   Conclusion

In this paper, we introduced the joint registration and segmentation algorithm using SSMs and compared our results

to the GAC algorithm presented in [3]. While we used the same way to obtain our SSM and used the same SSM for our experiments with both algorithms, [3] relies on evolving a level set curve which is merely restricted by the SSM information. Thus the shape information has the role of a guidance tool in the same way that curvature and external energy guide the evolution of the level set segmentation. This is the main difference to previous approaches - we do not merely guide our segmentation with the SSM, it is our segmentation tool. This means that the results are inside the anatomically possible configurations. The segmentation algorithm is combined with non-linear registration for additional guidance. We discovered that the level set curve in the GAC segmentations did not evolve far enough because of the restricting effect of the curvature. The presented algorithm does not rely on curvature and is thus able to segment anatomically correct features automatically. The joint algorithm combines the advantages of registration and segmentation and was able to score an even more accurate result. Future efforts could be focused on generating SSMs based on large databases to include more anatomic variety into the segmentation process.

Acknowledgement

The work has been carried out at and supervised by the Institute of Medical Informatics, Universität zu Lübeck.

5   References

[1] M. Nichols, N. Townsend, P. Scarborough and M. Rayner, *European cardiovascular disease statistics*. Technical report, European Heart Network, 2012.

[2] J. Ehrhardt, T. Kepp, A. Schmidt-Richberg and H. Handels, *Joint multi-object registration and segmentation of left and right cardiac ventricles in 4D cine MRI*.SPIE Medical Imaging, 90340M–90340M, 2014.

[3] M. E. Leventon, W. Eric, L. Grimson, and O. Faugeras, *Statistical shape influence in geodesic active contours*. Computer Vision and Pattern Recognition, 2000.

[4] A. Schmidt-Richberg, *Registration Methods for Pulmonary Image Analysis: Integration of Morphological and Physiological Knowledge*. Springer Science & Business Media, 2014.

[5] A. Tsai, et al., *A shape-based approach to the segmentation of medical imagery using level sets*. IEEE transactions on medical imaging, 137-154, 2003.

[6] T. S. Yoo, *Insight into images: principles and practice for segmentation, registration, and image analysis*. AK Peters/CRC Press, 2004.

The Schatten-p-Norm as a Distance Measure for Image Registration

K. D. Lux-Hoffmann[1], J. M. Lotz[3], and J. Modersitzki[2] [3]

[1] Medizinische Informatik, Universität zu Lübeck, kim.luxhoffmann@student.uni-luebeck.de
[2] Institute of Mathematics and Image Computing, jan.modersitzki@mic.uni-luebeck.de
[3] Fraunhofer Institute for Medical Image Computing MEVIS, {judith.lotz,jan.modersitzki}@mevis.fraunhofer.de

Abstract

There are many methods to fuse two images using image registration techniques. An even more complex problem is the registration of more than two images which occurs for example in a pathology context, where several tissue sections of one biopsy are made to analyze different biomarkers. Often it is necessary to register the images sequentially. In this paper, we present a distance measure based on the Schatten-p-Norm which is capable of registering several images at the same time. To accomplish this the Schatten-p-Norm is applied to the pixel-wise computed image gradients. Thus, linear dependency of the gradients is favored and consequently the matching of edges and similarity of the images. We compare it to well-known distance measures like Normalized Gradient Fields (NGF) and Mutual Information (MI). The experiments show, that the proposed distance measure yields comparable results to NGF and MI for two images and promising results for more than two.

1 Introduction

When talking about image processing one big topic is image registration. This methods is used, when it is necessary to fuse information from different images. For example in a pathology context it may be necessary to analyze different biomarkers of one biopsy. Therefore it would be useful to have an overlay of all biomarkers.

One part of a registration is to measure the distance between the two images that should be registered. A distance measure computes how similar two images are. Gradient normalized fields (NGF) and mutal information (MI) are two of them and will here be used as reference distance measures. NGF is based on the idea that two images are the same, when their gradients are linearly dependent [1]. Therefore the gradient fields of the images are computed and normalized. Subsequently the linear dependency of the gradient fields is measured over either the dot or the norm of the cross product of the gradient vectors of each pixel.

MI uses the entropy of images [2]. The entropy is a measure for the uncertainty of the value a random variable can take. For MI the joint entropy of two images is calculated. The random variable describes in this case the pixels intensities that occur in the images. Eventually the joint entropy will take on smaller values when homogeneous structures in the images match each other [3].

A major drawback of common distance measure is, that only the distance between two images can be measured. However, in case of the introduced registration problem it would be useful to measure the distance between more than two images. In this paper a distance measure is presented, which enables the registration of not only two, but three ore more images at the same time by using the Schatten-p-Norm.

2 Material and Methods

In image registration we try to solve

$$J[y] = D[T[y], R] + \alpha S[y - y^{ref}] \xrightarrow{y} min \quad (1)$$

where D is the distance measure, T and R are the template and reference image, $T[y]$ is the transformed template and S measures the reasonability of the transformation [4]. For this moment we are concentrating on the evaluation of the distance measure and ignore the transformation and the optimization process.

The presented distance measure is based on the Schatten-p-Norm. Therefore the Schtten-p-Norm is described first in this section. Afterwards the distance measure itself is explained.

2.1 Schatten-p-Norm

Given a matrix A the Schatten-p-Norm is calculated over the singular-values of A [5].

$$A = U\Sigma V^* \quad (2)$$

is the singular-value decomposition, where U and V consist of the singular vectors of A and Σ contains the singular values $\sigma_1, ..., \sigma_r$, with r as the rank of A. The Schatten-p-Norm is then defined as

$$||A||_p = \left(\sum_{i=1}^{r} \sigma_i^p \right)^{\frac{1}{p}} . \qquad (3)$$

To facilitate the optimization process in a future registration scheme, we use the Schatten-p-Norm to the power of p:

$$||A||_p^p = \sum_{i=1}^{r} \sigma_i^p \qquad (4)$$

This can be done, because monotone operations do not change the position of the extremum.

2.2 The Distance Measure

With the introduced Schatten-p-Norm, a distance measure can be derived as follows. We define R as the reference image and $T_1, ..., T_n$, with $n \in \mathbb{N}$ as the template images. At first the gradients of the reference image and the template images are needed. We use the forward difference quotient given as

$$\partial_j^{h+} f(x) = \frac{f(x + he_j) - f(x)}{h} . \qquad (5)$$

In this case $f(x)$ is the value of a 2D image at point x, which is linearly interpolated. e_j is the j-th unit vector, i.e. with j you can specify the direction of the gradient. In this case j is 1 for the gradient in x-direction or 2 for the y-direction. h is the distance between two measured points, which is here simply 1.
With this, the gradients of R and T are calculated. We get

$$g_R^j(x) \equiv \begin{pmatrix} g_R^1(x) \\ g_R^2(x) \end{pmatrix} = \partial R(x) \qquad (6)$$

for the reference image and

$$g_{T_i}^j(x) \equiv \begin{pmatrix} g_{T_i}^1(x) \\ g_{T_i}^2(x) \end{pmatrix} = \partial T_i(x) , \; i = 1, ..., n \qquad (7)$$

for the template images, with $j = \{1, 2\}$.
If the gradients of two images at a point x point into the same direction, we assume the edges in the images to be aligned, in other words, we assume the images to be similar. Thus, we want the image gradients to be linearly dependent and measure this by writing them into a matrix on which the Schatten-p-Norm is calculated. When two gradients are linearly dependent only one singular values is greater zero, otherwise more then one are greater zero. Hence we build a matrix out of the gradients on which the Schatten-p-Norm can be calculated. This is done for each pixel separately:

$$A_x = \begin{pmatrix} g_R^1(x) \; g_{T_1}^1(x) \; ... \; g_{T_n}^1(x) \\ g_R^2(x) \; g_{T_1}^2(x) \; ... \; g_{T_n}^2(x) \end{pmatrix} \qquad (8)$$

On this Matrix the Schatten-p-Norm is computed according to (4). In the last step the Schatten-p-Norms of all pixels are added together:

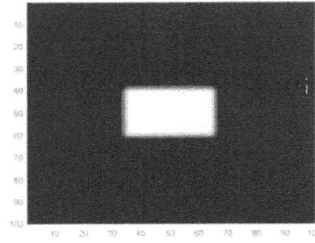

Figure 1: Synthetic data is used in the experiments: An image of size 100×100 pixels with a rectangle of size 30×20 pixels in it, where a gaussian smoothing was applied to.

$$D^{SPN}(R, T_1, ..., T_n) = \sum_{k=1}^{m} ||A_k||_p^p \qquad (9)$$

with m as the number of pixels. With this we get one value for the distance between all images.

2.3 Experiments

In this section three experiments are explained, to a) find a p-value for the Schatten-p-Norm suitable for image registration and b) evaluate the behavior of the distance measure with given transformations.
For the experiments synthetic data were generated. In an image of size 100×100 pixels a rectangle of size 30×20 pixels was placed in the middle. Afterwards a gaussian smoothing was applied to spread the information and create larger areas for the gradients (s. Fig. 1).
In the first two experiments it is necessary to explore if the new distance measure is robust for the normal use case. Therefore we use only one template image and a reference image, even though the distance measure was developed for more images. This is done, because it is the simplest case and a comparison with other distance measures is possible. At first we need to find a suitable value for p. Hence, the template image was shifted horizontally while the reference image was at a fixed position. Seen from the center of the reference image, the template image was shifted from -50 pixels to 50 pixels. For every shifted position the distance was measured. This was done for $p = 0.1$, $p = 0.5$, $p = 1$ and $p = 3$.
In the second experiment we evaluate the behavior of the distance measure if not only a horizontal shift but also vertical shift or a rotation is applied to the template image. The procedure was the same as in the previous experiment. The template image was shifted vertically from -50 pixels to 50 pixels, seen from the center of the reference image and rotated from $0°$ to $360°$.
In the last experiment the step to three images is taken. The setting is the same as in the second experiment, with the difference, that there is a second template image. The first template image is shifted as before, but at each position the second template image is shifted from -50 pixels to 50 pixels. This was done for two horizontal shifts, one horizontal

(a) $p = 0.1$ (b) $p = 0.5$

(c) $p = 1$ (d) $p = 3$

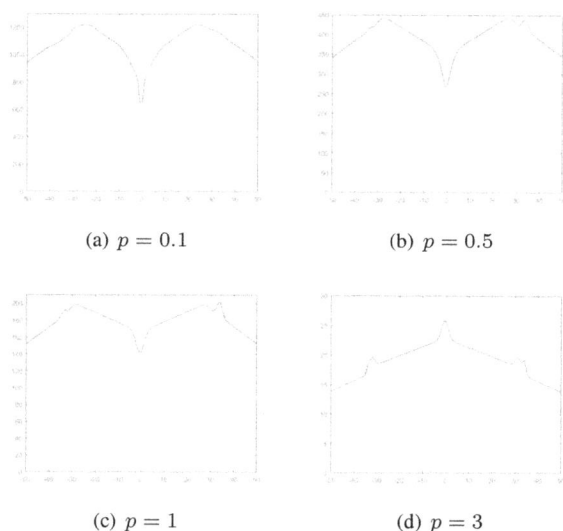

Figure 2: Here the results of the p-value experiment are diagrammed. The x-axis denotes the shift and the y-axis the value of the distance measure. The graph of $p = 0.1$ has global minimum in the middle and is very smooth. The graphs at 2(b) and 2(c) are similar to 2(a), but the global minima are not as distinct as in 2(a). And for $p = 3$ the graph shows a maximum at a zero shift. Therefore we choose $p = 0.1$.

and one vertical shift, and one horizontal shift and a rotation.

3   Results and Discussion

In this section we have a closer look at the experiments and discuss the results.

In the first experiment we were looking for a p-value for the Schatten-p-Norm, so that the distance measure shows a global minimum at zero shift and is smooth.

In Fig. 2 the results of the first experiment are shown. The graph in Fig. 2(a), where a p-value of 0.1 was used, has a huge minimum at zero shift. Also the slope near by the minimum is very steep, which is favorite behavior with respect to optimization.

The graph of $p = 0.5$ in Fig. 2(b) is similar to the first one. However the minimum is not as distinct as before. Also the slope is flatter near the minimum. The drop of the graph at the edges, which also occurs in Fig. 2(a) and Fig. 2(c), are only problematic for registration, if the starting guess is not well chosen.

When choosing $p = 1$ the graph in Fig. 2(c) still has a pronounced minimum at zero shift, but it is not as distinct as in the first two graphs.

The last graph, where a p-value of 3 was chosen, has no minimum but a maximum at zero shift (s. Fig 2(d)). Optimization is possible with this kind of graph, but is more difficult than optimize towards a minimum.

On the basis of this observations for the following experi-

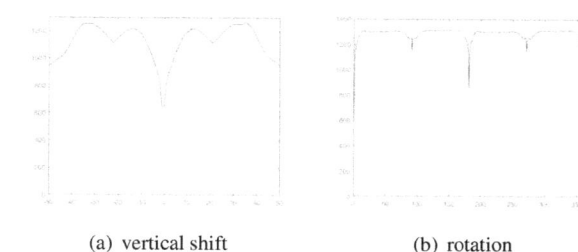

(a) vertical shift   (b) rotation

Figure 3: This figure displays the values of the distance measure for a vertical shift an a rotation. The graph of the vertical shift shows a global minimum at zero shift, but also two local minima. These are around the point, where the two rectangles start and end shifting over each other. The graph of the rotation has three big minima. One at $0°$ and one at $360°$, where the rectangles lay perfectly over each other, and the last at $180°$.

ments a p-value of 0.1 was chosen.

The next experiment is concerned with the behavior of the proposed distance measure when the template image undergoes a vertical shift and a rotation instead of a horizontal shift.

Fig. 3(a) shows the result of the vertical shifting. Like in the horizontal shifting a global minimum can be seen where the images match. Two additional smaller local minima occur at a shift of -20 and 20 pixels. These arise because the Schatten-p-Norm is computed on the gradients of the images. At a shift of -20 and 20 pixels the gradients in y-direction of the rectangles are aligned, which leads to a minimum. This also would be seen when a horizontal shift is applied, but the edges are shorter in that case.

In the second figure of this experiment the results of the rotation are displayed (s. Fig. 3(b)). The most distinct minima occur at a rotation of $0°$, $360°$ and $180°$. Due to the symmetry of the rectangles, this was expected. At these rotation angles the template image lays directly over the reference image. The local minima appear around a rotation of $90°$ and $270°$, when the rectangles lay vertical over each other. For comparison NGF and MI are used as distance measures and Fig. 4 shows the results of a horizontal shift. A distinct minimum can be seen at zero shift, with a rise of the function, the further away the two rectangle move from each other.

The same behavior can be observed when using the distance measure based on the Schatten-p-Norm. Therefore it can be said, that the proposed distance measure can be used theoretically for the registration of two images. With this in mind an analysis of the measurement between three images using the proposed distance measure is justified.

The last experiment deals with the actual problem the distance measure was designed to tackle. Instead of one template image, two template images where used. The distance measure was tested for two horizontal shifts of the template images, one horizontal and one vertical shift, and for one horizontal shift and a rotation.

The results of this experiment are shown in Fig. 5. A com-

(a) NGF   (b) MI

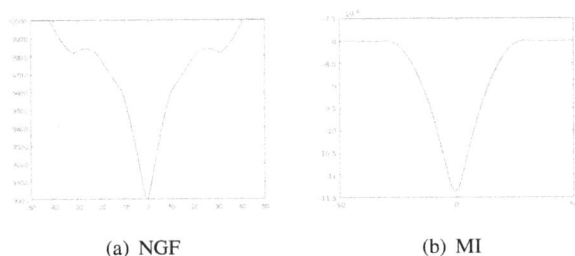

Figure 4: This graph shows the measurements of NGF and MI for a rectangle that is shifted horizontal over another rectangle.

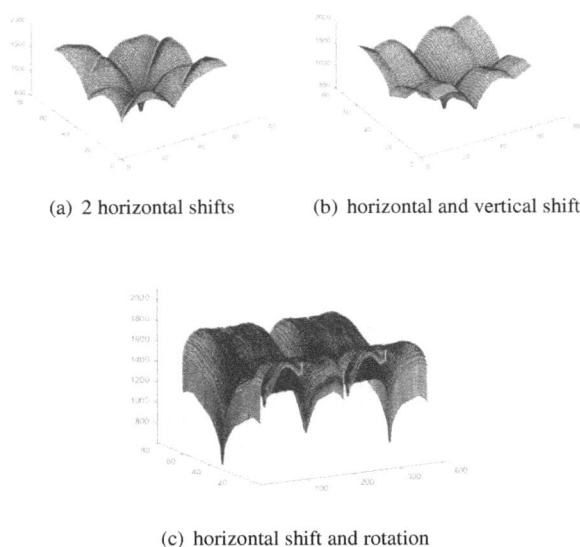

(a) 2 horizontal shifts   (b) horizontal and vertical shift

(c) horizontal shift and rotation

Figure 5: The results of the third experiment are displayed in this figure. All three graphs have their global minimum at the point, where all three rectangles have a perfect overlay. The valleys occur at areas, where at least one template rectangle lays over the reference rectangle and leads to the minimum.

parison with another distance measure is not possible, because to the best of our knowledge there is no distance measure, which can measure the distance between more than two images.

Fig. 5(a) and Fig. 5(b) look quit similar. There is a global minimum when all three rectangles match and local minima and valleys occur at shifts, where at least two rectangles are aligned.

The graph of the concurrent horizontal shift and rotation experiment in Fig. 5(c) looks like a combination of the graph in Fig 2(b) and Fig. 3(b). Like in the other two graphs of this experiment the minima are at spots, where two or more rectangles lay over each other.

4   Conclusion

In this paper a new distance measure based on the Schatten-p-Norm for image registration approaches was presented. It is capable of measuring the distance not only between two images, but between three or even more.

In the first step the gradients of the reference image and the template images have to be computed. Afterwards for each pixel the gradients of all images were written in to a matrix. On this matrix the Schatten-p-Norm was computed. In the end the norm values of all pixels were added together to achieve the final value of the distance measure.

We did three experiments to evaluate the robustness and behavior of the proposed distance measure. In a general use case with two images it showed comparable results to standard methods as NGF and MI. In the special use case of more than two images distinct minima occurred at locations, where a huge overlay of gradients was detected. This leads to the conclusion, that it is possible to measure the distance between three or more images and use the proposed distance measure for image registration of at least three images.

Future work includes experiments with real data and the development of an optimization method to enable the registration of more than three images.

Acknowledgement

The work has been carried out and supervised by the Institue of Mathematics and Image Computing at the Universität zu Lübeck.

5   References

[1] E. Haber and J. Modersitzki, "Beyond mutual information: A simple and robust alternative," in *Bildverarbeitung für die Medizin 2005*, pp. 350–354, Springer, 2005.

[2] P. Viola and W. M. Wells III, "Alignment by maximization of mutual information," *International journal of computer vision*, vol. 24, no. 2, pp. 137–154, 1997.

[3] H. Handels, *Medizinische Bildverarbeitung: Bildanalyse, Mustererkennung und Visualisierung für die computergestützte ärztliche Diagnostik und Therapie.* Springer-Verlag, 2009.

[4] J. Modersitzki, *FAIR: flexible algorithms for image registration*, vol. 6. SIAM, 2009.

[5] S. Lefkimmiatis, J. P. Ward, and M. Unser, "Hessian schatten-norm regularization for linear inverse problems," *IEEE transactions on image processing*, vol. 22, no. 5, pp. 1873–1888, 2013.

Propagation of landmarks with a registration graph

F. Zell[1] and M. P. Heinrich[2]

[1] Medizinische Ingenieurwissenschaft, Universität zu Lübeck, fenja.zell@student.uni-luebeck.de

[2] Institute of Medical Informatics, Universität zu Lübeck, heinrich@imi.uni-luebeck.de

Abstract

The generation of anatomic atlases may consume a lot of time but is necessary to produce large amount of data to represent the high variability of the human body. Therefore an automatic propagation of segmentations or landmarks is desirable which is achieved with the help of a registration graph. The algorithm starts with the registration of each image, using the optical flow and the subsequent calculation of their similarities. They are determined using the DICE metric. The results are stored in a matrix $\mathcal{W}$, based on which the registration graph is calculated iteratively. To quantify the error of the algorithm, the transformed landmarks were compared with the results of the "Coherent Point Drift (CPD)". The evaluation showed that the algorithm generated acceptable results but the precision has to be increased.

1　Introduction

Image registration is a fundamental task in medical image processing. Over the years, a lot of techniques referring to registration problems have been developed. The goal is to find correspondences between a set of images and to estimate a transformation which maps one image or a set of landmarks to the other. Image registration is considered an important step in image analysis. It is crucial for monitoring tumour growth or comparing the patient data with anatomical atlases. The generation of these atlases is extensive work. Therefore the automatic propagation of information from one atlas to another is a desirable goal. With the availability of automatic and reliable algorithms for atlas generation, further anatomical information can be extracted from medical scans. Especially machine learning methods and statistical shape models may benefit from accurate correspondences in atlas scans.

However, multiple reasons complicate the registration problem. Due to individual variation in human anatomy the propagation of specific landmarks remains an unresolved issue.

Multi-atlas based approaches are the most popular methods for dealing with this variability. Wolz et al. [1] describe a method for the automatic propagation of a set of manually labeled atlases to various sets of images. Therefore a manifold is learned, which helps finding similar images and transferring an initial set of atlases to all images with multi-atlas segmentation steps.

Another solution to handle the problem of the anatomic variability is explained from Krause et al. [2]. The key step is the use of a *pose graph* to determine objects with similar poses.

This work contains a combination of both approaches to propagate anatomic landmarks in CT-Scans. Therefore a *pose graph* is generated to represent correlations and similarities between individual atlas scans. Based on this graph, anatomic landmarks of one image are pairwise transferred to those scans which are most similar to them.

Once, a set of images with landmarks is generated, the calculation of a statistical shape model is straightforward [5]. These models are an approach for the segmentation of medical images. They are less likely to produce errors because the models contain information about the expected shape of an interesting structure. Based on this model, shape correspondences are detected automatically.

There is a major reliability problem regarding biological objects because there is a significant natural variability. Therefore it is important that enough shape variability from trainings data is included in the shape model. A proven way to solve this problem is to learn a mean shape and their variations from a collection of training samples. A large selection of data is essential for conclusive results. Therefore a reliable and automatic propagation of landmarks is advantageous [5].

The results of this propagation will be evaluated and compared with the "Coherent Point Drift (CPD)" from Myronenko and Song [3]. The CPD algorithm is a point set registration method for rigid or nonrigid transformations. The alignment of two points is considered a probability density estimation problem which is solved by maximizing the likelihood.

2　Material and Methods

2.1　Data

In this work, a random selection out of 82 abdominal contrast enhanced 3D CT scans from National Institutes of Health Clinical Center is used [4]. The total data set of

CT scans includes 53 male and 27 female subjects, ranging from 18 to 76 years. 17 of those were healthy, whereas the rest suffered from major abdominal pathologies.

The CT scans are acquired on Philips and Siemens MDCT scanners with a resolution of 512×512 pixels and a varying pixel size. Additionally, a slice thickness between 1.5 - 2.5 mm was chosen.

The atlases include manually performed slice by slice segmentation of the pancreas as ground truth.

Based on the varying pixel sizes preprocessing is unavoidable. Performing this step, the pixel size is resampled to $1\,mm^3$ and a region of interest is defined which contains the pancreas structure. Depending on this region the data is trimmed to reduce the dimension and improve the registration. As an example, one slice of the preprocessed image and its segmentation is shown in Fig. 1.

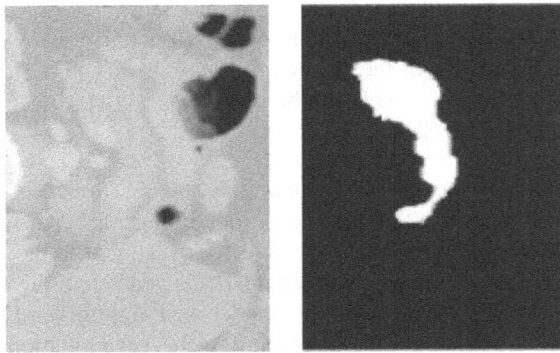

Figure 1: Preprocessed image of the pancreas and its related segmentation.

2.2   Overview of the method

To be able to propagate a number of landmarks for a set of images, the most appropriate transformation must be found. In this work, optical flow is used to determine displacement fields which describe the transformation from one image to another. This registration method is a suitable choice because of the assumption that small movements could be compared to the anatomic variability.

In the first instance, each image is registered to all the other ones with the optical flow, followed by using the DICE metric to compute the most similar one. This metric describes similarities between every single one of the images. The construction of a registration graph based on this metric is necessary for propagating landmarks to all images.

This procedure leads to the creation of a shape model which enables the segmentation of pancreas in unknown images.

2.2.1   Determine the displacement fields

The optical flow belongs to the non-linear registration methods and estimates displacement fields between a fixed image and a moving image. We defined $\mathcal{F} : \mathbb{R}^3 \to \mathbb{R}$ to be the fixed image and $\mathcal{M} : \mathbb{R}^3 \to \mathbb{R}$ the moving one with support in domain $\Omega_{\mathcal{F}}$ and $\Omega_{\mathcal{M}}$.

The aim is to transform the moving one to the fixed image, resulting in a high resemblance between both of them. Therefore the energy function, which contains the squared difference between the two images, is minimized. In equation (1) the energy function with the displacement field $u : \mathbb{R}^3 \to \mathbb{R}^3$ is defined by

$$E(\mathcal{F}, \mathcal{M}) = \frac{1}{2}(\mathcal{M}(x + u)) - \mathcal{F}(x))^2. \qquad (1)$$

The optical flow method is based on the assumption that the grey values of image objects do not change and only small displacements are possible. Using the first order Taylor expansion and a suitable regularization, the minimization problem can be solved. For more detailed information the work from Bruhn et al. [6] could be consulted.

In this work, the moving image $\mathcal{M}$ and fixed images $\mathcal{F}$ are considered to be the signed distance map of the atlases $\mathcal{A}$ to avoid different grey values between adjacent structures. Every possible allocation combination from moving and fixed images of $\mathcal{A}_1$ to $\mathcal{A}_{10}$ are examined, generating a total of 90 different displacement fields.

2.2.2   Measurement of similarity

After the generation of the displacement fields, the similarities between every moving and fixed images must be determined. Given the fact that atlases are binary objects, the DICE could be used to measure the similarities of the transformation. Similar images are represented by small weights. Therefore $1 - DICE$ was chosen for the calculation. The results are listed in a matrix $\mathcal{W}$ see with entries

$$\mathcal{W}_{i,j} = 1 - DICE(\mathcal{A}_{i_{fixed}}, \mathcal{A}_{j_{moving}}). \qquad (2)$$

Each row i of the matrix represents a fixed atlas and every column j a moving one. The diagonal of the matrix is defined as infinite.

2.2.3   Registration graph construction

Based on the matrix $\mathcal{W}$ a graph $\mathcal{G}$ is created, which represents images as nodes, using edges $(\mathcal{A}_i, \mathcal{A}_j)$ to connect similar objects. For the construction of this graph disjoint minimum spanning trees are calculated iteratively [2].

A minimum spanning tree is a subset of an undirected graph and connects all nodes without cycles. In addition, the sum of the weights of all edges in the tree must be as small as possible. This kind of graph was chosen for registration because only the most similar images are connected and therefore included in the registration process.

First of all, k single minimum spanning trees can be found in the matrix $\mathcal{W}$, when each weight is set to infinite after its utilization. As a result of the consecutive calculation, no more minimum spanning trees are left.

This procedure finishes by combining all the computed minimum spanning trees to produce a new graph $\mathcal{G}$. A major advantage of this method is its significant robustness, due to the fact, that every node is connected to at least k others.

2.2.4 Propagation

The final step is the transfer of landmarks along the graph. Therefore the most frequently connected node defined the starting point. This image has the highest degree in graph $\mathcal{G}$ as well as the greatest resemblance to the others. Based on the initial point, a set of landmarks is propagated to all adjacent images. For the transformation of landmarks, the previously calculated displacements fields are used.

As long as there are still images left without landmarks, the algorithm chooses the node with the largest number of adjacent images with landmarks available. This node then receives its landmarks next, calculating the median of each adjacent images transformation.

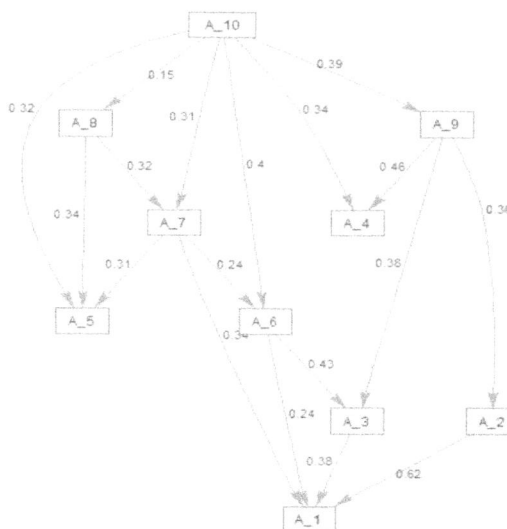

Figure 3: Registration graph $\mathcal{G}$ was merged, using found minimum spanning trees.

3 Results and Discussion

Applying the algorithm described in section 2 with ten atlases from the pancreas, two disjointed minimum spanning trees are determined. Therefore the smallest three values from the rows and columns of the weight matrix $\mathcal{W}$ are used to increase the precision. The two found spanning trees are illustrated in Fig. 2. On each edge the weight $w \in \mathcal{W}$ of similarity between the single images are visualized.

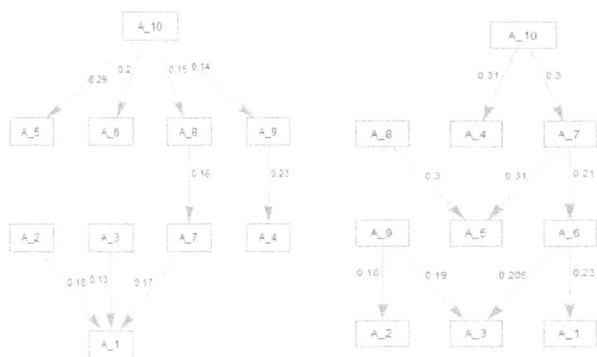

Figure 2: Disjointed minimum spanning trees generated based on matrix $\mathcal{W}$.

Combining these minimum spanning trees, the registration graph $\mathcal{G}$, shown in Fig. 3, is created. In this case, the starting point with the highest degree in $\mathcal{G}$ is image $\mathcal{A}_{10}$.

To propagate a set of landmarks from the starting point to all adjacent images $\mathcal{A}_4$ to $\mathcal{A}_9$, 100 random landmarks of image $\mathcal{A}_{10}$ were generated. These points represent the shape of the segmented pancreas, shown in Fig. 4.

To evaluate the accuracy of the determined landmarks there has to be an error calculation, containing the distance between the landmarks and the surface of the given structure. Therefore the signed distance map of the related atlases is used. The average absolute error for each image is presented in Table 1.

Figure 4: 100 randomly generated landmarks of a pancreas structure.

3.1 Comparison between direct and alternative route

To evaluate the error propagation through graph $\mathcal{G}$, a direct and an alternative way are compared. In the first instance, 100 landmarks from $\mathcal{A}_{10}$ are propagated directly to $\mathcal{A}_7$, resulting in a mean error of 3.92.

As well as the direct way, the alternative one also starts at $\mathcal{A}_{10}$ taking a detour $\mathcal{A}_8$ to $\mathcal{A}_7$. Going this way, the mean error is 3.69.

The new generated landmarks of both ways are seen together in Fig. 5, showing the landmarks of the direct way as crosses and the other ones as points.

Repeating this procedure with $\mathcal{A}_{10}$, $\mathcal{A}_9$ and $\mathcal{A}_4$, the difference between the average errors could be considered similar. The error of the direct way amounts 4.02, slightly varying to the alternative one with 3.92.

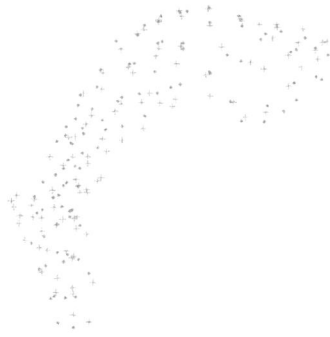

Figure 5: Propagating the landmarks from $\mathcal{A}_{10}$ to $\mathcal{A}_7$ directly (points) and with a detour (crosses) to $\mathcal{A}_8$.

3.2 Registration graph vs. CPD

To evaluate the registration graph method, the proven CPD algorithm was chosen. Therefore a set of 200 random landmarks of image $\mathcal{A}_{10}$ is generated and propagated to all other ones, using the CPD method. This procedure was repeated 30 times to find the minimal mean error, listing the results in Table 1. An example of the propagated landmarks from $\mathcal{A}_{10}$ to $\mathcal{A}_9$ with both approaches is presented in Fig. 6, indicating the landmarks from the registration graph method as points and the CPD created ones as crosses.

Table 1: Comparison of the mean error

Image	Registration graph	CPD
1	5.26	2.05
2	4.14	2.39
3	4.88	2.64
4	4.02	1.54
5	4.54	1.79
6	4.69	1.83
7	3.92	2.23
8	2.95	2.10
9	3.31	2.22

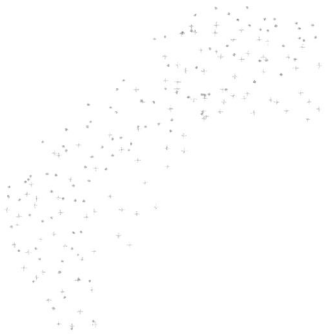

Figure 6: Comparison of the results from the CPD approach (crosses) and the registration graph (points).

3.3 Discussion

The results indicate that a registration graph orientated propagation with only one set of landmarks generates acceptable results but especially the comparison with the CPD algorithm underlines that the precision has to be increased. One option to optimize this is the usage of more data to rise the similarity between each image. Outliers, such as shown between the nodes $\mathcal{A}_2$ and $\mathcal{A}_1$, could be avoided this way. Another notable aspect is the small error propagation, indicating that a propagation through a highly complex graph could be possible.

4 Conclusion

In this work, the automatic propagation of landmarks with a registration graph was described. By using this algorithm, it is possible to generate a large set of data containing a wide variety.

Thinking one step ahead is the usage of these automatic generated landmarks to calculate statistic from models to find searched structures in unseen images.

Acknowledgement

The work has been carried out at the Institute of Medical Informatics, Universität zu Lübeck and supervised by Jun.-Prof. Dr. M. Heinrich.

5 References

[1] R. Wolz et al. *LEAP: Learning embeddings for atlas propagation.* NeuroImage, 49.2, pp. 1316-1325, 2010.

[2] J. Krause et al. *Fine-Grained Recognition without Part Annotations.* Proceedings of the IEEE Conference on Computer Vision and Pattern Recognition, pp. 5546-5555, 2015.

[3] A. Myronenko and X. Song, *Point set registration: Coherent point drift.* IEEE transactions on pattern analysis and machine intelligence 32.12, pp. 2262-2275, 2010.

[4] Cancer image archive, 2014. Available: https://wiki.cancerimagingarchive.net/display/Public/Pancreas-CT [last accessed on 2017-01-09].

[5] T. Heimann and H. Meinzer, *Statistical shape models for 3D medical image segmentation: a review.* Medical image analysis 13.4, pp. 543-563, 2009.

[6] A. Bruhn, , J. Weickert and C. Schnörr. *"Lucas/Kanade meets Horn/Schunck: Combining local and global optic flow methods.* International Journal of Computer Vision 61.3, pp. 211-231, 2005.

Groupwise Affine Registration of Medical Images with Missing Correspondences

H. Uzunova[1], H. Handels [2], and J. Ehrhardt[2]

[1] Medizinische Informatik, Universität zu Lübeck, hristina.uzunova@student.uni-luebeck.de

[2] Institut für Medizinische Informatik, Universität zu Lübeck, {handels, ehrhardt}@imi.uni-luebeck.de

Abstract

The registration of medical images becomes challenging when, due to pathological structures or anatomical variations, the images miss correspondences. This paper presents a fast and robust method for groupwise affine registration based on the RASL algorithm, which uses a sparse and low-rank decomposition to formulate the registration problem [1]. We adapt RASL for the alignment of 3D images and introduce a stochastic optimization scheme to enable the computational tractability. Additionally, a normalization scheme to generate more plausible and unique transformations is introduced. The algorithm has been applied to various medical images and proves its suitability for medical image registration. This approach shows advantages especially in the presence of pathologies and outperforms iterative groupwise registration based on ITK. The stochastic optimization scheme generates a significant acceleration allowing for a groupwise affine registration of ten 3D CT images in $\sim$ 5 minutes compared to $\sim$ 200 minutes with the naive approach.

1 Introduction

Image registration is a crucial part of various scientific fields such as computer vision and medical image analysis and processing. The robust registration of medical images is particularly challenging when pathologies hinder correspondence detection. The usual definition of image registration uses a template image, which is then aligned to a chosen reference image by finding proper transformations. However, it is often necessary to be able to align multiple images at the same time – a *groupwise* registration, e.g. for producing an atlas out of various images for the atlas-based segmentation.

In this work, we present an efficient and robust algorithm for groupwise affine registration of medical images in the presence of pathological structures. Our algorithm is based on a registration method named RASL (*Robust Alignment by Sparse and Low-rank decomposition for linearly correlated images*), which has been used in previous work to align non-medical 2D images [1]. The authors of [1] have shown the efficacy of this algorithm over a wide range of image corruptions and realistic misalignments. A drawback of RASL is the high computational demand required to solve the underlying optimization problem which prevents the application to 3D (medical) data. The contribution of our paper is threefold: (1) The algorithm has been expanded for aligning 3D images and a stochastic sampling scheme was integrated leading to a significant speedup allowing for efficient 3D registration. (2) We introduce a normalization scheme that results in more plausible transformations. (3) The algorithm has been applied in various medical settings, to prove its suitability for medical image registration and to show its advantage in the presence of pathologies.

2 Material and Methods

In this section, we first present the optimization problem formulated in [1] and then our suggested changes, which eventually lead to more plausible transformations and faster computation.

2.1 Robust image alignment by sparse and low-rank decomposition

Suppose we are given n 2D grayscale images $I_1, \ldots, I_n \in \mathbb{R}^{w \times h}$ and $vec : \mathbb{R}^{w \times h} \to \mathbb{R}^m$ is an operator which stacks the pixels of an image as a vector. Under the assumption that aligned images are linearly correlated, the data matrix

$$A = [vec(I_1)| \ldots |vec(I_n)] \in \mathbb{R}^{m \times n} \qquad (1)$$

should be approximately *low-rank* if the images are well-aligned. However, even small misalignments will lead to an increased rank of the matrix A. The misalignment of the input images is modeled by transformations τ_i, such that $I_1 \circ \tau_1, \ldots, I_n \circ \tau_n$ are well-aligned. In this work, we choose to limit the domain of τ to affine transformations.

In practice, images are often corrupted or happen to have differing pixels to each other – in medical images, those can be various pathologies, artifacts or anatomical variations not modeled by the transformation class. Thus, it is convinient to assume that each image I_j has a corresponding additive error e_j such that $I_1 \circ \tau_1 - e_1, \ldots, I_n \circ \tau_n - e_n$ are well aligned. The above observations can be written as

$$D \circ \tau = A + E, ||E||_0 \leq k \qquad (2)$$

where $E = [vec(e_1)|\ldots|vec(e_n)]$, $D = [vec(I_1)|\ldots|vec(I_n)]$, τ is the set of n transformations $\tau_1,\ldots,\tau_n$ and $||\cdot||_0$ denotes the l^0-norm. Because errors usually affect only a fraction of the image, this modelling yields sparse vectors e_j and hence E is a *sparse matrix*.

The above components can now be used to obtain the optimization problem

$$\min_{A,E,\tau} \ \mathrm{rank}(A) + \gamma||E||_0 \quad \text{s.t. } D \circ \tau = A + E \quad (3)$$

where $\gamma > 0$. While (3) follows naturally from the problem definition, this problem cannot be directly solved [2]. Therefore, a relaxed convex optimization problem is formulated using nuclear norm $||\cdot||_*$ and $\mathcal{L}^1$ norm:

$$\min_{A,E,\tau} \ ||A||_* + \lambda||E||_1 \quad \text{s.t. } D \circ \tau = A + E. \quad (4)$$

To solve (4) w.r.t. the transformations τ, the constraint $D \circ \tau = A + E$ is linearized using Taylor's formula

$$D \circ (\tau + \Delta\tau) \approx D \circ \tau + \sum_{i=1}^{n} J_i \Delta\tau \epsilon_i^T$$

where $J_i = \frac{\partial}{\partial \xi} vec(I_i \circ \xi)|_{\xi=\tau_i}$ is the Jacobian of the i-th image with respect to the transformation parameters and ϵ_i denotes the standard basis for $\mathbb{R}^n$. Based on this linearization, the authors of [1] propose the iterative RASL algorithm to solve the given optimization problem. The algorithm uses an outer loop to update the transformations $\tau_1,\ldots,\tau_n$ and an inner loop to minimize (4) with respect to A, E and $\Delta\tau$ using the Alternating Direction Method of Multipliers (ADMM) [3].

2.2 Restriction to plausible transformations

A significant problem in the existing algorithm is the fact, that the calculated transformations $\tau_1,\ldots,\tau_n$ are not restricted and computed independently. This can lead to implausible results such as zooming all images in or out up to a certain area of one color. Such transformations are, indeed, solutions of the minimization problem, but are not desirable in practice. The authors propose an intensity rescaling step in each iteration to avoid this problem, however, in our observations the resulting transformations do not correspond to the intuition. The results of this can be seen in Fig. 1(b), where all registered images are scaled by the same factor. Such transformations may lead to better solutions of the optimization problem, but can and should be cancelled out due to the fact that they do not contribute to the registration accuracy. Therefore, we propose the additional constraint

$$\tau_1 \circ \tau_2 \circ \cdots \circ \tau_n = Id \quad (5)$$

where $\circ$ denotes the concatenation of affine transformations. Note that (5) implicitly registers all images to the average shape (Fig. 1).

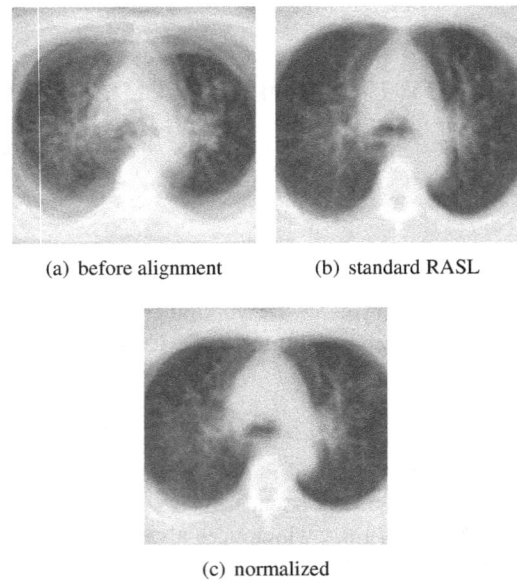

(a) before alignment   (b) standard RASL

(c) normalized

Figure 1: Mean of ten lung CT images (a) before alignment, (b) after RASL alignment and (c) after RASL alignment using the transformation constrain in (5).

2.3 Solving RASL efficiently by stochastic optimization

In previous work, RASL is particularly used for rigid, affine and projective alignment of 2D photographic images [1]. The extension of the algorithm to 3D image data is straightforward by using a vectorization operator $vec : \mathbb{R}^{w \times h \times d} \to \mathbb{R}^m$. The RASL registration of images with $200 \times 200 \times 100$ voxels then requires multiple times the ADMM optimization of (4) for matrices with 4 million rows. One possible solution is to reduce the image size in a multi-resolution mode. However, this might contradict the sparsity assumption. Instead, we propose to use stochastic sampling in the outer loop of the RASL algorithm to reduce the matrix size in (4) to be solved in the inner loop. Therefore, we select randomly a percentage $p \in (0, 1]$ of pixel positions in each iteration to compute the transformation updates $\Delta\tau$. Because only a small number of transformation parameters needs to be computed (e.g. 12 per image for affine transformations), we assume that a robust registration is possible even for small values p. This assumption has been confirmed by multiple experiments, some of which presented in this work. The stochastic RASL optimization is shown in Alg. 1. Note that the matrices A and E are not needed within the outer while loop and have only $p \cdot m$ rows.

The stochastic optimization leads to a necessary adaption of the weighting parameter λ depending on the parameter p. According to the suggestion in [1], we use the following adaptation rule: $\lambda = \frac{\lambda_0}{\sqrt{p}}$ where λ_0 is the weighting parameter for the full resolution images ($p = 1$). Note that the Log-Euclidean formula in [4] is used for the normalization in step 5.

Algorithm 1 Stochastic RASL optimization

INPUT: Images $I_1,\ldots,I_n \in \mathbb{R}^{w \times h \times d}$, initial transformations $\tau_1,\ldots,\tau_n$, weights $\lambda > 0$, percentage of used rows $p \in (0,1]$.

Step 0: Vectorize and stack the images:

$$D \leftarrow [vec(I_1)| \ldots |vec(I_n)]$$

while not converged **do**

　Step 1: Select $p \cdot m$ rows of D and save them in $\hat{D}$ corresponding to the reduced images $\hat{I}_i$

　Step 2: Compute Jacobian matrices w.r.t. transformation:

$$J_i \leftarrow \frac{\partial}{\partial \xi} \left(vec(\hat{I}_i \circ \xi) \right) \Big|_{\xi = \tau_i}, i = 1,\ldots,n$$

　Step 3: Transform the images:

$$\hat{D} \circ \tau \leftarrow \left[vec(\hat{I}_1 \circ \tau_1)| \ldots |vec(\hat{I}_n \circ \tau_n) \right]$$

　Step 4: Solve the linearized convex optimization with ADMM (inner loop):

$$(A^*, E^*, \Delta\tau^*) \leftarrow \arg\min_{A,E,\Delta\tau} ||A||_* + \lambda||E||_1,$$

$$\text{s.t. } \hat{D} \circ \tau + \sum_{i=1}^{n} J_i \Delta\tau \epsilon_i \epsilon_i^T = A + E$$

　Step 5: Update and normalize transformations:

$$\begin{aligned} \tau &\leftarrow \tau + \Delta\tau^* \\ \tau_i &\leftarrow \tau_i \circ \bar{\tau}^{-1} \quad \text{with } \bar{\tau} = exp(\tfrac{1}{n}\textstyle\sum_i log(\tau_i)) \end{aligned}$$

end while

Step 6 (optional): Compute $D \circ \tau$ and solve (for fixed τ):

$$(A^*, E^*) \leftarrow \arg\min_{A,E} ||A||_* + \lambda||E||_1, \text{ s.t. } D \circ \tau = A + E$$

OUTPUT: Transformations $\tau_1,\ldots,\tau_n$ (optional A^* and E^*) as solution of (4)

3　Results and Discussion

The first experiment evaluates the influence of the percentage p of the used image pixels on the registration accuracy and run-time. Table 1 shows the results for an affine registration of ten 3D CT images of the thorax without visible pathologies ($200 \times 200 \times 180$ voxels). Registration accuracy is measured by average pair-wise Dice overlaps of the lung masks. Computation time is measured for the complete groupwise registration process on a quad-core

Table 1: Sample size and its effect on speed and quality of alignment, measured using Dice-coefficient.

$p =$	Time (min.)	Dice
100%	202.23	0.7815
50%	91.85	0.7815
10%	21.48	0.7818
1%	05.72	0.7805
initial		0.7175

2.67GHz Xeon® CPU W3520. A speed-up of ≈ 35 can be achieved by using only 1% of the voxels in each iteration without degrading registration accuracy significantly. Note that the computation time does not seem to scale linearly with respect to p. This occurs due to the fact, that the algorithm performs certain operations on the whole image, such as reading, writing and warping.

The following experiments assess the registration accuracy of the stochastic 3D RASL algorithm in the presence of pathologies using four different data sets:

- *Lung CT Tumor:* Ten 3D CT images of the thorax with large lung tumors, $200 \times 200 \times 180$ voxels.

- *Lung CT Diverse:* Ten 3D CT images of the thorax with diverse lung pathologies like fibrosis or lung edema, $210 \times 210 \times 190$ voxels.

- *Brain MRI Tumor:* Ten MRI images with brain tumors, $200 \times 200 \times 80$ voxels.

- *Brain MRI Lesion:* Ten MRI images of the head with artificially generated stroke lesions, $140 \times 180 \times 140$ voxels.

In all experiments, an affine transformation model with $p = 0.01$ and $\lambda_0 = \frac{1}{\sqrt{m}}$ is used where $m = w \cdot h \cdot d$ is the number of voxels. Registration accuracy is measured in terms of average pair-wise Dice overlaps of lung or brain masks and compared to the iterative groupwise registration approach proposed in [5].This approach was implemented using the intensity-based affine registration method implemented in ITK. Here we use 3-level multiresolution, regular gradient descent optimizer and SSD as distance measurement. Note, that no multi-resolution scheme is used for RASL. The quantitative results are summarized in Table 2. As shown in Fig. 2 and 3, the sparse component contains predominantly pathological structures besides the residual alignment errors, except for the *Brain MRI Lesion* data set. Here, the lesions do not affect the registration accuracy due to slight intensity differences to healthy brain tissue and they are consequently not present in the sparse component. In contrast to that, RASL strongly improves the registration accuracy for data sets with prominent space-consuming pathologies like *Lung CT Diverse* and *Brain MRI Tumor*, where the groupwise ITK registration even misalignes the images compared to their initial state.

Table 2: Registration accuracy measured in Dice coefficients for four different 3D data sets. Compared are the initial alignment, the results of the groupwise ITK-based registration and the proposed stochastic RASL algorithm.

data set	Initial	Groupwise ITK-Reg	Stochastic 3D RASL
Lung CT Tumor	0.7175	0.7915	0.8135
Lung CT Diverse	0.6397	0.6692	0.7586
Brain MRI Tumor	0.9502	0.8966	0.9741
Brain MRI Lesion	0.9017	0.9568	0.9576

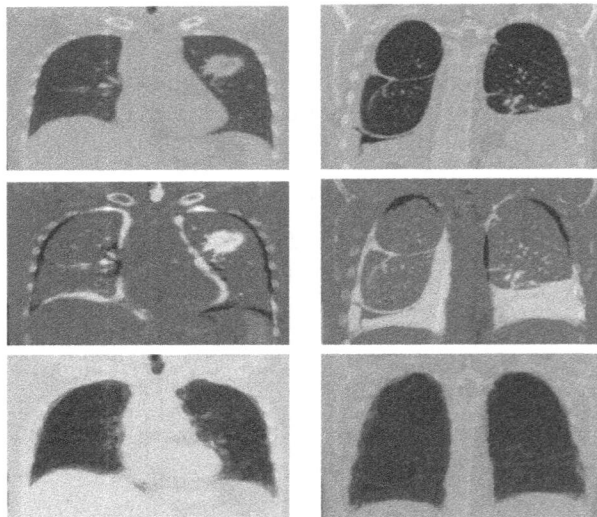

Figure 2: Results of the groupwise stochastic 3D RASL registration for the *Lung CT Tumor* (left column) and *Lung CT Diverse* (right column) data sets. Shown are the transformed images (top row) as well as sparse (second row) and low-rank components (third row)

4 Conclusion

We presented a stochastic groupwise robust registration method and applied it to 3D medical images. In our experiments, we compared the registration results of RASL to the ones of a standard intensity-based registration algorithm, using Dice overlaps of segmentation masks. The RASL algorithm shows to be run-time efficient and yields same or higher Dice coefficients as the standard method. Best results are achieved particularly for images with missing correspondences caused by space-consuming pathologies.

The RASL algorithm can be applied to 3D images because of significant acceleration generated by a stochastic optimization scheme. The normalization of the transformations shown in Sec. 2.2 results in a registration to the group average shape and therefore increases the plausibility of the resulting transformations.

Acknowledgement

The work has been carried out and supervised by the Institute of Medical Informatics, Universität zu Lübeck.

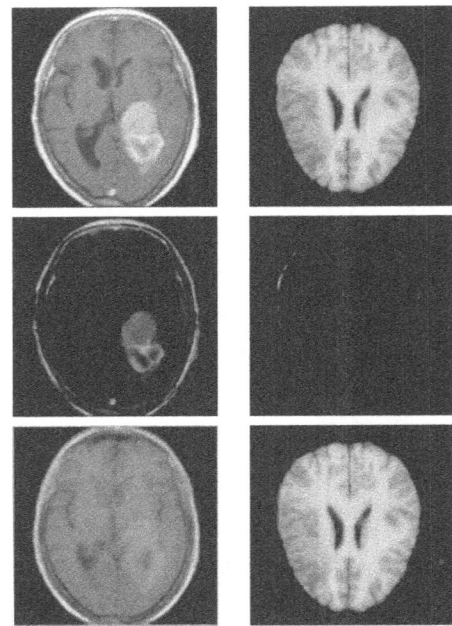

Figure 3: Results of the groupwise stochastic 3D RASL registration for the *Brain MRI Tumor* (left) and *Brain MRI Lesion* (right) data sets. Shown are the transformed images (top row), the sparse components (second row) and the low-rank components (third row).

5 References

[1] Y. Peng, A. Ganesch, J. Wright, W. Xu, and Y. Ma, "RASL: Robust alignment by sparse and low-rank decomposition for linearly correlated images," *Pattern Analysis and Machine Intelligence, IEEE Transactions*, vol. 34, pp. 2233 – 2246, Jan 2012.

[2] J. Wright, A. Ganesh, S. Rao, Y. Peng, and Y. Ma, "Robust principal component analysis: Exact recovery of corrupted low-rank matrices via convex optimization," in *Advances in neural information processing systems*, pp. 2080–2088, 2009.

[3] S. Boyd, N. Parikh, E. Chu, B. Peleato, and J. Eckstein, "Distributed optimization and statistical learning via the alternating direction method of multipliers," *Foundations and Trends® in Machine Learning*, vol. 3, no. 1, pp. 1–122, 2011.

[4] V. Arsigny, O. Commowick, N. Ayache, and X. Pennec, "A fast and log-Euclidean polyaffine framework for locally linear registration," *Journal of Mathematical Imaging and Vision*, vol. 33, no. 2, pp. 222–238, 2009.

[5] A. Guimond, J. Meunier, and J.-P. Thirion, "Average brain models: A convergence study," *Comput. Vis. Image Understand.*, vol. 77, no. 2, pp. 192–210, 2000.

Registration of thermographic Images and three dimensional Point Clouds

L. Schulz [1], B. Stender [2] and H. Handels

[1] Medizinische Ingenieurwissenschaft, Universität zu Lübeck, linda.schulz@student.uni-luebeck.de

[2] Drägerwerk AG & Co. KGaA, Lübeck, birgit.stender@draeger.com

[3] Institut für Medizinische Informatik, Universität zu Lübeck, handels@imi.uni-luebeck.de

Abstract

The breathing rate is a frequently neglected vital signal due to limitations of clinically established measurement techniques, which require attachment of sensors. Camera based monitoring of patients' breathing could solve this problem due to contactless data acquisition. Therefore, this paper examines if it is possible to map two dimensional thermal information to three dimensional depth information in order to extract a breathing signal. To register the information, intrinsic and stereo camera calibrations were carried out on two image sets with 50 images each. The results show low reprojection errors. The mapped depth information is slightly misaligned to the temperature information. Mapping thermal information to depth data is possible, but there still has to be some work carried out to get a more accurate mapping.

1 Introduction

The instantaneous respiratory rate (IRR) is one of the biomedical signals of the human body, that gives information about a patients' health. For example, a rise of the respiratory rate can have its cause in stress, fever or a sepsis [1]. Furthermore, the measurement of the instantaneous respiratory rate is important to detect sleep apnoea. In clinical routine monitoring the respiratory rate is often neglected due to limitations of clinically established measurement techniques, which require attachment of sensors.

Low cost color cameras combined with depth acquisition sensors, in short RGB-D cameras, can offer an alternative way to observe patients' respiratory rate. Depth information makes it possible to track human body parts and movements, which could potentially be used to determine the activity level of patients or to recognise sleep phases. In addition depth information can be used to measure the through breathing induced motion of the thorax [2]. The temperature modulation around mouth and nostrils due to exhaled air and the movement of the shoulders can be used to measure the breathing rate [3]. The aim of this paper is to investigate if it is possible to map thermal information acquired by a small thermal imaging camera, called CompactPRO, to the depth information acquired by a Kinect V2 sensor to detect the respiratory rate of patients. For this mapping, intrinsic and extrinsic calibration have to be carried out, to get a general transformation between three dimensional depth information and two dimensional thermal information. The conceptual idea is shown in Fig. 1.

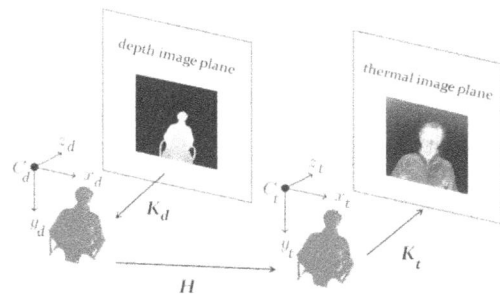

Figure 1: The three dimensional point cloud in the camera coordinate system C_d of the depth camera has to be transformed to the camera coordinate system C_t of the thermal camera and projected on to the thermal image plane.

2 Material and Methods

In this section, the cameras selected and their specifications are presented. Furthermore, it is demonstrated how the cameras are calibrated with acquired images, to enable the registration of thermographic information with depth information.

2.1 Camera Setup

In the following, the camera setup and the data recording procedure are presented.

RGB-D sensor

The Kinect V2 of *Microsoft Corporation, Redmond, WA, United States* provides a high resolution color sensor combined with a near-infrared (NIR) sensor, see Fig. 2. With the color camera it is possible to capture color information with a maximal resolution of 1920×1080 pixels. For NIR imaging the scene is illuminated actively by three infrared projectors and thus can obtain infrared and depth information with a maximal resolution of 512×424 pixels and an angular view of $70.6° \times 60.0°$. The distance measurement is carried out based on the time-of-flight principle, detecting the travel time of NIR pulses. By knowing the speed of light in air, the distance to an object can be computed indirectly. The acquisition can be obtained with a maximum of 30 frames per second [4].

Thermal sensor
The CompactPro of *Seek Thermal Inc., Santa Barbara, CA, United States* is a small thermal imaging camera, that was originally designed for smartphone use. It provides temperature data with a resolution of 320×240 pixels and a frame rate of roughly 11 Hz. The detection distance ranges from 6 inches to 1800 feet (0.1524 m to 548.64 m) and the camera has a manual zoom [5].

Attachment
The thermal camera is attached to the RGB-D camera to provide validity of stereo calibration in the calibration process explained in section 2.2. Because of its small size, the thermal camera is mounted on the RGB-D camera with a 3D printed rig, see Fig. 2. The housing around the thermal camera ensures, that it cannot be moved in either direction relative to the RGB-D camera.

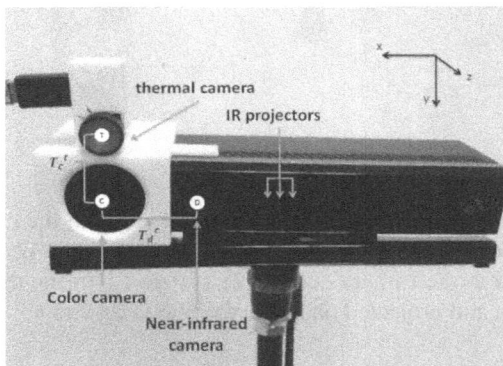

Figure 2: The attachment of thermal camera to the RGB-D camera ensures the validity of the stereo calibration.

Data Acquisition
Both cameras are connected to the computer via USB. To get data from the RGB-D camera, a collection of tools and libraries, called *iai_ kinect2* [6], is used, which offers a connection between the *libfreenect2* software from the OpenKinectProject and the *Robotic Operating System (ROS)*. An in-house developed recording tool makes simultaneous recording of both sensors possible. Thermal images are saved to hdf5 files, RGB-D information is saved to bag files. Both bag and hdf5 files can be read with *MATLAB 2016a, The MathWorks Inc.,Natick, Mass., United States* and images can be extracted for calibration.

2.2 Calibration Procedure

In this section, the calibration procedure is explained. First, the image distortion throughout the image acquisition and the projection parameters of a three dimensional scene to the image plane are determined, called the intrinsic camera parameters. Afterwards, the relative pose of the camera coordinate systems to each other are estimated, called stereo parameters.

The calibration is carried out with the MATLAB Computer Vision Toolbox, based on Zhangs camera calibration algorithm [7]. Several images of a planar pattern at different orientations and distances from the camera have to be taken. For chessboard patterns MATLAB provides an automatic corner point detection. Based on these points an initial guess of the intrinsic camera parameters is determined and the relative pose of the calibration target to the camera are determined in a closed-form solution. Afterwards, the distortion coefficients are estimated by solving linear least-squares. All parameters are then refined by minimizing the reprojection error p_e. The reprojection error is the distance between a pattern keypoint detected in a calibration image and the corresponding world point projected into the same image with the computed parameters.

The intrinsic calibration in this paper will lead to two intrinsic camera matrices in the following form:

$$K = \begin{bmatrix} f_x & \alpha & c_x \\ 0 & f_y & c_y \\ 0 & 0 & 1 \end{bmatrix} \tag{1}$$

where f_x and f_y is the focal length in pixels, c_x and c_y is the principal point in pixels and α, a skew coefficient which is non-zero if the image axes are not perpendicular. For simplicity, the parameter α is set to zero [8]. This intrinsic matrix describes how three dimensional scene points are projected to a two dimensional image plane. Furthermore, radial distortion coefficients are computed, which describe how the image is radially distorted due to the lens.

In order to provide a calibration target for the thermal calibration, a thermal active calibration target is required. This can be realized by a varying thermal emission. Therefore, an aluminium plate is sprayed with baking enamel, forming equally sized squares. This way, the sprayed squares have higher emittance when the board is cooled down under room temperature and appear darker in the thermal image. Unfortunately, the thermal calibration target is only visible in the color images of the RGB-D sensor, but not in the NIR images. The aluminium causes artifacts due to reflections of the active NIR illumination and the contrast between the aluminium and the baking enamel is very low in NIR images. An indirect calibration procedure is undertaken, where the relative pose between the two sensors and the color sensor are determined, respectively. The intrinsic camera parameters and the relative pose of the NIR and the color sensor of the RGB-D camera can be obtained with a laser-printed chessboard. This printed chessboard is fixed to an even board and thus can be mounted onto a tripod. Two different sets of images are taken: image set 1, which contains color and NIR images of the laser-printed chessboard and image set 2, which contains several color and thermal images of the thermal active chessboard. Image set 1 is used to calibrate the NIR sensor intrinsically and extract pose information between the NIR and color sensor. Image set 2 is used to calibrate the thermal sensor intrinsically and to get the pose information between color and thermal sensor. For both image sets, about 50 images are taken varying in orientation and distance to the cameras. For stereo calibration, it is important that the images of one image set from both cameras each show the same scene, so stereo calibration can be run. The stereo calibration result in the relative rotation R and translation T between two camera coordinate systems, which can be concatenated to the transformation matrix

$$H = \begin{bmatrix} r_{11} & r_{12} & r_{13} & t_x \\ r_{21} & r_{22} & r_{23} & t_y \\ r_{31} & r_{32} & r_{33} & t_z \\ 0 & 0 & 0 & 1 \end{bmatrix}, \qquad (2)$$

that describes the transformation between two camera coordinate systems. For further information see [9] and [8]. In the following, the intrinsic matrices will be notated as K_d (intrinsic matrix of the NIR camera) and K_t (intrinsic matrix of the thermal camera). A transformation matrix that describes the coordinate transformation from reference system b to reference system a will be notated as H_b^a.

To map the depth information captured by the NIR camera to the thermal information, we have to undergo four steps:

1. Undistort the given images with the computed distortion coefficients.

2. Convert the depth information from the two dimensional image into a three dimensional point cloud in the depth reference system with the information of the intrinsic matrix K_d as shown in [10].

3. Transform the point cloud to the reference system of the thermal camera with $H_d^t = H_c^t \cdot H_d^c$.

4. Project the point cloud into the image plane of the thermal camera with the information of K_t.

3   Results and Discussion

In this section, the calibration and registration results are presented and discussed.

3.1   Results

The results for the intrinsic calibration and the mean reprojection errors in pixels (px) are shown in Table 1.
The focal length of the NIR camera is around 365 pixels, the focal length of the thermal camera is around 548 pixels. The principal point of NIR camera is estimated as $c_d = [256.4, 208.6]$ pixels, and the principal point of the thermal camera is $c_t = [161.3, 139.3]$ pixels. The mean reprojection error of the depth images is 0.13 pixels, the mean reprojection error of the thermal images is 0.18 pixels.

Table 1: Intrinsic Camera Parameters and mean reprojection errors

	K_d	K_t
f_x	364.7 px	547.8 px
f_y	364.7 px	548.7 px
c_x	256.4 px	161.3 px
c_y	208.6 px	139.3 px
p_e	0.13 px	0.18 px

Fig. 3 shows the result of the stereo calibration of image set 2 and the computed relative pose between the two cameras. Camera 2 displays the thermal camera, camera 1 displays the color camera. The result of the relative translation between the color sensor and the thermal sensor is $T_t^c = [3.8, 43.1, -22.9]$ mm, which means that the color camera is shifted approximately 43 mm in y direction. The result of the relative translation between the NIR sensor and the color sensor is $T_c^d = [-50.0, 0.1, -0.4]$ mm, which means that the NIR camera is shifted approximately 50 mm in negative x direction.

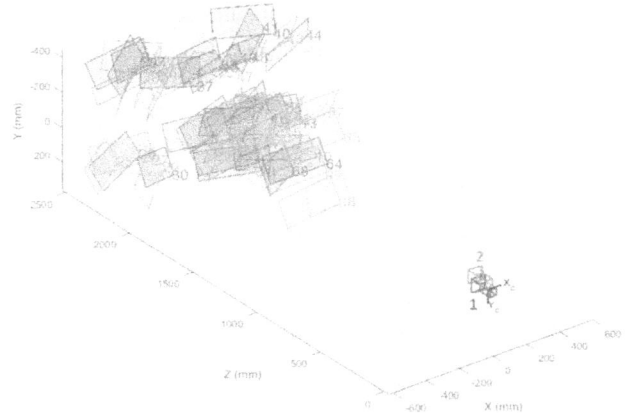

Figure 3: The stereo calibration results of image set 2 show the relative rotation and translation between the two cameras, camera 2 is the thermal camera, camera 1 is the color sensor of the Kinect. Further, the different poses of the calibration target relative to the two cameras are indicated.

The result of mapping a human point cloud to its corresponding thermographic image with the computed parameters is shown in Fig. 4. The human point cloud has been segmented for better differentiation between human and background. A small misalignment is visible, especially observable on the right arm and shoulder.

3.2   Discussion

Comparing the computed intrinsic camera parameters of the NIR camera to [6] and to the manufacturing default parameters of the Kinect shows, that the computed parameters are reasonable. There are no comparison data available for the CompactPro. The mean reprojection error of the thermal camera is higher than the mean reprojection error of the

Figure 4: Overlay of the mapped depth information and temperature image for a human point cloud.

NIR camera. This might be the consequence of its lower resolution (thermal image: 320×240 compared to depth image: 512×424) and the higher noise level of the thermal camera. Comparing the outcome of the translation vectors between the cameras and the positions in real world in Fig. 2, the translation vectors seem accurate, except for $T_t^c(z) = -22.9$ mm. The thermal camera is in z direction aligned with the Kinect housing and thus has a maximal shift of 10 mm.

An error source for the misalignment in the registration might be the computation of the intrinsic camera parameters of the thermal camera or the transformation between the camera reference systems defective. Due to the low resolution of the thermal image, even small errors in rotation and translation can have severe effects on the registration. Two different calibration targets are used instead of executing only one stereo calibration directly between the thermal camera and the NIR camera. The concatenation of the two rotations and translations with even small errors can multiply the overall error. This leads to a higher error propagation because of the two stereo calibrations.

4   Conclusion

In this work, the procedure for intrinsic and stereo calibration of the thermal camera CompactPro and the Kinect RGB-D camera is presented. Further, it was shown that mapping depth information to thermal information is possible. The calibration of the thermal camera is afflicted with higher errors and thus results in a misalignment of the mapped images.

The aim of future work will be to eliminate this misalignment by carefully examining every step of the calibration procedure, starting with the chessboard patterns used and ending with the computation of the camera parameters. One approach might be to examine the effect of only optimizing the focal length in the least squares optimization of the intrinsic parameter estimation.

Acknowledgement

The work has been carried out at Drägerwerk AG & Co. KGaA, Lübeck.
I would like to thank everyone, who helped me while working on the project and who supported me with this paper.

5   References

[1] E. Gultepe, J. P. Green, H. Nguyen, J. Adams, T. Albertson, and I. Tagkopoulos, "From vital signs to clinical outcomes for patients with sepsis: a machine learning basis for clinical decision support system," *J Am Med Inform Assoc.*, vol. 21, pp. 315–325, 2014.

[2] A. Prochazka, M. Schätz, F. Centonze, J. Kuchynka, O. Vysata, and M. Valis, "Extraction of breathing features using ms kinect for sleep stage detection," *Signal, Image and Video Processing*, vol. 10, no. 7, pp. 1279–1286, 2016.

[3] C. B. Pereira, X. Yu, M. Czaplik, V. Blazek, B. Venema, and S. Leonhardt, "Estimation of breathing rate in thermal imaging videos: a pilot study of healthy human subjects," *Journal of Clinical Monitoring and Computing*, 2016.

[4] P. Fankhauser, M. Bloesch, D. Rodriguez, R. Kaestner, M. Hutter, and R. Siegwart, "Kinect v2 for mobile robot navigation: Evaluation and modeling," in *International Conference on Advanced Robotics (ICAR)*, pp. 388–394, IEEE, 2015.

[5] Seek Thermal , http://www.thermal.com/products/compactpro/ , accessed: 13 January 2017, 10:51.

[6] T. Wiedemeyer. https://github.com/code-iai/iai_kinect2/tree/master/kinect2_calibration, accessed: 20 January 2017, 12:08.

[7] Z. Zhang, "A flexible new technique for camera calibration," technical report msr-tr-98-71, Microsoft Research, 1998.

[8] R. Szeliski, *Computer Vision: Algorithms and Applications*. New York, NY, USA: Springer-Verlag New York, Inc., 1st ed., 2010. pp. 39–54.

[9] Mathworks, https://de.mathworks.com/help/vision/ug/camera-calibration.html, accessed: 13 January 2017, 14:47.

[10] E. Lachat, H. Macher, M.-A. Mittet, T. Landes, and P. Grussenmeyer, "First experiences with kinect v2 sensor for close range 3d modelling," *International Archives of the Photogrammetry, Remote Sensing and Spatial Information Sciences*, vol. Volume XL-5,W4, pp. 93–100, 2015.

Automatic left ventricle segmentation on axial cine cardiac MR Imaging
– Phase 1 –

S. Kaminsky[1], W. Kristanto[2], J.P. Aben[2], and A. Mertins[3]

[1] Medizinische Ingenieurwissenschaft, Universität zu Lübeck, steffen.kaminsky@student.uni-luebeck.de
[2] Pie Medical Imaging,BV, {Wisnumurti.Kristanto, JeanPaul.Aben}@pie.nl
[3] Institute for Signal Processing, Universität zu Lübeck, mertins@isip.uni-luebeck.de

Abstract

This project focuses on the segmentation of hearts left ventricle. With producing an algorithm to segment ventricle, new possibilities arise in axial MR Imaging to calculate structure and volume of the heart. Automatically left ventricle segmentation was used mainly in short axis MRI, but just rarely in axial MRI. Image processing was used to determine a region of interest in the images to remove bland areas. By tracking back the ascending aorta in the left ventricle, the boundaries were approximatively segmented with this algorithm. Little artifacts still need to be removed in further researches.

1 Introduction

Cardiac cine short axis magnetic resonance imaging (MRI) has been used extensively in the clinical practice to evaluate heart function [1]. Given the spatial resolution limitation of the early clinical MRI devices, this imaging orientation promoted the best view to analyze ventricle function and morphology. Due to advances in MRI image acquisition technology, it is now possible to scan the whole heart in axial slice orientation with much more isotropic spatial resolution which promotes more accurate measurements. Recent studies were developed for left ventricle image segmentation more focussed on short axis MRI [8],[9] than on axial MRI[7].

Basically, it is of interest to segment the cardiac ventricle automatically, because there is a huge number of images for every patient. In cine cardiac MR Imaging there are produced around 1000 images for each patient. So it would take too much time to segment the heart area manualy. And due to this segmentation of 4D dataset there are usefull information to get from: the ejection volume of the heart, blood-capacity of heart in systolic and diastolic phase and thickness of myocardium. The aim of this project is to develop an automatic algorithm to segment left ventricle on the cardiac MR images dataset; axial MR Imaging.

2 Material and Methods

2.1 Images

Cardiac Magnetic Resonance (CMR) imaging was performed using a Philips Ingenia 1.5T system (Philips Med-ical Systems, Best, The Netherlands). A cine volumet- ric dataset was acquired in axial direction using Balanced Turbo Field Echo (sBTFE) sequence at 5 mm slice thickness, 2.5 mm interslice distance, and a matrix of 144 x 144 resulting in 2.43 x 2.43 mm pixel size. Repition-time of acquisition refers to 2.7690 ms and echo-time refers to 1.3590 ms. Twenty phases over the cardiac cycle were reconstructed retrospectively. In order to cover the whole heart and the ascending aorta, acquisition was performed twice, first over the upper part and then lower part of the heart with 10 mm overlapping between them. Afterwards, the two acquired datasets were manually combined with the redundant overlapping slices removed. In total, the complete dataset consists out of 960 Images with 48 slices and 20.

2.2 Automatic segmentation

Basically the algorithm contains three steps (Fig.1) and will be performed with Matlab 2011b (Mathworks, Natick, MA, USA).

2.2.1 Heart localization

The first step of the algorithm is, where roughly the heart is located in the image. For this purpose, the advantage of the heart movement is used. The pixels with much movement will have high variation in signal intensities over time. Calculating the standard deviation of signal intensity of every pixel over time, the highest values will appear within and around (the surrounding tissue of heart is also in passive movement, effected from the heart movement) the heart.

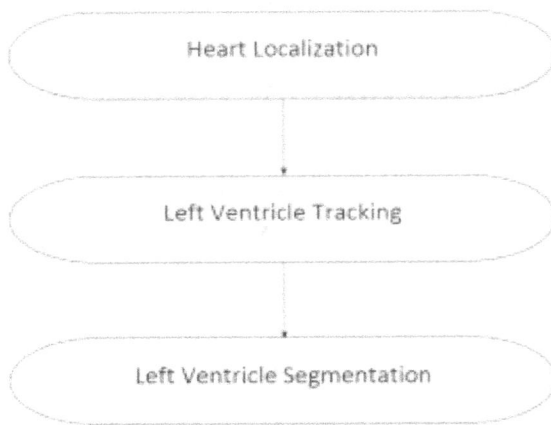

Figure 1: Flowchart of algorithm.

Static structure like bones and furthermore tissue should have low signal values. Nevertheless there are some artifacts in calculating standard deviation due to inhomogeneity in magnetic fields during measurement and the resulting variations in signal intensity for same pixels in different frame. After blurring with a Gaussian filter (25x25 mask, Standard-Deviation 5) and thresholding at 85 percent of total signal intensity most of artifacts can be eliminated [5]. To get rid of the few remaining artifacts its necessary to calculate the biggest connected parts of pixels in binary image. This leads always to the region of heart, because most pixels with high standard deviation, which are close together and even connected in an 8-pixel-neighbourhood can belong with high probability to region of heart. The remaining binary image need to be dilated, because otherwise there is a little probability that a small areas of the heart fell out of the region of interest (ROI) [3]. Computing the area of standard deviation had some kind of average behaviour between systolic and diastolic phase. Thus plotting standard deviation without image-dilation can sometimes segment too small for diastolic phase, because the ROI is plotted to every frame of its slice. Finally a convex hull was calculated around the dilated binary image. With this binary mask, the region of interest can be plotted in the 4D Images. Examples for superior, middle and inferior areas are shown at Figure 3.

2.2.2   Ventricle tracking

The next step is to locate the left ventricle. The idea to segment the left ventricle is, to detect in the superior images the ascending aorta and track it back in the left ventricle. The aorta possesses a circular shape in axial view. By applying a Hough transform [2],[4] this circular structure can be detected with a circle detection. To rule out other circular structures detected with this method, such as the descending aorta or pulmonary artery, the ascending aorta is detected by selecting the circular structure with the biggest diameter (Fig.2). With the knowledge of the position of ascending aorta, the ROI Images can be tracked back to the

left ventricle by thresholding them at 40 percent of total signal intensity and afterwards with the use of 3D connected pixels. That means if a pixels has another pixel in his 26-connected neighborhood [2], then they are recognized as connected pixels.

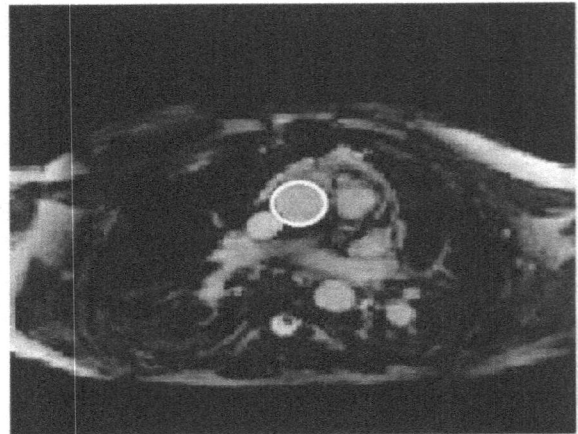

Figure 2: Circle detection: Ascending Aorta detected using Hough transformation.

2.2.3   Left ventricle segmentation

There was a problem in segmentation of left ventricle: In diastolic phase, when the myocardium is relaxed and the ventricle hold the biggest dilatation, the myocardial septum that separates left and right ventricle become so thin that, due to partial volume effect, the left and right ventricles appears to be connected. Still due to this fact, the left atrium which is located directly above the disjunction of left and right ventricle and posesses similar signal intensity in MR images as the two ventricles, seems also to be connected and is not removable by just thresholding the image. For separation this connected area it was applied an image erosion with a 3x3 mask with value 1, followed by an image-opening. This step eliminates the connection to the right ventricle. Afterwards an image-dilation with the same mask as before was applied. The thresholding step also leaves some holes inside the ventricles from papillary muscles which has lower pixel values than the surrounding blood area in the heart. These holes were closed using image-filling methods [2].

3   Results and Discussion

The algorithm provides good first approximation of left ventricle segmentation (Fig. 4). Note that in Figure 4 the small connected part of left and right ventricle in the middle of image near to the descending aorta. The segmented area is independent of the connected area because of the image-opening function as mentioned before. Nevertheless some frames were still remain some artifacts. The first problem is caused by the presence of papillary muscles, when they are located at boundary area from ventricle to myocardium the

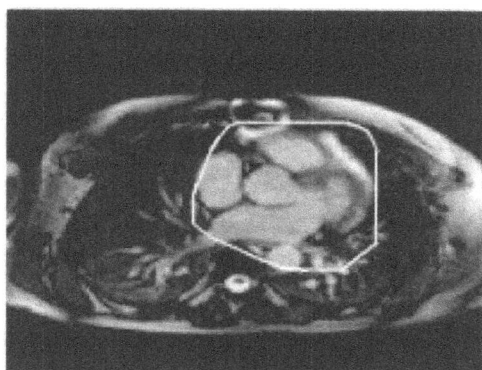

(a) Superior part of the heart

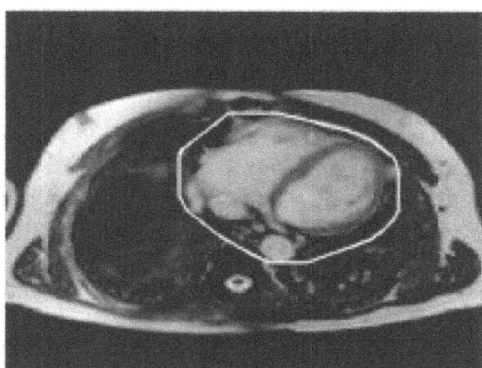

(b) Middle part of the heart

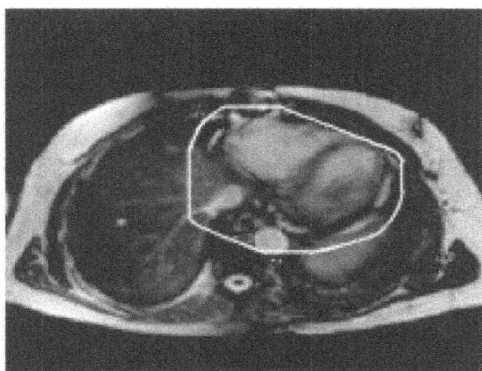

(c) Inferior part of the heart

Figure 3: Segmented Region of Interest (ROI) in axial MRI data

remaining gaps via image processing cannot be replaced by image-filling, because these gaps are opened. So the ventricle will not be marked at the boundary to the myocardium.

This problem can also be seen on Figure 4. Note that on the right side of the segmented area, the segment is a little bit too small. This results from the papillary muscle when the image-erosion is processed, because of the intensity gap. It can not be corrected afterwards by image-dilation.

The next problem resulted from data acquisition. To measure the whole heart column in z-axis, it is necessary to perform two measurements, one the superior and one for the inferior part of heart, with overlap between those two parts. At the slice positions where the transition between these two acquisition take place, which occurs in the middle of heart, there appears to be signal loss that makes these slices to have lower signal intensity distribution than the rest of the slices. This is problematic for the segmentation, as applying the same thresholding step as the rest here failed to segment the left ventricle properly.

Also keep in mind that a possible weakness of this algorithm acquires, if the patient has a deasease, e.g. aorta possesses not the biggest diameter in the images or has not a circular structure. Then the whole algorithm could fail. Thus alternatives need to be implemanted for such situations. Thus the algorithm has to be tested to different data to verify its reliability.

3.1 Outlook of this project

The work described in this article is the result of the first part of this project to segment the left ventricle in axial cardiac cine MRI images. The next part of the project will be to finalize the left ventricle segmentation. Specific problems encountered in the first part will also be addressed in the next part. Some algorithms need to be developed to compensate for signal loss artifact at overlapping slice locations.

The next problem to be addressed is to move away from setting a fixed threshold and make it adaptive, similar to Otsu's method [6]. It gives the oportunitiy to determine threshold value for every single image automatically. So far the threshold value is determined manually, because Otsu's method often thresholds too low. An automatically estimation makes it more reliable for other patient images. Finally, the following part of the project will be used for reduc- ing papillary muscles influences on the segmented images and determination of upper and lower ends of ventricle, to have the possibility to calculate whole volume of ventricle. To determine the upper end there is a good chance to take advantage of finding the aortic valve, because it provides boundary of left ventricle and aorta. For the lower part the influence of the myocardium is required, because of a drop in signal intensity compared to the blood filled ventricle. Otherwise the user has to determine upper and lower boundary manually.

4 Conclusion

On balance it is a progress in getting the possibility of segmentation the approximate area of left ventricle with ascending aorta. It is a good basement for improvements in

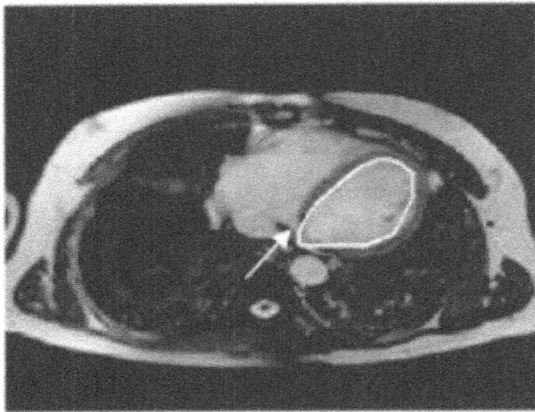

Figure 4: Segmentation of left ventricle with overlap (arrow) to the right ventricle and papillary muscle on the boundary to myocardium

continuative researches in this field image segmentation in axial view. The biggest goal of this work was to separate left ventricle from right ventricle due to the appearance of left atrium in some slices. Nevertheless improvements are necessary to eliminate artifacts of papillary muscles at the boundaries of the segmented image or signal intensity lose in middle slices resulting out of data acquirement.

Another important advantage of this project was the localization of heart area to receive ROI. With removing parts of the image located at the outer of ROI, the focus on postprocessing was on the heart area and could not be influenced by high signal intensities from fat/muscle tissue. Due to the fact (as mentioned in 2.2.1) that the left ventricle is detected by finding the biggest area of connected pixels, there arise a problem with high intensities values of fat/muscle tissue. Note e.g. in Fig. 3 (b) torax muscles give a bigger connected area than the heart structer. ROI removes this influence by surrounding the heart. Another removed negative influence is the circular structures of spinal canal in vertebral column. Without ROI the algorithm can lead to difficulties in circle detection for aorta. Nervetheless computing of ROI take some time, because of huge data volume of 960 images, but it is neccessary though.

Acknowledgement

This work has been produced at PIE Medical Imaging,BV, company in Maastricht/ Netherlands specialized for solutions in cardiovascular analysis, and was supervised by Institute for Signal Processing, University of Luebeck.

5   References

[1] D. Pennell et al., *Clinical Indications for Cardiovascular Magnetic Resonance (CMR): Consensus Panel Report*. In: Journal of Cardiovascular Magnetic Resonance Vol. 6, No. 4, pp. 727–765, 2004

[2] A. McAndrew, *Introduction to digital image processing with Matlab*. Course Technology, Press Boston, MA, United States 2004, ISBM: 0534400116.

[3] C.A.Cocosco et al.,   *Automatic cardiac region-of-interest computation in cine 3D structural MRI*. In: International Congress Series 1268 (2004) 1126–1131

[4] C. Petitjean, *A review of segmentation methods in short axis cardiac MR images*. In: Medical Image Analysis 15(2011) 169-184

[5] I.T. Young, *Fundamentals of image processing*. Cip-Data Koninklijke Bibliotheek, Den Haag

[6] N. Otsu, *A threshold selection method from gray-level histograms*. In: IEEE Trans. Sys., Man., Cyber. 9 (1): 62–66 (1979)

[7] S. H. James et al., *Accuracy of Right and Left Ventricular Functional Assessment by Short-Axis vs Axial Cine Steady-State Free-Precession Magnetic Resonance Imaging: Intrapatient Correlation with Main Pulmonary Artery and Ascending Aorta Phase- Contrast Flow Measurements*. Canadian Association of Radiologists Journal, Volume 64, Issue 3, August 2013, Pages 213-219. doi: 10.1016/j.carj.2011.12.016.

[8] L. Wang et al., *Left ventricle: fully automated segmentation based on spatiotemporal continuity and myocardium information in cine cardiac magnetic resonance imaging (LV-FAST)*. BioMed research international 2015. doi: 10.1155/2015/367583

[9] H. Hu et al., *Automatic Segmentation of the Left Ventricle in Cardiac MRI Using Local Binary Fitting Model and Dynamic Programming Techniques*. PLOS ONE 9(12): e114760 (2014). doi: 10.1371/journal.pone.0114760

A Review of Neural Network based Semantic Segmentation for Scene Understanding in Context of the self driving Car

J. Niemeijer[1], P. Pekezou Fouopi[2], S. Knake-Langhorst[2], and E. Barth[3]

[1] Medizinische Informatik, Universität zu Lübeck, Joshua.Niemeijer@student.uni-luebeck.de
[2] German Aerospace Center, Braunschweig, {Paulin.Pekezou,Sascha.Knake-Langhorst}@dlr.de
[3] Institute of Neuro- and Bioinformatics, Universität zu Lübeck, barth@inb.uni-luebeck.de

Abstract

This paper tackles the challenge of scene understanding in context of automated driving. To react properly to the conditions given by the surrounding scene, the car has to understand its environment. Further the real time capability of a method solving this task is essential. For scene understanding the car has to detect and classify its surrounding objects. For this purpose a semantic segmentation can be employed to assign a class label to every pixel. In this paper we evaluate the state of the art methods for the semantic segmentation and perform tests on the FCN-8 architecture. Due to hardware limitations, we train the FCN-8 on a downscaled version of the Cityscapes Dataset, containing urban traffic scenes. The evaluation of the results shows, the necessity to train the FCN-8 on the original size City Scapes Dataset. We conclude that we need to purchase a better hardware.

1 Introduction

Scene understanding is an important task in the context of automated driving. In order to react properly to the conditions given by the surrounding scene, the self driving car has to understand its environment. For example information about the state of other traffic participants or objects are needed to predict their behaviour and avoid collisions. Besides the quality of scene understanding, the real time capability of a method solving this task is essential, because of the fastly changing traffic conditions. Amongst other things the self driving car perceives its environment through a camera. By performing a semantic segmentation on the recorded images, one receives labels for every pixel, describing the object it is part of. Hence an understanding of the present objects and their location is created. These information in turn are the basis of further scene analysis. In this paper we review a variety of pixel wise semantic segmentation Methods based on fully convolutional neural networks. In addition we show the results of training and testing the FCN-8 architecture from the paper [1] on the Cityscapes Dataset [2].

2 Material and Methods

Semantic segmentation can be understood as the pixel wise labeling of an image, to declare which class-label each pixel belongs to. An example can be observed in Fig. 1. The top picture shows the image picturing a traffic scene to be segmented, the image on the left side is the groud truth semantic segmentation and the image on the right side is

Figure 1: On top the downscaled version of the original Cityscapes validation image can be seen. The image on the left side is the ground truth and the image on the right is the result of the network trained on the Cityscapes Dataset.

the semantic segmentation computed by a neural network. The different gray values indicate the classes the pixels belong to (e.g. car or road etc.). There are many algorithms to tackle this task, however recently artificial neural networks have offered the best results. In general there are two main architectures of artificial neural networks to cope with the task. Bounding Box approaches first detect the object and then classify all the pixels belonging to this object. However recent challenges showed that methods which are based on the pixel wise segmentation architecture proposed in [1], perform generally better. Hence in this paper we take a closer look at these methods.

The general architecture, as it is presented in [1] can be ob-

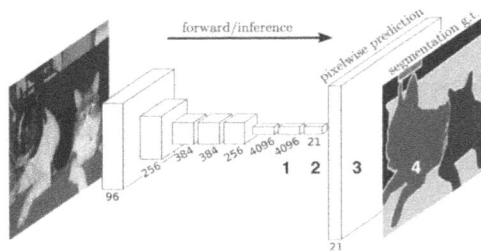

Figure 2: The general architecture of pixel wise segmentation is presented. The image information is processed to a feature map of the shape (1). From this feature map a score map for each of the (in this case 21) classes is derived (2). These score maps are upsampled to the original image size (3) and for every pixel the class with the highest score is chosen (4). The image was taken and modified from [1].

served in Fig.2. The architecture is divided up into four stages. In the first stage the image data is processed, by extracting the most important features. The output of this data processing is a downscaled feature map, in which every pixel represents a receptive field in the original image and hence holds the information of this area. Based on this information in the second stage a score map Sc is computed for every class. So the output of this layer, the score map (Sc), is of the shape $lengthFeatureMap \times widthFeatureMap \times numberClasses$. The score values can be interpreted as a probability of a pixel, to belong to a semantic class. In the third stage, the now coarse score map is interpolated to the size of the original image. For interpolation commonly the bilinear interpolation is used. Here an upsampled score value is weighted by the inverse distance to its four closest neighbours in the lower-resolution score map. Now for every pixel in the original image for every class there is a score. To determine the class label of an pixel (x, y), in the fourth stage the class with the maximum value over all score values is computed, making it the pixel's class label CL.

$$\forall (x, y) \; CL = \arg \max_{z} Sc(x, y, z) \qquad (1)$$

Approaches that are based on this architecture try to improve the data processing, the creation of the score layer or the interpolation stage. For example the approach presented in [3], tries to improve the interpolation stage through an architecture based on the Laplacian Pyramid. In this approach a low-resolution score map is refined by using higher frequency details derived from higher-resolution feature maps. The idea is based on the assumption that lower-resolution feature maps have bigger receptive fields and hence more context information, leading to better predictions in the score map. Whereas higher resolution feature maps contain more information about local structures. So refining a low-resolution score map, which is computed from a lower-resolution feature map, with features derived from a high resolution score map saves details, hence creating confident predictions with a high level of details.

Conditional random fields are used in the creation of the score map. The approach in [4] is to capture the spatial context of a pixel and thus improving its classification, by using conditional random fields. The conditional random fields is thereby constructed, so that for every spatial position in the coarse feature map, there is a node. The pair wise connections between a node and all other nodes are drawn within a certain range around the node. By choosing the range spatial relations like above or under can be modelled. An other possibility is, to improve the creation of the feature map. In [5] a new shallower architecture of the residual network is used to create better features and thus improve the creation of the score map. Residual networks are neural nets that include neurons that map their own input to their output (residual units). The advantage is that these nets are easier to optimize. This new approach is based on the finding that paths in the residual nets that are deeper than the effective depth (number of residual units) aren't trained fully end to end. So by reducing the effective depth the net becomes fully end to end trainable. Empirical results show that these networks outperform deeper residual networks [5]. The approach in [6] also tries to improve the feature map. A pyramid scene parsing network is proposed, which combines global context information and local cues to improve the pixel wise predictions. However its architecture is slightly different from the baseline approach. In this case not the score map, which is created by a residual network, is upsampled, but directly the feature map. The pyramid pooling module is applied to the upsampled feature map and fuses features under four different scales. It thus separates the feature map into different sub regions and forms pooled representation for different locations. Then these new features are upsampled to the size of the original feature map and concatenated with it. Subsequently the pixel wise predictions are made by an convolutional layer.

Table 1: Model performance on the Cityscapes Dataset [2]

Model	mean class IoU	mean class iIoU
ResNet-38 [5]	80.6	57.8
PSPNet [6]	80.2	58.1
LRR-4x [3]	71.8	47.9
Adelaide-context [4]	71.6	51.7
FCN 8s [1]	65.3	41.7

Table 1 shows an overview of the presented methods. As it can be seen the FCN-8 performs worse than the methods inspired by its architecture (information about the dataset and the metric in Results and Discussion). The main reason we decided to use the FCN-8 was that at the time it was already available as a Caffe model. In addition to this the models from [5] and [6] were published only recently and developing a new architecture was not the goal of this work.

2.1 FCN-8

Fig. 3 shows the architecture of the FCN-8, which we used to perform experiments on the Cityscapes Dataset. This network is a fully convolutional network. Thereby the spatial information of the image is preserved. By employing a fully connected layer this information would be destroyed [1]. The creation of the feature map is done in the layers Conv1 to Conv6-7. This network is a neural network for classification trained on the same dataset as the semantic segmentation without the classification layer. In this case the network is the VGG-16 [7]. The FCN-8 combines the output of fine and coarse layers, as it can be observed in Fig. 3. This is done by first scoring the output of these layers after pool3, pool4 and conv6-7 and then aligning the score maps through upsampling (interpolation) and cropping. After the score maps are brought to the original image size, they are merged by a 1×1 convolution, summing the score maps. By combining the information of coarse and fine layers the model makes local predictions with respect to global structures. The interpolation is initialized as a bilinear interpolation. During training the whole model is fine tuned end to end.

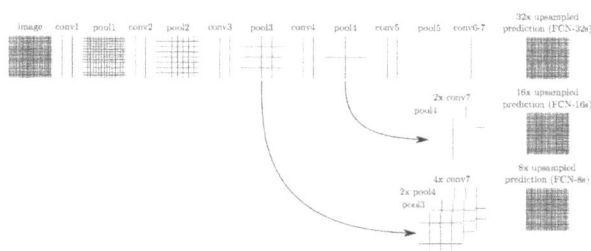

Figure 3: The figure (from [1]) shows the architecture of the FCN-8. (Conv = convolutional layer/pool=pooling layer)

3 Experiments

3.1 Dataset and quality metric

The Cityscapes Dataset contains 5000 images of street scenes with fine annotations for the purpose of semantic segmentation. An example for these images is pictured in Fig. 1. The images are taken in 50 cities under different conditions, seasons and daytimes. Annotations include 30 classes, which are divided up into the categories flat, human, vehicle, construction, object, nature, sky and ground. The images have got a HD resolution of 2048×1024 pixels. The training set includes 2974 images, the validation set includes 501 images and the test set includes 1525 images. Due to the fact that we do not have the ground truth images of the test set and hence can not evaluate results on this test images, we performed the tests on the validation set.

To evaluate the results of the experiments we use the quality metric IoU and iIoU, as is done in the cityscapes benchmark. These metrics are scalar values $\in [0, 1]$ computed per class over the whole test set to evaluate the segmentation. IoU (intersection-over-union) is defined as follows:

$$IoU = TP/(TP + FP + FN) \qquad (2)$$

In which per class TP stands for true positive, FP for false positive and FN for false negative. This could also be interpreted as dividing the intersection of the segmentation

with the ground truth by the union of the segmentation and the ground truth. IoU is biased toward object instances that cover large image area. To address this problem $iIoU($ instance-level intersection-over-union) is used:

$$iIoU = iTP/(iTP + FP + iFN) \qquad (3)$$

iTP and iFN are computed by weighting the contribution of each pixel by the ratio of the class' average instance size to the size of the respective ground truth instance.

3.2 Training

Training the FCN-8 on the Cityscapes Dataset was done using a server with a Tesla K-20 GPU. This GPU has got 5 GB of RAM, which limited the maximum size of the images due to the fact that the image information and the size of the derived feature maps increase with the size of the image. Accordingly we downsampled the images from 2048×1024 pixels to 512×256 pixels, which offered a good RAM Capacity utilization. We already had a caffe model (https://github.com/shelhamer/fcn.berkeleyvision.org.git) of the FCN-8 trained on the Pascal Dataset to initialize the FCN-8 with for finetuning. We made the assumption that the features derived from the Pascal Dataset are well suited for the Cityscapes dataset. This can be assumed due to the fact that these dataset are similar and in general the features derived are mostly some kind of edge detection. We also adopted the interpolation from the given model, assuming it would be well suited due to the fact that the changes of the interpolation parameters are very small and the datasets are similar. Thus we retrained the layers that create the class-scores. For training we used the train IDs from the Cityscapes Dataset, encoding the 19 classes that can be observed in table 1. The training was done for 100000 iterations with a learning rate of 1e-14 and a batch size of three (again due to the low RAM).

3.3 Results

Fig. 1 shows example of the semantic segmentation performed on the rescaled validation set. On top the image to be segmented is pictured, on the left side the ground truth and on the right side the result of the FCN-8 on the rescaled image. It can be observed that in general the main structures of the picture, as cars, the street and the vegetation are well captured. In general these structures tend to occupy more space in the segmentation, than they actually occupy in the ground truth image. However the model struggles to segment finer structures as e.g. the rider.

In Table 2 the results of the test on the rescaled validation data are displayed relating to the classes (rescaled). On the left side of these scores the scores of the FCN-8 trained on the full sized Cityscapes Dataset (orig) can be seen. These scores are computed on the full size Cityscapes Test Set.

In general it can be observed that the FCN-8 trained on the rescaled Cityscapes Dataset performs on the validation dataset worse than the FCN-8 trained on the full sized

Table 2: Performance of the FCN-8 on the Cityscapes Dataset. orig. FCN-8 trained on the full size Dataset; rescaled FCN-8 trained on the rescaled dataset

Class	IoU orig.|rescaled	iIoU orig.|rescaled
building	89.2 | 69.4	/ | /
fence	44.2 | 2.8	/ | /
pole	47.4 | 10.9	/ | /
road	97.4 | 81.9	/ | /
sidewalk	78.4 | 34.1	/ | /
sky	93.9 | 70.1	/ | /
terrain	69.3 | 19.5	/ | /
traffic light	60.1 | 3.6	/ | /
traffic sign	65.0 | 15.5	/ | /
vegetation	91.4 | 74.6	/ | /
wall	34.9 | 0.8	/ | /
truck	35.3 | 1.1	22.2 | 1.1
train	46.5 | 1.0	26.7 | 3.54e-04
rider	51.4 | 0.3	33.4 | 0.4
person	77.1 | 30.7	55.9 | 29.9
motorcycle	51.6 | 0.3	31.1 | 0.1
car	92.6 | 65.4	83.9 | 50.1
bus	48.6 | 1.7	30.8 | 0.5
bicycle	66.8 | 28.7	49.6 | 23.7
average	65.3 | 27.0	41.7 | 13.2

Cityscapes Dataset and tested on the test dataset (65.3 vs. 27.0 IoU and 41.7 vs. 13.2 iIoU). Especially those classes which make up a low portion of the rescaled training data have a much lower IoU and iIoU scores. For example the classes truck and train have a prior portion of 0.0046 and 0.0016 and the IoU scores fall from 35.3 to 1.1 and from 46.5 to 1.0. Whereas the class car has a prior portion of 0.052 and the IoU scores only fall from 92.6 to 65.4.

The deployment time for both Models is 0.5 seconds or lower. The FCN-8 orig. was deployed on the TitanX (16 GB RAM) and the FCN-8 rescaled on the Tesla K20 GPU.

3.4 Discussion

The evaluation of FCN-8 trained on the rescaled dataset on its own training dataset showed similar results as the evaluation on the rescaled validation dataset (29.2mean IoU/13.35mean iIoU). It can be assumed, that no Overfitting took place but possibly Underfitting. Although the loss function converged it was still high, leading to the assumption that training for more than 100000 iterations might bring an improvement.

That the FCN-8 trained on the rescaled Dataset offers results, that are a lot worse is caused by the fact, that the information available for training the network is only 1/16 of the original information. The rescaling of the images leads in particular to the fact, that classes that make a small portion of the original training data, are not well enough represented for FCN-8 to offer good results. If the FCN-8 is trained on the Cityscapes Dataset in its original size, this model doesn't fulfil the real time requirements (runtime less than 100 ms) since it takes around 0.5 seconds to segment an image. However deploying the model on dedicated hardware like Nvidia DPX2 (http://www.nvidia.com/object/drive-px.html) can considerably reduce the run time.

4 Conclusion

The results of this work show that training the FCN-8 on the rescaled Cityscapes Dataset leads to a low detection rate, while deploying the FCN-8 on the full size Cityscapes Dataset doesn't fulfil the real time requirements. However for deploying the FCN-8 in a self driving car, a good detection rate is necessary. Hence we need to train the the FCN-8 on the full sized City Scapes dataset. For that we will need a stronger GPU with enough RAM, like the Nvidia Titan X for training or the Nvidia DPX2 for the deployment of the model. Another goal should be to improve the semantic segmentation by employing better architectures and training the models to distinguish between object instances.

The methods of semantic segmentation presented in this paper, aren't limited to the use in context of self driving cars, but can also be used in a medical context. An application example is the detection and monitoring of tumors.

Acknowledgement

The work has been carried out at the German Aerospace Center, Braunschweig (Institute of Transportation Systems) and was supervised by the institute of Neuro- and Bioinformatics at the Universität zu Lübeck.

5 References

[1] E. Shelhamer, J. Long, and T. Darrell, "Fully convolutional networks for semantic segmentation," *IEEE Transactions on Pattern Analysis and Machine Intelligence*, vol. PP, no. 99, pp. 1–1, 2016.

[2] M. Cordts, M. Omran, S. Ramos, T. Rehfeld, M. Enzweiler, R. Benenson, U. Franke, S. Roth, and B. Schiele, "The cityscapes dataset for semantic urban scene understanding," *CoRR*, vol. abs/1604.01685, 2016.

[3] G. Ghiasi and C. C. Fowlkes, *Laplacian Pyramid Reconstruction and Refinement for Semantic Segmentation*, pp. 519–534. Cham: Springer International Publishing, 2016.

[4] G. Lin, C. Shen, A. van den Hengel, and I. D. Reid, "Exploring context with deep structured models for semantic segmentation," *CoRR*, vol. abs/1603.03183, 2016.

[5] Z. Wu, C. Shen, and A. van den Hengel, "Wider or deeper: Revisiting the resnet model for visual recognition," *CoRR*, vol. abs/1611.10080, 2016.

[6] H. Zhao, J. Shi, X. Qi, X. Wang, and J. Jia, "Pyramid Scene Parsing Network," *ArXiv e-prints*, Dec. 2016.

[7] K. Simonyan and A. Zisserman, "Very deep convolutional networks for large-scale image recognition," *CoRR*, vol. abs/1409.1556, 2014.

Deep learning for spinal centerline extraction

N. Vogt[1], C. Lorenz[2], T. Brosch[2], and M.P. Heinrich[3]

[1] Medizinische Informatik, Universität zu Lübeck, nora.vogt@student.uni-luebeck.de
[2] Philips GmbH Innovative Technologies, Research Laboratories, D-22335 Hamburg
[3] Institute for Medical Informatics, Universität zu Lübeck

Abstract

In the field of spinal diagnosis and therapy planning, an accurate extraction of the spinal centerline is of great importance. This work investigates the suitability of two different convolutional neural network architectures for the task of spinal centerline segmentation in computed tomography images. While the first architecture is based on the processing of two-dimensional image patches and classifies the central pixel of each patch, the second network directly infers voxel-wise predictions of three-dimensional volumes. Consisting of multiple pathways, the model simultaneously extracts local and global features on different input scales and resolutions. In a post-processing step, a region growing algorithm is applied on the network's class probability outputs. Detected centerline point candidates are subsequently connected to a final line segmentation. An evaluation over 50 medical cases resulted in an overall segmentation accuracy of 2.70 mm mean distance to the ground-truth centerlines.

1 Introduction

This work is motivated by recent successes of convolutional neural networks (CNNs) in image classification tasks and aims to study the performance of deep CNNs for the application of spinal centerline extraction in 3D computed tomography (CT) images. An accurate, fast and automatic segmentation of the centerline is of interest for clinical diagnosis and therapy planning and can serve as a basis for further processing steps in medical applications [1]. For this reason, two different CNN approaches were applied, evaluated and compared with respect to their accuracy and computational requirements. Finding the centerline was formulated as a voxel-wise binary classification problem, where a voxel was assigned the label c_1 if the center line passed through that voxel, and c_0 otherwise.

In a first setup, 2D CNNs were trained to yield patch-wise predictions. To obtain 3D probability outputs, a 2D patch was extracted for each voxel of the input and subsequently classified, depending on wheter the central pixel of the patch belonged to the centerline or the background class. The fully convolutional neural network (FCNN), as an alternative approach, obtained 3D image segments as inputs and facilitated the direct inference of voxel-wise class predictions. By considering different scales and resolutions of an image in the feature learning process, it produced accurate 3D probability outputs. At the same time, memory and computational requirements were kept practicable, when using a graphics processing unit (GPU). To get an impression of alternative spinal centerline segmentation algorithms and recent achievements concerning CNNs, the reader is referred to [1] and [2].

2 Methods

This section concentrates on the introduction of multi-scale 3D FCNNs. A brief introduction to 2D CNNs is given in [3]. The network architecture used throughout the experiments of this paper was proposed by Brosch et al. As its topology has not been published yet, the basic ideas will be introduced by referring to a strongly related network of Kamnitsas et al. [2].

2.1 Fully Convolutional Neural Networks

The multi-scale 3D FCNN presented in [2] is capable of simultaneously performing class predictions of all voxels of an image segment in one pass through the network. The processing of whole images is usually constrained by available GPU memory. The processing of segments is therefore a compromise between patch-based and whole image voxel-wise segmentation approaches. Furthermore, it provides a flexible handling of varying input sizes, given that all images can be divided into segments of the predefined size. Fig. 1 shows a shallow version of the proposed 11-layers deep network of [2], which consists of two parallel, non-interactive, convolutional pathways. Provided with an input image of given resolution, each path extracts a segment to learn a mapping from a voxel's adjacent intensity values to its ground-truth class label. While the upper pathway processes a segment having the same resolution as the input image, the lower path extracts a second segment that is centered at the same location but has a larger scale and lower resolution than the first one. Like in common CNNs, each convolutional layer learns a set of kernels, also referred to as "channels". Convolving those kernels with the layer's

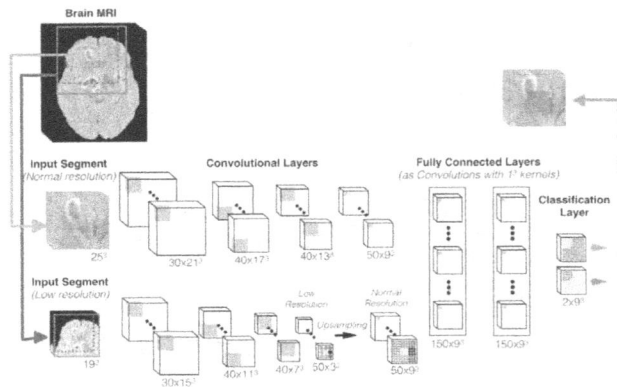

Figure 1: 3D multi-scale FCNN of [2], introduced for brain lesion segmentation. It consists of two pathways (each five convolutional layers deep, using kernels of size 5^3) and two fully connected layers to combine the FMs of both paths. Each path extracts an image segment of different resolution and scale (here the upper segment has a native resolution while the lower one is downsampled by a factor of three). To match the dimensions of the upper path's output, the lower path's output is upsampled before reaching the final layers. The FM numbers and sizes are depicted as Number × Size (for details, see image source: [2])

input yields a volume of feature maps (FMs) for each layer. To combine the outputs of both paths in two fully connected layers, the feature map (FM) volume of the last layer of the low-resolution path needs to be upsampled to match the upper path's one. Doing so, the authors manage to process images at multiple resolutions and scales simultaneously. Thus, the inclusion of both local and larger contextual information in the feature learning process is achieved. The network finally outputs a 3D softmax probability map, representing the class predictions for each voxel of a target segment (see Fig. 1 at the upper right). Note, that Fig. 1 does not represent the exact architecture used in the experiments of this paper.

3 Material and Experimental Setup

3.1 Data and pre-processing

The CNNs were evaluated on 350 training, 50 validation, and 50 test 3D sagittal CT spine images. To reduce the computational demands, the data was downsampled to an isotropic voxel size of $3\,\mathrm{mm}^3$. In a further pre-processing step, the data was standardized to have zero mean and unit variance. The ground-truth spinal centerlines were provided by the extraction algorithm proposed in [1].

3.2 Experimental setup

2D patch-wise classification For the binary classification task of assigning one of the labels c_0 or c_1 to each central patch pixel, 2D patches of size 35×35 were extracted (see Fig. 2(a)). In the training setup, all possible centerline

centered patches were extracted and stored with the class label c_1. The patches of class c_0 were randomly sampled in the background, while keeping a minimum distance of 5 mm to the centerline. This way, training sets of multiple class distributions could be easily extracted and studied in the two-dimensional setup. The result depicted in Fig. 2 was obtained by a 5-layer 2D CNN, containing three convolutional layers (with 32, 48, and 64 channels) and two fully connected layers (with 96 and 2 neurons for the binary classification), respectively. The network's parameters resulted from a cross entropy cost function optimization, using a minibatch size of 18, a momentum of 0.9, and a learning rate of 0.018 for the first and 0.009 for the last ten epochs. Besides max-pooling layers, rectified linear units (ReLUs) and softmax activations were applied. Experiments with strongly unbalanced class distributions hypothesized limitations of the cross entropy cost function for the detection of the greatly underrepresented centerline patches. The problem of dominating background predictions has been recently discussed in [4]. As a solution, the authors introduced a more robust training criterion, which weighs the squared error for the classes c_0 and c_1 as follows:

$$C = r \frac{\sum_{i=1}^{n} (GT(i) - P(i))^2 GT(i))}{\sum_{i=1}^{n} GT(i)} + \\ (1-r) \frac{\sum_{i=1}^{n} (GT(i) - P(i))^2 (1 - GT(i)))}{\sum_{i=1}^{n} (1 - GT(i))}, \quad (1)$$

with $GT(i)$ denoting the binary ground-truth class (value 0 or 1) of the ith sample and $P(i)$ being the posterior probability (value range $[0, 1]$) of sample i belonging to the foreground class. The (user-defined) parameter r weighs the influence of errors made respecting foreground (first term) and background (last term) predictions. This way, the sensitivity and specificity behavior of the network can be greatly influenced by the parameter r. Since the background class is over-represented in the application of centerline segmentation, it is advisable to penalize the specificity error more strongly (e.g. by choosing $r = 0.1$).

3D voxel-wise classification Similar to the network described in Section 2.1, a multi-pathway 3D FCNN was applied to obtain voxel-wise class predictions for 3D centerline segmentations. To facilitate the processing of arbitrary sized images, 3D segments of the largest possible size (fitting into all training, validation and test volumes) were classified in one forward pass. First, "shallow" FCNN experiments were performed with filter sizes of 7^3, later two more layers were inserted to obtain deeper FCNNs, reducing all filter sizes to 3^3 (following the advice of [2]). To handle the strong under-representation of centerline voxels in the 3D images, the sensitivity and specificity loss function (1) was applied throughout all experiments with a fixed ratio of $r = 0.1$ (note, that $GT(i)$ now denotes the ground-truth label of voxel i). To obtain 3D predictions for a whole 3D test volume, the image was divided into non-overlapping segments, being successively tested by the FCNN. The resulting 3D probability maps (obtained for each segment) were then concatenated to reconstruct a 3D output matching the size of the test image. It can be assumed, that a less abrupt formulation of the ground-truth mask could lead to more accurate probability outputs. Therefore, further experiments were carried out with distance surrogates replacing the binary labels during the training process. In this setup, a voxel's Euclidean distance to the closest centerline point $d(i, L) \in [0, \infty)$ was mapped to a value $f(i) \in [1, 0)$, with:

$$f(i) = \frac{1}{1 + d(i, L) \times 0.1\,\text{mm}}. \qquad (2)$$

3.3 Implementation details

All experiments were implemented using the 'The Microsoft Cognitive Toolkit' (CNTK) [5], an open-source toolkit for deep learning algorithms. Due to huge memory loads, which were caused by the extraction of largely overlapping test patches, the evaluation for the 2D CNNs was limited to a visualization of single reconstructed probability map slices. While the training of a 2D CNN with 196 306 parameters took about 6 hours on the CPU, the time could be significantly reduced to 30 minutes training on a NVIDIA GeForce GPU. The GPU training of a "shallow" FCNN with kernel size 7^3 and 741 265 parameters took three times longer than the training of a deeper FCNN (the kernel sizes of 3^3 reduced the parameter number to 183 601), which reduced the training time to 10 minutes. Depending on the image size, inference and concatenating image segments of one 3D case required about 2 to 5 seconds on a GPU and about one minute on a CPU.

3.4 Post-processing

Binary segmentation results could be easily obtained by simply thresholding a network's probability output. However, thresholded centerline segmentations exceed the size of ground-truth centerlines by far, given that the lines are represented by single connected landmarks. To yield a similar set of most probable landmarks from a network's probability output, a repeated region growing with intensity criterion and sphere boundary was applied as a post-processing step. While the sphere radius was fixed to 20 mm in all experiments, the intensity threshold had to be optimized for each network independently. The evaluation was finally performed by comparing ground-truth and segmented centerlines, both being represented by a set of connected landmarks.

4 Results and Discussion

The evaluation over all 50 test images was carried out for 3D FCNNs only, since the storage of largely overlapping patches in the patch-based approach emerged to overburden the memory requirements. Still, a 2D CNN example result is given for visual inspection.

4.1 Segmentation results

2D patch-wise classification Fig. 2(d) shows an example of a reconstructed probability map (obtained by concatenating all patch predictions) for one sagittal CT test slice.

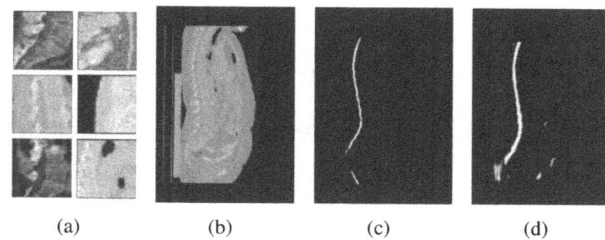

(a)   (b)   (c)   (d)

Figure 2: (a) 2D training patches belonging to the classes c_1 and c_0. (b) Sagittal slice of a downsampled 3D test image, (c) corresponding ground-truth mask, and (d) probability map obtained by the 2D CNN approach.

Network	mask_s	mask_d	dist_s	dist_d
MD	**2.53**	2.70	4.09	4.07
MD_overlap	1.51	**1.35**	1.64	1.71
DC	0.47	**0.52**	0.43	0.43

Table 1: Evaluation of shallow (s) and deep (d) 3D FCNNs optimized on either binary masks (mask) or distance surrogates (dist). The distance measures MD and MD_overlap are specified in mm.

Since negative samples with a distance to the centerline of less than 5 mm had not been represented in the training set, the centerline estimation appeared wider than the ground-truth centerline mask (depicted in Fig. 2(c)). Aside from some high prediction outliers, the network detected the centerline quiet accurately.

3D voxel-wise classification The average performance of four different FCNN architectures has been summarized in Table 1. Shown are evaluations of a "shallow" FCNN with kernel sizes of 7^3, as well as a (two layers) deeper FCNN with kernel sizes of 3^3. Each of the networks was trained using the binary mask, as well as the distance surrogate map (denoted as mask_s, mask_d, dist_s, and dist_d, respectively). While the Dice coefficient (DC) serves as a measure of the volumetric overlap of segmentation and ground-truth centerline, the mean symmetric distance (MD) measures the average Euclidean distance between both lines. The MD_overlap computes the mean symmetric distance without including distances of "non-overlapping" parts (that are the segmentation parts, that exceeded the ground-truth line at the border landmarks and vice versa). The best performance regarding DC and MD_overlap were obtained by a deep FCNN trained on a binary mask (mask_d). Incomplete segmentations and landmark outliers (mainly arising from difficulties in choosing an overall optimal threshold for the region growing algorithm) led to the comparably worse performance of the deep network, dist_d. The most accurate results, which were obtained by two "deep" 3D FCNNs, are shown in Fig. 3. The segmentations, depicted in Fig. 3(c) and 3(e) represent the connected outputs of the region growing algorithm, performed on the probability maps of the networks mask_deep and dist_deep (Fig.3(b) and 3(d)). The training on a binary mask led to slightly su-

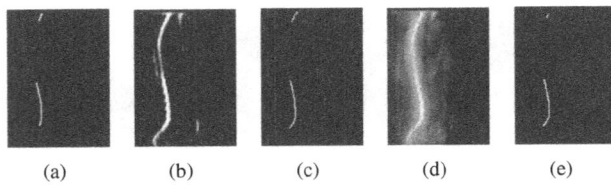

Figure 3: Sagittal example slice showing one of the best results. (a) Ground-truth centerline, (b) mask_deep probability output, (c) corresponding post-processed segmentation, (d) dist_deep probability output, and (e) segmentation.

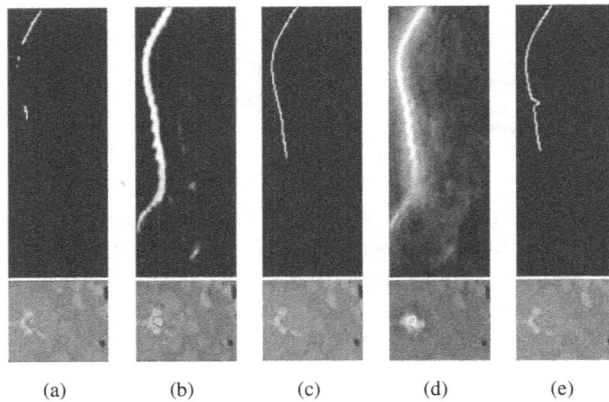

Figure 4: Sagittal example slice of a problem case. The image arrangement (a)-(e) is consistent with the one of Fig. 3. The bottom shows an axial example slice of the same images, overlaid on the gray value CT image.

perior results (DC=0.72 and SMD=1.46 mm) than the one on a distance surrogate map (DC=0.68 and MD=1.84 mm). A test case, that yielded one of the worst accuracies, is shown in Fig. 4. The post-processed segmentations of both networks mask_d (DC=0.24 and MD=3.23 mm) and dist_d (DC=0.28 and MD=3.24 mm) exhibited some deviations from the ground-truth line. While Fig. 4(c) shows a more reasonable mismatch to the ground-truth line, the clear line outliers of Fig. 4(e) represent a common failure of the region growing algorithm applied on the distance surrogate probability maps. Even if the network's probability maps appeared visually more accurate than the ones obtained by the training on binary masks (a fact that is difficult to see in the gray value representation of this paper), the threshold criterion of the region growing algorithm caused problems of either disconnected centerlines (in case of a low threshold) or problems of outliers detected as landmarks (in case of a high threshold).

4.2 Discussion

Large memory requirements and test times, exceeding 2 hours per test case, prohibited a detailed evaluation of the 2D CNN approach. As initial experiments indicated improved accuracies of FCNNs, the performance of 2D CNNs was not further investigated. 3D multi-scale FCNNs not only considered additionally convolutional information of the third image dimension, but also included different resolutions and scales of the input to improve the networks feature discrimination abilities. The evaluation over all test cases proved the network to be accurate, while requiring only a few seconds for the processing of each 3D test image. Since the DC is known to be especially "strict" for small object segmentation, it might be a less appropriate metric for the centerline extraction evaluation. While the region growing performed especially well on the probability maps of "deep" FCNNs trained on binary masks, it faced some problems in the distance surrogate setup. Given that the probability outputs of dist_d networks visually appeared to be more precise than mask_d networks' ones, it is possible, that a modification of the post-processing step can further improve the classification accuracy.

5 Conclusion

In this paper, the performances of 2D CNNs and 3D multi-scale FCNNs were compared with respect to the application of spinal centerline extraction. Being fast in training and inference, the 3D FCNNs clearly out-performed the common 2D CNNs regarding memory and computation requirements. Post-processing the network's probability outputs with a region growing algorithm yielded accurate segmentations of the centerline with a mean symmetric distance of 2.70 mm and a Dice coefficient of 0.52. It is assumed that performances can be further improved by increasing the depth of the networks, extending the training set by data augmentation and modifying the post-processing step. In future research, the spinal centerline segmentation task is furthermore going to be extended to the labeling of centerline sections.

Acknowledgment

The work has been carried out at Philips GmbH Innovative Technologies, Hamburg.

6 References

[1] T. Klinder et al., *Automated model-based vertebra detection, identification, and segmentation in CT images.* Elsevier, 2009.

[2] K. Kamnitsas et al., *Efficient Multi-Scale 3D CNN with Fully Connected CRF for Accurate Brain Lesion Segmentation.* Elsevier, 2016.

[3] M. Nielsen, *Neural Networks and Deep Learning.* Available: http://neuralnetworksanddeeplearning.com [last accessed on 2017-01-07].

[4] T. Brosch et al., *Deep convolutional encoder networks for multiple sclerosis lesion segmentation.* Springer, 2015.

[5] A. Agarwal et al., *An Introduction to Computational Networks and the Computational Network Toolkit.* Microsoft Technical Report MSR-TR-2014-112, 2014.

Development of concepts for high quality image reconstruction based on adaptive grid sizes

A. Pfahl[1], M. Wagner[2], A. Bieberle[3], and T. M. Buzug[4]

[1] Medizinische Ingenieurwissenschaft, Universität zu Lübeck, annekatrin.pfahl@student.uni-luebeck.de
[2] AREVA Endowed Chair of Imaging Techniques in Energy and Process Engineering, Technische Universität Dresden, Michael.Wagner3@tu-dresden.de
[3] Institute of Fluid Dynamics, Helmholtz-Zentrum Dresden-Rossendorf, a.bieberle@hzdr.de
[4] Institute of Medical Engineering, Universität zu Lübeck, buzug@imt.uni-luebeck.de

Abstract

In this work an alternative data processing concept is investigated for the correct reconstruction of slice images acquired by an ultrafast electron beam X-ray tomography scanner mainly used for analyzing multiphase flows. Currently, image reconstruction is performed on regular pixel grids by filtered back projection to achieve rapid data processing performances but leading to non-optimal image qualities. To accomplish an improved image quality the usage of irregular reconstruction grids and iterative reconstruction methods is analyzed considering the geometric arrangement and, thus, the real spatial resolution of the ultrafast CT scanner. Finally, a two-stage reconstruction approach is proposed reducing the required amount of computer memory as well as the computational time and inserting prior knowledge about the object of interest. First simulations of different irregular grids provide a promising basis for further successful implementation of the proposed two-stage reconstruction concept.

1 Introduction

For high dynamic multiphase flow investigations a worldwide unique ultrafast electron beam X-ray tomography scanner, called ROFEX, is operated in the department of Experimental Thermal Fluid Dynamics at Helmholtz-Zentrum Dresden-Rossendorf (HZDR). This non-invasive imaging system allows studies in optically inaccessible technical devices and fluids, e.g. presented in [1].

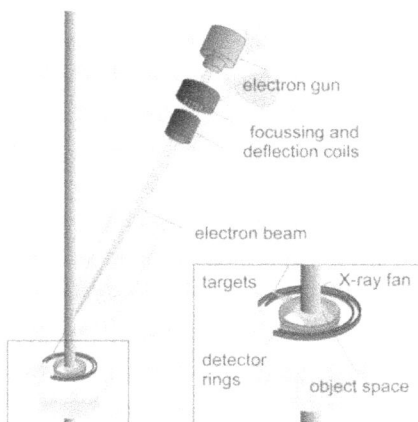

Figure 1: Principle sketch of the ultrafast electron beam X-ray tomography scanner [2].

The principle setup is shown in Fig. 1. In contrast to usual CT scanners, an electron beam, generated by an electron gun, is focused and deflected onto a circular target covering the object to obtain projection data from different angular positions. In Table 1 relevant geometric parameters of the scanner are compiled. Additional technical information is described in [3]. The CT scanner provides a maximal

Table 1: Selected geometric parameters of the ROFEX CT scanner.

Parameter	Description	Value
rSrc	radius of target ring	182.5 mm
nSrc	number of source elements respectively projections	500
dSrc	X-ray spot size	1 mm
rDet	radius of detector ring	108 mm
nDet	number of detector elements	432
dDet	width of detector element	1.3 mm
rTest	radius of investigated object	60 mm

scanning frequency of $f_{deflect}$ = 8,000 fps. The statically and circularly positioned detectors are periodically sampled with f_{samp} = 1 MHz. Thus, the number of projections can be calculated as $nSrc = f_{samp}/f_{deflect}$.

Currently, standard computed tomography reconstruction technique, namely Filtered Back Projection (FBP), is employed for image reconstruction on regular pixel grids with a typical pixel size of 0.25 mm^2. Due to the huge amount of data, delivered by the CT scanner, this rapid reconstruction method is usually preferred. By using parallel processing hardwares, like Graphic Processing Units (GPUs) [4], the reconstruction frequency can achieve the maximal scanning rate of the CT scanner whereby on-line CT scanning can be performed.

Adversely, the fast reconstruction method includes several influencing factors that affect the resulting image quality negatively. To analyze these effects quantitatively, this work contributes to develop a new reconstruction approach that, in fact, will not achieve high-speed performances, but may improve the image quality. Therefore, irregular grids will be used a) involving the correct geometric arrangement of the CT scanner and b) including information about the object of investigation. Furthermore, the use of Algebraic Reconstruction Techniques (ART) will be tested, obligatory required for image reconstruction on irregular grids.

2 Material and Methods

2.1 Data Processing

To demonstrate influencing factors that reduce the resulting image quality, an analysis of data processing was carried out (see Fig. 2).

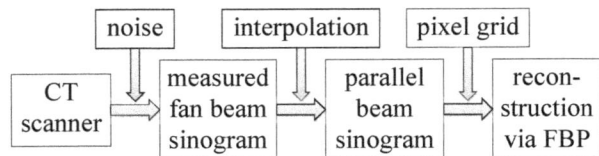

Figure 2: Current data processing schema based on regular pixel grids.

Because of the electron deflection procedure, the positions of the X-ray spots pSrc are not equidistantly distributed onto the target. For this reason and furthermore due to the fan beam geometry of the projections, an extensive interpolation must be conducted on the measurement data to obtain parallel beam projections from equidistant positions. This allows the application of an one-dimensional filter onto each projection for the FBP that is addressed to high data throughput instead of a computational-intensive two-dimensional filtering, required for the originally acquired fan beam data set. In addition to the interpolation process, a high sensitivity to noise and an interpolation on user-defined, not CT-system-defined reconstruction pixel grids have been stated.

To eliminate occurring interpolation errors a new data processing concept would be required, like schematically illustrated in Fig. 3. The objective is to reconstruct the image directly from the measured fan beam projection data set onto an irregular grid that perfectly fits to the acquisition geometry (spatial resolution) and may include prior knowledge about the object of investigation. By using ART the influence of noise can also be considered during reconstruction. Subsequently, to obtain information on structures of irregular reconstruction grids, a literature review was carried out.

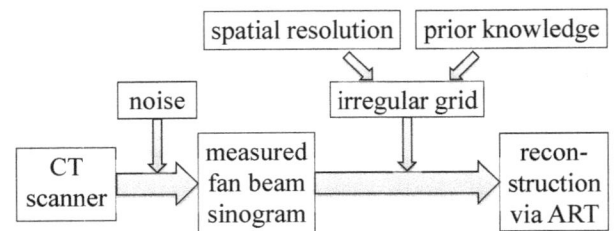

Figure 3: Proposed data processing schema based on irregular grids.

2.2 Image Discretization

For discretization of images different approaches were presented in the past. Finally, the following three approaches are most relevant to this work:

- Regular grids, e.g. pixel grids, hexagonal grids [5]
- Adaptive grids, e.g. grids which are determined by a segmentation that is done during the reconstruction process [6] or which are subdivided into finer segments in each iteration step [7]
- Exact grids, e.g. projection intersection grids [8], polygon grids

In 2007 Knaup et. al [5] tested hexagonal grids. Here, the number of pixels could be reduced by 13% and the reconstruction process correspondingly accelerated by 12%. Unfortunately, by using the FBP algorithm the image quality could not be enhanced compared to pixel grids.

Buyens et. al [6] focused on ART and divided the reconstruction grid into irregular rectangles that are adapted to the reconstructed image iteratively. Therefore, a segmentation process is performed in alternation to the reconstruction iterations. As a result the ART was accelerated but no improved image quality could be reported.

The work of Kim et. al [7] followed a similar strategy. Here, a so called octree was implemented to adapt the reconstruction grid to the image. The reconstruction interval of the ART was reduced but, again, a higher image quality was not mentioned.

A promising approach was shown by Goswami et. al [8]. They introduced so-called projection intersection grids (PI-grid) used for measurements of multiphase flows. Reconstructing the image via ART or the Method of Moving Asymptotes (MMA) the image quality of both simulated and experimental data was enhanced. Because of this, PI-grids are the focus of this work.

The structure of a PI-grid follows the intersection points between all projection lines that are assumed to be as infinitesimal small. Fig. 4 (left) demonstrates the generated PI-grid for a simulation of three projections for a simplified fan beam geometry. Another type of exact grids adapts the structure of the PI-grid to the geometric arrangement of the CT scanner. Here, the infinite thin pencil beams are replaced by realistic expanded beams, which result of the X-ray spot size and the active area of the detector elements.

As can be seen in Fig. 4 (right) the crossing points then become intersection polygons reflecting the real spatial resolution of the imaging system. Therefore, this grid represents the most precise reconstruction grid and is of importance for this analysis as well.

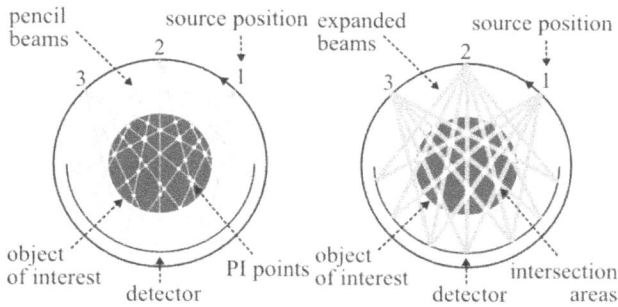

Figure 4: Principal structure of PI-grid (left) and polygon grid (right).

Both, PI- and polygon grid were generated by the simulation of X-ray fan beams using the following data processing steps. Geometrical parameters (see Table 1) were defined first to generate and visualize all relevant elements. Subsequently, permitted beams that fulfill a given angle criteria were simulated between the source spots and detector elements. Afterwards, all intersection points/polygons inside the object could be calculated.

3 Results and Discussion

3.1 Polygon Grid and Spatial Resolution

Simulating expanded X-ray fan beams during the above-mentioned processing algorithm, an irregular grid of intersection polygons is generated. Since those areas exactly correspond to the spatial resolution of the CT scanner, it may be the most precise reconstruction grid. Nevertheless, the simulation of projection rays is challenging since computing time exponentially increases with the number of source and detector elements as shown in Table 2. Con-

Table 2: Computing time for various spatial resolution map configurations.

nSrc	nDet	Number of proj. lines	Computing time
5	4	11	2 s
15	13	98	33 s
25	22	275	30 min

sequently, a full polygon grid for typical ROFEX scanner data acquisition parameters cannot be performed this way. An approach that follows the concept of parallel computing could be pursued prospectively but was not implemented so far.
Exemplary, Fig. 5 illustrates the grid simulating fan beam projections between 25 X-ray source positions and 22 detector elements where all components are equidistantly distributed onto the target ring (rSrc) and detector ring (rDet).

For better representation, the X-ray spot size dSrc and the width of detector elements dDet were raised by the factor of 10. This map totally includes 26,052 intersection polygons. Thereby, the larger the polygon's area the brighter they are. 9.24% of all polygons have smaller areas than 0.25 mm^2. Respecting the scaling factor, this percentage raises for smaller values of dSrc and dDet. Thus, the real spatial resolution is much higher than currently assumed.

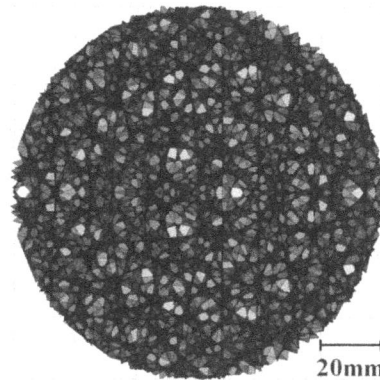

Figure 5: Spatial resolution map simulated with 25 X-ray source positions and 22 detector elements.

3.2 Projection Intersection Grid

In the next step the PI-grid was generated analogous to the polygon grid procedure (see Fig. 6). The resulting grid contains totally 6,235 intersection points. It can be easily seen that the spacing between single grid points differs tremendously. Thus, higher and more sparsely sampled regions occur. Another important fact is the enormous amount of grid points. Using ART for reconstruction the system of equations, to be solved, becomes very large and the reconstruction time increases. This computational effort might not be manageable by the available computer memory.

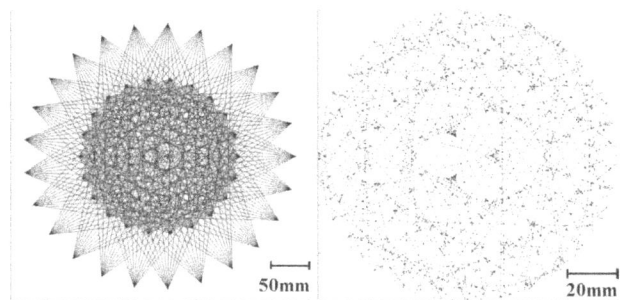

Figure 6: Simulation of pencil beams with 25 X-ray source positions and 22 detector elements (left) and resulting PI-grid using the crossing points (right).

3.3 Novel Reconstruction Approach

A comparison between the PI-grid and a regular pixel grid demonstrates its decisive advantage: for the design of a regular pixel grid that accomplishes the spatial resolution of the calculated PI-grid, a single pixel hypothetically must have an area of $7.4 \cdot 10^{-5}$ mm^2 which corresponds to the smallest grid point spacing within the PI-grid. Then, the regular pixel grid would contain $1.5 \cdot 10^8$ pixels – 24.000 times higher than the PI-grid.

However, under real conditions (nSrc = 500; nDet = 432) the number of PI-grid points also increases enormously. This leads to a huge system of equations for reconstruction. Thus, a new concept of reconstruction and data processing is proposed. To minimize the computational effort within the algebraic reconstruction, prior knowledge about the object of interest is included. To do so, the following steps have to be performed. First, fast reconstruction is conducted with FBP on a regular pixel grid providing an initial image. Second, a segmentation process (e.g. thresholding or levelsets) recognizes all elements inside the object. Transition areas between the element edges can be defined where subsequently the exact PI-grid/polygon grid is calculated for. So, the huge number of grid elements is restricted to a small area. An additional reconstruction with algebraic methods is finally performed on the exact grid within the transition areas (see Fig. 7). After all, both reconstruction results are merged correspondingly.

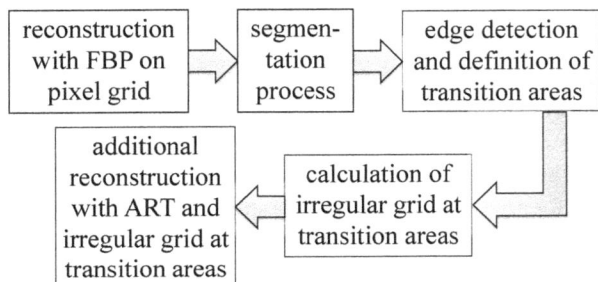

Figure 7: Developed data processing schema based on irregular grids and two-stage reconstruction.

So far, the new concept was not implemented completely. First simulations have already demonstrated that the reconstruction time can be reduced to a third compared to a reconstruction of the entire image with ART. For this reason, the approach will be continued in further studies where the resulting image quality of different algebraic methods, like Kaczmarz' algorithm and Simultaneous Algebraic Reconstruction Technique (SART), is compared.

4   Conclusion

This work presented a new two-stage CT image reconstruction concept based on irregular exact grids to accomplish an improved image quality. A literature review and an analysis of the geometric arrangement of the ultrafast CT scanner ROFEX had shown that PI-grids and polygon grids are suitable to include information of the data acquisition setup. Both types of irregular grids were successfully implemented and analyzed for variable parameter settings. Furthermore, the use of algebraic techniques may allow the integration of prior knowledge about the investigated ob-

ject. Although their application leads to higher computational efforts, the novel concept reduces these extensive calculations onto a limited region of interest.

Acknowledgement

The work has been carried out at the Department of Experimental Thermal Fluid Dynamics, Institute of Fluid Dynamics, Helmholtz-Zentrum Dresden-Rossendorf and supervised by the Institute of Medical Engineering, Universität zu Lübeck.

5   References

[1] J. Zalucky, T. Claußnitzer, M. Schubert, R. Lange, and U. Hampel, "Pulse flow in solid foam packed reactors: Analysis of morphology and key characteristics," *Chemical Engineering Journal*, vol. 307, pp. 339–352, 2017.

[2] M. Wagner, J. Zalucky, M. Bieberle, and U. Hampel, "Hydrodynamic investigations of bubbly flow in periodic open cellular structures by ultrafast X-ray tomography," in *10th Pacific Symposium on Flow Visualization and Image Processing*, 2015.

[3] F. Fischer, D. Hoppe, E. Schleicher, G. Mattausch, H. Flaske, R. Bartel, and U. Hampel, "An ultra fast electron beam x-ray tomography scanner," *Measurement Science and Technology*, vol. 19, no. 9, p. 094002, 2008.

[4] T. Frust, M. Wagner, J. Stephan, G. Juckeland, and A. Bieberle, "Rapid data processing for ultrafast X-ray computed tomography using scalable and modular CUDA based pipelines," *Computer Physics Communications*, submitted 2016.

[5] M. Knaup, S. Steckmann, O. Bockenbach, and M. Kachelrieß, "CT image reconstruction using hexagonal grids," in *2007 IEEE Nuclear Science Symposium Conference Record*, vol. 4, pp. 3074–3076, IEEE, 2007.

[6] F. Buyens, M. Quinto, and D. Houzet, "Adaptive mesh reconstruction in X-ray tomography," in *MICCAI Workshop on Mesh Processing in Medical Image Analysis*, 2013.

[7] S. Kim, A. Sakane, Y. Ohtake, H. Suzuki, Y. Nagai, K. Sato, H. Fujimoto, M. Abe, O. Sato, and T. Takatsuji, "Efficient iterative CT reconstruction on octree guided by geometric errors," in *6th Conference on Industrial Computed Tomography, Wels, Austria (iCT2016)*, 2016.

[8] M. Goswami, A. Saxena, and P. Munshi, "A new grid-based tomographic method for two-phase flow measurements," *Nuclear Science and Engineering*, vol. 176, no. 2, pp. 240–253, 2014.

A syntactic approach to wreck pattern recognition in sonar images

M. Constapel[1], T. Teubler[1] and H. Hellbrück[1,2]

[1] Fachbereich Elektrotechnik und Informatik, Fachhochschule Lübeck,
  manfred.constapel@stud.fh-luebeck.de, torsten.teubler@fh-luebeck.de, horst.hellbrueck@fh-luebeck.de
[2] Institut für Telematik, Universität zu Lübeck, hellbrueck@itm.uni-luebeck.de

Abstract

Autonomous underwater vehicles (AUV) capable of doing robust online wreck recognition are key for inspection and monitoring scenarios either including one or multiple AUVs. Therefore, a technique for recognition of wrecks in pre-recorded sonar images by means of compter vision and syntactic pattern recognition is described and illustrated. Our framework-agnostic approach utilizes for implementation the widely known Open Source Computer Vision library (OpenCV) for image processing. Based on an example we approximate the shape of a wreck by a convex polygon whereupon some of the vertices of the polygon are converted to digestable features using Menger curvature providing processable data in a syntactic pattern domain.

1 Introduction

This paper illustrates a method for recognition of wrecks in sonar images. Sonar images are generated by a sonar device. On one hand sonar devices emit acoustic waves repetitive at certain power levels and usally high frequency, on the other hand they receive echos from objects and the sea bed caused by reflection.

Clear differences between acoustic and optical images are the amount of time to spend for generating an acoustic image. This is because of lower propagation speed of waves in water at roughly 1500 meter per second, neglecting salinity, temperature and pressure, and the feasible resolution with regard to the lower frequency of acoustic waves compared to visible light. Nevertheless, in our work images containing depth information taken by an optical camera come into being as proxies for the serving acoustical sonar images.

Fig. 1b shows a sonar image enhanced by two sobel operators to reveal some of the details of the wreck covered in the gray-scale image on the left-hand side for visual inspection.[1] In average the wreck is 15 meter below the sonar device at a depth of 17 meter. The wreck has an overall length of 80 meter with pointing up superstructures. The bow and parts of the anchorage gear are located in the upper part, the stern and main parts of the superstructure can be seen in the lower part. At all the sonar image in Fig. 1a covers an area of 100 meter in length and 50 meter in width. We will refer to the wreck in Fig. 1a throughout this paper to illustrate our approach of syntactic pattern recognition.

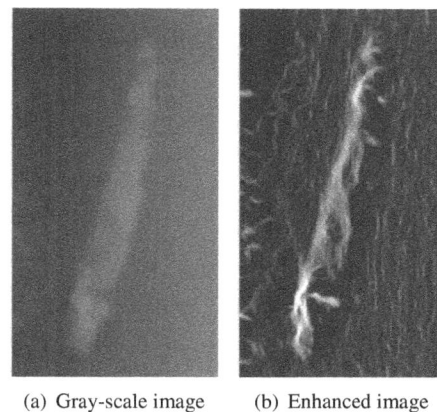

(a) Gray-scale image   (b) Enhanced image

Figure 1: Sobel operators applied to a gray-scale sonar image create an image suitable for visual inspection

1.1 Related work

We are going to use a variant of an embedded multi-beam sonar feature extraction for online AUV control [1]. Although their work is focused on a workflow and processing chain for AUV control, a masking technique based on a Canny algorithm for shape recognition of wrecks is included. We put that basic idea of an expert system able to do object classification into a different and more narrow direction by means of syntactic pattern recognition.

In medical applications [2] syntactic pattern recognition is common. Various other applications such as letter recognition or fingerprint recognition [3] are supported by syntactic pattern recognition. Syntactic pattern recognition is

---

[1] Raw data were taken by an *Imagenex 837 DeltaT multi-beam sonar* in the port of Vancouver at position 49° 19.4' N 122° 54.5' W in March 2005.

a form of pattern recognition, in which each object can be represented by a variable amount of symbols as nominal features. In our work we will show suitability of that kind of pattern recognition for recognition of wrecks, therefore we rest upon syntactic pattern recognition [4] in this work.

1.2 Assumptions

Interference of coincident waves and multi-path effects have to be taken into account when dealing with acoustic waves. In this work we assume:

- processing of non-distorted sonar images

- sonar images taken by a down looking sonar device perpendicular to the sea bed as far as possible and aligned to earth's gravitational vector respectively

- suspected wrecks situated on the sea bed not be fractionized

1.3 Preliminaries

Since we need an image-like representation for applicability to computer vision algorithms by OpenCV [?] we have to convert sonar raw data to a gray-scale sonar image as shown in Fig. 1a. Thus, intrinsics of an inherent optical camera with regard to scale are to be known (e.g. aperture angle) for mapping the pixels of the image plane as seen in Fig. 1a to the snapped sea bed gathered from the sonar raw data.

The gray-scale values of the image shown in Fig. 1a encode depth information in a range from 2 meter (white pixels) to 20 meter (black pixels) in 256 discrete steps. Such a gray-scale image is a starting point for our implementation.

2 Material and methods

This part is divided into six sections each one describing a major step of our implementation. Sections 2.1 and 2.2 can be treated as pre-processing, namely normalization and edge detection. Section 2.3 introduces the impact of convex polygons in our work and it puts everything to the starting line. Sections 2.4 and 2.5 lay the foundation in terms of vertices and features for the culminating final section 2.6 describing our syntactic approach.

2.1 Normalization

In the normalization step gray-scale values are lifted and dropped linearly in order to make usage of the full gray-scale range. Additionally we apply a gaussian blur for smoothing and expand the lift and drop about 10 percent above and below the extremes to simply set the brighter one-tenth to bright and the darker one-tenth to dark. This procedure cancels out irregularities of the sea bed and amplyfies the height differences of large objects to emphasize their shapes as perceivable in Fig. 2a. This leads to a more robust edge detection.

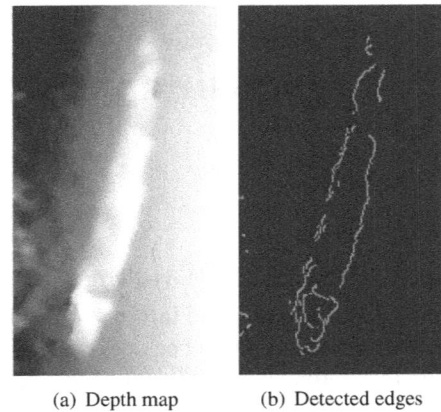

(a) Depth map   (b) Detected edges

Figure 2: A normalized depth map makes life a bit easier for edge detectors to provide robust results

2.2 Edge detection

We apply an adaptive edge detector based on the established Canny algorithm to the normalized depth map for extraction of edges in the image. The adaptive edge detector determines the arithmetic mean μ and standard deviation σ of the normalized depth map by simply taking all pixels gray-scale value. A lower and upper threshold is established by putting μ and σ together with a factor c, in the first place by substracting and thereafter by adding: $thres = \mu \pm c\sigma$.

Fig. 2b shows the result after passing the thresholds and the normalized depth map to a Canny edge detector. As a consequence the self-adjusting thresholds give rise to the edge detection so it yields more reliable results regardless a given depth map.

2.3 Convex polygon

Presumed the wreck not be parted a convex polygon is a good fit for the shape we would expect to see from the perspective of the sonar device and our fictional camera respectively. Prior to estimation of the polygon an outlier discussion by segmentation is performed for removal of edges not belonging to the wreck hinted with aid of three parallel lines in Fig. 3a. Except for special-purpose watercrafts, regular ships (e.g. merchant vessels) exhibit a longish shape by their hulls. Therefore, edges detected beyond a certain distance widthwise are assumed to have no dependency to the hull we want to expose.

For this purpose we determine pixelwise the center of mass and a linear regression line passing the center of mass in order to bisect congregations of pixels concerning the pixels showing an edge only. All pixels not within the outermost lines shifted away by a certain extend from the intermediate regression line are assumed to be outliers. Due to truncation of outliers, thus, only remaining pixels are considered for estimation of the polygon.

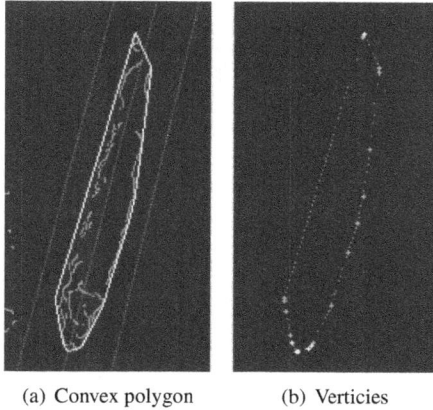

(a) Convex polygon   (b) Verticies

Figure 3: Polygon yields a set of vertices addressable via (x, y)-coordinates

2.4 Vertex extraction

We are going to pass the convex polygon in Fig. 3a encircling an object suspected to be a wreck to a Hough transform for generating vertices. In particular the Hough line transform is a transform mainly used to detect straight lines. In order to generate vertices we take the midpoints of all detected lines. The dotting effect after application of our Hough line-to-vertex transform by drawing dots for corresponding vertices is visualized in Fig. 3b. As usual in image processing coordinates of vertices are referenced vertically by y and horizontally by x within a vertically flipped Cartesian grid originated in the upper left corner.

For further processing we need to have all vertices either in a clockwise or counter-clockwise order. A comparsion algorithm for all vertices is described in Algorithm 1 to permit sorting for the subsequent feature extraction.

Algorithm 1: Vertex comparison
input: vertices a, b of convex polygon,
        center c convex of polygon
result: wether a is left of or further away than b
if (a_x - $c_x \geq 0$) and (b_x - $c_x < 0$): true
if (a_x - $c_x < 0$) and (b_x - $c_x \geq 0$): false
if (a_x - $c_x = 0$) and (b_x - $c_x = 0$):
    if (a_y - $c_y \geq 0$) or (b_y - $c_y \geq 0$):
        if ($a_y > b_y$): true
        if ($a_y < b_y$): false
    if ($b_y > a_y$): true
    if ($b_y < a_y$): false
d = (a_x - c_x) (b_y - c_y) - (b_x - c_x) (a_y - c_y)
    if $d < 0$: true
    if $d > 0$: false
d1 = (a_x - c_x) (a_x - c_x) + (a_y - c_x) (a_y - c_y)
d2 = (b_x - c_x) (b_x - c_x) + (b_y - c_y) (b_y - c_y)
    if $d1 > d2$: true
    if $d1 < d2$: false

2.5 Feature extraction

In this step an ordered set of vertices gathered by vertex extraction as shown in Fig. 3b becomes vital for further processing. Hence, we define a set of vertices $v_i \in V$ to wrap the vertices highlighted in Fig. 3b we consider as feature. A triple of consecutive vertices $w_i = (v_{i-1}, v_i, v_{i+1})$ is defined as the vertices span a triangle we will frequently refer to. We are going to utilize all w_i for assigning feature attributes to $v_i \in w_i$. For this purpose the Menger curvature [5] takes place. A *leg* is assigned to v_i in case of low Menger curvature. Legs are more likely a straight line than a circular segment. Contrary at high Menger curvature a triple w_i seems to be more like a curve than a line. In such case we assign a *turn* to v_i.

The Euclidean lengths $a_i = |v_{i-1} - v_i|$, $b_i = |v_i - v_{i+1}|$ and $c_i = |v_{i-1} - v_{i+1}|$ of a triangle spanned by w_i are obtainable via difference of position vectors of two adjacent vertices. The semi-perimeter $s_i = \frac{a_i+b_i+c_i}{2}$ is used for calculation of the area A_i covered by w_i by Heron's formula.

$$A_i = \sqrt{s_i(s_i - a_i)(s_i - b_i)(s_i - c_i)} \qquad (1)$$

The Menger curvature c_i is reciprocal to the radius of a circle passing all vertices of w_i. The smaller a circle the higher its curvature and vice versa. This curvature is used for assigning feature attributes to vertices as shown in Fig. 3b; the thicker dots represent the set of vertices, the white ones are *turns*, the gray ones remain as *legs*.

$$c_i = \frac{4A_i}{a_i b_i c_i} \qquad (2)$$

We are going to assign feature attributes to all verticies v_i by calculation of c_i. Our feature extraction would yield an uniform distributed set of vertices for a perfect circle. For wrecks which are modelled by a polgon we expect to get higher densities of vertices and higher Menger curvatures at bow and tail. Therefore, by comparing values of c_i to the arithmetic mean $\bar{c}$ we establish our condition for assignment of a feature attribute to a vertex: If $c_i > \bar{c}$ then v_i is a *turn* else it is a *leg*. An example of the Menger curvature is shown in Fig. 4. Due to our condition, for vertex $v_i = (95, 44)$ the Menger curvature is $c_i = 0.12$, therefore it is a *leg*, because it is less than $\bar{c} = 0.16$. In other words v_i is considered to be more straight than curved.

2.6 Pattern recognition

We propose an alphabet A with two symbols "leg" and "turn" as well as a sequence $S \supseteq A$ listing a finite count of those symbols. Symbols are ordered in relation to their vertices, either clockwise or counter-clockwise from the point of view of the underlying polygon.

Furthermore we propse a simple grammar G just consisting of two existential quanitifiers namely "at least one" and "exactly two", and a concatenation operator "is followed by", very much alike the arithmetic operator for addition without commutative and distributive properties. Such operators augmented with directional attributes are used very often in

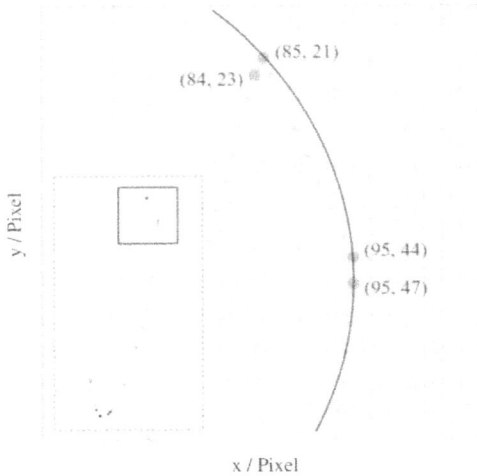

Figure 4: A Menger curvature applied to a vertex at position (95, 44) by fitting a circle through its neighbors

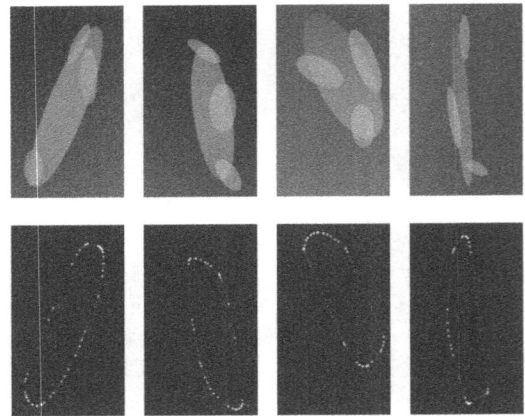

Figure 5: Four examples of artifical sea beds and wrecks generated randomly by gradients and parametrized ellipses (upper row) and their corresponding results from feature extraction (lower row)

syntactic pattern recognition. In our work are going to use them in a slightly simplified way.

$$A = \{L : "leg", T : "turn"\}$$
$$Q = \{\exists_{\geq}^{1} : "at\ least\ one", \exists_{=}^{2} : "exactly\ two"\}$$
$$O = \{\rightarrow : "is\ followed\ by"\}$$

To check wether a sequence S contains a polygon representing a wreck the following pattern is proposed.

$$\exists_{=}^{2} [\, \exists_{\geq}^{1} (L) \rightarrow \exists_{\geq}^{1} (T) \,]$$

In order to provide operability of the pattern we must have at least four vertices $|S| \geq 4$ forming a convex polygon. The pattern yields either true or false. It examines the existence of two relatively small roundings for bow and tail and the occurence of symmetry we expect to see looking at a longish shape of some ship from bird's-eye. Due to varying amount of symbols in S a desireable property of the proposed pattern is its invariance to the cardinality of S. Prior knowledge about a potential wreck in terms of size (1) could be used as a preceding filter.

3 Evaluation

Due to lack of sonar images we decided to generate a set of artifical sonar images for evaluation. A subset of such sonar images is shown in Fig. 5. At all we randomly generated 30 unique sea beds each containing a hypothetical wreck. One main ellipse and three minor ellipses are used to shape a wreck. Minor ellipses are generated in a way they always overlap the main ellipse but do not interfere themselves. To keep the resulting sonar images realistic we put constraints in for absolute gray-scales, gray-scale changes for gradients as well as rotations and scales for all ellipses. Our syntactic pattern as proposed in 2.6 was able recognize all wrecks in our set of randomly generated artifical sonar images. As a result we showed the applicability of our approach for syntactic pattern recognition of wrecks in sonar images.

4 Conclusions

We proposed a leightweight and easy-to-implement technique for wreck recognition by means of syntactic pattern recognition. For this purpose we showed in stepwise manner an abstract implementation of our algorithms.

Further evaluation to verify versatility of our approach deduced from robustness and reliability must to be done in future work, especially for cobbled sea beds and small wrecks. Therefore, more raw data of multi-beam sonars or convenient sonar images showing wrecks are essential.

Acknowledgements: This work was funded by the Federal Ministry for Economic Affairs & Energy (035SX361C, BOSS).

5 References

[1] M. Sion, T. Teubler, and H. Hellbrück, "Embedded multibeam sonar feature extraction for online auv control," in *OCEANS 2016-Shanghai*, pp. 1–4, IEEE, 2016.

[2] M. R. Ogiela, R. Tadeusiewicz, and L. Ogiela, "Image languages in intelligent radiological palm diagnostics," *Pattern Recognition*, vol. 39, no. 11, pp. 2157–2165, 2006.

[3] B. Moayer and K. S. Fu, "A syntactic approach to fingerprint pattern recognition," *Pattern Recognition*, vol. 7, no. 1, pp. 1–23, 1975.

[4] K. S. Fu, *Syntactic methods in pattern recognition*, vol. 112. Elsevier, 1974.

[5] I. Hahlomaa, *Menger curvature and Lipschitz parametrizations in metric spaces*. Univ., 2005.

[6] J. Sklansky, "Finding the convex hull of a simple polygon," *Pattern Recognition Letters*, vol. 1, no. 2, pp. 79–83, 1982.

Object tracking for accurate irritation-free 3D shape measurements of human faces and body parts

O. Rost[1,2], A. Brahm[2], P. Dietrich[2], I. Schmidt[2], P. Kühmstedt[2], G. Notni[2,3] and R. Huber[4]

[1] Medizinische Ingenieurwissenschaft Universität zu Lübeck, oliver.rost@student.uni-luebeck.de
[2] Fraunhofer Institute IOF, 07745 Jena {anika.brahm,patrick.dietrich,ingo.schmidt,peter.kuehmstedt}@iof.fraunhofer.de
[3] Technical University Ilmenau,98693 Ilmenau, gunther.notni@tu-ilmenau.de
[4] Institute of biomedical optic, Universität zu Lübeck, robert.huber@bmo.uni-luebeck.de

Abstract

For detailed measurements of human faces we use resolution three-dimensional (3D) scanners. However these sensors have a small field of view. To be able to scan people, it is therefore necessary to track the face to be scanned. In this paper we describe a setup for an additional sensor which we used for tracking the face and we introduce several algorithmic methods and describe one in detail which we found to suit the problem best. This sensor is a classical passive stereo setup; comprising of two calibrated cameras and generates position and feature data of human faces or body parts in motion. An more accurate determination of tracked positions can be necessary for future robot assistance systems. Test measurements were made, to ensure that the algorithm meets the lateral resolution requirement for the position tracking.

1 Motivation

Three-dimensional (3D) measurements of object surfaces with high accuracy have become important tasks in industrial applications. In particular, accurate 3D reconstruction of human faces and body parts, or body movements offers new applications in medicine, gesture detection and many more. Several optical methods have been reported [1][2], as they provide non-destructive, contactless and precise measurements. We are using an active triangulation stereo 3D-Scanner, based on near infrared (NIR) illumination, with a field of view of 40 cm x 40 cm for detailed face scanning. The scan volume of this scanner is about 60 cm deep [3]. For applications where people move, the scanner needs to be moved to keep the face inside the scan volume. In this paper we evaluate methods to generate the necessary tracking information. Two angels are required to describe the necessary 3D-scanner orientation.

This can be achieved with a single camera. However, with a single camera the distance can only be measured if the object is of known size. For the general case, where we do not have such a priori information a second camera is required for triangulation.

We use a passive stereo camera setup to get the depth information. With this setup, the depth coordinate can be directly calculated without knowledge about the object size and therefore, we use this setup for our described algorithm and measurements. We describe different algorithmic methods to determine the depth information in measurement scenarios.

To clarify, we are using 2 sensor systems on the same mount

(Fig. 1.): The high resolution 3D-Scanner and the passive stereo tracking system. The 3D-Scanner has a high accuracy resolution in a range of $100~\mu$m [3] to detect all uneven surfaces. The passive stereo scanner only needs to produce coarse depth information, because it is only there to adjustment the 3D-Scanner. One possible future application for the method described in this paper is using the tracking information to control a robot arm on which the scanner unit is attached.

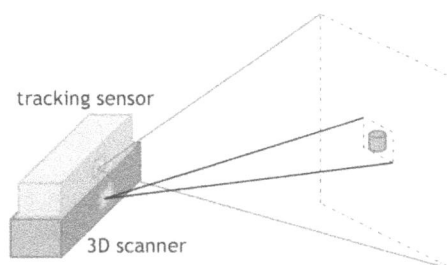

Figure 1: This figure shows a tracking setup with two 3D measurement systems visible: The stereo setup with a large field of view, the 3D scanner with a small field of view.

2 Material and Methods

In this section we present the basic principles for the development of the tracking sensor. Prior to that, we start in section 2.1 on general knowledge of face detection approaches to find the best detection algorithm for our application to get their two-dimensional coordinates (x,y) in the selected scene. We present also the passive 3D sensor technology to explain the principle of depth information in a stereovision setup. Thereby, we give an introduction of two different methods to determine the depth information. In 2.2 we show the general hard- and software setup to generate all missing tracking coordinates and in section 2.3 we describe our full algorithm structure and explain the chosen chronologic order for the final algorithm.

2.1 Basic principles – tracking approaches

We evaluated the application of two different algorithms for the object detection. For the first method a template (e.g. from a database) of an object has to be created [4]. To find the best match for the template it must be compared with the camera images. The next method based on machine learning. This algorithm is called *Viola Jones Algorithm* and is described in detail in [5]. Basically, it can be generated from many pictures of an object to a generalized picture of the object. The generalized picture is used in classifiers which were connected in series and processed by a majority principle. The classifiers searched through each image section and selected a face if the majority of all classifiers returned a positive value for the searched image section.

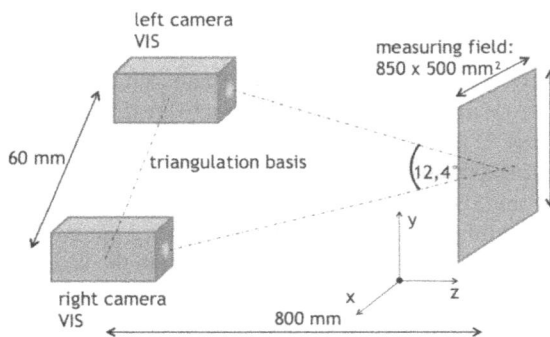

Figure 2: Experimental stereo-vision setup to calculate the depth values without knowledge about the object size.

The next path described existing different measurement methods to determine the depth information in measurement scenarios. These methods based on a classical passive stereo-vision setup and are shown in Fig. 2. A depth coordinate can be directly calculated without any knowledge about the sample. To measure the real values physically, a constant distance between both cameras and a camera calibration is necessary. The line between both cameras is called triangulation basis and can be identified accurately with a Fraunhofer internal calibration software. During the calibration, the program determines all relevant intrinsic and extrinsic parameters of the camera system. The intrinsic parameters describe that each fictive beam of a camera can represented as one point in the original picture. The extrinsic parameters describe the orientation of the cameras to a world coordinate system of the scene. Both, intrinsic and extrinsic parameters, contributes the object transfer from a world coordinate system to a camera

coordinate system. After all parameters were found, the mathematical concept of triangulation can be used to calculate the desired depth information of the scene (z-coordinate) [2].

One concept that we have implement is called *disparity*. The basic concept of the disparity is shown in Fig. 3 and described by [7]. This algorithm was able to find general corresponding points in both camera pictures and works with the Fraunhofer internal calibration software.

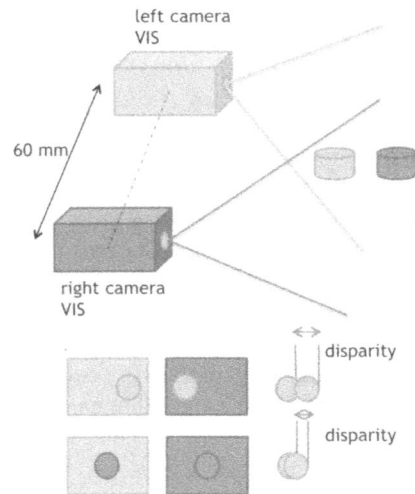

Figure 3: Principle of the disparity: objects which are far away in the scene corresponds to a smaller disparity. The receive disparity from rectify images returns the depth value informations about an object.

The other concept is called *feature* and must be able to find certain pre-defined edges (often called *keypoints*) in a picture that characterize it best, and further is able to find these keypoints in altered pictures of the same scene. The most commonly used algorithms for this feature detection are SIFT and SURF (Scale Invariant Feature Transform and Speeded Up Robust Features) (see [8][9]) but they were to slow for real-time application. A good alternative offers the fast and free feature detection algorithm ORB (Oriented FAST and Rotated BRIEF) [6]. This type of feature detection finds the searched keypoints by the operation with a circle detection which checks the image for edges and also calculates the keypoint orientations.

2.2 Hardware and software setup

In order to improve and simplify the created algorithm structure, we used the free library of programming functions for computer vision which is called *OpenCV*. The programming language for the development of the tracking algorithm was Python. The hardware setup which was used for first experiments consisted of either one or two uEye CMOS cameras with a resolution of 1280x1024 pixels and a frame rate of 60 Hz (at full frame). Further, a Pentax TV-Lens with a fixed focal length of 8.5 mm was used as objective. The interface of the cameras was an USB 3.0 standard and thus, easy to integrate with *OpenCV*.

2.3 Algorithmic structure

For our algorithm the initial data is a camera frame and the aim is to find a face in the image i.e. calculate the two-dimensional coordinates plus an average distance value of the face to the camera. In section 2.3.1 we describe two algorithms to find the face and in 2.3.2 we show two algorithms to get the z value.

2.3.1 Face detection

For simplicity reasons, we call the two different approaches for locating the face in the image *matchTemplate* and *Viola Jones Algorithm*.

The *matchTemplate* algorithm require a pre generated image which it compare to the camera frames. We use the *OpenCV* function *matchTemplate* [4] to find the best matches between the template image and the camera frames. A match is the position of a potential face as a single x,y coordinate. Each match is accompanied with a corresponding fit value on which we apply a threshold of 0.5, i.e. we only use those matches with a fit value above 0.5.

For the *Viola Jones Algorithm*, we load a training file with a generalized image of faces and use the *OpenCV* function *detectMultiScale* [4] to detect the face in a camera frame. As training data, we used the material from the University of Augsburg [10]. All further calculations were done with the *Viola Jones Algorithm*.

2.3.2 Measuring the distance between camera and Sensor

After finding the face in a camera image, we expand the tracking algorithm with two methods, *disparity map* and *feature detector*, to get a z-coordinate, i.e. the distance between the sensor and the face.

The *disparity map* method works with the disparity function described in section 2.1. We rectify the camera images first (using *stereoRectify* from *OpenCV*) and after that, we use the *OpenCV* function *StereoSGBM* [11] to generate a map of disparities. This function has many adjustment options described in [11]. We convert the map into a depth map using the calibration information. The average depth value is calculated over a small selected region in the front of the face. We implement a rectangle (Rect1) which is smaller one being inside of the face detection rectangle (Rect2). The location of Rect1 is the exact center of Rect2 with each edge being exactly 60% in length of the counterpart edge in the Rect2. All received depth values in the smaller part were sort and the median was calculated. This obtained value is our z value. A small selected region in front of the face with the calculated median as depth

value is shown in Fig. 4.

The *feature detection* algorithm uses the ORB feature detector [6] to match features in both cameras. The general principle is described in section 2.1. The image rectangle containing the face (which we know from the face detection algorithm) and limits the field in which we apply the feature detector, i.e. we only apply the feature detector to the face region. The feature detector is applied on both images and could find the same feature in the second camera frame. After that, both features in the same camera scene are matched by the standard method of nearest-neighbor-criterion. All matches have to be lie on the same epipolar line. To reduce false-positive matches, all matches with a distance from more than 10 pixels on the y axis are deleted. For calculation of the depth value, we sort all detected feature matching points and determine the first ten results. Next, we calculated the average of the first ten matching points and use the receive value for our distance between the camera and the face. A feature detection example is shown in Fig. 5.

3 Results and Discussion

Of the two face detection algorithms, the *Viola Jones Algorithm* yields better results for our application. The matchTemplate based algorithm is sensitive to ambient light, which has a disturbing influence on the detection robustness and produces a lot of false positive matches. The *Viola Jones Algorithm* is less sensitive to ambient light. It is stable and works with different backgrounds in the measurement scene. With the readily available face detection training data, the usage of the algorithm is simple and fast to apply.

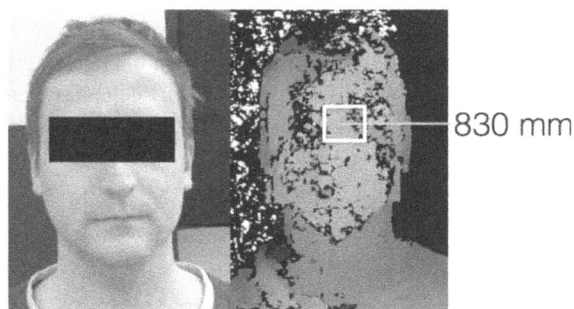

Figure 4: (left) A picture of male person. (right) The related *disparity map* with depth information (z).

The lateral resolution of the tracked depth information (z-value) was determined in the *disparity map* and in the feature detection algorithm to a measuring range of 100 mm. As a result of the disparity map algorithm, we received the depth information in form of z-values in a 2D *disparity map* of the measured scene. One example is shown in Fig. 4 where a face of a male person could be tracked. The face is shown in grey values (left) and the *disparity map* with depth information is presented with grey colored areas (right). Each grey colored area stand for a depth information. Objects which are far away have a darker gray value and correspond with a smaller disparity. As opposed to closer objects which have a high disparity and correspond with a brighter grey value. The calculated depth information in the selected small rectangle in front of the face fulfill the required 100 mm lateral resolution. But there are certain errors in the *disparity map* as well. Their false color depends on inaccurate calculated camera parameters or on the homogeneous background. Further development is necessary.

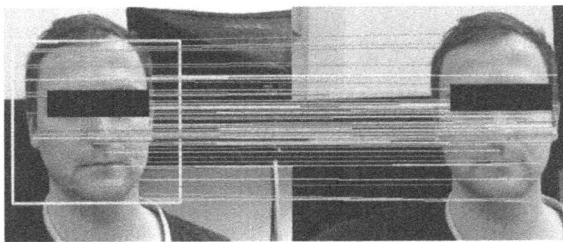

Figure 5: (left) Feature detected image of on camera - the face of the male person is framed. (right) Image of the second camera as reference image for feature matching.

The result of the feature detection algorithm is shown in Fig. 5. The left image shows a detected face in grey values, framed by a rectangle to reduce the tracking area for the feature detection. To get the depth information of a scene both cameras were needed for the feature matching. The lines between both images describe the connection between the matched features in form of keypoints in both images.

In comparison, both algorithms can be used to track the depth information (z) of faces in the scene. The advantage of the *disparity map* is the determination of the z value for each pixel in the disparity image. Furthermore, the *disparity map* allows to differentiate between background and face by comparing the disparity (or depth) values, whereas the feature detection algorithm does not provide this additional information. The advantage of the ORB detection is the faster calculation.

4   Conclusion

In conclusion, we described different tracking setups and methods for the future application with a high-resolution 3D sensor. We introduced that the *Viola Jones Algorithm* is able to calculated the x and y coordinates of a scene. Further, we presented two different approaches for the detection of the missing depth information; the disparity map and the feature detection. Both work with a stereo-vision setup of two cameras. The detection algorithms we tested are in general stable enough for the required testing conditions, e.g. environmental conditions and tolerance of 100 mm for the depth resolution (z-values). More investigations are necessary to compare the accuracy of both approaches. The aim of the next investigation is to improve the accuracy of the depth information values for both methods of disparity map and feature detection, as well as to reduce the background influences and realize a fast processing time without latency.

Acknowledgement

The work has been carried out at Fraunhofer Institut of Applied Optics and Precision Engineering IOF, Jena, supervised by the Institute of Institute of biomedical optic, Universität zu Lübeck. We acknowledge financial support from Federal Ministry of Education and Research in the project IRESTRA (Research Grant No. 16SV7208K).

5   References

[1] J. Bayerer, F.P. Frese, *Machine Vision, Methods of image Acquisition*. Springer, Berlin, 2016

[2] S. Zhang, *Handbook of 3D Machine Vision*. CRC Press, New York, 2013

[3] S. Heist, P. Lutzke, I. Schmidt, P. Dietrich, P. Kühmstedt and G.Ñotni, *GOBO projection-based high-speed three-dimensional shape measurement*. Proc, Jena, 2016

[4] 2017. *The OpenCV Tutorials*. Available: http://docs.opencv.org/2.4/opencv_tutorials.pdf - CallforPapers.aspx [last accessed on 2017-01-05]

[5] P. Viola and M. Jones. *Rapid Object Detection using a Boosted Cascade of Simple Feature*. COMPAQ, Cambridge, 2001.

[6] E. Rublee et. al. *ORB: an efficient alternative to SIFT or SURF*. Willow Garage, Menlo Park, 2011

[7] K. Takaya. *Real-time Stereo Disparity Map for Continuous Distance Sensing Applications - A Method of Sparse Correspondence*. University of Saskatchewan, Canada, 201, pp. 475-486

[8] D.G. Lowe. *Distinctive image features from scale-invariant keypoints*. International Journal of Computer Vision, 2009

[9] H. Bay, T. Tuytelaars and L. Van Gool. *Surft: Speeded up robust features*. European Conference on Computer Vision, 2006

[10] 2017. *opencv haarcascades*. Available: https://github.com/opencv/opencv/tree/master/data CallforPapers.aspx [last accessed on 2017-01-10]

[11] 2017. *StereoSGBM*. Available: http://docs.opencv.org/master/d2/d85/ classcv_1_1 StereoSGBM.html CallforPapers.aspx [last accessed on 2017-01-10]

Evaluation of Bluetooth Positioning for Medical Device Tracking

T. Kirchmann[1], M. Pelka[1] and H. Hellbrück[1,2]

[1] Fachbereich Elektrotechnik und Informatik, Fachhochschule Lübeck, thore.kirchmann@stud.fh-luebeck.de, mathias.pelka@fh-luebeck.de, horst.hellbrueck@fh-luebeck.de

[2] Institut für Telematik, Universität zu Lübeck, hellbrueck@itm.uni-luebeck.de

Abstract

Health care applications benefit from geo-information. One example is to keep track of medical devices, e.g. wheelchairs or ventilators. Medical devices are expensive and not easy to replace. In order to prevent devices from being stolen or misplaced, geo-information, e.g. coordinates, is beneficial. Geo-information is obtained with positioning systems. This paper investigates three Bluetooth low-cost positioning methods to determine the position of such devices. This paper discusses positioning methods, it presents the measurement setup and analyzes the positioning results. For the evaluation the positioning methods are implemented using Bluetooth Low Energy (BLE) modules. All methods are compared based on their accuracy, precision and empirical cumulative distribution function. In the experimental evaluation accuracies better than 6 m in 95% of all measurements are achieved. This shows that positioning based on Bluetooth devices is possible and are able to determine the correct room or corridor.

1 Introduction

Positioning systems consist of *tags*, which refers to the devices with unknown coordinates, as well as *reference points* or *anchors*, which denote devices with known coordinates. Positioning systems are not necessarily easy to deploy, neither they are cost-effective. This work aims to build positioning systems with a minimum amount of sensors based on cost-effective Bluetooth Low Energy (BLE) beacons.

The contributions of this work are a review of theoretical positioning methods along with an evaluation of three positioning methods based on BLE. It is shown that BLE is a feasible standard for positioning.

This paper is structured as follows: Section 2 gives an overview of existing indoor positioning technologies. Section 3 discusses lateration, proximity and fingerprinting as positioning methods. Section 4 provides implementation details and in Section 5 an evaluation is presented. Section 6 provides a summary and an outlook to future work.

2 Related Work

An overview of indoor positioning technologies is given by Mautz in [1]. Common positioning technologies are based on radio frequency systems, e.g. WiFi, which employs time-of-flight (TOF), direction or angle based positioning methods. To achieve precise positioning, special developed networks are used e.g. nanoNET based on Nanotron's nanoLoc receiver [2]. Other examples for positioning methods employ laser range finders or ultra sound [1]. This work focuses on cost-effective and easy to deploy positioning methods. Bluetooth is a common standard for wearable devices with low energy consumption, while the devices are cost-effective. Bluetooth as an indoor positioning system is discussed by Faragher and Harle [3].

Bluetooth is inapplicable to some approaches, e.g. TOF and direction based positioning methods. Therefore this work investigates lateration, as suggested by Rod et al. in [4], proximity, discussed by Brida et al. [5] and fingerprinting, discussed by Bilke [6].

3 Overview of Positioning Methods

In this section an overview on different positioning methods, applicable with Bluetooth devices, is provided.

The Friis transmission equation describes how the received signal strength (RSS) decreases depending on the distance d to the antenna [7]:

$$\frac{P_r}{P_t} = A_r A_t \frac{\lambda^2}{d^2} \tag{1}$$

where P denotes the power and A the antenna parameters respectively for the transmitter (subscript t) or receiver (subscript r). λ denotes the wavelength of the radio signal, which is inverse proportional with increasing frequency. For Bluetooth, operating in the 2.4 GHz ISM band, this equals $\lambda \approx 12,5$ cm. The Friis transmission equation is accurate when $d > 2\lambda$.

Lateration

Determining the position based on distance estimation is called lateration [8]. The distance d to the unknown coordinate $\mathbf{r}$, shown in Figure 1, is

$$d = |\mathbf{r} - \mathbf{r}_i| = \sqrt{(x - x_i)^2 + (y - y_i)^2 + (z - z_i)^2}. \quad (2)$$

Eq. (2) is a nonlinear problem with n measurements and n equations. The unknown position is denoted by $\mathbf{r} = [x, y, z]^T$. Respectively the coordinates of the ith reference points are given as $\mathbf{r}_i = [x_i, y_i, z_i]^T$. The distance d between the unknown tag position and a reference point is estimated by Eq. (1). Eq. (2) is solved with the Gauss-Newton algorithm.

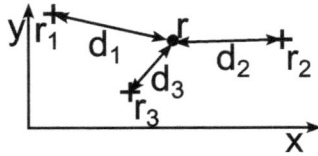

Figure 1: Lateration.

Proximity

To determine the position $\mathbf{r}$ based on proximity, a number of reference points $\mathbf{r}_i$ is required. The device is assigned to a position $\mathbf{r}_i$, if a connection between reference point and tag is established. Since the Bluetooth standard is able to transmit data with ranges up to 100 m, it is necessary to limit the range to decrease the positioning error. Therefore a threshold RSS_{min} is chosen. As soon as the measured RSS exceeds the threshold, the tag is assumed to be in *in range* or in *proximity*, this is expressed by $D_{\mathbf{r}_i}$:

$$D_{\mathbf{r}_i} = \begin{cases} 1, & \text{if } RSS_r \geqslant RSS_{min} \\ 0, & \text{if } RSS_r < RSS_{min} \end{cases} \quad (3)$$

The position of the tag is estimated as the average of all $\mathbf{r}_i$ in range:

$$\mathbf{r} = [x, y, z] = \left[\frac{1}{n} \sum_{i=1}^{n} x_i, \frac{1}{n} \sum_{i=1}^{n} y_i, \frac{1}{n} \sum_{i=1}^{n} z_i \right] \quad (4)$$

Eq. (4) is called centered proximity. Other solutions exists as well, for instance *weighted proximity* [5].

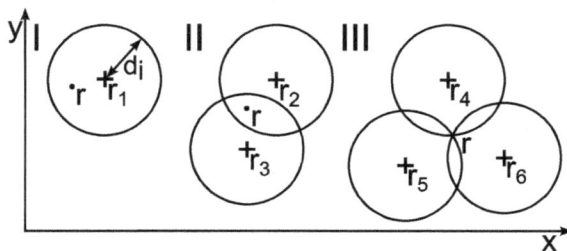

Figure 2: Proximity-based positioning approach.

Figure 2 shows positioning based on proximity for three cases. In the first case the tag in proximity to one reference point. Therefore $\mathbf{r}$ is assumed to be equal to the corresponding coordinate of the reference point $\mathbf{r}_i$. In the second and

third case the tag is in range of more than one reference point. For the second case the Eq. (4) results in one position. The third case states the general case with at least three measurements.

Fingerprinting

In Figure 3 a grid of fingerprints $\mathbf{f}_{i,j}$ is visualized. During the *offline phase* a database with fingerprint is created The fingerprints are based on RSS measurements over all reference points at a given position. The fingerprint $\mathbf{f}_{i,j}$ is in our case the mean of a number of RSS measurements and stored in the fingerprint database. In the database the fingerprints are stored and are represented by the fingerprint database $\mathbf{H}$ in Eq. (5). The fingerprint database has size k times l and is denoted as:

$$\mathbf{H} = \begin{bmatrix} f_{1,1} & f_{1,2} & \cdots & f_{1,l} \\ f_{2,1} & f_{2,2} & \cdots & f_{2,l} \\ \vdots & \vdots & \ddots & \vdots \\ f_{k,1} & f_{k,2} & \cdots & f_{k,l} \end{bmatrix} \quad (5)$$

$\mathbf{H}$ consists of the elements

$$\mathbf{f}_{i,j} = [\overline{RSS_1}, \overline{RSS_2}, \ldots, \overline{RSS_n}], \quad (6)$$

where $\overline{RSS}_m$ is the averaged RSSI measurement at one position for a reference point, obtained during the *offline phase*.

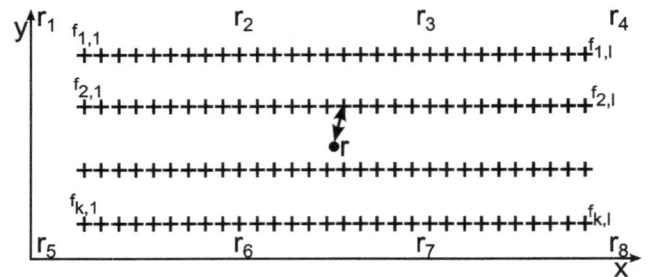

Figure 3: Fingerprinting based positioning approach, where $\mathbf{f}_{i,j}$ denotes a fingerprint.

During the *online phase* live measured RSS values are compared with $\mathbf{H}$ to determine the position. The tag measurement vector is given as

$$\mathbf{y} = [RSS_1, RSS_2, \ldots, RSS_n]. \quad (7)$$

The vector $\mathbf{y}$ is compared with each element of $\mathbf{H}$. Bilke et al. [6] suggests the *Euclidean distance* and the *cosine similarity* to compare $\mathbf{y}$ with the database.

Euclidean distance

The Euclidean distance is denoted as

$$d(\mathbf{y}, \mathbf{f}_{i,j}) = \|\mathbf{y} - \mathbf{f}_{i,j}\| = \sqrt{\sum_{m=1}^{n} (y_m - f_{(i,j)m})^2}. \quad (8)$$

$d(\mathbf{y}, \mathbf{f}_{i,j})$ is the distance between the measured vector $\mathbf{y}$ and the fingerprint $\mathbf{f}_{i,j}$. Eq. (8) is calculated overall $\mathbf{f}_{i,j}$ to determine the minimum $d(\mathbf{y}, \mathbf{f}_{i,j})$, which is chosen as best actual position.

Cosine similarity

The cosine similarity compares the angle between two vectors

$$\underset{\forall i,j}{\alpha(\mathbf{y}, \mathbf{f}_{i,j})} = \frac{\mathbf{y} \cdot \mathbf{f}_{i,j}}{\|\mathbf{y}\| \|\mathbf{f}_{i,j}\|}. \tag{9}$$

With $-1 \leqslant \alpha(\mathbf{y}, \mathbf{f}_{i,j}) \leqslant 1$. If $\alpha(\mathbf{y}, \mathbf{f}_{i,j}) = 1$ the compared vectors are pointing in the same direction, $\alpha(\mathbf{y}, \mathbf{f}_{i,j}) = -1$ results for vectors pointing into the exact opposite direction. The position is determined by maximizing Eq. (9), therefore the entry with the best similarity.

4 Implementation

This section presents the measurement hardware composed of a smartphone and BMD-300 Bluetooth Low Energy modules, as well as the Android based measurement software. The BMD-300 modules are based on the nRF52832 from Nordic Semiconductor. These are mounted on an evaluation kit, built by the company Rigado [9]. A HTC One mini Android smartphone records the RSS data from the Bluetooth beacons.

The software is developed for the HTC One mini smartphone and is able to connect with up to six Bluetooth devices. It measures the RSS value for the n reference points every 75 milliseconds. The data is evaluated with MATLAB.

5 Evaluation

In the following, lateration, proximity and fingerprinting are evaluated based on accuracy and precision. Additionally, the empirically cumulative distribution function of the measurement error is evaluated. The tag was moved along the ground truth.

The *positioning error* ε is defined as the Euclidean distance between ground truth $\mathbf{r}$ and estimated position $\hat{\mathbf{r}}$:

$$\varepsilon = |\mathbf{r} - \hat{\mathbf{r}}| \tag{10}$$

The *accuarcy* of the positioning error is the mean of the positioning error

$$\mu = E\left[\varepsilon\right] = E\left[\|\mathbf{r} - \hat{\mathbf{r}}\|\right]. \tag{11}$$

The *precision* is given as the standard deviation of the positioning error

$$\sigma = \sqrt{E\left[\varepsilon^2\right] - E\left[\varepsilon\right]^2}. \tag{12}$$

Additionally the *cumulative distribution function (CDF)* is denoted as $F(\varepsilon) = P(\varepsilon \leqslant x)$ with $0 \leqslant x \leqslant 1$. With the CDF the positioning error is determined in percentiles, e.g. the 95 percentile.

Measurement setup

For the measurement setup are four reference points employed, mounted at an altitude of 2.4 m. The measurement area has a size of 13 times 4 meters and is divided into a grid with a spacing of 1 m, therefore $k = 13$ and $l = 4$. The fingerprints consists of $n = 100$ RSS measurements for each position in the grid. The coordinates (in [m]) of the reference points are $\mathbf{r}_1 = [0, 0]^T$, $\mathbf{r}_2 = [3.04, 0]^T$, $\mathbf{r}_3 = [0, 13]^T$ and $\mathbf{r}_4 = [3.95, 13.5]^T$. The BMD-300 Bluetooth module has a maximum power consumption of 0.072 W [9]. This results in an energy consumption of 0.631 kWh per module per year.

Lateration

The RSS is measured and the distance based on Friis transmission equation Eq. (1) is calculated. Then the position is estimated with Eq. (2), and solved by the Gauss-Newton algorithm [8].

In Figure 4a the result of the Gauss-Newton algorithm is shown. Dots mark the reference points, the solid line represents the ground truth, whereas the crosses indicate positions estimated by the Gauss-Newton algorithm. The position estimation is shifted towards negative x-coordinates, this indicates a systematic measurement error. This measurement error is caused by the RSS measurements, which are prone to changes in the measurement angle, geometry and antenna characteristics, as well as interferences along with other radio technologies. The empirical CDF $F(\varepsilon)$ for the positioning error ε, is shown in Figure 4b.

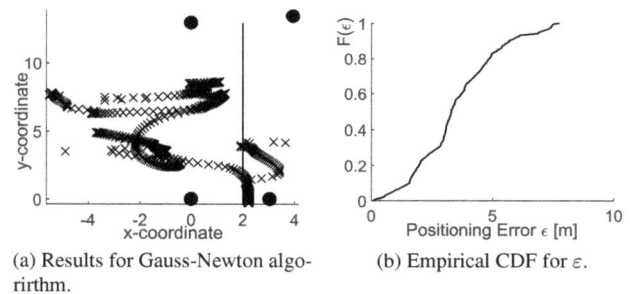

(a) Results for Gauss-Newton algorirthm.   (b) Empirical CDF for ε.

Figure 4: Results for Lateration.

In the experimental evaluation an accuracy of 3.5 m and a precision of 1.66 m, is obtained. Based on Figure 4 positioning errors ε less than 6.9 m, in 95% of all measurements, are determined.

Proximity

The proximity is measured with Eq. (3). The RSS_{min} values are chosen for each reference point to correspond to a distance of 7 m. This divides the measurement area into three parts. Due to Eq. (4), proximity assume the tag position to be in the center of these areas. The centers of the areas are at the coordinates $[1, 3.5]^T$, $[1, 7]^T$ and $[1, 10.5]^T$.

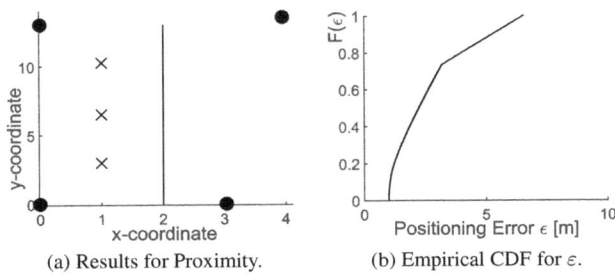

(a) Results for Proximity.   (b) Empirical CDF for ε.

Figure 5: Results for Proximity.

In Figure 5a the positioning results are shown. The dots represent the reference positions, the line shows the ground truth and the crosses indicate the estimated position. Figure 5b shows the empirical CDF of the positioning error ε. An accuracy of 2.7 m, a precision of 1.7 m, are obtained and the 95 % percentile is smaller than 5.9 m.

Fingerprinting

To estimate the position based on fingerprinting $\mathbf{H}$ is compared with the live measurement. Therefore either Euclidean distance Eq. (8) or cosine similarity Eq. (9) is employed.

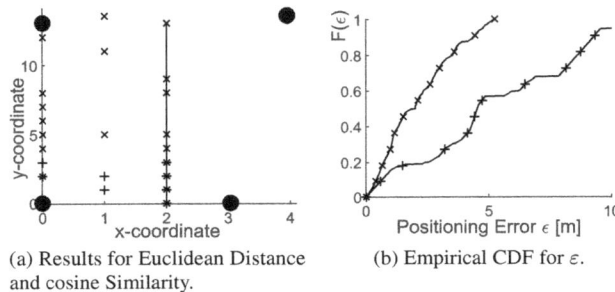

(a) Results for Euclidean Distance and cosine Similarity.   (b) Empirical CDF for ε.

Figure 6: Results for Fingerprinting.

The results for fingerprinting are given in Figure 6. Plus markers belong to the results of Euclidean distance, cross markers to cosine similarity. The solid line represents the ground truth and the dots the reference points. Euclidean distance results in an accuracy of 5.3 m, a precision of 3.2 m and a 95 % error over all measurements of 11 m. The cosine similarity leads to an accuracy of 2.1 m, a precision of 1.4 m and a 95 percentile over all estimates of 4.8 m.

6   Conclusion and Future Work

Table 1 summarizes the evaluation results. Lateration and proximity achieves similar results in terms of accuracy. It shows that cosine similarity performs better than Euclidean distance. The Euclidean distance achieves poor results and cosine similarity achieves the best positioning results in this experiment. Due to the given evaluation, the best results are accomplished by proximity and fingerprinting. A positioning error less than 6 m in 95% of all measurements is

achievable, which allows tracking of medical devices. Position estimation with low-cost Bluetooth devices is feasible and provide an easy to deploy solution. Given the achieved positioning accuracy, it was shown that with position estimation based on proximity and fingerprinting corridors and rooms are distinguished.

Approach	Accuracy	Precision	95% Percentile
Lateration	3.5	1.7	6.9
Proximity	2.7	1.7	5.9
Euc. Dis.	5.3	3.2	11.0
Cos. Sim.	2.1	1.4	4.8

Table 1: Evaluation results in [m].

In future work fingerprinting and proximity positioning methods are investigated in more detail.

Acknowledgments

This publication is a result of the research work of the Center of Excellence CoSA in the RosiE project which is funded by German Federal Ministry for Economic Affairs and Energy (BMWi) ZF4186102ED6.

7   References

[1] R. Mautz, "Overview of Current Indoor Positioning Systems," *Geodesy and Cartography*, 2009.

[2] Nanotron, "nanoNET Chip Based Wireless Networks," *White Paper*, 2007.

[3] R. Faragher and R. Harle, "An Analysis of the Accuracy of Bluetooth Low Energy for Indoor Positioning Applications," *Proceedings of the 27th International Technical Meeting of The Satellite Division of the Institute of Navigation*, 2014.

[4] M. Rodriguez, J. P. Pece, and C. J. Escudero, "In-building Location using Bluetooth," *Workshop on Wireless Ad-hoc Networks*, 2005.

[5] N. Patwari and A. O. Hero III, "Using Proximity and Quantized RSS for Sensor Localization in Wireless Networks," 2003.

[6] A. Bilke, "Ortung und Navigation mit Mobilen Geräten in Gebäuden," M.Sc. thesis, HTW Berlin, 2012.

[7] H. T. Friis, "A Note on a Simple Transmission Formula," *Proceedings of the IRE*, 1946.

[8] M. Pelka, G. Goronzy, and H. Hellbrück, "Impact of Altitude Difference for Local Positioning Systems and Compensation with Two-Stage Estimators," in *Localization and GNSS*. IEEE, 2016.

[9] *Data sheet: BMD-300 Series Module for Bluetooth 4.2 LE*, Rigado LLC, 11 2016, v1.7.

7

Signal Processing

Compressing Vector Fields by Using K-SVD

T. Jahner[1], T. Parbs[2], and A. Mertins[2]

[1] Medizinische Informatik, Universität zu Lübeck, jahner@student.uni-luebeck.de
[2] Institute for Signal Processing, Universität zu Lübeck, {parbs, mertins}@isip.uni-luebeck.de

Abstract

The compression of vector fields is in general interesting for medical uses. It is advantageous for all big data set applications like in functional magnetic resonance imaging. The K-SVD algorithm is one of the most widely used algorithms for dictionary learning and sparse representation. Strangely however, you can't find a lot of research with it involving vector fields. In this paper, we discuss the compression of constructed 64x64 pixel vector fields using sparse representation via K-SVD. We constructed a training data set of 100 vector fields and used the K-SVD with different input parameters on it. Then we tested the quality of the dictionary by reconstructing a test data set of 100 vector fields. We were able to show that it is possible to compress vector fields using K-SVD with a high reconstruction precision relative to the compression rate.

1 Introduction

The K-SVD algorithm is widely used for dictionary learning and sparse representation. It was subject of multiple scientific papers in the last years and is still used in research. It is an approach for the synthesis based sparse representation model [1], which sees signals $\mathbf{f}$ not as a vector, but rather as a linear combination of different bases. A coefficient vector $\mathbf{x}$ is here multiplied with an dictionary $\mathbf{D}$ so that the signal $\mathbf{f}$ can be represented as: $\mathbf{f} = \mathbf{Dx}$. Because we want to compress $\mathbf{f}$ by only transmitting $\mathbf{x}$, which should be sparse. The method is based on the assumption that a signal can be represented with only a small number of columns of $\mathbf{D}$. An interesting field which uses this method is compressed sensing. The idea of compressed sensing is that by exploiting signal redundancies and structures, fewer measurements are necessary than proposed by the Nyquist-Shannon sampling theorem. A lot of signals however are not 2D, but 3D in form of vector fields. Examples for this are ultrasonic imaging [2] and motion fields [3].

However the usage of K-SVD for vector fields is not well researched even when the importance and usability of sparse representation in compressing vector fields is seen [3] [2]. Synthesis based sparse representation is successfully used in image compression (for example JPEG2000 [4]) and are among the most effective algorithms for compression.

Another approach in the field of compressing vector fields uses the wavelet transformation. For example, the usage of multiwavelets has been shown to be very effective in compression [5]. However, this approach is as well a form of dictionary based compression. The advantage of predefined functions as dictionaries is its simplicity in comparison to learned dictionaries [4]. The method of predefined functions has the disadvantage that it needs to be suitable to represent the signal. Learned dictionaries on the other hand are always learned for the specific signal and have therefore sometimes better results.

On the following pages we will describe the K-SVD algorithm, its components and principle. We will demonstrate how we obtained our training and test data and the algorithm used to evaluate our result. Subsequently we will demonstrate our findings, discuss these and make a conclusion on using K-SVD for the compression of vector fields.

2 Material and Methods

We used the K-SVD algorithm, an algorithm to create the training and test data and an algorithm to test the difference between the reconstructed and the original image. All these algorithms are explained in this section.

2.1 Dictionaries

First we want to define and explain what a dictionary is. Mathematically, we can define a dictionary $\mathbf{D} \in \mathbb{R}^{N,K}$, $K > N$ as a matrix that satisfies

$$\mathbf{f} = \mathbf{Dx}$$
$$\text{or} \qquad\qquad (1)$$
$$\mathbf{f} \approx \mathbf{Dx}$$

where $\mathbf{f} \in \mathbb{R}^N$ is a signal and $\mathbf{x} \in \mathbb{R}^K$ is a coefficient vector. Note that $\mathbf{D}$ is overcomplete due to $K > N$. However it is obvious, that the problem in (1) has an infinite number of solutions due to the overcompleteness of $\mathbf{D}$. To obtain a good result for compression from this solutions we need to constrain $\mathbf{x}$ to obtain as many zero elements as possible. A common constrain to enforce sparsity is minimizing the l^0-norm $\| \cdot \|_0$, which counts the non-zero elements and search a solution for:

$$\min_{\mathbf{x}} \| \mathbf{x} \|_0 \text{ subject to } \mathbf{f} = \mathbf{Dx}. \qquad (2)$$

This way we will get the best coefficient vector $\mathbf{x}$ for compression, because it will only be very sparsely populated. There are different ways to obtain such a dictionary. The first way is to specify a dictionary by using a set of functions. This is an approach used by multiwavelets. The other approach is to learn the dictionary from a training data set, which is what the K-SVD does.

2.2 K-SVD

We used the K-SVD algorithm proposed in [4] to produce an overcomplete dictionary, which then is used for sparse coding. The idea behind the K-SVD algorithm is the combination of the k-means algorithm with the singular value decomposition (SVD). Mathematically the K-SVD algorithm can be defined as a minimization. Suppose that $\mathbf{f}_i$ is a part of a signal set $\mathbb{F} = \{\mathbf{f}_i \in \mathbb{R}^M | i = 1, \ldots N\}$, than we can compose a data set matrix $\mathbf{F} = [\mathbf{f}_1, \mathbf{f}_2, \ldots, \mathbf{f}_N]^T$. Given that we do the same with the coefficient vector $\mathbf{x}_i$, we get $\mathbb{X} = \{\mathbf{x}_i \in \mathbb{R}^K | i = 1, \ldots, M\}$ and $\mathbf{X} = [\mathbf{x}_1, \mathbf{x}_2, \ldots, \mathbf{x}_M]$. As a minimization we want to solve:

$$\min_{\mathbf{D}, \mathbf{X}} \| \mathbf{F} - \mathbf{DX} \|_F^2 \text{ subject to } \forall i: \| \mathbf{x}_i \|_0 \leq T_0 \qquad (3)$$

where $\mathbf{D} \in \mathbb{R}^{N,K}$, $K \in \mathbb{N}$ and $K > M$ is an overcomplete dictionary, $\mathbf{x} \in \mathbb{R}^K$ is a vector of coefficients, T_0 is the threshold of non-zero elements and $\| \cdot \|_F$ is the Frobenius norm. However because we want to use it to compute vector fields in complex representation, we have to define the variables in the complex domain, therefore $\mathbf{f} \in \mathbb{C}^n$, $\mathbf{D} \in \mathbb{C}^{N,K}$ and $\mathbf{x} \in \mathbb{C}^K$.

The K-SVD has two distinct stages. First it starts with the sparse coding stage utilizing a pursuit algorithm. In [4] the authors compared different pursuit algorithms, namely matching pursuit, orthogonal matching pursuit and basis pursuit. While we used the orthogonal matching pursuit in our implementation, all matching pursuit algorithms are viable. The pursuit algorithm is necessary to sparsify the coefficient vector $\mathbf{x}$ and is vital.

The second stage is the codebook update stage. In this stage the dictionaries are updated based on the SVD of an error matrix. The error matrix $\mathbf{E}_k$, $k = 1, \ldots, K$, is computed by

$$\mathbf{E}_k = \mathbf{F} - \sum_{i \neq k} \mathbf{d}_i \mathbf{x}_i^T \qquad (4)$$

where $\mathbf{d}_i$ is column of $\mathbf{D}$ so $\mathbb{D} = \{\mathbf{d}_i \in \mathbb{C}^N | i = 1, \ldots, K\}$ and $\mathbf{D} = [\mathbf{d}_1, \mathbf{d}_2, \ldots, \mathbf{d}_K]$. This basically means that the error matrix is computed by the signal minus the sum of all reconstructed elements, except the corresponding column of k. Then we restrict the columns of $\mathbf{E}_k$ to only the columns which correspond to the non zero elements of $\mathbf{x}_i$ in $\mathbf{E}_k$, which is then $\mathbf{E}_k^R$. Now the SVD is used to retrieve

$\mathbf{U}$ and $\mathbf{V}$:

$$\mathbf{E}_k^R = \mathbf{U} \triangle \mathbf{V}^H, \qquad (5)$$

where $\mathbf{U}$ is a $N \times N$ unitary matrix, $\triangle$ is a diagonal $N \times M$ non-negative matrix, $\mathbf{V}$ is a unitary $M \times M$ matrix and $\mathbf{A}^H$ is the hermitian of $\mathbf{A}$. With $\mathbf{U}$ and $\mathbf{V}$ we can now update the dictionary column $\mathbf{d}_k$ by the first column of $\mathbf{U}$ and update the coefficients $\mathbf{x}_i^R$, which are the non-zero entities of $\mathbf{x}_i$, by the first column of $\mathbf{V}$. In our implementation, the sparse coding stage and the codebook update stage are repeated for a fixed number of iterations.

2.3 Orthogonal Matching Pursuit

The orthogonal matching pursuit (OMP) is one of the fastest and an often used matching pursuit algorithm [6]. It is stated as followed:

Data: Signal $\mathbf{f}$, dictionary $\mathbf{D}$
Result: coefficient vector $\mathbf{x}$
Initialization;

- the residual $\mathbf{r}_0 = \mathbf{f}$

- a counter $n = 1$

- a subset of selected column indices's $C = \{\}$

- a stop condition $n_{\max}$

Procedure;

1. Find the index λ_n, that solves the optimization

$$\lambda_n = \arg \max_{j=1\ldots K} |\langle \mathbf{r}_{n-1}, \mathbf{d}_i \rangle| \qquad (6)$$

2. Add λ_n to C and the chosen atoms $\mathbf{Z}_n = [\mathbf{Z}_{n-1} \mathbf{d}_{\lambda_n}]$. Where $\mathbf{Z}_0 = \{\}$

3. solve

$$\mathbf{x}_n = \arg \min_{\mathbf{x}} \| \mathbf{Dx} - \mathbf{f} \|_2, \qquad (7)$$

where $\| \cdot \|_2$ is the l^2-norm.

4. Calculate the new approximation and residual:

$$\begin{aligned} \mathbf{a}_n &= \mathbf{Dx}_t \\ \mathbf{r}_n &= \mathbf{f} - \mathbf{a}_t \end{aligned} \qquad (8)$$

5. $n = n + 1$ repeat from 1. if $n < n_{\max}$.

The stop condition is in our implementation the number of columns in the signal, as proposed in the implementation shown in [1]. It is a form of greedy algorithm and is one of the fastest available algorithms for this task. We use this algorithm in the sparse coding stage of the K-SVD and to obtain the coefficients for the reconstruction of the training data.

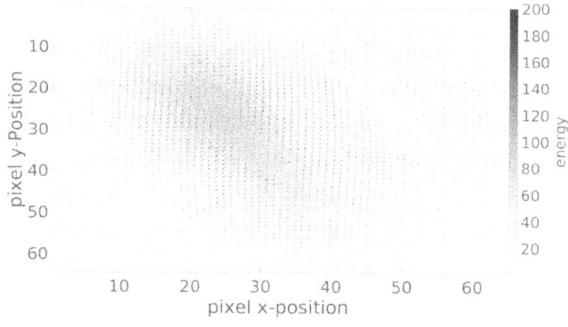

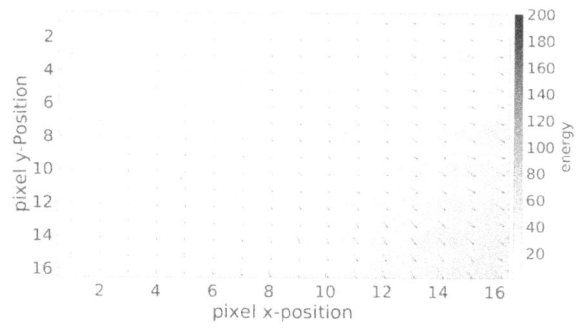

Figure 1: Example of our constructed training and test data. The black arrows are realizations of the vector field, however they are hard to see. In the background you can see the underlying field used to generate the vector fields.

Figure 2: This is a segment of Fig. 1 for a better understanding of the general look of the dataset, note that this is only for illustration purposes.

2.4 Training and Test Data

We use the K-SVD on a constructed vector field. This vector field is computed by spatial derivative of a field containing several gaussian modes, whose parameters change according to an autoregressive-moving average models to create smooth elastic motion fields. Due to time constraints, we used a training vector of a hundred images of size 64×64, which we partitioned into patches of different sizes. See Fig. 1 for an example image and corresponding vector field of our constructed training and test data. Fig. 2 is a segment of Fig. 1 and should demonstrate better how the vector fields look like, because they are hard to see in Fig. 1 due to the size of the image and vectors. Fig. 1 is composed of two parts. First there is the background image, which is the corresponding field used to generate the vector field shown as arrows.

2.5 Sum of Euclidean Distances

As a quantitative measurement of similarity between original and reconstructed image we use the sum of euclidean distances, which we use on normalized data for better comparability. Further, because we have used the complex annotation for the vectors, these have to be split now. Therefore we define two functions $\mathrm{Re}(\mathbf{x})$, which returns the real part of $\mathbf{x}$ and $\mathrm{Im}(\mathbf{x})$, which returns the imaginary part of $\mathbf{x}$. Given a signal set $\mathbb{F}$ from our test data and a reconstructed signal $\hat{\mathbf{f}}$, where $\mathbb{F} = \left\{ \hat{\mathbf{f}}_i \in \mathbb{R}^M | i = 1, \dots N \right\}$ is the signal set and $\hat{\mathbf{F}} = [\hat{\mathbf{f}}_1, \hat{\mathbf{f}}_2, \dots, \hat{\mathbf{f}}_N]^T$ is the signal set matrix, the sum of euclidean distances is here defined as

$$
\begin{aligned}
\mathbf{g} &= \frac{\mathbf{f}}{\| \mathbf{f} \|} \\
\hat{\mathbf{g}} &= \frac{\hat{\mathbf{f}}}{\| \mathbf{f} \|} \\
p_i &= \mathrm{Re}(g_i), q_i = \mathrm{Im}(g_i) \\
s_i &= \mathrm{Re}(\hat{g}_i), t_i = \mathrm{Im}(\hat{g}_i) \\
e &= \sum_{i=0}^{M} \sqrt{(p_i - s_i)^2 + (q_i - t_i)^2}
\end{aligned} \tag{9}
$$

The sum of euclidean Distances is a well known and understood principle of distance and error measurement [7].

3 Results and Discussion

We tested the K-SVD algorithm with three thousand parameter combinations on a constructed testing data set of a hundred images. The parameters for the K-SVD are the number of dictionary elements K, the number of linear combinations per patch L, the number of iterations I and the patch size P. We used different value ranges for each as follows:

$$
\begin{aligned}
K &= 20, 40, \dots, 180, 200 \\
L &= 4, 8, \dots 36, 40 \\
I &= 100, 200, \dots, 900, 1000 \\
P &= 2^3, \dots, 2^5
\end{aligned} \tag{10}
$$

In the following analysis, we only show the maximum number of iterations $I = 1000$ because they are the most relevant.

In Fig. 3 you can see the sum of l_2-norms for our test data

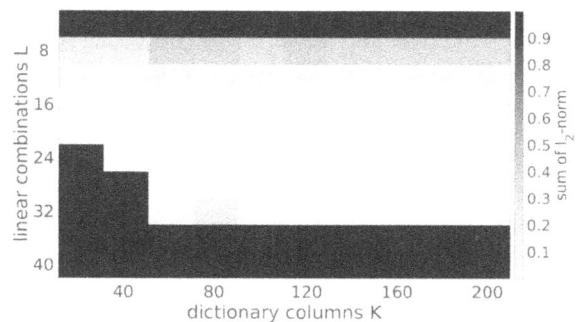

Figure 3: Graphic representation of the evaluation of K-SVD on vector fields with 8x8 patches.

and a reconstructed image using a learned dictionary on test data on a patch size of 8x8 and in Fig. 3 you can see the same on a patch size of 16x16. The minimum in Fig. 3 is at

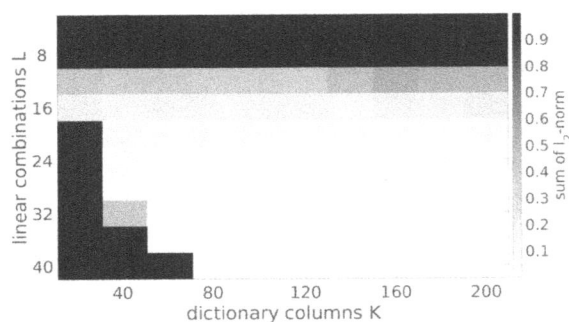

Figure 4: Graphic representation of the evaluation of K-SVD on vector fields with 16x16 patches.

$K = 180$, $L = 32$ with a value of $E = 2.8617 \cdot 10^{-4}$. The black parts of Fig. 3 and Fig.4 are values that are bigger than 1. These values come to be on the one hand from a to small number of linear combinations in the upper part and on the other hand due to overfitting for a high number of linear combinations. The minimum in Fig. 4 is at $K = 200$ and $L = 40$ with a value of $E = 0.0015$.

These results appear to be pretty good at first glance. It seems that the dictionaries perform really well, while a smaller patch size is to be preferred. However, under the viewpoint of space reduction the result is not as good, because in the case of 8x8 patches we only save 50% of the space complexity in comparison to the original signal, while the bigger patch size saves nearly 85%. Secondly it can be seen, that the error gets worse fast with a lower number of linear combinations and therefore a reduced space complexity. On the other hand it is really interesting for compressed sensing, because a relatively small number of bases can be used, while still maintaining a relatively high precision, which is a demand of compressed sensing.

4 Conclusion

In this paper we discussed the usage of K-SVD to compress vector fields We tested how the K-SVD would work on vector fields and pinpointed interesting and uninteresting parameter combinations for K-SVD. We could show that the K-SVD is an option for compression of vector fields. We could further reveal that it may be interesting to look more into K-SVD and machine learning for compressed sensing of vector fields. Future work is required to show that these findings can be reproduced on functional magnetic resonance images and to show if these findings can be reproduced for compressed sensing.

Acknowledgement

The work has been carried out at and was supervised by the Institute for Signal Processing, Universität zu Lübeck, Lübeck.
Special thanks to Marco Maaß for his help with technical issues.

5 References

[1] R. Rubinstein, T. Peleg and M. Elad, *Analysis K-SVD: A Dictionary-Learning Algorithm for the Analysis Sparse Model.* IEEE Transactions on Signal Processing, vol. 61, no. 3, pp. 661-677, 2013.

[2] J. S. Turek, M. Elad, and I. Yavneh, *Clutter Mitigation in Echocardiography Using Sparse Signal Separation.* International Journal of Biomedical Imaging, vol. 2015, Article ID 958963, 18 pages, 2015.

[3] G. Ottaviano, P. Kohli, *Compressible Motion Fields.* IEEE Conference on Computer Vision and Pattern Recognition, pp. 2251-2258, Portland, OR, USA, 2013.

[4] M. Aharon, M. Elad and A. Buckstein, *K-SVD: An Algorithm for Designing Overcomplete Dictionaries for Sparse Representation.* IEEE Transactions on Signal Processing, vol. 54, no.11, pp. 4311–4322, 2006.

[5] J. E. Fowler and L. Hua, *Wavelet Transforms for Vector Fields Using Balanced Multiwavelets.* IEEE Transactions on Signal Processing, vol. 50, no.12, pp. 3018–3027, 2002.

[6] J. A. Tropp and A. C. Gilbert, *Signal Recovery From Random Measurements Via Orthogonal Matching Pursuit.* IEEE Transactions on Information Theory, vol. 53, no. 12, pp. 4655-4666, 2007.

[7] J. C. Gower, *Euclidean Distance Geometry.* Math. Scientist, no. 7 pp. 1–14, 1982

A Comparison of RIDE and ICA for the Decomposition of ERPs in EEG-data

C. Baumgart[1]
[1] Medizinische Informatik, Universität zu Lübeck, carolin.baumgart@student.uni-luebeck.de

Abstract

Residue iteration decomposition (RIDE) is a method used in cognitive neuroscience for analyzing event-related potentials (ERPs). It was developed for decomposing ERPs into several component prototypes, which represent different processes taking place in the brain after task initiation. In early studies other methods, especially the independent component analysis (ICA), were used for that purpose. RIDE is a highly specialized approach and thus it is very difficult to compare its outcome with general approaches like ICA. The developer of RIDE showed that ICA is outperformed in specific simulated scenarios, but so far there is no ground truth available for ERPs in real EEG data. In this work, I will motivate how the original ERPs can be reconstructed to compare both approaches and to check if a combination of both methods would improve the results.

1 Introduction

In contrast to spontaneous EEGs, event-related potentials (ERPs) provide a temporal high-resolution insight in mental processes [8]. They occur from stimuli and are thereby very important for research in neuroscience [4]. Through decomposition the ERPs can be separated into different components which are partly crucial for psychologists like the P1, N1 and P2, which are influenced in case of attention and the N2 and P3, which are influenced by the purpose of trials [8]. The names of the components are descriptive. N-components have negative amplitudes and P-components represent positive potentials. Moreover, the names number the components corresponding to their chronological order. The most common methods for decomposition are the principal component analysis (PCA) and the independent component analysis (ICA), which work on the basis of frequency distribution of the amplitudes [8]. These methods are established since even superimposed components can be separated. However, it is pretty difficult separating the requested components and therefore only used in single cases by now. Moreover, some components have trial-based latencies, which blurs the waveform through averaging over all trials. This prior knowledge could not be integrated in ICA. These problems were solved by the so called RIDE algorithm developed by Ouyang et al. [4]. In this work the RIDE algorithm and the ICA will be explained. Furthermore, both methods will be compared with respect to the previous work from Ouyang et al. [6] and it will be shown if these algorithms can be combined to gain even better results.

2 Material and Methods

For the analysis, real EEG data sets kindly provided by Dr. Verleger were used and preprocessed with the Brain-Vision Analyzer 2. For the further analysis, MATLAB and the MATLAB-toolboxes EEGLAB [1], FastICA [2] and the RIDE toolbox [5] were used. For the RIDE method the unchanged data sets exported from BrainVision Analyzer 2 were used. They consist of an array with data points for the time courses of every single-trial for each electrode used for measurement. For the ICA these time courses were concatenated to avoid biased results caused by averaging.

2.1 RIDE

The residue iteration decomposition (RIDE) is an iterative method to estimate ERP component prototypes in data sets with different variabilities and temporal overlap of these components [4]. Besides that, latencies of the components are estimated for single-trials and then used for finding functional differences at specific test conditions [4]. To reach that goal the ERPs with the same conditions are typically averaged. But caused by different latencies and amplitudes the result is very blurred and less informative, which can be seen in Fig. 1(a). This can be avoided by using the RIDE method since it separates the ERPs into latency-invariant component prototypes through latency estimation [5]. Usually, RIDE is used to calculate three different prototypes S, C, and R, which represent different mental processes. The S prototype corresponds to stimulus-locked components of the ERP containing neuronal processes like visual perception, R corresponds to response-locked components, which contain motor-related processes and C corresponds to the intermediate latency-variable components containing other cognitive processes [6]. The computation of this prototypes needs their latency information. For S the stimulus-onset is used as latency (L_S), the latencies for R (L_R) are the reaction times (RTs) from single-trials but C has unknown latencies (L_C). RIDE estimates L_C and the prototypes using the following steps:

1. **Initial latency estimation:** Prototype waveforms are initially set to zero by $S = C = R = 0$. Afterwards, L_C is calculated using Woody's algorithm [9]. It computes the latency by calculating the cross-correlation between the averaged ERP (used as template) and each single trial waveform. Thereby, the single-trials are averaged and this average waveform is used as a new template for further calculations. Through iterations the latencies get adjusted until the template does not change anymore.

2. **Prototype separation:** The prototypes are estimated based on three sets of latencies [6]. In a specific time window, prototypes S and R are subtracted from the single-trial ERP and the residual of all trials is aligned to the correlated latency L_C [5]. Afterwards, the median of residual is used as the new prototype C. Prototypes S and R have to be updated as well and therefore C and the other prototype are subtracted from the single-trial ERP in the same time window. The median of the resulting residual waveforms is now used as the new prototype. The time windows for every prototype are predefined by the user and determined by the visually inspected average ERP. The L1 norm minimization is used instead of L2 norm minimization for preventing distortions and a stronger robustness against noise [6].

3. **Latency re-estimation:** The latencies of prototype C are again calculated for every iteration with high precision. For each electrode/ channel the cross-correlation between the template and the single-trials is computed [5]. Prototype C is used as template. After that, the results are averaged and the maximum of cross-correlation time course is used as the new latency of the correlated single-trial.

Step 2 and 3 are repeated until the prototypes and latency L_C do not change anymore. Through this process overlapping components can be reduced.

4. **Baseline adjustment:** After computation of the three components the baseline has to be adjusted. The baseline of S is placed in the early time window [0,200ms] to the values of the original ERP and the baseline of C and R are placed to zero in this time window [6].

The outcome of this processing consists of three important parts. First, we get the patterns of the three component prototypes obtaining the topographies and the waveforms for each electrode [6]. Second, the new estimated ERP, which is the reconstructed summation of all latency-locked components. Thus, it is less blurred than the original ERP. These two ERPs are shown in Fig. 1(a) as the grand-average which

means that the results are averaged over all channels. It becomes clear that many overlaps occur in the original ERP and after shifting the components depending on their latencies the components are clearer visible in the resulting ERP [5]. Thirdly, we get the variability information out of the algorithm. This information contains the amplitude information of all prototypes and the latency information for prototype C.

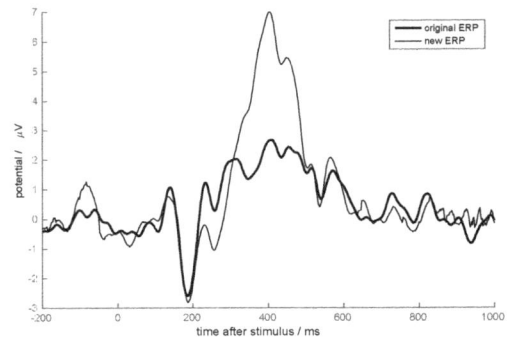

(a) Grand-average of original ERP and newly calculated ERP.

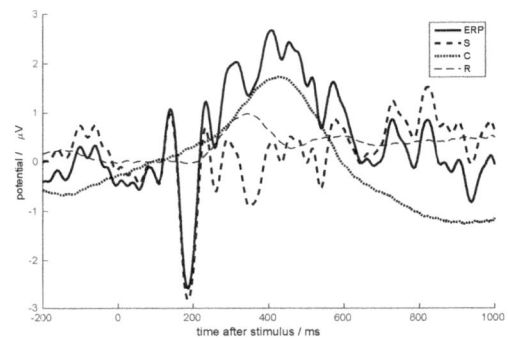

(b) Original ERP with stimulus-locked prototypes.

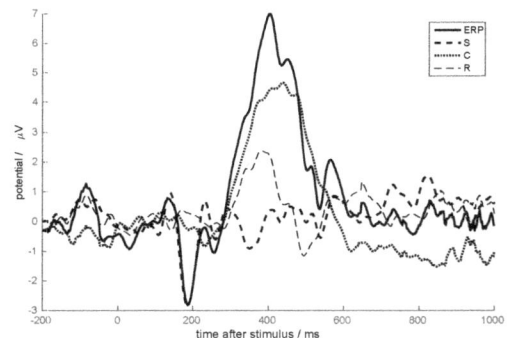

(c) New ERP with latency-locked prototypes.

Figure 1: ERPs and corresponding component prototypes.

In Fig. 1 the estimated component prototypes are figured with the ERPs. In Fig. 1(b) the original ERP is shown with the stimulus-locked prototypes and in contrast Fig. 1(c) shows the new calculated ERP without overlapping components and the latency-locked component prototypes. It becomes clear that the summation of component prototypes amounts exactly to the ERPs and that RIDE separates the ERP into components with the same roles e.g. the lock to the same event [4].

2.2 ICA

The independent component analysis (ICA) can be used for component separation, too. It estimates as many statistically independent signal sources from the electrodes as channels are used for measurement [6]. Based on the work of Hyvärinen [3], ICA can be described by the following steps:

1. **Preprocessing:**

   (a) **Centering:** The input data has to be centered by subtracting its mean value from all data points.

   (b) **PCA and Whitening:** Before whitening the data needs to be decorrelated by PCA. Therefore, the matrix of covariance becomes the identity matrix, which can be easily set to variance 1 by normalizing each component by its eigenvalue.

   (c) **Dimensional reduction:** The eigenvectors can get sorted by the quantity of their eigenvalues. Thus, the dimension of the data set can be reduced by deleting the components with the smallest eigenvalues to avoid overamplification of noise by the whitening.

2. **Component extraction:** The direction for the independent components (ICs) is found iteratively to maximize the measure of non-Gaussianity by a nonquadratic nonlinearity function. In this way, the unmixing matrix can be calculated and then used for matrix multiplication with the dimensional reduced data matrix to reconstruct the independent sources of a mixed signal.

This time, the order of the calculated components cannot be rated, it is random.

2.3 Comparison

First, it has to be said that a direct comparison between both methods is fairly difficult because they are based on completely different assumptions and theoretical frameworks. Some ICA assumptions cannot be fulfilled in this case. The mixture of different sources at the electrodes is not purely linear, the signal transduction through tissue should be constant and most importantly the sources have to be statistically independent, which is not true for most cognitive processes, of course [8]. Furthermore, statistical processes are often liable to aberration and the number of components is normally unknown [6]. Whereas, RIDE has strict assumptions as well, which makes the interpretation of results quite difficult e.g. the calculated prototypes should reveal a linear superposition [4]. Certainly this assumption leads to a bias if the component prototypes interact strongly nonlinear. RIDE is a fairly fast algorithm and is quite robust at

noisy data but too strong noise or artifacts may cause incorrectness at computation of latencies and therefore likewise in the resulting prototypes. A further great advantage is the feature that artifacts are not enhanced by this method and that it does not use global optimizations [4]. Nevertheless, the preprocessing is very important for both methods to improve the results and there are still some possibilities for further improvement. The last problem of both methods is the number of computed components. For ICA there is the possibility of determining the number of components beforehand by handpicking from all ICs, which are usually many. Whereas, RIDE can normally calculate just up to three components (S, C and R) or if necessary separate C further in two prototypes. So, any further measured process will become part of any component.

2.4 ICA and RIDE Combined

RIDE estimates C as the residue after extracting S and R components from the signal. Obviously, component C might be consisting of even more sub-components, a problem already discussed by Ouyang et al. [4]. The proposed ICA variant on the other hand might extract the underlying spatial independent components very well, but so far we have not estimated trial-wise latencies within the individual trials. Thus, averaging over each trial can only be done stimulus-locked, which again leads to blurring of potentially extracted components. In a first approach, we tested whether or not we can find the RIDE components in any of the calculated ICs and to what extend we can reconstruct these components with these ICs. The purpose of this trial was to get components repeated in every single-trial but without distortions caused by inaccurate latencies. In Fig. 2 there are different components from two selected electrodes (FCz placed in the middle of forehead and POz placed in the middle of occiput) and their reconstructions with selected ICs is shown.

You can see that the reconstructions of component prototypes S and R correspond to the RIDE prototypes. By contrast, the reconstruction of C component does not work well. This might be caused by the way C is computed: Since the latency of C in unknown at the beginning and has to be estimated first, the waveform of C has to be estimated as well [7]. Therefore, the calculated components S and R are subtracted from the ERP. Certainly the ERP of real data is not free of noise. As a consequence, the calculated residue may contain this noise and therefore component prototype C becomes noisy as well. At first sight it may seem like the reconstruction does not work well, but indeed it makes it look like the ICA enables the possibility of improving the results of RIDE algorithm, especially for a better estimation of the C complex.

3 Conclusion

In Fig. 1 it became clear that there is a strong effect whether or not the latencies receive attention since the stimulus-locked prototypes are blurred compared to the latency-

(a) Component S at channel FCz.   (b) Component S at channel POz.

(c) Component C at channel FCz.   (d) Component C at channel POz.

(e) Component R at channel FCz.   (f) Component R at channel POz.

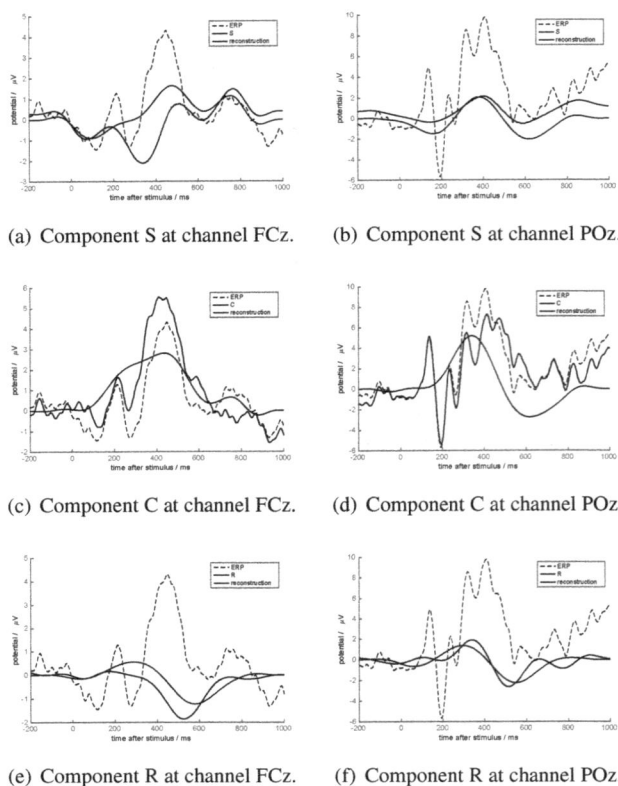

Figure 2: RIDE Components and their reconstructions with the reconstructed ERPs from selected Channels.

locked ones. Nonetheless, real data reveals that the latencies calculated by RIDE method may not be reliable since some prototypes are displaced to implausible positions, e.g. the R component appears to be ahead of the C component in single trials. Thus, it seems to be important to reconstruct the components on a single trial level by ICA and thereby to improve the results of RIDE. In contrast to the comparison Ouyang did in [6], in this work the component prototypes estimated by RIDE were not just compared against single independent components computed by ICA but against combinations of these independent components. Therefore, the results of ICA look much more like the RIDE-prototypes and perhaps the results are even better. As a first result, it cannot be confirmed that RIDE works much better than ICA at analyzing ERPs so far. However, the results of ICA can only improve by using RIDE components as a priori information. At this point, we are able to say, that the results from the combination of the residue iteration decomposition and the independent component analysis looks promising and that further investigations have to be conducted to evaluate if the results are actual improving.

Acknowledgement

The work has been carried out at Institute for Neuro- and Bioinformatics, Universität zu Lübeck and supervised by PD Dr. rer. nat. Amir Madany Mamlouk, Institute for Neuro- and Bioinformatics, Universität of Lübeck and Prof. Dr. Rolf Verleger, Department of Neurology, University Medical Center Schleswig-Holstein, Lübeck.

4 References

[1] A. Delorme and S. Makeig, *EEGLAB: an open source toolbox for analysis of single-trial EEG dynamics including independent component analysis.* Journal of Neuroscience Methods, vol. 134, pp. 9–21, 2004

[2] A. Hyvärinen, *Fast and robust fixed-point algorithms for independent component analysis.* IEEE TRANSACTIONS ON NEURAL NETWORKS, vol. 10, no. 3, pp. 626–634, 1999

[3] A. Hyvärinen, J. Karhunen and E. Oja, *Independent component analysis.* John Wiley & Sons, 2001.

[4] G. Ouyang, G. Herzmann, C. Zhou and W. Sommer, *Residue iteration decomposition (RIDE): A new method to separate ERP components on the basis of latency variability in single trials.* Psychophysiology, vol. 48, pp. 1631–1647, 2011.

[5] G. Ouyang, W. Sommer and C. Zhou, *A toolbox for residue iteration decomposition (RIDE) – A method for the decomposition, reconstruction, and single trial analysis of event related potentials.* Journal of Neuroscience Methods, vol. 250, pp. 7–21, 2015.

[6] G. Ouyang, W. Sommer and C. Zhou, *Updating and validating a new framework for restoring and analyzing latency-variable ERP components from single trials with residue iteration decomposition (RIDE).* Psychophysiology, vol. 52, pp. 839–856, 2015.

[7] G. Ouyang, W. Sommer and C. Zhou, *Reconstructing ERP amplitude effects after compensating for trial-to-trial latency jitter: A solution based on a novel application of residue iteration decomposition.* Psychophysiology, vol. 109, pp. 9–20, 2016.

[8] J. Seifert, *Ereigniskorrelierte EEG-Aktivität.* Pabst Science Publishers, Lengerich, 2005.

[9] C. D. Woody, *Characterization of an adaptive filter for the analysis of variable latency neuroelectric signals.* Medical and Biological Engineering, vol. 5, pp. 539–554, 1967.

A Neural Network for Single-Channel EEG Prediction and Encoding of Attention Selection

S. Grosnick[1], L. Fiedler[2], and J. Obleser[3]

[1] Medizinische Informatik, Universität zu Lübeck, simon.grosnick@student.uni-luebeck.de
[2] Insitut für Psychologie, Universität zu Lübeck, lorenz.fiedler@uni-luebeck.de
[3] Insitut für Psychologie, Universität zu Lübeck, jonas.obleser@uni-luebeck.de

Abstract

A persons focus of attention in a multi-speaker environment (*cocktail party problem*) can be encoded from single-channel EEG. In a listening task, two audio books were presented simultaneously to eighteen participants. They were asked to either attend to the male or the female speaker while a 64-channel EEG was recorded. A forward encoding model based on a neural network was trained to process two speech stimuli in order to predict a single-channel EGG-response. Comparing the predicted EEG-response to the actual measured one provides information about the subjects focus of attention. Confirming prior results, we have shown that a single-channel EEG-response can be predicted based on two presented stimuli and that the participants focus of attention can be encoded. Further, we observed that the neural network model can achieve a better classification accuracy than a linear model that is based on ridge regression.

1 Introduction

Listeners are able to attend to one speaker while ignoring distracting sound sources, which is the so called *cocktail party problem* [1]. It is a popular research scenario and motivated several studies in the past decade. Their results have shown that it is possible to predict a persons focus of attention based on his EEG-response [2]. Fiedler et al. presented a forward encoding model using ridge regression [3]. Different representations of speech were tested, achieving a $77,13\%$ classification accuracy in identification of the attended speech stream.

The main goal of this paper is to examine whether a more complex model, based on a neural network, is able to outperform the linear approach in [3]. Therefore the exact same task is performed. Processing two input stimuli, the model predicts an EEG-response which will be used to encode the subjects focus of attention. A short overview of this forward model can be seen in Fig. 2.

2 Material and Methods

As mentioned before, this work is based on the approach of Fiedler et al. and will be applied to the same data. Therefore this chapter will first give some brief information about their experiments. A more detailed description can be found in [3] and [4]. Afterwards the new encoding model and network design will be explained. The entire model was implemented using the *Neural Network Toolbox* for *Matlab (MathWorks, Inc.)*.

Participants and Task Eighteen subjects (23-68 years) participated in the study. A *cocktail party scenario* was created by simultaneously presenting two audiobooks spoken by a female and male speaker. Twelve trials of five minute length were presented to each participant and their task was to either attend the male or the female voice.

EEG-Data Acquisition and Preprocessing During the experiment a 64-channel scalp-EEG was recorded. In a preprocessing, the data were highpass-filtered at $f_c = 2$ Hz, lowpass-filtered at $f_c = 8$ Hz and downsampled to 125 Hz. Both, the EEG- and stimulus data were split up into samples of one minute length, resulting in 60 samples for each participant (42 training-, 9 testing- and 9 validation samples).

Representation of speech The effect of several ways to represent the speech stimulus was evaluated within the experiments in [3]. Two of those representations were also examined in this work and will be explained shortly.

First, the envelope of the continuous speech signal serves as an input for the model. As shown in Fig. 1, the envelope is an approximation of the signals amplitude. It can be obtained by calculating the absolute values of the signals hilbert transform. To examine whether the neural network benefits from an input with more detailed information, the sub-bands of a cochleogram served as inputs as well [5]. This approach is motivated by the processes in the outer and middle ear, where the audio signal is split up into different frequencies. In this work, each audio signal was represented by 24 frequencies.

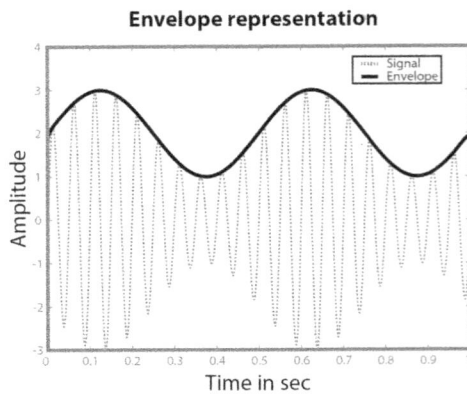

Figure 1: The envelope is an approximation of the signals amplitude and can be calculated using the absolute values of the signals hilbert transform.

2.1 Encoding Model

A neural network was chosen to replace the linear approach in [3]. It requires two inputs X_1, X_2 and is processing the stimulus data to predict the participants EEG-response on *Cz*. The network was trained in a way that it always expects the first input to be the attended stimulus while the ignored stimulus came along with the second input. Fig. 2 gives a brief overview of the implemented model, the actual design of the network will be explained more detailed later on.

The aim of this model is to classify the input stimuli into *attended* and *ignored*. Therefore the network predicts two EEG-responses Y_A and Y_B which were then compared to the measured EEG Y_M. Both EEG-responses are the results of two scenarios that are using the input stimuli S_1 and S_2 in a different way.

The first signal Y_A is obtained by setting the inputs as $X_1 = S_1$ and $X_2 = S_2$. As described above, the model is trained to expect the first input to be the attended stimulus. This configuration represents *scenario A*, which means that the subjects focus of attention was on S_1. Changing the inputs to $X_1 = S_2$ and $X_2 = S_1$ will lead to *scenario B*, where the subject was attended to S_2. This configuration generates the EEG-response Y_B.

Goodness of fit   Using the Pearson-correlation, both predicted EEG-responses Y_A and Y_B were correlated to the measured EEG Y_M. Calculating $r_A = corr(Y_A, Y_M)$ and $r_B = corr(Y_B, Y_M)$ will obtain two coefficients that can be compared. The greater (i.e. more positive) value will indicate the *true* scenario.

Classification accuracy   The neural networks accuracy will be defined as the percentage of correctly classified trials. A trial was classified correctly if there is a higher correlation between the actual measured EEG-response and the signal that was predicted when the attended stimulus was set as X_1.

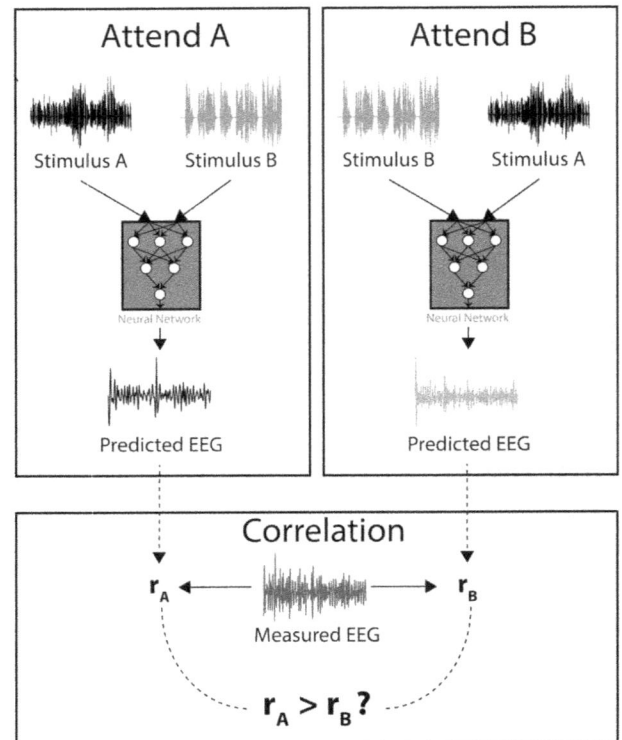

Figure 2: A neural network processes the input stimuli to predict two single-channel EEG-responses which will be compared to the actual measured EEG-response, using the Pearson-correlation. The greater (i.e. more positive) correlation coefficient indicates the stimulus that was attended by the subject.

2.2 Network design

The human brain uses large clusters of neurons to solve problems and process information. Each neuron is connected to many others and uses a function that combines all of its inputs to create an output signal. Neural networks are a mathematical approach to model this process of the human brain. All networks consist of three parts. The first one is the input layer, representing the information that should be processed. The area of *hidden layers* is located behind the input and can contain multiple layer with a variable amount of neurons. Within each neuron an activation function (e.g. *sigmoid function*) is applied to combine the neurons inputs, resulting in a single output per neuron. The results of the last hidden layer will be processed in the output layer to form the networks final output.

The learning algorithm   The Scaled Conjugate Gradient Backpropagation (SCG) [6] is an algorithm for fast supervised learning and can be applied to train a neural network. In comparison to other common learning algorithms like BFGS [7], the SCG is faster and does not rely on any user-dependent parameters. It also can handle large input data, which makes is it a good choice for this work.

Hidden layers Selecting the number of neurons in each hidden layer is a process of finding an acceptable balance between performance and accuracy. We determined the number of neurons and layers by evaluating the classification accuracy for different configurations. Therefore, up to three layers with different amounts of neurons $n \in \{5, 15, 25, 50, 87\}$ were chosen. Each configuration was used to train a network for three participants. The four best performing configurations were then used to train networks on all eighteen participants. In the end, an overall performance per network was denoted.

Time-lags The brains reaction to a stimulus is delayed and the first negative peak can be seen around $100ms$ after the stimulus is presented, followed by a positive peak at $200ms$. The entire event-related potential (ERP) can last for $\sim 700ms$. Therefore a delay was added to the model, allowing the network only to predict a value of the EEG-response based on the past $700ms$ of the stimulus. An overview of the entire network is visualized in Fig. 3.

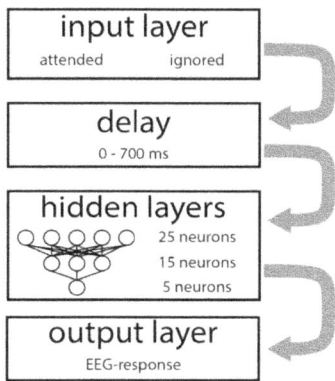

Figure 3: The best performing network configuration consists of three hidden layers with 25, 15 and 5 neurons. A delay area was added to the network to model the time-lag between a presented stimulus and the resulting ERP in the human brain.

3 Results and Discussion

The results have shown that the neural network benefits from a more detailed input. A better classification accuracy for nearly all tested network configurations could be achieved when the stimulus was represented by its cochlea sub-bands. Therefore only the results for this representation will be presented in the following.

3.1 Determination of network configuration

The final network architecture was selected among 20 test configurations with up to three hidden layers. These 20 configurations were trained on three participants to calculate a mean classification accuracy per configuration. In Fig. 4 the results for all tested options with a maximum of two hidden

layers can be seen. A brighter square indicates a higher classification accuracy. We can see, that a more complex configuration leads to a tendentially improved accuracy. In this context *complex* means that a configuration has a larger number of neurons in its highest layer. These trends are still present if the number of hidden layers is increased to three.

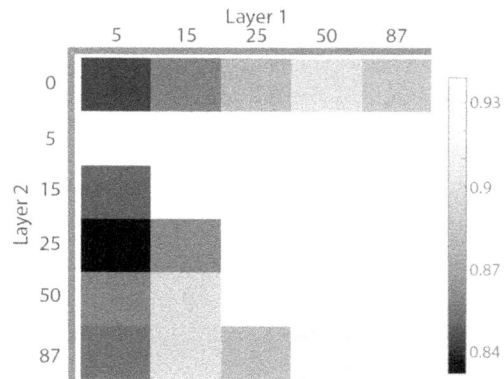

Figure 4: This figure visualizes the mean accuracy of all tested network configurations with a maximum of two hidden layers. Each square represents a configuration with different amount of neurons in its hidden layers. A brighter square indicates a better accuracy.

The four best performing network configurations were selected to be trained on all eighteen participants. In Fig. 5 we can see that three configurations achieve a slightly better performance than the linear approach while two networks are close to the linear model but still with a lower accuracy.

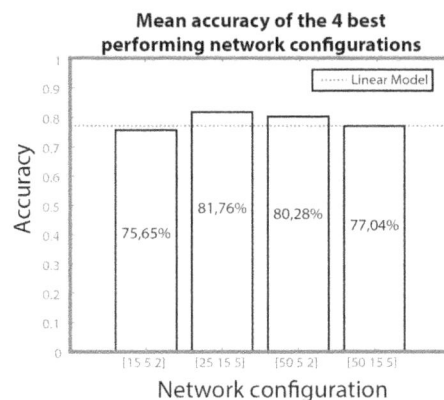

Figure 5: Two network configurations achieved a higher overall classification accuracy than the linear approach.

Network two, containing 25, 15 and 5 neurons, achieved the best overall performance, therefore its classification accuracy per participant was plotted against the linear model (see Fig. 6). With 81.78% accuracy, the neural network model is able to outperform the linear model in [3].

3.2 Analysis of best performing network

The brain is responding to a stimulus with an ERP which can be measured with an EEG. If the neural network is able to create plausible output signals, its response to a simple stimulus should look similar to an ERP. Therefore a test input was created, using a rectangle impulse on all frequencies of either the attended or ignored stream. Both results are illustrated in Fig. 7. The test stimulus on the attended stream leads to an EEG-response that looks similar to an ERP, which indicates that the network is capable of predicting plausible outputs. For the test stimulus on the ignored stream we can not notice such a strong ERP in the predicted output. This might indicate that the first input, which is supposed to be the attended stimulus, is better represented or higher weighted in the network.

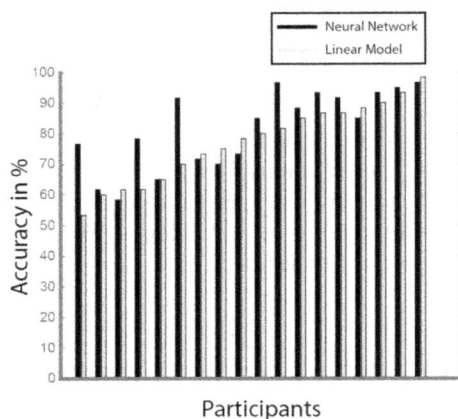

Figure 6: The classification accuracy of the best performing network is plotted against the linear model. Especially for participants with bad accuracy in the linear approach, the neural network seems to perform better.

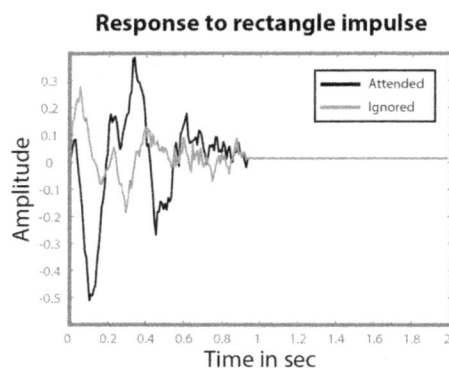

Figure 7: The neural networks response to a test signal on the attended stream (black line) looks similar to an event-related potential and confirms that the network is capable of predicting a plausible output.

4 Conclusion

A simple network architecture was implemented to examine whether it is worth applying a machine learning approach to the problem of attention selection and signal reconstruction. The results confirmed that it is possible to encode a subjects focus of attention based on a predicted single-channel EEG-response. Further, the neural network benefits from a more detailed stimulus representation and is able to achieve a better classification accuracy than the linear approach in [3]. In the next steps we will extend our network using current deep learning trends like convolutional layers or a long short-term memory network. Therefore more efficient deep learning toolboxes, implemented in *Python*, will be applied. In addition to the network changes, we consider using data that was collected with an in-ear-EEG instead of a traditional scalp-EEG. Nowadays in-ear-EEG systems can be implemented into comparably small hearing aids and if the in-ear-EEG also provides information about the participants focus of attention, the development of future hearing aids might benefit from these results.

Acknowledgement

The work has been carried out and supervised by the Institute of Psychology I, Universität zu Lübeck.

5 References

[1] E. C. Cherry, *Some experiments on the recognition of speech, with one and with two ears*. In: Journal of the Acoustical Society of America 25, pp.975–979, 1953.

[2] J. A. O'Sullivan et al., *Attentional Selection in a Cocktail Party Environment Can Be Decoded from Single-Trial EEG*. In: Cereb. Cortex 25, pp.1697–706, 2015.

[3] L. Fiedler et al., *Scalp EEG predicts listeners' attentional focus and attentional demands under continuously varying signal-to-noise ratio* (poster). Advances and Perspectives in Auditory Neuroscience (APAN), San Diego, 2016.

[4] L. Fiedler et al., *Single-channel in-Ear-EEG predicts the focus of auditory attention to concurrent tone streams and mixed speech*. bioRxiv, 2016.

[5] G. J. Brown and M. P. Cooke, *Computational auditory scene analysis*. In: Computer Speech and Language 8, pp.297-336, 1994.

[6] M. F. Møller, *A scaled conjugate gradient algorithm for fast supervised learning*. In: Neural Networks Vol. 6, No. 4, pp.525-533,1993.

[7] R. Battiti and E. Masulli, *BFGS Optimization for faster and automated supervised learning*. INCC 90 Paris. International Neural Network Conference. 2, pp.757-760, 1990.

Structure the Noise!
Machine Learning for Alarm Diagnosis

Ilja Stechmann[1], Andreas Weng[2], and Philipp Rostalski[3]

[1] Medizinische Informatik, Universität zu Lübeck, ilja.stechmann@student.uni-luebeck.de

[2] Drägerwerk AG & Co. KGaA, Lübeck, andreas.weng@draeger.com

[3] Institut für Medizinische Elektrotechnik, Universität zu Lübeck, philipp.rostalski@uni-luebeck.de

Abstract

Alarms are the most important source for clinical personal to identify a life-threatening condition of a patient. Since the requirements in hospitals to improve the quality of treatment and minimize patient risk are changing, it gets more important to understand the alarm situation at different stations in a hospital. Currently a lot of devices are saving their alarm history and from the beginning of 2016 Dräger has also started maintaining a central alarm history database for their devices. The monitoring devices are also saving medical trend data. This paper evaluates if it is possible to determine if an alarm is relevant by combining the trend data with an alarm classification and an alarm cause. For this analysis, neural networks are used. With the results of this work it might be possible to provide a service to reduce alarms on stations in future.

1 Introduction

As shown in [1] in 1997, a lot of alarms are produced on clinical stations at the hospital, and most of them are not clinical relevant. This problem has grown in the last 20 years since more and more devices have their own alarm system and more patients get connected to such devices. Since the alarm systems are not standardized, there are a lot of inconsistent alarm functions and characteristics. This might lead to duplicate alarm conditions and makes it hard to obtain data from the systems to build consistent alarm data. Because of this and other problems, a huge alarm burden is created and may cause an alarm fatigue [2].

This alarm fatigue can lead to serious patient risk. The following case described in [3] should demonstrate it:

A 23-year-old healthy male presented for a laparo-scopic bilateral inguinal hernia repair. It was the first such procedure this hospital had performed, so they decided to film it. It was an uneventful general anesthesia induction, and the patient was paralyzed with atracurium and maintained with isoflurane gas. At some point the anesthesiologist left the head of the bed so that he could watch the procedure on the video monitors and chat with the film crew. When the surgeon paused momentarily to switch sides, the anesthesiologist returned to the head of the bed and announced to everyone that the patient was in cardiopulmonary arrest. Unnoticed, the breathing circuit had become disconnected at the Y-connector under the drapes. All the alarms were flashing on the anesthesia machine, but they had apparently been silenced. This patient sustained severe permanent brain damage.

As shown alarm management is a serious and important matter. Therefore the *joint commission national patient safety* of the USA set up a *National Patient Safety Goal (NPSG)* on clinical alarm safety. The first two points of this *NPSG* require hospitals to make an organizational priority for alarm safety and depending on their own internal situation the most important alarms had to be identified by 2014. The third point calls hospitals to establish specific components of policies and procedures to manage the identified alarms by 2016 [2].

Since 2016 Dräger provides an alarm history database, which saves all medical alarms of connected Dräger monitors inside a hospital. Furthermore, all Dräger devices are saving medical trend data of the connected patients for at least 24 hours. In addition, Dräger has software components that allows to send data to a central database.

In order to complete the first two points of the *NPSG*, the data of the alarm history database must be evaluated. This evaluation could be a visualization which is accessible by clinical personal. To provide services to accomplish the third point of the goal, the alarm data must be further analyzed. Since the monitors provide medical trend data, the combination of these trends and the alarm message must be evaluated to determine if there are any pattern or criteria that allow to consider an alarm as false positive. This will be the goal of this paper. With this knowledge it would be possible to give hospitals a tool to detect stations with many irrelevant clinical alarms.

Since all the data can be saved to a central database, the alarm causes can be further analyzed and service like alarm limit recommendations can be provided to reduce not relevant alarms.

2   Material and Methods

In this section the available data, the used methods of the analysis and also the alarm situation of a hospital will be described.

2.1   Alarm situation in hospital

To get an understanding of the situation of the alarm distribution in hospitals, data from a hospital was allocated by means of the alarm history database. The data included alarms of one week. Also the information about the care unit the alarm came from, the alarm limits, the alarm duration and the alarm cause were covered. Table 1 shows the distributions of the alarms. The data was obtained from 21 beds and two different stations. The average alarm duration, until a response was received, was 20 seconds and the maximum response time was more than one hour. The abbreviations systolic (S), mean (M) and diastolic (D) are used [4].

Table 1: Alarm distribution for one week

alarm cause	number of alarms
Arterial Pressure M	2078
Pulse Oxygen Saturation	1395
Peak inspiratory pressure	1300
Heart Rate	980
Arrhythmia	255
Arterial Pressure S	220
End-tidal Carbon Dioxide	197
Arterial Pressure D	26
Non-Invasive Pressure M	19
Central Venous Pressure	13
Respiration Rate	15
Generic Pressure M	15
Pulse Rate	5
Temperature	2
Generic Pressure S	1
Right Ventricle Pressure M	1

2.2   Alarm data set

The analyzed data set was collected by the authors of [5] to characterize alarm patterns in general. Therefore they collected patient monitoring alarms during 25 cardiac surgeries. Additionally, about 124 hours of medical trend data and about 7500 alarms were gathered. These alarms were annotated by at least two physicians. In case of differing opinions about an alarm, a third physician had to decide about the final annotation. The alarm annotation categories were relevant, not relevant and alert.

As described in [6] and [7] the pulserate, the pulse oxygen saturation and the blood pressure belong to the most important alarms. The authors of [8] who worked with the same data set additionally considered the heart rate for further analysis, which will be done for this analysis, too. Considering Table 1, this solution would make it possible to an-

alyze more than 70% of the captured alarms. Also alarms which did not have a full trend data set were excluded from the analysis. Table 2 shows the distribution of alarms used for analysis.

Table 2: Distribution of annotated alarms

anotation	number of alarms
alert	703
relevant	1124
not relevant	945

2.3   Analysing tools

Since finding rules and patterns in such complicated data sets without medical knowledge is difficult, an automated method shall be used to classify the given alarms. As described in [9] a large number of different classification algorithms are available with different benefits and disadvantages. Statistical pattern recognition and machine learning methods were compared by classifying the same data. One of the data sets the authors of [9] tested was similar to the alarm data set in terms of amount of training samples, number of features and classes. Machine learning methods like *Neural Network* (NN), *Random Decision Forests* and *Predictive Value Maximization* had performed best on this kind of data. Therefore, this paper will investigate if it is possible to create a NN which can determine the corresponding alarm class. The NNs have been implemented in Python using the pybrain library [10]. NNs are inspired by the biological nervous system, learns by examples and can be configured for data classification [11].

In order to train the NN, 75% of the data will be used and 25% will be used as a test data set. To find a good generalization, different architectures will be tested. Also the learning rate and the momentum will be varied. The learning rate describes the change of the parameters into the direction of the gradient. The momentum adds information of the last training step to the current one. To prevent overfitting a L2 regulator is added and set by the weight decay parameter [10].

To test the ability of the NN, different architectures have been designed. Table 3 shows the created NN. All architectures include 37 input neurons and three output neurons. The first input neuron describes the condition (e.g. lower limit violation), the second neuron the alarm type (e.g. heart rate) and the other 35 provide the medical trend data. The seven observed parameters pulse rate, SpO2, systolic, diastolic and median blood pressure, heart rate and temperature are provided to the network with a value 4 and 8 seconds before and after and the value when the alarm occurred. Since the network is supposed to classify the alarm in three categories, there are three output neurons.

3   Results and Discussion

The created and tested networks are shown in Table 3. The table shows the architecture (arch), learning rate (η), momentum (m), weight decay (L2), number of trainings epochs (n) and best result for the classification on the test data in percent (E). E shows the test error.

Table 3: Network architectures and test results

arch	η	m	L2	n	E/%
37-15-3	0.0001	0.4	0.1	100	38,29
37-15-3	0.0001	0.01	0.1	500	35,11
37-15-3	0.0001	0.01	0.5	500	38,15
37-15-3	0.00001	0.01	0.1	1000	40.61
37-15-3	0.00001	0.01	0.5	1000	39,17
37-1000-3	0.0001	0.4	0.1	200	30.92
37-1000-3	0.0001	0.01	0.1	200	27,02
37-1000-3	0.0001	0.01	0.5	150	31.35
37-1000-3	0.00001	0.01	0.1	1000	28.76
37-1000-3	0.00001	0.01	0.5	450	35,26
37-20-40-3	0.0001	0.4	0.1	100	40.32
37-20-40-3	0.0001	0.01	0.1	50	41,33
37-20-40-3	0.0001	0.01	0.5	50	58.82
37-20-40-3	0.00001	0.01	0.1	50	57.95
37-20-40-3	0.00001	0.01	0.5	100	52.17
37-300-30-3	0.0001	0.4	0.1	150	37.86
37-300-30-3	0.0001	0.01	0.1	150	39.45
37-300-30-3	0.0001	0.01	0.5	100	52.46
37-300-30-3	0.00001	0.01	0.5	500	32,51
37-300-30-3	0.00001	0.01	0.1	1000	29.77

As Table 3 shows, 20 different types of networks with four different architectures were tested. The networks with just one hidden layer proved to be more effective with a big hidden layer. This is interesting, since there was only a small input layer and a small output layer. This could imply that the hidden layer is not supposed to have too many neurons as shown in [9] for similar data sets. Even with small learning rates the big layered network needed less trainings epochs and had a 10% lower test error.

Same goes for networks with two hidden layers. It was tested if it is more effective to build a NN with two small layers where the first layer is smaller than the second one. This net was compared to a NN with bigger layers where the first layer was 100 times bigger and the second layer was 10 times bigger than the output layer. The latter also had an about 10% lower test error.

3.1 Training

Most architectures had a common training trend. Fig. 1 shows exemplary the training progress of the third 37-1000-3 architecture. Most nets trained quite fast and got a good result after 10 - 30 training epochs. Further training resulted in worsening the result again. The training error had the same trend. The training and test results were also often close to each other which is untypical for NN. This implies these architectures are not suited to classify the data.

Networks with small learning rates had comparable training and test trends but needed more epochs to get to the best training point. Fig. 2 shows the training of the fourth 37-1000-37 architecture. The curve was wider and better test

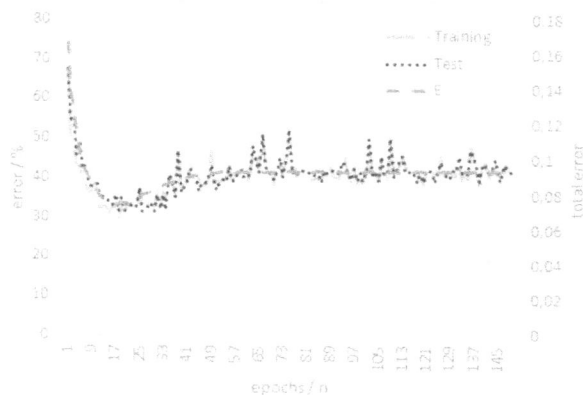

Figure 1: Training trend of 37-1000-3 architecture with η = 0.0001, m = 0.01, L2 = 0.5. The NN trains fast to get a good result. Afterwards the results get worse again.

results were achieved. Also the training error was much smaller than the test error. That could point to an overfitting but as the training proceeded, the total error and the test error became smaller. The best total error was calculated after about 800 training epochs through the test error still become smaller after further training.

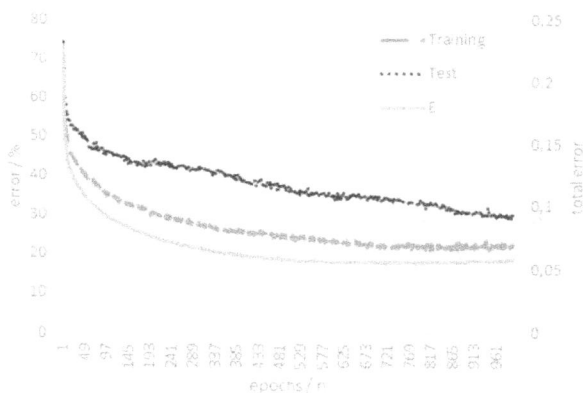

Figure 2: Training trend of 37-1000-3 architecture with η = 0.00001, m = 0.01, L2 = 0.1. The NN trains slowly but has good test results.

3.2 Validation

The validation data set consisted of 175 alerts, 281 relevant and 236 not relevant alarms. Fig. 3 shows the result of the fourth 37-1000-3 architecture for the test data set. The correct classification class is presented on the x-axis. The bars show how the training samples were classified in percent. The network performed badly on classifying alerts correctly. Only 55% were recognized correctly. Most not relevant alarms also were not classified in a proper way. Just 65% were analyzed precisely. On the other hand, the network performed well for identifying relevant alarms. 84% of the relevant alarms were matched correctly.

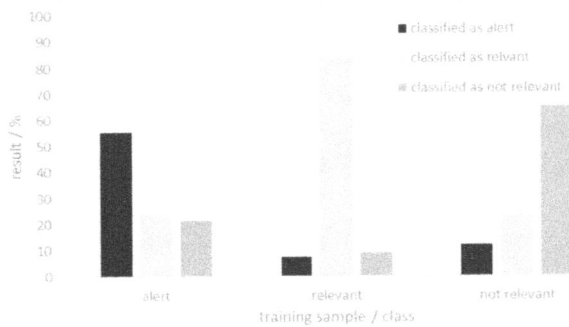

Figure 3: Classification result of 37-1000-3 architecture with $\eta = 0.00001$, m = 0.01, L2 = 0.1. The NN shows the best classifying result for the relevant class. This class had the most training samples.

4   Conclusion

With the best test error of 27%, the implemented networks are not applicable for analyzing alarms yet. On the other hand more than 70% of the alarms were classified correctly, therefore this paper points out the general potential to determine an alarm relevance by combining medical trend data with an alarm cause.

As the test results showed, the relevant alarms were predicted most appropriately. Due to the fact, that there were about 60% more relevant alarm samples available than alert samples and the validation results for classifying relevant alarms correctly were about 30% higher, for further analysis larger and more balanced data sets should be used.

The used data sets taken from surgeries included intentionally caused periods of asystole. So the used data set might not be applicable for alarms which occurred during other kinds of surgeries or on care units. Therefore data from varied intensive care units should also be analyzed since the alarm causes and alarm patterns might be different.

Also there are still a lot of ways to increase the performance of neural networks that were not used here. Dropout could be used to reduce overfitting as shown in [12]. Also there are more regularization methods available like L1 or MaxNorm. Additionally, more information can be used to determine the significance of an alarm. The used data set had no details about alarm limits or general information about the patient like gender or weight which could help to detect different alarm criteria.

For further analysis the mentioned points should be considered. Also other machine learning methods like *Random Decision Forests* and *Predictive Value Maximization* should also be investigated in the future.

Acknowledgement

The work has been carried out at Drägerwerk AG & Co. KGaA, Moislinger Allee 53, 23558 Lübeck and supervised by the Institute for Electrical Engineering in Medicine, Universität zu Lübeck.

5   References

[1] C. L. Tsien and J. Fackler, "Poor prognosis for existing monitors in the intensive care unit.," *Crit Care Med. 25(1997), pp 614 -619*, 1997.

[2] M. Vockley, "Clinical alarm management compendium," *AAMI Foundation*, 2015.

[3] A. S. Lofsky, "Turn your alarms on!," *The Official Journal of the Anesthesia Patient Safety Foundation*, 2004.

[4] A. Noble *et al.*, "alarms 7 days 2 mu," *Dräger Intern*, 2016.

[5] F. Franz *et al.*, "Analysis alarm annotation final," *Dräger Intern*, 2013.

[6] M. Wallace and N. Dunbar, "User of observation charts to identify clinical deterioration," *Australian commission on safety and quality in healthcare*, 2009. https://www.safetyandquality.gov.au/wp-content/uploads/2012/02/UsingObservationCharts-20091.pdf.

[7] M. Elliott and A. Coventry, "Critical care: the eight vital signs of patient monitoring," *CRITICAL CARE*, 2012.

[8] F. Schmid, M. S. Goepfert, F. Franz, D. Laule, B. Reiter, A. E. Goetz, and D. A. Reuter, "Reduction of clinically irrelevant alarms in patient monitoring by adaptive time delays," *Springer Science+Business Media Dordrecht 2015*, 2015.

[9] S. M. Weiss and I. Kapouleas, "An empirical comparison of pattern recognition, neural nets, and machine learning classification methods," *Department of Computer Science, Rutgers University, New Brunswick, NJ 08903*, 1989. http://ijcai.org/Proceedings/89-1/Papers/125.pdf.

[10] T. Schaul, J. Bayer, D. Wierstra, Y. Sun, M. Felder, F. Sehnke, T. Rückstieß, and J. Schmidhuber, "PyBrain," *Journal of Machine Learning Research*, 2010.

[11] S. Haykin, *Neural Networks and Learning Machines*. Pearson Prentice Hall, 3 ed., 2009.

[12] N. Srivastava, G. Hinton, A. Krizhevsky, I. Sutskever, and R. Salakhutdinov, "Dropout: A simple way to prevent neural networks from overfitting," *Journal of Machine Learning Research 15*, pp. 1929 – 1958, 6 2014.

Classification of Solitary Pulmonary Nodules using Deep Learning Networks

F. Kanter[1], R. Wiemker[2], T. Brosch[2], T. Klinder[2], and A. Mertins[3]

[1] Medizinische Ingenieurwissenschaft, Universität zu Lübeck, frederic.kanter@student.uni-luebeck.de
[2] Philips GmbH Innovative Technologies, Research Laboratories, D-22335 Hamburg
[3] Institute of Signal Processing, Universität zu Lübeck, mertins@isip.uni-luebeck.de

Abstract

Solitary pulmonary nodules (SPN) are often detected during computed tomography (CT) of the lung. Widespread use of thin-slice CT and growing interest in lung cancer screening lead to a frequent detection of small lung nodules, which creates a need for computer aided detection (CADe) and diagnosis (CADx). The range of SPNs includes benign causes as well as malignant ones (lung cancer and metastases), which makes a differential diagnosis necessary. As one building block towards a CADx system, we implemented an algorithm to classify SPNs into the three main types *solid*, *subsolid* and *mixed*, which are used by many malignancy risk models. For our classification task the methods of HU thresholding, a k-nearest neighbour algorithm, and our network were implemented. The methods we have tested so far could not solve the problem at a satisfactory accuracy level, possibly due to insufficient training data.

1 Introduction

Lung cancer is the leading cause of cancer deaths in the United States of America and 20%-30% of the patients diagnosed with lung cancer also have detected SPNs [3]. Imaging-based diagnoses of solitary pulmonary nodules (SPNs) are improving because of the advances in CT technology, increasing the detection rate of SPNs during screening programs [3]. Computer aided diagnosis systems led to an improvement of a radiologist's receiver operating characteristic curve (ROC) for detecting pulmonary nodules [1]. Furthermore, trials showed that screenings using low dose computer tomography (CT) reduces mortality from lung cancer by 20% [2]. Apart from detection of nodules, CT also allows characterization of nodules. The most important types are solid, subsolid, and mixed nodules. They are characterized through their attenuation in CT. For example subsolid nodules have a component with ground-glass attenuation, which is higher than the attenuation of lung parenchyma and lower than that of soft tissue [3]. Differentiating the nodule types helps estimating the malignancy likelihood. Computer aided diagnosis systems are being developed to assist in lung cancer screenings. Primarily, computer systems that automatically *detect* nodules have been developed [1]. By using different approaches for *classifying* the found nodules, we want to further improve the diagnosis of lung cancer. Therefore, the classification was carried out by using a HU threshold method, a k-nearest neighbour algorithm, and especially designed deep learning network. All algorithms were tested on the same set of CT images.

2 Material and Methods

2.1 Data and preprocessing

This work uses CT images from the National Lung Screening Trial (NLST) [5]. The NLST was a trial to determine whether screening for lung cancer with low-dose helical CT reduces mortality from lung cancer in high-risk individuals compared to the conventional method of chest radiography. CT images from the NLST collection come from over 25,000 participants and 3 screens per person over an interval of 3 years. Each screening exam contains a collection of one localizer image and 2 to 3 axial reconstructions of a single helical CT scan [5]. For this work, a subset of 300 participants is used. A list is available in which all nodules found in each image are recorded. However, no machine readable annotations for the nodule locations are available. For each nodule only a slice number and a lobe location is available. To generate nodule locations we applied a CADe algorithm [8] and matched the resulting candidates to the NLST markings, using heuristics based on geometric distance and lobe location. Afterwards each image was divided into patches of 30x30mm, 40x40mm, and 50x50mm spacing, where every patch contains only one nodule. All following classifiers were tested on the resulting 216 nodules, which contain 168 nodules of type solid, 34 of type subsolid and 14 of type mixed. Fig.1 shows nodules of each type.

2.2 Hounsfield Threshold Classification

SPNs are mainly characterized by their density, which leads to different Hounsfield unit in the images. The idea of a

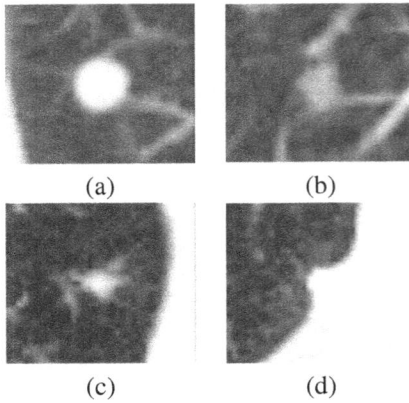

Figure 1: (a) Example for a solid pulmonary nodule, (b) Example for subsolid nodule, (c) Example for mixed nodule, (d) Hard to classify nodule labeled with type mixed.

simple threshold classification algorithm is based on the assumption that SPNs can be separated by comparing the Hounsfield values and finding a fixed border. In this approach the comparison was done by using the mean value, the median, or the mean value of the largest 5% quantile. First a threshold based on the knowledge of characteristic HU values of lung parenchyma and soft tissue such as pulmonary vessels was chosen. Later on, an interval of HU values for each type of nodule was determined by evaluating the results of the first classification.

2.3 k-Nearest Neighbour Classification

The k-nearest neighbour (kNN) algorithm is a well known tool for pattern classification [4]. The algorithm calculates the distance between the input vector it is given and classifies an example by the majority of the labels among the k-nearest neighbours in the training set [4]. For this work the input vector is the Hounsfield histogram of a not classified nodule and the distance metric is the Euclidean distance. Mainly four different configurations of the input vector are used such as raw, cumulative sum, normalized and normalized cumulative sum. This was done to avoid a huge distance between nearly similar histograms that just differ from each other by a small shift of the bins. Each configuration was tested with 5, 10 and 15 neighbours.

2.4 Deep Learning with CNNs

The principal function of a neural network (NN) is to minimize the error between a desired output and the calculated network output. Every neuron in each layer has two main attributes called weights and bias. The weights expressing the importance of the neurons inputs to the desired output and the bias is the negative threshold for activation. During the training all weights and biases are progressively updated by using the stochastic gradient descent (SGD). So called convolutional neural networks (CNN) are widely used for image classification tasks. Current trends lead to increasingly deeper networks, which are able to learn complex

features for classification. Fully connected layers are inefficient for detecting image structures. Most important patterns, e.g., edges appear several times in one image, thus some patterns could be learned repeatedly by several neurons. CNNs use concept of shared filters and local connectivity to solve this problem of NNs. In the forward pass a convolutional layer's input is convolved with a filter kernel using a sliding window approach. The spatial positions of a feature are represented in the filters response, which are called feature maps. More convolutional layers are able to learn more complex features, since each layer combines the patterns of all channels of the previous layers. To be used for classification tasks, fully connected layers have to be stacked at the top of convolutional layers to map the activations to neurons of the same number as the number of classes to differentiate. Often the softmax function is used to convert the network's predictions to probabilities in the range of 0 to 1.

2.4.1 CNTK

To create the networks of this work the Microsoft Cognitive Toolkit (CNTK) by Microsoft Research was used. CNTK is an open-source toolkit for deep learning algorithms. It provides many components and functions to create all kind of networks like NN or CNN. Furthermore it is compatible with widely used programming languages like C++, thus it is possible to embed your network in a desired programming environment.

2.4.2 Network Architecture

In this section, the used network will be described shortly and most important attributes will be mentioned. The used network contains 2 convolutional layers and 6 fully connected layers, each layer containing 32 neurons. The input patches are images of the order of 1,225 pixel or for training with volumes in order of 42,875 voxel. The schematic architecture of the network is shown in Fig.2. As mentioned before the network was trained with patches of 30x30mm, 40x40mm, and 50x50mm picture cutouts around one nodule. For the cost function two different approaches were made. First the cross-entropy (ce) and secondly a sensitivity-specitivity-ratio (ratio) were tested. The cross-entropy is defined as

$$C = -\frac{1}{n}\sum_{z}[y\ln a + (1-y)\ln(1-a)], \qquad (1)$$

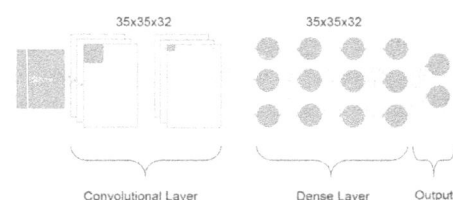

Figure 2: Schematic structure of the network.

where n is the total number of items of training data, x is the number of training inputs, y is the corresponding desired output, a is the output of the neuron and b is the bias. The cross entropy coupled with softmax outputs was described working much better for classification problems than the traditionally used quadratic cost function [6]. Furthermore the sensitivity-specificity-ratio was chosen, since it should work better with unbalanced training sets [7]. It is defined as

$$C = w \left(\frac{r_p}{r_p + f_n} \right) + (1 - w) \left(\frac{r_n}{r_n + f_p} \right), \quad (2)$$

where w is a adjustable ratio, r_p and r_n are the right positive and right negative classified objects, whereas f_n and f_p denote the false negative and false positive classified objects.

2.4.3 Regularisation

One of the biggest problems when working with NN or deep learning networks is overfitting. After enough epochs of training a network can learn a problem to perfection. But the validation shows that often the network just adjusts to the training data and does not generalize well. To solve this problem the $L2$ regularisation, dropout regularisation and a combination of both methods were tested. $L2$ regularisation is defined as

$$C = C_0 + \frac{\lambda}{2n} \sum_w w^2, \quad (3)$$

where C_0 denotes the cost function without regularisation, n is the number of weights and w are the weights. The dropout method indicates the excluding of randomly selected nods during training. As parameters the percentage of nods and the epoch to start with dropout can be chosen.

2.5 Receiver Operating Characteristics

In this work the receiver operating characteristic (ROC) method is used to analyse and optimize a two class classification problem. For different parameters the resulting probabilities in form of true positive rate and false positive rate are determined. Setting the true positive rate as y-axis and the false positive rate as x-axis a plot is generated. The resulting curve serves as an indicator of the classifier's performance. The ideal form of a ROC curve rises vertical and a diagonal form indicates a pure random decision process. By calculating the area under curve (AUC) the ROC curve can be used as a quality metric. The value lies in an interval of 0 to 1, where 0.5 is the worst value which is the result of a random process. Values lower than 0.5 emerge from wrong interpretation of the used data.

3 Results and Discussion

In this section the results for classifying the solid pulmonary nodules, the most common class, are shown. The results for the other mentioned classes are nearly as good as the solid results concerning classification error and area under curve.

Table 1: Results HU thresholding

Mean Value	Sensitivity	Specificity	AUC
			0.55
-800 HU	0.75	0.15	
-575 HU	0.62	0.5	
-375 HU	0.05	0.98	

Median	Sensitivity	Specificity	AUC
			0.52
-800 HU	0.77	0.12	
-575 HU	0.49	0.59	
-375 HU	0.03	0.95	

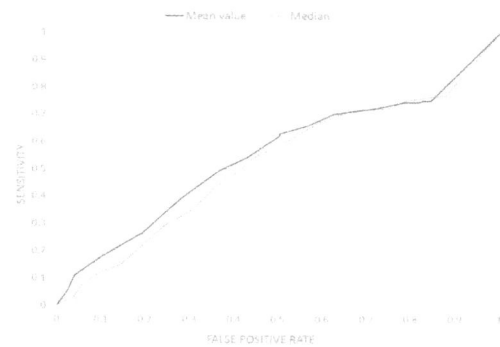

Figure 3: ROC curves for HU thresholding method.

3.1 HU Thresholding

The results of the simple HU thresholding method are more or less a random process result as expected. In table 1 it is shown that the algorithm does not work well for this task. It can either yield a high sensitivity or specificity but not both. The threshold was varied between -800 HU and -375 HU and was changed by a step size of 25 HU. Fig.3 displays the resulting ROC curves and the AUC are in the range of 0.55 for mean value and to 0.52 for median. This result underlines that the algorithm is not suited to solve the given classification task.

3.2 k-Nearest Neighbour

The kNN method seems to provide a slightly better results than simple thresholding. The best results were made by using 5 neighbours, this is shown in table 2. Increasing the number of neighbours leads to inferior results. It has to be considered that the sample for mixed nodules is quite small, so that too many neighbours lead to false classifications. An overview of the resulting ROC curves of all configurations for the kNN algorithm is presented in Fig.4. To sum up

Table 2: Results kNN algorithm

Number of Neighbours	Histogram	AUC
5	raw	0.62
5	norm	0.52
5	cumulated	0.57
5	norm+cumulated	0.5

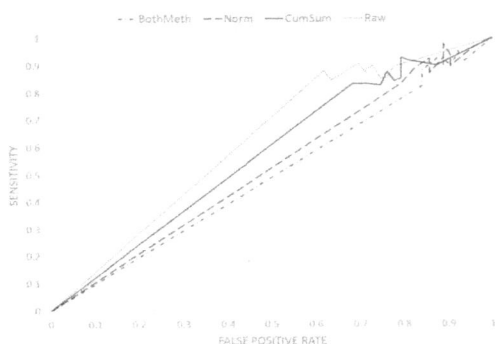

Figure 4: ROC curves for kNN method.

the results the kNN algorithm works slightly better than the thresholding approach, but the results are still no satisfying classification.

3.3 Convolutional Neural Network

One of the problems using deep learning networks is that they have a broad spectrum of tuning parameters. During this work many different configurations have been tried and the best achieved results are represented in this section. Furthermore the different configurations of training sets are compared based on the results of the best working configurations. Fig.5 displays the ROC curves with the highest AUC in all tested cases. The configuration set which contained data normalization, the cross entropy as cost function and using slices delivered the best results. Fig. 5 shows that

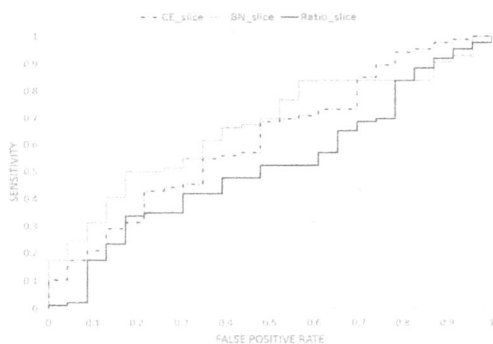

Figure 5: ROC curves for deep learning network.

a network with normalized inputs and without regularisation works best for our problem. Considering the results in

Table 3: Results neural network

Norm	Regularisation	patch	cost function	AUC
yes	none	slice	ce	0.67
no	L2+Dropout	slice	ce	0.62
no	L2+Dropout	slice	ratio	0.52
yes	none	volume	ce	0.61
no	L2+Dropout	volume	ce	0.61
no	L2+Dropout	volume	ratio	0.5

table 3 the statement holds true that our network performs better compared to the other approaches. But the classification is still not satisfying for a practical use in a CAD system.

4 Conclusion

The trained convolutional network has not yet achieved satisfactory performance as a classifier for pulmonary nodules. Comparison to other methods shows that this task is far from trivial. Due to the fact of only subtle differences in HU values and a broad variety of possible shapes for each class, even radiologists often disagree on the correct classification. However, there are indications that deep learning networks might have the highest practical potential to solve this problem if sufficient data becomes available.

Acknowledgement

The work has been carried out at Philips Research Labs Hamburg.

5 References

[1] J. P. Ko and M. Betke, *Chest CT: Automated Nodule Detection and Assessment of Change over Time*. Radiology, vol. 218, pp. 267-273 2001.

[2] A. McWilliams et al. *Probability of Cancer in Pulmonary Nodules Detected on First Screening CT*. The New England Journal of Medicine, 2013.

[3] M. T. Truong et al. *Update in the Evalution of the Solitary Pulmonary Nodule*. RadioGraphics, vol. 34, pp. 1658-1679, 2014.

[4] K. Q. Weinberger, J. Blitzer and L. K. Saul, *Distance Metric Learning for Large Margin Nearest Neighbor Classification*. University of Pennsylvania, 3330 Walnut Street, 2009.

[5] National Cancer Institute, *Cancer Data Acess System*. Avialable: https://biometry.nci.nih.gov/cdas/learn/nlst/images/ [last accessed on 2017-01-17].

[6] X. Glorot, Y. Bengio *Understanding the difficulty of training deep feedfoward neural networks*. DIRO, Universite de Montreal, 2010.

[7] T. Brosch, Y. Yoo, L. Y. W. Tang, D. K. B. Li, A. Traboulsee and R. Tam *Deep convolutional encoder networks for multiple sclerosis lesion segmentation*. Springer, 2015.

[8] M. Bergholdt, R. Wiemker and T. Klinder, *Pulmonary Nodule Detection Using a Cascaded SVM Classifier*. SPIE Medical Imaging, 2016.

Design and implementation of a noise reducing filter solution for biosignals derived from phonocardiography and beat-to-beat finger plethysmography

E. Hachgenei[1], U. Limper[2] and P. Rostalski[3]

[1] Medizinische Ingenieurwissenschaft, Universität zu Lübeck, enno.hachgenei@student.uni-luebeck.de
[2] Deutsches Zentrum für Luft- und Raumfahrt, Institut für Luft- und Raumfahrtmedizin, Köln, ulrich.limper@dlr.de
[3] Institute for Electrical Engineering in Medicine, Universität zu Lübeck, philipp.rostalski@uni-luebeck.de

Abstract

Cardiac systolic time intervals are used to monitor the performance of the human heart. A suitable filter option for phonocardiogram (PCG) and arterial pressure data will be the foundation of a robust detection algorithm, for the estimation of Systolic Time Intervals (STIs). STIs are a non-invasive technique, to monitor the performance of the left ventricle. In order to find the required filter characteristics the data of three subjects were used to calculate mean waveforms of both heart sounds and the pulse wave (FPG), which then were used to perform a frequency analysis. The derived frequency ranges were applied to filter parabolic flight PCG and FPG data. Tested on noise afflicted data, good results were achieved to enable an efficient processing of the data. On the basis of the filtered data it is now possible, to develop an automated analysis software including a detection algorithm.

1 Introduction

The necessity of long-term monitoring of the cardiovascular changes of astronauts exists, because of the prolongation of space missions.

Echocardiography is the current in-hospital standard for the monitoring of the left ventricular function. Because of the needed medical training to properly use a medical ultrasound and evaluate its results, the ultrasound is not a potential option for the usage in space missions. An alternative, simple and mobile non-invasive monitoring technique, is the technique of measuring STIs. The measurement system consists of electrodes to record an one-lead electrocardiogram (ECG), a microphone to record a PCG and a pressure sensor to record an arterial pulse wave.

STIs have been widely used in the 1960s and 1970s to determine the function of the left ventricle [1]. Since, they have been replaced by echocardiography, which provides more detailed information. But for the requirements of an easy to use long-term monitoring technique the STIs are well suited.

The aim of this project is, to find an appropriate filter combination, to enable a robust detection of heart sounds and characteristic points in the pulse wave. This filter combination has to be implemented in Matlab. Main requirements for the filters are, to increase the signal to noise ratio in both, the PCG and the arterial pressure data, to remove high frequency noise and any other artefacts consisting of different frequency components than the signals of interest. Based on the results of this study an analytic algorithm, to determine

the length of the mechanical systole, will be realised. For this algorithm to work well, it is crucial to detect important points in the data precisely, for example the onset of the second heart sound. The result of this project is supposed to be a filter which enables this precise detection. The derived filter combination will then be tested on existing data, which was recorded on a parabolic flight campaign, to validate the usability of the filter.

2 Material and Methods

The data, which was retrospectively analysed to optimize the filters, consisted of an ECG, a PCG and arterial pressure measurements [6]. In order to be able to draw conclusions about the characteristics of the recorded data, it was acquired in a quiet environment, to minimize the impact of any noise and artefacts. All biosignals were recorded with a sampling frequency of 2 kHz. The test subjects were three healthy adults, one female, 30 years old and two male subjects, 29 and 54 years old. The data was acquired within the context of a zero-g parabolic flight, where it served the purpose of baseline recordings.

To obtain the required filter specifications a frequency analysis was performed in Matlab. In order to further reduce the impact of any kind of noise or artefacts on the result of the frequency analysis, prior a mean waveform of both relevant heart sounds and the pulse wave was calculated. To generate a mean waveform the first and the second heart sound were detected and distinguished in the PCG and the pulse wave was detected in the arterial pressure data. The detected heart sounds and pulse waves were divided in data arrays of the same length to allow the calculation of a mean waveform.

2.1 Detection and processing of heart sounds and pulse waves

To detect the R-peaks in the prefiltered ECG a simple threshold detection was used, by that the R-peak position was determined and could be used to create a time window in which the first heart sound was expected. The first heart sound was localized by a maximum search within the time window of 0 - 5 ms following a R-peak in the ECG. The full heart sound was extracted out of the PCG by cutting out a data frame of 0,165 s around the peak. That is the maximum duration of the first heart sound [2]. The second heart sound was localized by a maximum search in the time interval between the end of the first heart sound and the following R-peak. The length of the second heart sound was set to 0,145 s, the maximum duration of the second heart sound [2].

To generate an averaged waveform, all heart sounds were normalized, then all waveforms were averaged. Fig. 1 b) shows the result of the extraction and normalization of the first heart sounds (S1) of all subjects, the black curve represents the mean waveform. Fig. 1 c) shows the results for the second heart sounds (S2) for all subjects. Based on this mean waveform a Fourier analysis was performed. To check for interindividual differences a Fourier analysis for every averaged waveform of each subject was performed. Those steps were executed for both, the first and second heart sound.

Detecting the onset and the end of a pulse wave was done by a minimum search, within the time window between two R-peaks, in the arterial pressure signal. The end of the previous pulse wave was at the same time the onset of the following pulse wave. The pulse wave was then cut in a frame of 0,725 s, this was found to be the shortest duration of a pulse wave in the recorded data. Fig. 1 a) shows all pulse waves of all subjects in grey and the mean pulse wave in black. After all pulse waves were separated, the same steps as for the heart sounds were performed. Normalized mean waveforms were calculated, and for each subject as well as for the overall mean waveform a Fourier analysis was performed.

2.2 Filter selection and implementation

The results of the frequency analysis were used to design one filter for the PCG and one filter for the arterial pressure data in Matlab. In both cases an Infinite Impulse Response (IIR) Butterworth filter was chosen. An IIR-Filter was chosen because of its lower complexity than a Finite Impulse Response (FIR) filter and therefore lower computational cost [3]. A phase shift caused by an IIR-Filter is non crucial, because zero-phase filtering was used. This method prevents the occurence of a time shift, by filtering the data

from both directions, this way the same time shift occurs in both directions and cancels each other out. A Butterworth filter was used, because its amplitude response is particulary flat within the pass band and a very steep transition between stop and pass band is not required [4]. Additional to the found filters, moving average filters were used, to further smoothen the results. This was necessary because, it was of interest to generate and use the first, second and third derivative of the filtered data.

The found filters were tested on data which was recorded during a zero-g parabolic flight.

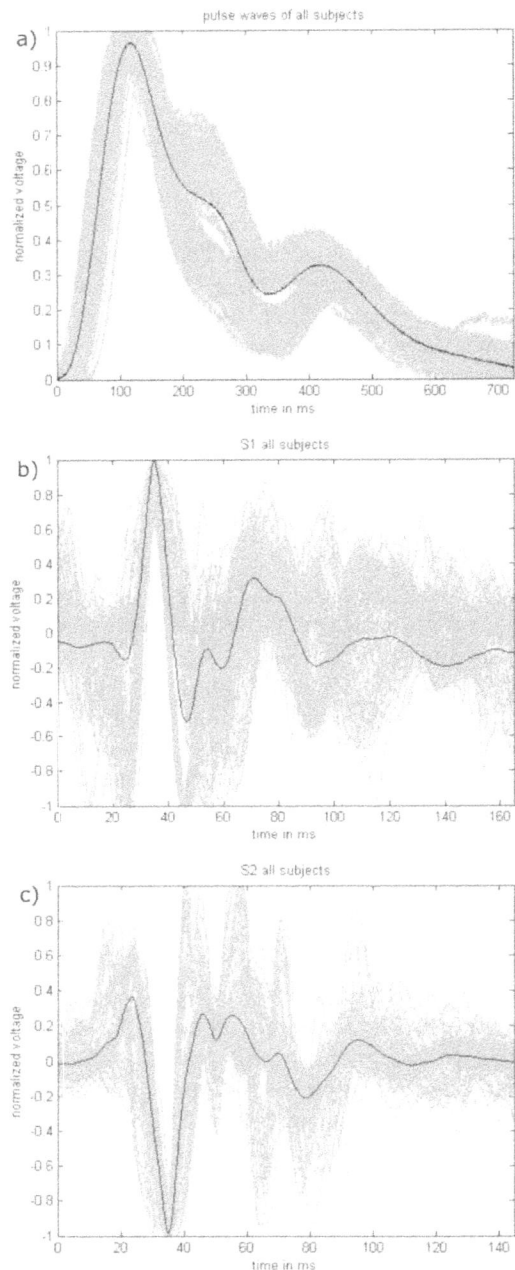

Figure 1: Waveforms of all recordings in grey and the resulting mean waveform in black for the a) pulse wave, b) first heart sound (S1) and c) second heart sound (S2).

3   Results and Discussion

The results of the frequency analyses of the mean waveforms shown in Fig. 1 are illustrated in Fig. 2. The spectrum of the pulse wave only consists of low frequency components, the highest contributing frequency lies between 7 Hz and 8 Hz. The spectrum of the first heart sound consists of frequencies between 0 Hz and 80 Hz with a prominent peak around 25 Hz. The spectrum of the second heart

sound has a frequency range from 0 Hz to 140 Hz, with its peak between 20 Hz and 45 Hz. The frequency ranges of the heart sounds differ from those in the literature, which span from 0 Hz to 300 Hz [5].

Based on the results of the frequency analyses the filter specifications are chosen. For the PCG, which consists both heart sounds, a Butterworth-bandpass-filter with a passband from 35 Hz to 140 Hz is applied, because the highest frequency of the second heart sound is 140 Hz and the microphone, which was used to record the PCG is designed to record frequencies of 35 Hz to 3500 Hz. Both transition bands are 10 Hz wide and the stop band attenuation is 60 dB, with these specifications good results were obtained.

For the arterial pressure data a Butterworth-lowpass-filter is used. It has a cut-off frequency of 15 Hz, a transition band width of 10 Hz and a stopband attenuation of 60 dB.

In Fig. 4 the results of the filtering of the PCG are illustrated. Fig. 4 a) and c) both show raw PCG data. The data of Fig. 4 a) has been recorded during a zero-g, Fig. 4 c) during a hyper-g phase of a parabolic flight. The data of the zero-g phase contains a lower amount of noise, because in this phase of the parabolic flight there was no contact between the subject and the plane. Fig. 4 b) and d) show the corresponding filtered graphs to Fig. 4 a) and Fig. 4 c). It is easy to see in Fig. 4 b), that the chosen filter works good to improve the signal to noise ratio of the zero-g data. Both heart sounds can be distinguished, the second heart sound in general has a lower amplitude than the first heart sound, but it still can be recognized. In the more noisy data of the hyper-g phase it is difficult to distinguish all second heart sounds, even after filtering, because of their small amplitude in comparison to the noise level. Fig. 4 d) shows, that the filter improves the signal to noise ratio and enables all first heart sounds to be detected, even in noisy data. In comparison, the data amplitude of the hyper-g phase data decreases more after filtering than the amplitude of the zero-g data.

The filtered arterial pressure data has been cleared of any high frequency noise (Fig. 3), which opens up the opportunity to calculate derivatives as basis for a detection algorithm.

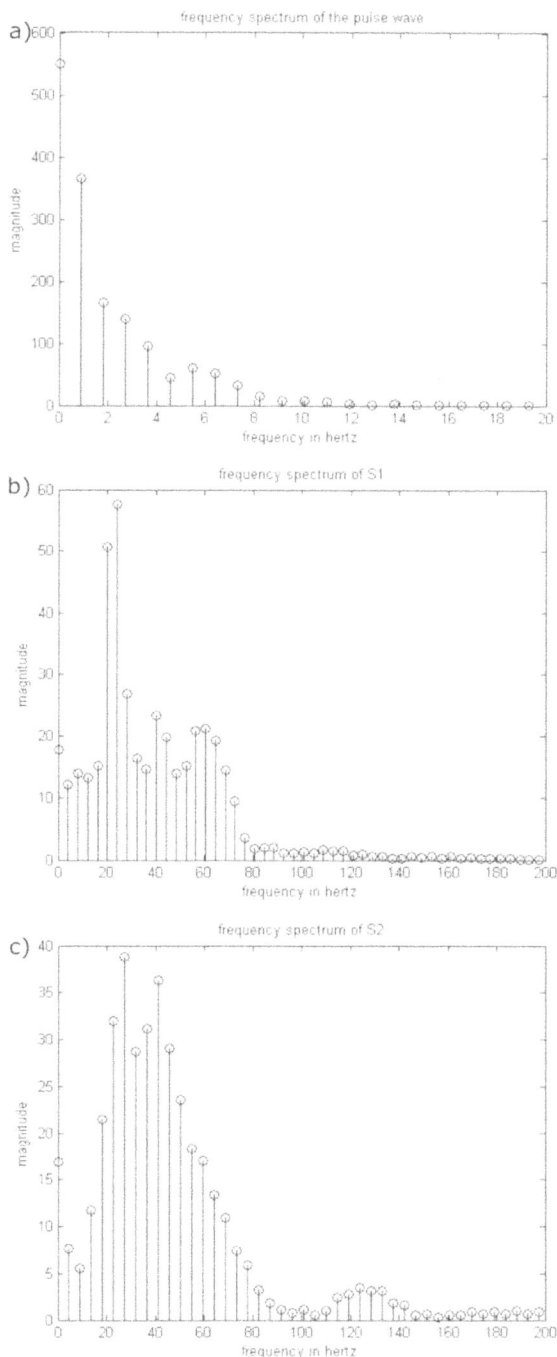

Figure 2: Frequency spectra of the a) mean pulse wave, b) mean first heart sound (S1) and c) mean second heart sound (S2).

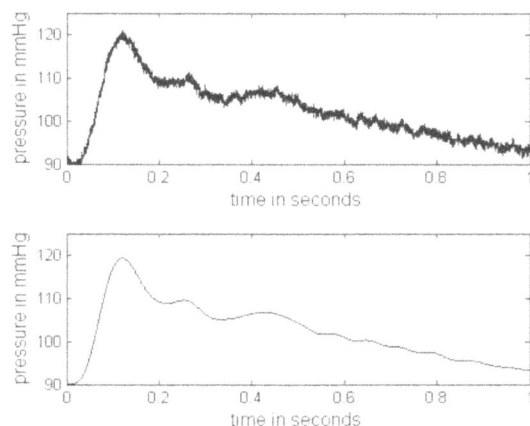

Figure 3: Raw pulse wave (top) compared to filtered pulse wave (bottom).

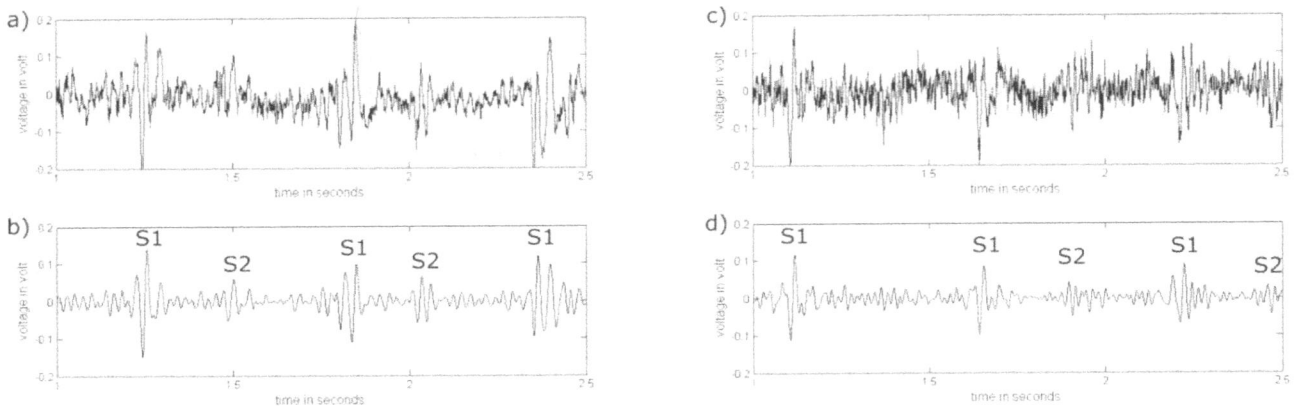

Figure 4: Raw and filtered PCG data in comparison: a) raw PCG measured in zero-g conditions and b) the corresponding filtered data, c) raw PCG measured in hyper-g conditions and d) the corresponding filtered data.

3.1 Discussion

Frequency analysis led to an optimized filtering of the PCG and the arterial pressure signals, this made STI estimation more stable. But even after filtering it is not possible to extract all the heart sounds of each heart beat out of noisy data. Mostly second heart sounds, which have a lower amplitude than first heart sounds, sometimes can not be distinguished from noise. But the field of application of measuring STIs is not supposed to be a noisy environment, the filtering of the hyper-g data only answered the purpose of testing its robustness. In the data of the calmer environment all second heart sounds could be recognised.

It should be considered to try to improve the isolation of the microphone, this way the amount of noise could be limited and it would maybe be possible to extract a greater number of the second heart sounds.

Fig. 1 b) and c) show, that the waveforms contributing to the mean waveform differ quiet a bit from each other, this is due to interindividual differences between the subjects, it could be informative to conduct a study in which a larger amount of subjects are probed to check for differences between the frequency spectra of their heart sounds. If there is found to be a large interindividual variance it could be discussed, if it is reasonable to realise an individual filter for each subject, based on a baseline measurement.

4 Conclusion

The derived filter combination produces, even used on noisy measurements, data, which is a very good basis for a detection algorithm. The next steps are, to implement such an algorithm which automatically analyses changes in the STIs. After that, a testing of a long time monitoring of the systolic function of the heart through STI detection can be realized. This method could not only be applied for astronauts, it could also be an useful technique for a home monitoring device used on high risk patients.

To further evaluate the findings of this project, it would be of interest, to test the impact of interindividual differences in the frequencies of the heart sounds between different subjects and different subject groups, by realizing a project in which a larger subject group is probed.

Acknowledgement

The work has been carried out at the German Aerospace Center, Cologne and supervised by the Institute for Electrical Engineering in Medicine, Universität zu Lübeck.

5 References

[1] K. Tavakolian, *Systolic Time Intervals and New Measurement Methods*. Cardiovascular Engineering Technology, 2016.

[2] A. A. Luisada, F. Mendoza and M. M. Alimurung, *The duration of normal heart sounds*. British Heart Journal, 1949.

[3] M. Werner, *Digitale Signalverarbeitung mit Matlab*. Vieweg + Teubner, 2012.

[4] M. Meyer, *Signalverarbeitung, Analoge und digitale Signale, Systeme und Filter*. Springer, 2014.

[5] A. K. Abbas, R. Bassam, *Phonocardiography Signal Processing*. Morgan and Claypool Publishers, 2009.

[6] S. Moestl, *Development, Verification and Practical Application of a Signal Analysis Program for Determining Systolic Time Intervals in Parabolic Flight*. Master Thesis, 2016.

Development of a ball balancing robot
–Modularity and control concept–

A. Ibbeken [1], G. Männel [2] and P. Rostalski [2]

[1] Medizinische Ingenieurwissenschaft, Universität zu Lübeck, ibbeken@student.uni-luebeck.de

[2] Institute for Electrical Engineering in Medicine, Universität zu Lübeck, {ge.maennel, philipp.rostalski}@uni-luebek.de

Abstract

This project deals with the further development of a ball balancing robot (BallBot) as a testbed for complex, modular, hierarchical and distributed control strategies. A BallBot requires fast feedback control to fulfill the task to balance on a ball by itself. The developed control concept uses a hierarchical control scheme in which certain parts of the control algorithm are outsourced to the motor driver modules. The overall control scheme is intended to be able to accommodate various motor driver modules with different requirements without requiring manual retuning of controllers. In this context current step response experiments for each motor driver shield were performed for the purpose of identifying the range of relevant motor dynamics. First tests lead to the conclusion, that the motor current measurement is not accurate enough for a sufficient control. Therefore the concept needs adaption. Instead of controlling the resulting motor torques a speed control is developed. Finally the closed-loop system response of a controlled motor driver module is compared to the open loop response.

1 Introduction

In order to develop and test new control algorithms a Ball Balancing Robot (BallBot) was developed in [1]. Based on this recent project, this paper deals with its further development.

A BallBot is a system whose task is to autonomously balance on a ball. Since the system itself is unstable and has highly nonlinear dynamics, advanced and sophisticated control algorithms are necessary, which also could be used in similar fashion for issues in modern medical controlling engineering. The control of a BallBot is related to the control of an inverted pendulum.

So far mainly the assortment of the required hardware components and its setup has been realised. The hardware setup includes three motor units each containing a motor driver shield, a quadrature encoder to obtain the actual wheel speed and direction, a Faulhaber DC motor and an omni-wheel which is in physical contact to the ball to balance on [1]. Those three functional modules are connected with an acrylic disc, in a manner that they can be placed on a basketball. To be able to obtain information about the position and orientation of the body an Inertial Measurement Unit (IMU) is used, including an accelerometer and a gyroscope. As local processing unit each of the three motor driver modules is equipped with an Infineon XMC4800 microcontroller. Those microcontrollers are linked over EtherCAT, a bussystem that allows real-time communication. The EtherCAT communication takes place between a master and one or more slaves according to the EtherCAT protocol. Control schemes used in other BallBot constructions concen-

trated on controlling with full state feedback using a Linear Quadratic Regulator (LQR) [2],[3],[4]. Since one goal of this project is to develop a distributed and modular system, a different control architecture is required. A first approach is done by cascading several proportional-integral (PI) controllers similar to an approach in [5] for the respective controlled variables of the system, whereby controllers for motor speed and torque are implemented locally on the modules' own microcontrollers. The aim of this control architecture is to reach a higher flexibility of the setup. For example one should be able to change single hardware components without adjusting the entire control system. This offers the advantage, that in the process of authorisation of for example a medical product or device it might be possible to avoid having to undergo an authorisation procedure for the entire product again after carrying out certain changes of the modules. In order to test the exchangeability of components this project uses two different motor driver shields. To characterise the two different systems, the step response from pulse width modulation (PWM) to motor speed and to the motor current when the motor is stalled, is recorded for both motor driver shields with motors. Consequently the dominating time constant and the static gain were estimated. For each motor driver module a PI controller for the motor speed was tuned, with the objective to achieve similar closed-loop responses. In the following the developed control concept is presented and the results of the executed tests of the different motor driver shields will be depictured and discussed. Finally the implementation of a speed control for the two differing systems is tested and compared to the systems without implemented control.

2 Material and Methods

The aim of the developed current control concept is to keep the setup on top of the ball. Therefore current state information needs to be available to the control like the angular displacements and velocities between the upright position of the setup and the actual tilted position. This measurement can be made with the help of an IMU. To prevent the construction from tilting the controller calculates the needed PWMs for the motors to apply a torque to raise the setup (Fig.1).

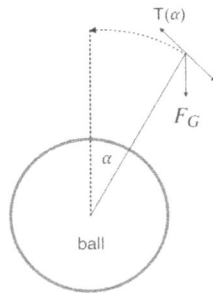

Figure 1: Due to the gravitational force F_G a torque acts on the setup and forces it to tilt (tilting angle α or for another tilting direction β). To get the tilted setup back to an upright position a counter torque $T(\alpha)$ needs to be applied to the ball.

In order to develop the control the following assumptions were made. As there are three possible planes in which an angular displacement can take place, the control law is split into separate controllers for each plane using the assumption that each plane can be considered independent. To simplify the system the angular displacement in the horizontal plane — more specifically the rotation around the z-axis — is ignored as this movement does not describe a tilting motion, but simply the setup turning around itself. In the first place the mentioned angular displacements of the two vertical planes shall be controlled to an upright position which means that the reference value for the angles α and β are set to zero.

Other BallBot projects often used model-based control with state feedback [2],[3],[4]. A different approach which is applied here is to use a cascaded PI controller and to empirically determine the control parameters.

The structure of the cascade is illustrated in Fig.2. As displayed, the master control consists out of two individual control loops both receiving input values from the IMU.

The aim of the master control consists of computing a torque vector $(T(\alpha), T(\beta))$ that acts on the ball to bring the setup in an upright position and to transfer this vector to the related motor torques $(T(1), T(2), T(3))$.

The outer part of the master control (Fig.2: Control 1) is responsible to regulate the angular position of the robot. The difference between the actual measured values and the set values is equal to the control error $(e(\alpha), e(\beta))$. Using a PI controller, which combines a fast reacting proportional control with the precise control of an integrative controller,

a resulting velocity vector $[\dot{\alpha}, \dot{\beta}]$ is calculated which contains information about the angular velocity that is needed to get the system to its initial position.

The inner controller of the master, also a PI controller (Fig.2: Control 2) is responsible to compute resulting torque vectors in reference to the vertical planes with the angular velocity errors $[e(\dot{\alpha}), e(\dot{\beta})]$ and the output of the outer control of the master as input. The torque vector $[T(\alpha), T(\beta)]$ contains the direction and amount of the torques that are needed as counterforces to the torques induced due to gravitational forces F_G in each regarded plane. Since a torque can only get applied to the ball with the help of the motors it is important to transfer and split the computed total torque to separated partial torques acting on each motor $[T(1), T(2), T(3)]$.

Until this point, all of the computing took place on the master. The next step, controlling the actual motor torques is done on each slave module. The set values are transmitted over the EtherCAT bussystem and on each microcontroller a separate feedback loop is implemented to control the motor torques of the single motors and to compute the resulting PWMs (Fig.2: Slave controller).

For this purpose one can measure the actual motor currents by using the current sense voltage output pin of the motor driver shield. The measured voltage is proportional to the motor currents, which is proportional to the motor torques. This signal is used as feedback for the torque controller.

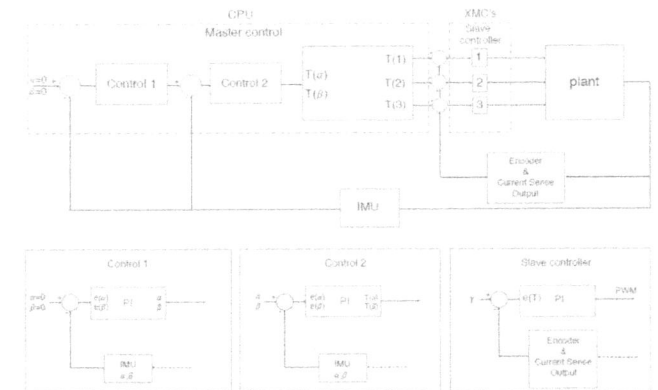

Figure 2: The control of the entire system (upper graphic) is structured into several independent controllers. Each used controller is depicted in detail in the lower graphics. While 'Control 1' and 'Control 2' are implemented on the master (CPU), the implementation of the three slave controllers takes place on each XMC microcontroller.

The advantage of this modular control is on the one hand that CPU resources are saved, due to the external processing of the final motor parameters on the XMC microcontrollers. On the other hand a modular setup is of interest because it offers the benefit to get a more flexible system architecture. Due to the controllers of the slaves being independent of the master control, changes in the hardware setup can be made more easily with only adjusting the controller on the module controller.

2.1 System Parameter Estimation

To be able to characterise the systems, a step response of the motor currents was taken. Therefore different PWMs were chosen as input parameters (V_{IN}) for the motors and, while holding the wheels in position by hand, the current output (I_{OUT}) was measured by using the current sense voltage output pin. As we assume the system to be a first order system one can use the following transfer function to characterize the system [8].

$$\frac{I_{OUT}(s)}{V_{IN}(s)} = \frac{\kappa}{1 + s\tau} \qquad (1)$$

where κ is the gain of the system, that indicates about what factor a unit step in the input is amplified and τ is the time constant, that describes the bandwidth of a first order system. With the help of the recorded data τ and κ can be estimated.

3 Results and Discussion

To characterise the systems of the two different types of motor driver shields, the step response of the motor currents was taken and the time constants τ (Fig.3) as well as the gains κ (Table 1,2) of the systems were estimated. The two shields under comparison are the Pololu motor driver shield 33926 [6] — capable to deliver up to 3A of continuous motor current — and the Pololu motor driver shield 5019 [7], delivering up to 12A of continuous motor current.

The time constant can be determined graphically by applying the tangent at the time when an input step is applied to the system. The elapsed time at the point of intersection between the tangent and the stationary value equals the time constant. For the motor driver shield 33926, the time constant τ is approximately 0.85 ms (Fig.3a), while for the motor driver shield 5019 the time constant τ is approximately 0.7 ms (Fig.3b), which means that the second system is reacting about 16% faster than the first system.

Table 1 and Table 2 show computed gains for varying input voltages for both motor driver shields. By computing the mean value one receives $\kappa = 1.69$ A/V for the motor driver shield 5019.

For the motor driver shield 33926 the computed gains for varying input voltages are strongly differing from one another with a standard deviation of $\sigma = 0.128$ and a measurement gain from 0.155 A/V up to 0.41 A/V. This indicates an inaccuracy of the measurement as usually the gain for a linear first order system should be the same for any input. According to the datasheet [6] for an output motor current of 1.5 A the related feedback current of the current

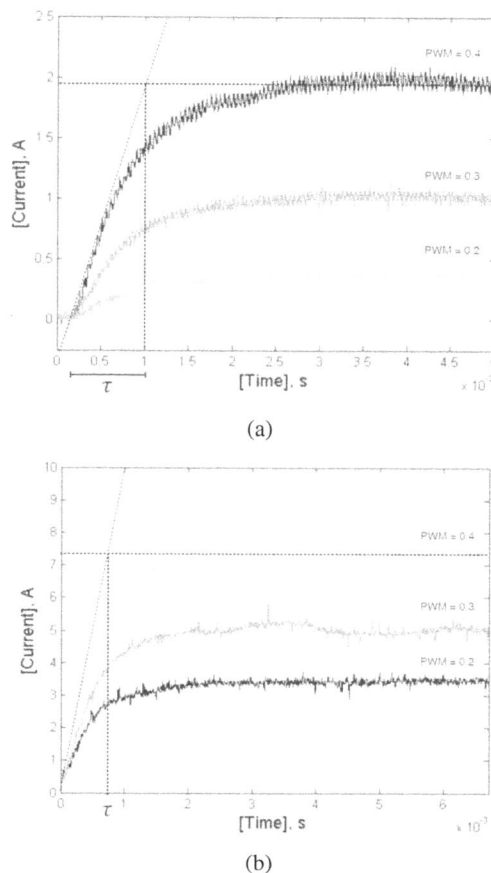

(a)

(b)

Figure 3: Step response of the motor driver shields 33926 (a) and 5019 (b) for different PWM inputs. The dotted lines were used to read out the respective gain τ

sense voltage output pin can be in a range from 2.86 mA to 4.28 mA and that the accuracy of the measurement is supposed to be better than 20%. Those statements imply that the measurement of motor currents is not very reliable for the motor driver shield 33926. One can conclude that in this case probably the currents were not measured accurate enough for a precise result.

Another point of contradiction is the measured current amplitude of the driver shield 5019. It is striking, that the computed currents seem to be constantly to high. The measured values reach up to 8 A even though the used laboratory bench is just capable of approximately 5 A output current. To transfer the measured voltages to currents a conversion factor from the reference manual of the motor driver shield was used (140 mV/A). The conversion factor of the motor driver shield 5019 is just an approximate value which can differ up to 53 % [7]. The range of the conver-

	Table 1: Gain κ for Driver Shield 5019	
Stationary value [A]	Input Voltage [V]	Gain [A/V]
3.98	2.4	1.66
5.88	3.6	1.63
8.63	4.8	1.79

	Table 2: Gain κ for Driver Shield 33926	
Stationary value [A]	Input Voltage [V]	Gain [A/V]
0.37	2.4	0.155
1.02	3.6	0.28
1.98	4.8	0.41

sion value is between 98 mV/A up to 214 mV/A. By using an adjusted conversion factor of 214 mV/A the measured values get more realistic. Still it is not for sure if the conversion factor is correctly chosen. Due to the huge inaccuracy of the measurement it is not recommended to use this motor current measurement in order to calculate the applied motor torque. Hence a PI controller was implemented to control the speed of the motor instead of the torque. The advantage of this approach is that the motor speeds can be determined more precisely by the encoder. The control parameters were empirically determined.

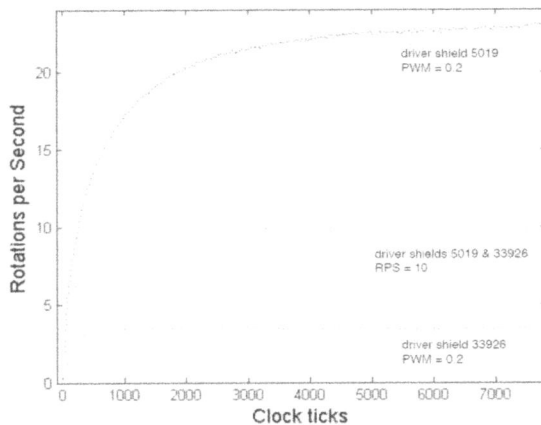

Figure 4: Comparison of the two different systems without a regulation of speed (dotted plots) for an input PWM of 0.2 compared to an implemented speed control (input: Rotations per Second (RPS)).

In Fig.4 the system responses of the differing motor driver shields on the one hand for an open loop system using a PWM as input (dotted plots) and recording the resulting step response are shown. On the other hand the system behaviour for speed regulated systems using a target wheel speed as input are depicted. By the means of those graphs one can see that the open loop systems react very different to the same chosen input step. By applying a speed control, the system dynamics get accommodated to each other, so that the closed loop behaviour of both systems can be considered as similar even though the open loop behaviour is not. That way the master control does not have to be adjusted if a slave module is exchanged. Only the control of the slave itself needs to be rewritten and accommodated.

4   Conclusion

A control architecture to achieve the balancing of a Ball-Bot with modular hardware components was introduced. The concept considered a cascade of different PI controllers which were distributed on a mastersystem and several slave modules. The main task of the controller on the master-system is to calculate the motor torques required to stay in upright position while the controllers on the motor units control the applied torque accordingly to the set values of the master. With this setup an independence of the used

hardware on the motor units seems possible. Two different actuation units were compared accordingly to their dynamic behaviour. Therefore the step responses from input voltage to stall current and therefore torque were measured. Those systems were characterised by estimating system parameters as the gain κ and the time constant τ. A relatively large deviation in the gain was recorded due to the low accuracy of the current measurement of the used hardware. To adapt the control system, instead of controlling the motor torque a speed controller was developed. With the help of this controller the closed loop responses of the different modules were very similar. The next step will be an adjustment in the control law running on the master system in order to balance the whole system. Furthermore an extension of the control to move through space is desired. Therefore a measurement of the spatial position of the robot in the room would be needed.

Acknowledgement

The work has been carried out at the Institute for Electrical Engineering in Medicine.

5   References

[1] S. Rettmer, G. Männel, and P. Rostalski *Development of a Ball-Balancing Robot platform.* ‚Universität zu Lübeck , 2017.

[2] M. J. Bjärenstam, M. Lennartsson, *Development of a ball balancing robot with omni wheels.* , Lund University, 2012.

[3] P. Fankhauser, C. Gwerder, *Modeling and Control of a Ballbot.* , ETH Zürich, 2010.

[4] T. B. Lauwers, G. A. Kantor, R. L. Hollis, *A Dynamically Stable Single-Wheeled Mobile Robot with Inverse Mouse-Ball Drive.* , Carnegie Mellon University, 2006.

[5] U. Nagarajan, A. Mampetta, G. A. Kantor, R. L. Hollis , *State Transition, Balancing, Station Keeping, and Yaw Control for a Dynamically Stable Single Spherical Wheel Mobile Robot.*, Carnegie Mellon University, 2009.

[6] Pololu, *AUTOMOTIVE THROTTLE H-BRIDGE ACTUATOR/ MOTOR EXCITER 33926.* Available: https://www.pololu.com/product/2503 [last accessed on 26.01.2017].

[7] Pololu, *Automotive fully integrated H-bridge motor driver VNH5019A-E.* Available: https://www.pololu.com/product/2507 [last accessed on 26.01.2017].

[8] IOPScience, *A Simple Method for Estimation of Parameters in First order Systems.* Available: http://iopscience.iop.org/article/10.1088/1742-6596/570/1/012001/pdf [last accessed on 07.02.2017].

Development of a Ball-Balancing Robot platform

S. Rettmer [1], G. Männel [2], and P. Rostalski [2]

[1] Medizinische Ingenieurwissenschaft, Universität zu Lübeck, sven.rettmer@student.uni-luebeck.de

[2] Institute for Electrical Engineering in Medicine, Universität zu Lübeck, {ge.maennel, philipp}.rostalski@uni-luebeck.de

Abstract

In modern intensive care units and associated devices, increased connectivity and interoperability promises additional and more reliable functionality for the patients' benefit. As medical devices and control loops involve humans in the loop and are hence safety critical, surrogate devices of similar complexity can be developed, in order to test novel techniques under non-critical conditions. In this paper, a ball-balancing robot is designed in the way as to be able to act as such a surrogate device. In addition to details on the construction of the robotic device, a first attempt at an EtherCAT-based network interface is described, which is intended for later use in hierarchical and distributed algorithms.

1 Introduction

In order to allow for efficient and automated diagnosis and therapy using intelligent medical devices and assistance systems, interoperability needs to increase. A large range of different data can then be gathered, combined and distributed, in order to implement new therapy related functionality. Therefore complexity and functionality of medical devices is rapidly increasing. Also more and more different devices are interconnected to work together in order to provide a better functionality [6]. This growth in information and functionality holds a lot of potential but also requires appropriate control strategies and algorithms. As medical devices and control loops involve humans in the loop and are hence particularly safety critical, surrogate devices of similar complexity and requirements can be developed to serve as test benches for advanced and yet-to-be-validated methods. A ball-balancing robot (Ballbot) is such a surrogate device. It is a special form of the well-known example of the so-called inverted pendulum [2]. A robotic device hereby balances on a ball to stay in an upright position otherwise it would fall off the ball. That makes a Ballbot an interesting object for control theory and control algorithms. In accordance to the robots modular design, a hierarchical control structure is presumed.

Several Ballbots have already been developed. The first Ballbot was developed in 2006 at the Carnegie Mellon University (CMU) by Lauwers et al. The human-sized robot incorporates two arms intended to carry objects or help older humans to stand up. The bot is actuated by rollers driven by DC motors which are in contact with the ball [5]. Another Ballbot was developed at the Eidgenössische Technische Hochschule (ETH) in Zürich by Frankhauser et al. This robot is of smaller dimension than the one from CMU. Special wheels, so called omniwheels, drive the ball. These wheels are able to move perpendicular to their spinning direction via a series of barrels along the wheels' cicumfer-ence [2]. Also at the Lund University a Ballbot was developed by Bjärenstamm et al. The robotic device is of smaller dimension than the one from CMU and from ETH. It is also driven by omniwheels [1]. Another Ballbot has been developed by UFactory [3]. It is also small in dimension and driven by omniwheels but the control algorithm does not rely on a mathematical model.

2 Material and Methods

The development of the Ballbot is separated in the construction of a prototype and the programming of the microcontroller.

2.1 Construction of the prototype

Two ways to actuate the robotic device are discussed here. One can be done by putting actuated rollers to the ball which move it, like in [5] and the other one by using three motors, which are rotationally fixed at 120°, with omniwheels like it is done in [1] and [3]. The approach with the actuated rollers would need at least two motors to move the robotic device. The body would then be constructed around the two rollers. In contrast to that is the approach with three motors and omniwheels. Here the wheels are positioned on the ball at a certain angle to the ball. Furhtermore the wheels are perpendicular to the surface of the ball. The second approach with the three motors and wheels is chosen because it is able to rotate around the z-axis. For this approach the motors have to react fast enough on the given motor torques by the control algorithm to keep the robotic device in an upright position. In [1] Faulhaber motors, with a maximum torque of 70 mNm, and a 3 : 1 transmission ratio are used. Therefore the same motortype is used for this project. Instead of a cog belt transmission, gear boxes are used. Fig. 1 shows the arrangement of the motors and

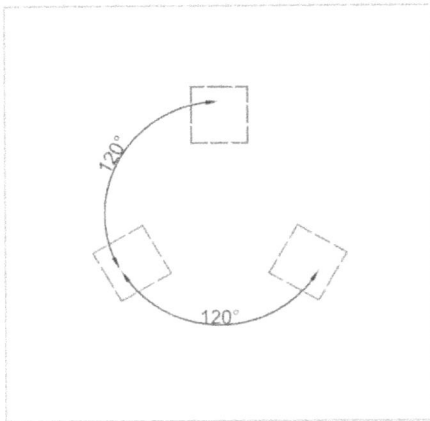

Figure 1: Top view of the robot with the arrangement of the aluminium mountings (dark grey).

their mountings to each other. Also omniwheels with two rows of barrels are used.

For the construction of the body a similar concept as in [1] and [3] is chosen. Both concepts use commercially available materials for the body and they seem easy reproduceable. The goals for the body of the prototype are: light in weight and low on costs. Also the material for the body should be rigid enough so it will not break or deform if the body falls of the ball. Therefore only plastics and light metals like aluminium are considered for the body.

When the ball is not moving, e.g. when it is clamped, the robotic device should be able to stay in an upright position on the ball without actuation due to a large support area. The support area indicates the area the wheel is circling around on the ball. It differs depending on the angle in which the wheels are positioned to the center of the ball. For example a low angle, e.g. 5°, between the wheels and the ball would cause a small support area. In contrast a higher angle, e.g. 50°, would cause a larger support area. The largest support area can be found at the equator of the ball.

The motor is applying its torque, over the wheels onto the ball. Hereby the transferred torque differs to the transmission ratio between the wheel and the ball. If the ratio is smaller than one, e.g. one revolution of the wheel to two revolutions of the ball (around the z-axis of the ball), there would be a loss in the applied torque by the motor. And if the ratio is bigger than one there would be a gain in the applied torque but at the same time the wheel is spinning faster than the ball. This means a transmission ratio of one or higher would be desired. Therefore the transmission ratio between the perimeter of the wheels and the ball has to be calculated. To consider a large variety of ratios, different ball and wheel sizes are considered. Also different angular positions of the wheels on the ball are considered. At each angle the wheels orbit with a different perimeter on the ball. Therefore at each angle a new perimeter has to be calculated. For this (1) is used where α is the chosen angle and d is the diameter of the ball used for the calculation:

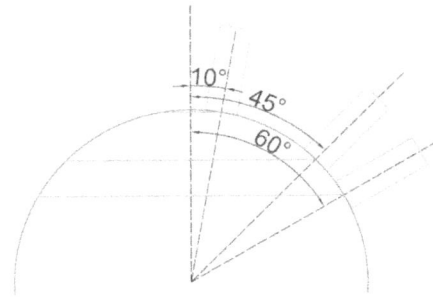

Figure 2: Ball and wheel locations for three different angles and the resulting perimeters (lines).

Table 1: Transmission ratio

45°	Wheel size	Ball size (Ø)
4.9 : 1	35 mm	244 mm
1.4 : 1	125 mm	244 mm
3.7 : 1	35 mm	181 mm
1 : 1	125 mm	181 mm
4.5 : 1	35 mm	223 mm
1 : 1.26	125 mm	223 mm

$$U = \pi \cdot d \cdot \sin(\alpha). \tag{1}$$

Both rows of the barrels on the omniwheels should be in contact with the ball otherwise it could happen that the ball will not move on one of the rows, leading to a limitiation of the movement of the ball perpendicular to the wheelorientation. This could cause disturbances and a loss of transmission.

In Fig. 2 three different angles between the ball and the wheels are shown. The lines (light grey) on the ball show the resulting perimeter at the specific angle. The support area and the transmission ratio between ball and wheel differ at each angle. At 10° the robot would circle a small support area and therefore the robotic device could fall off the ball easily. At 60° the support area is larger compared to the one at 10°. At 45° the support area is between the values of the other two angles.

Table 1 shows the results of the calculation for the angle of 45° with wheels of a diameter of 35 mm and 125 mm and balls with a diameter of 181 mm (U = 570 mm), ≈ 223 mm (U = 700 mm) and ≈ 244 mm (U = 765 mm). At 45° the perimeters changed from 570 mm to 403 mm, 700 mm to 495 mm and 765 mm to 541 mm. With the changed perimeters the ratio between the ball and the wheels are calculated. Keeping in mind that a transmission ratio of one or higher would be good every ratio smaller than one is neglected. The combination of the biggest ball with a diameter of ≈ 244 mm and the biggest wheel with a diameter of 125 mm is chosen because of the ratio being slightly abouve 1 : 1. Due to the lack of availability of wheels with a diameter of 125 mm wheels with a diameter of 100 mm are chosen and used for the construction of the prototype. Also hubs are needed to connect the wheels with the motor shaft. To keep the weight low perspex is used as the basis of the robot. For the mountings and the motorbrackets aluminium is used.

2.2 Microcontroller

The requirements on the microcontroller are that it should be fast enough to process complex control algorithms. Also the microcontroller should have enough interfaces for sensors and actuators and it should support a bussystem for the hierachical control structure.

The chosen bussystem is called Ethernet for Control Automation Technology, short EtherCAT, and it is a real-time capable bussystem first developed by Beckhoff Automation. EtherCAT uses the physical layer of Ethernet and operates on master-slave principle. The EtherCAT master sends a standard Ethernet packages to the slaves, which are are processed by the slave while they pass the slave reads the data which is designated for it and places its data to the package which is dedicated for the master. Due to this process the cycle times of EtherCAT are much shorter ($\leq$ 100 μs) than those of normal Ethernet (1 ms) [4]. Out of the few microcontroller supporting EtherCAT the XMC4800 by Infineon was chosen. Next to EtherCAT it supports common communicationprotocols such as CAN, I2C, PWM outputs and ADC inputs. It also features two hardware position interfaces (POSIF), capable of handling the signal from the motor encoders. Right now the Relax EtherCAT Kit is used. The Relax Kit supports Arduino hardware which allows the use of sensors and motorboards suitable for the Arduino. And the Relax Kit realises the physical layer for EtherCAT. In later setups the Relax Kit will not be neccessary.

For the estimation of the tilt angle of the robot and the speed of the motors suitable sensors are needed. In order to measure the motor speed optical encoders are used. For the tilt angles an inertial measurement unit (IMU) is used. With the IMU it is possible to measure the tilt angles and the angular accelerations of the robot around the x-, y- and z-axes. The IMU is connected to the microcontroller via I2C. The measurements of the IMU are later needed by the control algorithm to balance the robotic device on the ball.

For test purposes a simple EtherCAT network is established connecting one XMC4800 and a master running on a linux computer. The XMC thereby controls one motor. The encoder is connected to the pins of the POSIF, by which the data of the encoder is read out and converted into the speed of the motor. The speed of the motor is controlled with a simple PI-controller. Via EtherCAT the measured speed is send to the master and then displayed. Furthermore the set value for the motor speed can be changed on the master, which is then send to the slave, there the changes are processed and the speed of the motor is then adjusted.

With the robot resting on the perspex plate upside down, the actuation is tested. The motors are given speed and direction to simulate the movement.

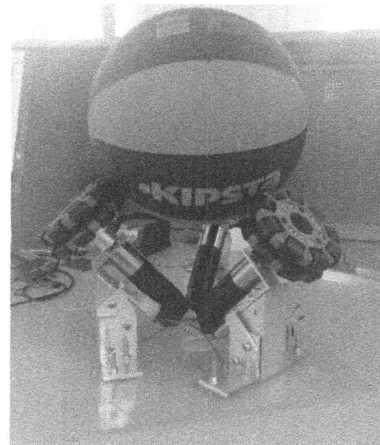

Figure 3: Final prototype, upside down, of the Ballbot.

3 Results and Discussion

Some goals for the prototype are achieved. The body of the robot weighs about 3.5 kg and is lighter than the bodies of [5] (55 kg), [2] ($\approx$ 10 kg) and presumably as light as the body in [1]. Also the costs for the used material i.e. the aluminium and the perspex are low (about 40 €). It is light enough so that it can move on the ball.

After first tests it turned out that the hubs make the whole actuators a bit longer then expected. This means that the wheels were not in full contact with the ball. Therefore the chosen angle of 45° is changed to 42°. The support area slightly decreases compared to the 45° but a better contact between wheel and ball is established. Due to the angle of 42° a transmission ratio of 1.63 : 1 is achieved. Which is near the calculated ratio of 1.4 : 1 but nearer to a ratio of 2 : 1. So far no negative side effects have occurred. The contact between the wheels and ball is good, such as both rows of the wheels have contact to the ball.

Fig. 3 shows the final prototype. It also shows the aluminium mountings and motor brackets. If the motors are turning at a fast pace the motors and the mechanical construction start to vibrate. It could be possible that the encoders are affected as well, meaning that the read out encoder data might differ. To avoid this further fixation of the motors would be neccesary.

The established EtherCAT network is capable to control one motor via EtherCAT. Also it is possible to read out the sensor data the IMU.

Fig. 4 shows one of the tested movements of the robot. The shown movement is the rotation of the roboter. In order to rotate the ball on the robotic device the three motors have to turn in the same direction. For this example the motors and the wheels, denoted by 1, 2 and 3 are turning clockwise. The result is that the robot is turning counter clockwise.

Table 2 shows all the tested movements for the robot while it is upside down. The table shows which motor has to turn and how it turns to simulate movement in a specific direction along the coordinate system in Fig. 4. Unfortunately the robot was not able to perform the movements on the

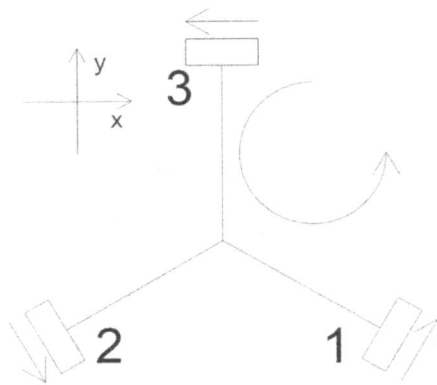

Figure 4: This figure shows one of the tested movements, rotation, of the robot. One can see the three wheels with the motors, denoted by 1, 2 and 3.

Table 2: Possible movements of the robot. To move along the negative y-axes (y-) motor 1 has to turn counter clockwise (c.c.w.), motor 2 has to turn clock wise (c.w.) and motor 3 does not turn at all (-).

Motor 1	Motor 2	Motor 3	Movement
c.w.	c.w.	c.w.	turning
c.w.	c.c.w.	-	y+
c.c.w.	c.w.	-	y-
c.w.	c.c.w.	c.w.	x+
c.c.w.	c.w.	c.c.w.	x-
c.w.	-	c.w.	y+, x+
-	c.c.w.	c.c.w.	y+, x-
-	c.w.	c.w.	y-, x+
c.c.w.	-	c.c.w.	y-, x-

ball. The used motor boards support a current of up to 3 A. So when two motors are connected to one motor board it does not provide enough current to actuate the robot on the ball. Therefore the motor boards should be replaced. To test if the motors are strong enough to move the robotic device on the ball, the motors were directly connected to a current supply with 4 A. With this the motors were capable of moving the robotic device.

4   Conclusion

So far a body for the robot has been constructed as well as a first EtherCAT network for test purposes has been established. While the robot can not balance on the ball the goals for a light weight and cheap materials are achieved. The established EtherCAT network is capable of controlling the speed of one motor via EtherCAT. It is possible to aquire the speed of the motor. One of the next steps would be the expansion of the network to use all three planed motors and the IMU. This could be done by adding another slave to the network. Instead on a laptop the EtherCAT master could be installed on a Raspberry Pi. Most of the hardware meets the requirements. One microcontroller has enough power to operate two motors or one motor and the IMU. The IMU

measures the angle and the angular velocity. Only the motor boards do not suit the requirments. As soon as the robotic device is on top of the ball the motors do not run. This is due to the maximum rating of the motor boards. The motor boards only support a current up to 3 A but to get the motors running at least 4 A are needed. Therefore the motor boards should be replaced.

A simple way to implement a controller for the robot would be to use the data of the sensors of the robot like it is done in [3]. For the design of a more complex controller a mathematical model of the robot should be established like it is done in [2]. The model in [2] could fit for the here presented robot by adjusting the parameters for weight, diameter etc. The mathematical model could also be used for a simulation of the system.

As soon as the robot is capable of balancing on the ball the movement in space should be considered as the next step. Therefore the change from a stationary power supply to batteries would be necessary.

Acknowledgement

The work has been carried out at the Institute for Electrical Engineering in Medicine, Universität zu Lübeck. I wish to acknowledge the help provided by Phillip Wendland, Morten Mey and Tobias Drewes.

5   References

[1] M. J. Bjärenstam and M. Lennartsson, *Development of a ball balancing robot with omni wheels*. Lund University, Lund, Sweden, 2012.

[2] P. Frankhauser and C. Gwerder, *Modeling and Control of a Ballbot*. Eidgenössische Technische Hochschule Zürich, Zürich, Switzerland, 2010.

[3] UFactory, *How to make a Ball Balancing Robot*. Available: http://www.instructables.com/id/How-to-make-a-Ball-Balancing-Robot/ [last accessed on February 10, 2017]

[4] EtherCAT Technology Group, *EtherCAT Technical Introduction*. Available: https://www.ethercat.org/en/technology.html [last accessed on February 10, 2017]

[5] T. Lauwers, G. A. Kantor and R. L. Holli, *A dynamically stable single-wheeled mobile robot with inverse mouse-ball drive*. In: Robotics and Automation 2006, ICRA 2006, Proceedings 2006 IEEE International Conference, pp. 2884–2889, 2006.

[6] J. D. Bronzino, *Medical Devices and Systems*. CRC Press, 2006.

8

Medical Imaging

Simulated Monochromatic X-ray Images by Forward Projection Utilizing the Lambert-Beer Law

K.L. Soika[1], S. Melnik[2] and M.P. Heinrich[3]

[1] Medizinische Informatik, Universität zu Lübeck, kira.soika@student.uni-luebeck.de
[2] Fraunhofer Institute for Production Systems and Design Technology (IPK), Berlin, steffen.melnik@ipk.fraunhofer.de
[3] Institute of Medical Informatics, Universität zu Lübeck, heinrich@imi.uni-luebeck.de

Abstract

X-ray computed tomography (CT) is a non-invasive technique using radiography for imaging inner structures of specimens. Three-dimensional CT scans are constructed from X-ray images recording the attenuation of X-rays leaving the X-ray tube. Common reconstruction algorithms assume that the attenuation of X-rays follows an exponential function over distance and can be described by the Lambert-Beer law. However, this assumption is only valid for a monochromatic spectrum of X-rays. In practice, X-ray tubes have a polychromatic spectrum, which leads to beam hardening artifacts in CT scans. Beam hardening is typically identified by false density gradients in regions of homogeneous matter and streaks between high attenuating structures. In this paper, we present an approach to simulate monochromatic X-ray images, which could be used for beam hardening correction. The approach uses the Lambert-Beer law to determine the attenuation of X-rays and thereby the intensity of each pixel in the simulated X-ray image.

1 Introduction

X-ray computed tomography (CT) is a non-invasive technique based on radiography for imaging inner structures of specimens [1]. Today CT imaging finds a wide range of applications. Possible application fields are structure determination, material testing, medical diagnostics and therapy. In the following, we will give a brief introduction into the physical principles, the problems arising due to artifacts and simulation of X-ray imaging.

Three-dimensional CT scans are constructed from two-dimensional X-ray images recording the attenuation of X-rays, whose source is the X-ray tube. X-rays are matter penetrating rays, which show an interaction with the crossed matter. For CT imaging, there are two relevant interaction effects: absorption, caused by the photoeffect, and scattering, caused by Rayleigh and Compton scattering. The different interaction processes affect the attenuation of X-rays. The probability of interaction is dependent on the X-ray energy and the mass density of the material the X-ray interacts with. The attenuation of the X-rays is measured by an image detector. The recorded attenuation is depicted in X-ray images. To construct a CT scan, several X-ray images with different projection angles are acquired. For the purpose of constructing a three-dimensional CT scan, a cone-beam is utilized for obtaining X-ray projection images.

CT imaging suffers from artifacts caused by inaccuracies during the image reconstruction. If X-ray images were to be acquired using a monochromatic energy, a high count of photons, an infinite image detector resolution, no scatter and no motion, a CT scan which perfectly reflects the reality could be reconstructed. Since this is not possible in practice artifacts appear in the reconstructed images [2].

An artifact, which is caused by the polychromatic spectrum of the X-ray tube, is beam hardening. In practice, the X-ray tube has a polychromatic spectrum, which causes the attenuation of X-ray energy to become a superposition of exponential functions. This is due to the fact that different energies are not attenuated in a uniform manner. Lower energies of the spectrum are attenuated more strongly. The inverse Radon transform assumes that the attenuation of X-ray energy can be described by the Lambert-Beer law [3]. This implies that the attenuation follows an exponential function over distance. However, this assumption is only valid for monochromatic energies. The diverse attenuation of polychromatic energy causes "hardening" of the X-ray beam, which produces inconsistencies in the measured X-ray images and beam hardening artifacts in the reconstructed CT scan. Beam hardening is typically identified by false density gradients in CT scans of homogeneous matter and streaks between high attenuating structures [1]. This is visualized in Fig. 1, which shows a CT scan of hip prosthesis in a human body.

Simulated X-ray images can be used to reduce artifacts in three-dimensional CT scans. In general there are two types of X-ray image simulation, the analytical and the statistical one. The analytical simulation constructs specimens using known analytical equations. An example is the forward

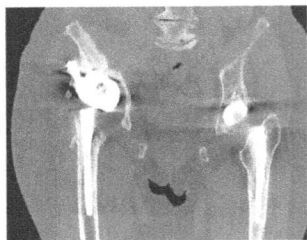

Figure 1: CT scan of hip prosthesis in a human body with beam hardening artifacts specially around the acetabulum.

projection of mathematical objects like cuboid, cone, ellipse and cylinder. A virtual specimen constructed from these geometrical objects is called phantom. The simulated X-ray images are generated based on the attenuation of line integrals along their path from X-ray tube to image detector. In contrast, the statistical simulation predicts the behaviour of the X-rays using a random number generator and the physical probability of interaction processes. For example the Monte-Carlo simulation for predicting scatter distribution is a statistical simulation [4].

In the past, different techniques were developed to reduce beam hardening, e.g. linearization [5], dual energy [6] and iterative post-reconstruction [7].
In this paper we present an approach, which performs an analytical simulation by using the Lambert-Beer law to determine the attenuation of the X-rays. Therefore, we show how to construct phantoms using mathematical objects (Section 2.2), calculate the interaction intervals of X-ray (Section 2.3) and how the Lambert-Beer law, presented in section 2.4, can be used to generate monochromatic X-ray simulation images (Section 2.5). Simulated monochromatic X-ray images are free from beam hardening and can be used for artifact correction.

2 Material and Methods

The simulation of monochromatic X-ray images is achieved by forward projection. A simulation by forward projection requires a known phantom and a known setup of the projection environment. Having a phantom in the simulation environment enables us to determine the attenuation of each X-ray beam. Before we explain the calculation of intervals, we briefly show how phantoms can be constructed. Afterwards, a short recap of the Lambert-Beer law is given. Finally, we demonstrate how to simulate monochromatic X-ray images by forward projection utilizing the Lambert-Beer law.

2.1 Forward Projection

As mentioned before, the presented simulation of monochromatic X-ray images is an analytical forward projection using cone-beam. Important for forward projection is prior knowledge of the phantom and how the X-ray tube and the image detector are positioned in the environment.

Fig. 2 shows the cone-beam projection environment and how the coordinate systems of the X-ray tube and the image detector are set with respect to a defined origin coordinate system. Coordinates can be easily converted from one system to another via transformation matrices. Knowing the setup of the environment, a line is drawn from the X-ray tube to the pixel centre of each pixel. For each line, the interval within the phantom needs to be known in order to calculate the pixel intensity. Therefore it is essential to know the projected three-dimensional phantom as well as the exact interval of interaction. These issues will both be addressed in the following sections.

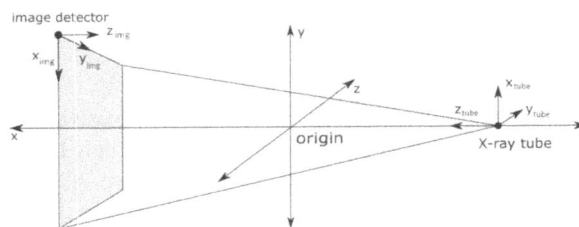

Figure 2: Cone-beam projection environment with origin coordinate system (middle), X-ray tube coordinate system (right) and image detector coordinate system (left).

2.2 Phantom Model

For simulation we are able to construct several three-dimensional phantoms by using geometric objects and logical operators. Cuboid, cone, ellipse and cylinder were used as geometrical objects. To construct more complex phantoms, logical and (AND) and logical difference (DIFF) operators are used to combine geometrical objects. In addition, the objects can be manipulated by scaling, rotation and translation. It is possible to construct a wide range of phantoms, having the basic geometric objects combined with AND and DIFF. Even phantoms like hip prosthesis can be build. In Fig. 3(a) a cupped quarter sphere is shown. It is constructed by an ellipse as basic object and two cuboids and a second ellipse as DIFF objects. Fig. 3 also shows a three-dimensional phantom of a hip prosthesis 3(b), which is generated by a number of basic objects combined with logical AND and DIFF objects.

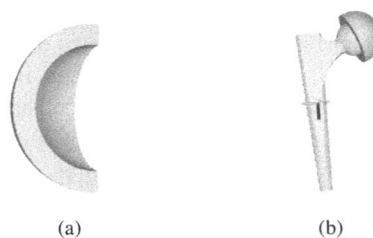

(a)   (b)

Figure 3: Phantom models generated by geometrical objects combined with logical operators. (a) shows cupped quarter sphere and (b) shows a hip prosthesis.

2.3 Interval Determination

In order to determine the interval where the X-ray interacts with the phantom, the intersection points of the X-ray and surface of the phantom are analytically calculated. The resulting three-dimensional intersection points can be easily converted into a one-dimensional line. In our approach the transformation from a three-dimensional intersection point into a one-dimensional line was done by taking the distance between the intersection points and the X-ray tube into account. Subsequently, the intervals of interest can be measured from the one-dimensional line. To calculate the intervals of interest for phantoms constructed via logical operator objects, the intersection points of all objects need to be identified. Having all intervals of the logical operation objects, the interval can be determined by building the conjunction (AND) or the difference (DIFF) of the basic object interval and all logical object intervals. Fig. 4 schematically visualizes this process.

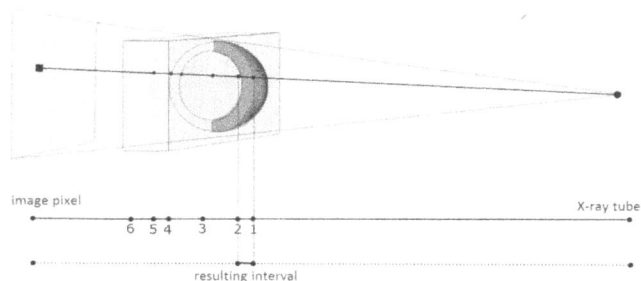

Figure 4: The upper part visualizes the forward projection environment with the X-ray tube, image detector, three-dimensional phantom and geometrical objects. The phantom is constructed by an ellipse as basic object and two cuboids (for readability, only the one which affects the interval is visualized) and a second ellipse as DIFF objects. A projection line from X-ray tube to a pixel centre of the image detector crossing the phantom is visible. The first line shows all intersection points of the projection line and the basic (intersection points: 1, 5) and DIFF solids (cuboid intersection points: 3, 6; ellipse intersection points: 2, 4) of the three-dimensional phantom on a one-dimensional line. The second line shows the resulting intersection points and the interval, which is taken into account for projection.

2.4 Lambert-Beer Law

The Lambert-Beer law states a linear relationship between the absorbance and the thickness of absorbing material [8]. On the basis of the Lambert-Beer law, the intensity of an attenuated X-ray, crossing through an absorbing object, can be describes as follows [8]:

$$I = I_0 exp^{-\int_L \mu(x)dx}, \qquad (1)$$

where I_0 is the X-ray energy before entering an object, $\mu(x)$ is the linear attenuation coefficient of the actual material and L is the length the X-ray interact with the object. I

describes the intensity of the attenuated X-ray after leaving the object.

2.5 Simulated Images

As previously described, a forward projection along the lines from the X-ray source to each pixel centre is used for the calculation of simulated X-ray images. During the forward projection the Lambert-Beer law is used to calculate the attenuated X-ray energy for every pixel. In order to use (1) to calculate the attenuated X-ray energy, the X-ray energy used for simulation, the linear attenuation coefficient and the length along which the X-ray interacts with the phantom needs to be known. The monochromatic energy used for simulation has to be chosen in a manner that no total absorption occurs. The linear attenuation coefficient can be obtained from the cross sections from the epdl97 database [9]. The length the X-ray interacts with the phantom can be derived as described in section 2.3. Having all variables of (1) defined, the attenuated ray energy can be calculated for every pixel and thus the simulated X-ray images can be generated.

3 Results and Discussion

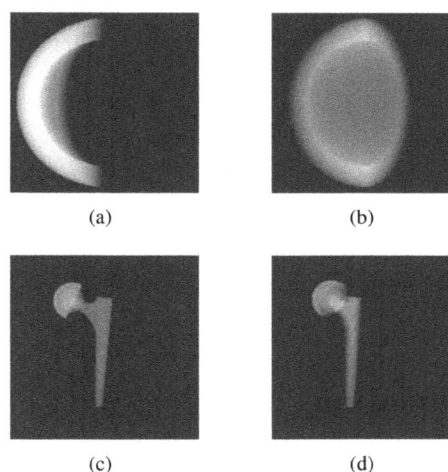

Figure 5: Simulated monochromatic X-ray images from cone-beam from different projection angles. X-ray images on top (a, b) show projections of a cupped quarter sphere phantom (see Fig. 3(a)). The absorbing material used for the cupped quarter sphere phantom was silver. X-ray images at the bottom (c, d) show projections of a hip prosthesis phantom (see Fig. 3(b)). For the hip prosthesis phantom, titanium was mainly used as absorbing material. Next to this, magnesium a lower absorbing material was used for the femoral head. Simulations were done with a monochromatic energy of 80 keV and a detector resolution of 256x256 pixel. In the simulated images white pixel represents high absorption and black pixel no absorption.

The outcome of our proposed approach for generating simulations yields simulated monochromatic X-ray images.

In Fig. 5 X-ray images of the cupped quarter sphere and the hip prosthesis phantom, which were visualized in Fig. 3, are shown. The X-ray images were simulated using a cone-beam from different projection angles and a detector resolution of 256x256 pixel. The monochromatic energy was set to 80 keV. In the resulting images, dark areas represent low attenuation and bright areas represent high attenuation. The different pixel intensities in Fig. 5(a) and Fig. 5(b) are caused by the different thickness of the phantom in ray direction. In Fig. 5(c) and Fig. 5(d) the different pixel intensities are caused by the different materials the phantom consists of as well as the different thickness.

As only primary rays with monochromatic energy were used for imaging, the resulting images are scatter and beam hardening free.

For evaluation X-ray images were simulated using Monte-Carlo simulation with a polychromatic spectrum. Only primary rays were considered. An example is shown in Fig. 6. The simulated monochromatic X-ray images were compared to full Monte-Carlo simulated X-ray images. The comparison showed a good matching of the pixel intensities and thus of the amount of attenuation.

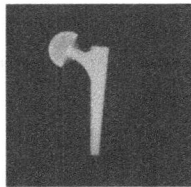

Figure 6: Monte-Carlo simulated polychromatic X-ray image from cone-beam respecting only primary rays. Image shows a simulation of the hip prosthesis phantom visualized in Fig. 3(b).

4 Conclusion

In this work monochromatic X-ray images were simulated by a forward projection which uses the Lambert-Beer law. Furthermore, a possible procedure for generating complex phantoms was presented. The approach allows the user to construct phantoms like hip prosthesis for simulation by simply using geometrical objects combined with logical operations. Once the phantom is placed in the projection environment,the interval, which has to be taken into account for attenuation, can be calculated for each ray. The intervals are calculated by converting the intersection points of ray and phantom surfaces into a one-dimensional line. On the basis of the interval, the attenuated energy of each ray is determined by the Lambert-Beer law. Subsequently, the intensity of each pixel can be identified. Since the simulation is performed with monochromatic energy and only primary X-rays, the resulting image is scatter and beam hardening free.

In further work the simulated monochromatic X-ray images can be used to generate correction images. Therefore the original image needs to be preprocessed, so that it is scatter free. For example this can be done by subtracting a scatter image generated by using a modified Monte-Carlo simulation. Having a scatter free original image and the simulated monochromatic X-ray image, the differences between the two images can be used to calculate a correction image. Subtracting the correction image from the original image before three-dimensional reconstruction will remove inconsistencies which cause beam hardening artifacts.

Acknowledgement

The work has been carried out at Fraunhofer IPK, Berlin, and supervised by the Institute of Medical Informatics, Universität zu Lübeck.

5 References

[1] J. J. Lifton, A. A. Malcolm, and J. W. McBride, "The application of beam hardening correction for industrial X-ray computed tomography," in *Proc. 5th Int. Symp. NDT in Aerospace*, 2013.

[2] F. E. Boas and D. Fleischmann, "CT artifacts: causes and reduction techniques," *Imaging in Medicine*, vol. 4, no. 2, pp. 229–240, 2012.

[3] J. D. Ingle Jr and S. R. Crouch, "Spectrochemical analysis," *Old Tappan, NJ (US); Prentice Hall College Book Division*, 1988.

[4] J. Hsieh, "Computed tomography: principles, design, artifacts, and recent advances," SPIE Bellingham, WA, 2009.

[5] G. T. Herman, *Fundamentals of computerized tomography: image reconstruction from projections*. Springer Science & Business Media, 2009.

[6] A. Coleman and M. Sinclair, "A beam-hardening correction using dual-energy computed tomography," *Physics in medicine and biology*, vol. 30, no. 11, p. 1251, 1985.

[7] O. Nalcioglu and R. Lou, "Post-reconstruction method for beam hardening in computerised tomography," *Physics in medicine and biology*, vol. 24, no. 2, p. 330, 1979.

[8] H. H. Telle, A. G. Ureña, and R. J. Donovan, *Laser chemistry: spectroscopy, dynamics and applications*. John Wiley & Sons, 2007.

[9] D. E. Cullen, J. H. Hubbell, L. Kissel, *et al.*, "Epdl97: The evaluated photon data library, 97 version," *UCRL-50400*, vol. 6, no. 5, pp. 1–28, 1997.

Failure detection of X-ray tubes for medical devices
– a case study –

M. Mecking[1] and T. Hipp[2]
[1] Medizinische Ingenieurwissenschaft, Universität zu Lübeck, marie.mecking@student.uni-luebeck.de
[2] Siemens Healthcare GmbH, Erlangen, tobias.hipp@siemens.com

Abstract

Unpredicted failures result often in downtimes which lead to profit loss, unsatisfied customers and risks for patients. Preventive maintenance is for X-ray tubes too cost-intensive, especially because there is a variance in the meantime to failure. Therefore, historical data from log-files were extracted, analyzed and prepossessed for the classification of failure indications. The two classification methods Random Forest (RF) and Support Vector Machine (SVM) were compared to select the best classifier. To forecast future parameter values a Autoregressive Integrated Moving Average (ARIMA) model was used. The failure prediction with RF had the lowest failure rate and achieved an average class failure of 5%. Because of the non-linear processes of the data set ARIMA wasn't a suitable model to forecast this data. Therefore, a hybrid forecasting model is suggested. However if the forecasting method is improved based on the results of the classification a failure prediction with this data is possible.

1 Introduction and Previous Work

The Failure of medical equipment can cause a big loss in time, money and endangering lives due to unpredictable downtimes. Predictive maintenance can in general reduce downtimes up to 45% and saves costs up to 12% [1]. The X-ray tube is one of the most critical part of a X-ray examination device and the clinical workflow is affected or even stopped in case of a failure. Currently reactive exchanges of the X-ray tube is mainly performed, because of high material costs, which make preventive maintenance unprofitable. However the X-ray tubes are already equipped with sensors that provide a great evaluation basis for the status of X-ray tubes, so the fundamentals for predictive maintenance exist and already to the customer delivered x-ray systems can be monitored. The sensor parameters are written to a log-file each time radiation is released. Log-files from the single systems are transferred to a data warehouse from where the data can be accessed.

There are many ways of a X-ray tube to fail e.g. the loss of vacuum, too much arcing, overheating or breakage of a focus filament. Those failures are not clearly observable from the sensor data, but fluctuations of the parameters can point to them. In [2], Fukuda et al. observed historical sensor data of a failing conventional X-ray tube. A noticeable fluctuation of the air kerma, a radiation quantity, were detected prior to a failure in comparison to three other tubes. The authors suggested to use the slope over time of air kerma to predict breakdowns. For prediction of system failures several machine learning algorithms can be used like Random Forest (RF), Support Vector Machine (SVM) or Neural Networks (NN). In areas e.g. electromechanical drive systems

the performance of RF is slightly better than SVM on the drive system data set in [3] in contrast to the comparatively bad results with NN. The measured data where besides the usual vibration rate, also voltage and current for the use in machine learning algorithm.

Forecasting short-term classified machine failures is in general possible with Autoregressive Integrated Moving Average (ARIMA) models, NN models or SVM models [4]. By combining models like ARIMA and NN better results in predicting linear and nonlinear time series are achieved [5]. Though in [6], Fan et al. achieved good results in using ARIMA for failure prediction. However further investigations in this topic will consider hybrid models to improve prediction reliability [9].

This case study analyzes the possibility of failure prediction with RF or SVM and ARIMA with a data set aggregated from X-ray tube log-files.

2 Materials and Methods

In this paper features, which are extracted from log-files of X-ray tubes will be analyzed, preprocessed and transformed to generate features. The two models, RF and SVM, were compared to determine the method for the highest sensitivity on machine failure prediction and to evaluate the important features. On that base ARIMA is implemented to predict tube failures half a month ahead. The whole process is presented in fig. 1.

From the data warehouse log-files in the format presented in Table 1 were extracted which includes measurements that describe the generation of the X-rays over a certain focus

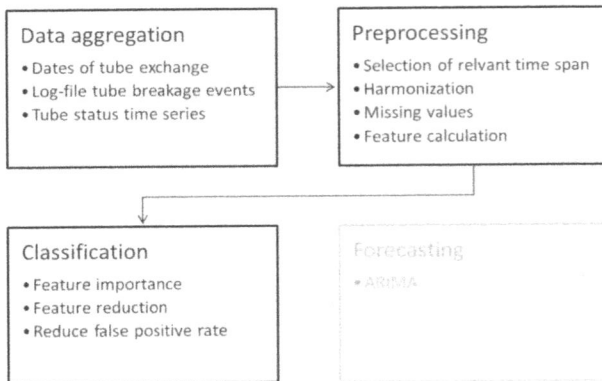

Figure 1: Process of data-preperation and modelling

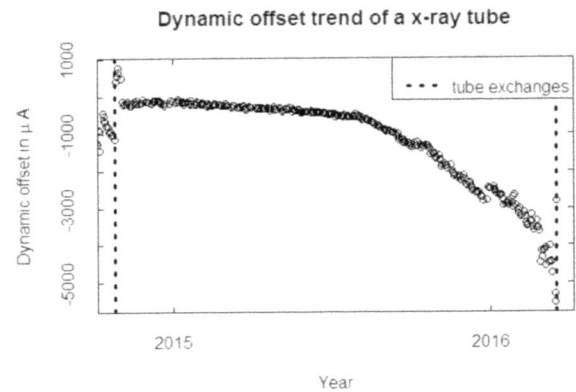

Figure 2: Qualitative trend of dynamic offset with marked downtimes over the lifetime of a X-ray tube.

filament, the run-time and the intensity of the radiation. In this first approach it seemed likely to observe mainly the failures of the filaments, because they are more correlated to those parameters than the other failures. The analyzed X-ray tubes have a small and a large focus and are operated in two different modes. The small focus is in average 4.5 times more often used than the large focus and has consequently a higher wear rate, which results in higher failure rates caused through the breakdown of the small focus. Additionally one mode with low load operation is in average 3.5 times more frequently used, especially in correlation with the use of the small focus, than the high load mode.

Table 1: Format of the data set.

Serial ID	Time	Mode	Focus	Measurements e.g dynamic offset
111111	Time 1	high	large	-374
111111	Time 2	low	small	-5472

In a first approach the data set was homogenized by analyzing just the events with small focus and lower load. As in fig. 2 the wear over time of the small focus is observable with the dynamic offset, a parameter which regulates the load on the thinning filament over time of operation; the thinner the filaments gets through heat induces material elimination the less load is necessary to generate the same scan conditions and the probability of breakage is higher. However defining a concrete threshold of the dynamic offset to predict filament failures were not sensitive enough because of the high variance of the dynamic offset values during breakage. In general the observed measurements are not mainly taken for predictive analysis unlike in other areas like electromechanical drive systems [3].

Before modeling, the data set was cleaned from missing values and just time series with enough data points were selected and fitted so that one time series contains just one tube failure. At the end 121 time series are included in the analysis. New possible features were calculated by the cumulative sum and the slope of the parameters over the time series as in [2] suggested. In the end 72 possible features

were generated from 24 parameters. The data set was reduced by hourly aggregation. Then the data set was labeled with three classes. The first class contains approximately 98% of the data set and represents the healthy status of the tube, the second class contains all data points within the 15 days time range, whereas the day of the breakdown is labeled with the third class. Dates of tube failure, the dashed lines in fig. 2, were determined from certain log events and service exchange dates. Due to the confidential character actual values of the features where replaced by fictive values.

2.1 Feature selection and failure prediction

The prepossessing and modeling was implemented with the free software environment R [7]. The sensor data set from the X-ray tube was classified with RF and SVM to evaluate the best failure prediction for this case. The general requirements for the model were to handle an imbalanced data set with few features in comparison to the number of observations. The true error of the models were estimated with cross validation. By comparison of accuracy and the average class error (ACE) the best model was selected.

RF is a ensemble method where many decorrelated trees are generated and averaged. An advantage of RF is the calculation of the feature importance, which is based on the accumulated partition performance of the features at each split in a tree [8]. This characteristic was mainly used to reduce the false positive rate on the second and third class, to reduce features and to prevent overfitting. A low false positive rate should prevent tube exchanges on a too early state.

SVM is a classification technique where the features are transformed into a high dimensional feature space and are separated by a hyperplane [8]. For this study the SVM has to seperate the data set into three classes which is implemented by solving several two-class problems.

2.2 Forecasting

For this study a forecasting method was needed which was suited for a non-stationary and non-seasonal data set for short-term forecasting up to a month. In [6] ARIMA is preferred for short-term time series analysis over NN, which is more effective on long-term time series. ARIMA integrates autoregressive with moving average methods by differencing consecutive time series values in dependence of the differencing order and then apply the methods on it. Differencing leads to a compensation of the non-stationarity.

3 Results and Discussion

Time series of 121 tubes were analyzed. On average a tube operated for less than 2 years, but there is a huge standard deviation of over one and a half year. Also at other features like the minimal dynamic offset at the time of failure, the deviation is about 40% in comparison to the average value. Therefore, no proper threshold for the features could be defined and machine learning algorithms were necessary. The results of a basic RF onto the data set with cross validation is relatively good regarding the accuracy, which is quite high because of the unbalanced data set. A better conclusion on the performance is the ACE which is also quite low with 8%. Class three has a high error rate with 20,6% through the low representation of the class in the data set. In comparison to RF, the basic SVM had a bad performance on the second and third class such that an ACE of 55% was reached.

For feature reduction the importance of the variables were observed. Based on the Gini index the features were selected. It is a rate of dispersion which is calculated for each node [8]. It was apparent that the accumulated features of the parameters and all features which contains data of the dynamic offset were the most important. The slope of the time series were the least important. The models were tested on their performance again with the selected features. The classification accuracy increased through the feature reduction. Nevertheless the ACE with 44% of the SVM are still unacceptable in comparison to the results of RF. There an ACE of 5% was achieved, mainly because of the dropped error rate of the third class from 20,6% to 11,8%. The results are listed in Table 2.

Table 2: Model comparison

Step	Model	Accuracy	ACE
Basic	RF	99,88%	8%
	SVM	97.93%	55%
Feature reduction	RF	99,89%	5%
	SVM	98,69%	44%

In contrast to the failure classification of the RF the results of ARIMA on the time series are not promising. At the end of operation time of the tube none of the forecasted values

of the features were a good approximation for a appropriated prediction and in general below real parameters. This is probably caused by the weaknesses of the ARIMA model, that includes just historical data in the model and assumes on that basis future values which are a linear function of the elapsed and considered time and errors [6]. So at the end ARIMA assumes that the underlying process is linear. This linearity assumption is apparent in fig. 3 where sometimes error predictions of up to 20% of the actual value were predicted through the trend to a linear factor. The error accumulates over all time series which are needed as features and that result in a classification. Therefore, a model is needed that can include prior data and is able to predict non-linear processes to improve the forecasting specially at the end of the operation time. During the normal lifetime the features are evolving quite linear. For such a case Khashei and Bihari [9] propose to combine the models to use the advantages of both. The advantages of NN are the consideration of prior information, the generalization and handling of non-linear processes. In their paper the hybrid model was successfully tested on three different time series in comparison to pure ARIMA and pure NN.

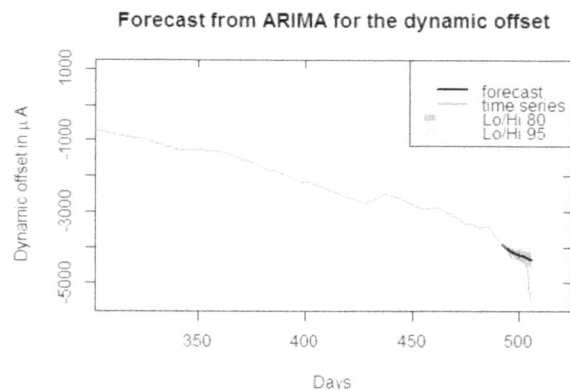

Figure 3: Qualitative forecast of the dynamic offset with ARIMA. It is apparent that huge differences between the forecast and the real value can occur.

4 Conclusion

This study analyzed the possibility of failure prediction with RF or SVM and ARIMA on the basis of X-ray tube parameters from log-files. The failure prediction of RF was significant higher with 5% as the average class error than with the SVM which achieved 44% average class error as can be seen from Table 2. The highest error was caused through the third class because of the small representation in the data set. During the feature engineering it was noticed that the slope over time of the parameters were not important for the classification of the system status as in [2] suggested. The most important features were the accumulated sums of the aggregated parameters and the dynamic offset.

To improve reliability of the model the data set can be expanded by the observations with a high load and with a small focus. The failure detection variety of the model can be expanded by applying the observations with a large focus with high and low load on to the model. Although the occurrence of tube failures caused through large focus defects are more infrequently, because it is less sensitive and used. Another possibility to expand the model is to search for more events in the log-files that indicate a specific failure and train those failures to the model. That would probably explain tube failures which appear quite early in comparison to the averaged operation time.

Given that the approach of using ARIMA to forecast the time series for failure prediction was unsuccessful, another approach is suggested. ARIMA performed well on the linear parts of the measurement but couldn't handle non-linearity in the long term. Therefore, further investigations should be done to include non-linearity. The combination of ARIMA and NN could be a solution.

Acknowledgement

This work has been carried out at Siemens Healthcare GmbH, Erlangen and supervised by the Institute for Neuro- and Bioinformatics, Universität zu Lübeck. We want to thank our colleagues at Siemens Healthcare GmbH for their support in the research.

5   References

[1] G. P. Sullivan, R. Pugh, A. P. Melendez and W. D. Hunt, *Operations & Maintenance Best Practices - A Guide to Achieving Operational Efficiency*. U.S. Department of Energy, Release 3.0, p. 52, August 2010.

[2] A. Fukuda, K. Matsubara and T. Miyat, *Estimation of the life expectancy of a filament of the conventional X-ray unit: a technical case study*. In: Radiological Physics and Technology, vol. 8, no. 5, pp. 107–110, 2015.

[3] C. Lessmeier, J. K. Kimotho, D. Zimmer and W. Sextro, *Condition Monitoring of Bearing Damage in Electromechanical Drive Systems by Using Motor Current Signals of Electric Motors: A Benchmark Data Set for Data-Driven Classification*. At: European Conference of the PHM Society 2016, Paderborn University, Paderborn, 2016.

[4] R. Adhikari and R. K. Agrawal, *An Introductory Study on Time Series Modeling and Forecasting*. LAP Lambert Academic Publishing, Germany, 2013.

[5] G. P. Zhang, *Time series forecasting using a hybrid ARIMA and neural network model*. In: Neurocomputing, vol. 50, pp. 159–175, 2003.

[6] Q. Fan and R. H. Fan, *Reliability Analysis and Failure Prediction of Construction Equipment with Time Series Models*. In: Journal of Advanced Management Science, vol. 3, no. 3, pp. 203–210, 2015.

[7] R Development Core Team, *R: A Language and Environment for Statistical Computing*. R Foundation for Statistical Computing, Vienna, 2008. Available: http://www.R-project.org [last accessed on 2017-01-16].

[8] T. Hastie, R. Tibshirani, J. Friedman, *The Elements of Statistical Learning*. Springer Verlag, 2. edition, pp. 241 f., 417f., 587 f., New York, 2009.

[9] M. Khashei and M. Bihari, *An artificial neural network (p, d, q) model for timeseries forecasting*. In: Expert Systems with Applications, no. 37, pp. 479–489, 2010.

Influence of patient motion on the depiction of microcalcifications in digital tomosynthesis

M. Eulers[1], A. Cordes[2] and J. Barkhausen[3]

[1] Medizinische Ingenieurwissenschaft, Universität zu Lübeck, mathias.eulers@student.uni-luebeck.de
[2] Institut für Medizintechnik, Universität zu Lübeck, cordes@imt.uni-luebeck.de
[3] Klinik für Radiologie und Nuklearmedizin, Universitätsklinikum Schleswig-Holstein Campus Lübeck, joerg.barkhausen@uksh.de

Abstract

Within the last decade digital breast tomosynthesis emerged as a new imaging modality for the early detection of breast cancer. Whereas the advantages have been demonstrated in many clinical trials, there are still controversies about the accuracy for the detection of microcalcifications. Our study aims to analyze the influence of motion on the visualization of microcalcifications. Since motion parameters and artifact-free ground truth data are often not available for patient data, a database for motion quantification in digital tomosynthesis is generated. Therefore, a microcalcification phantom is placed on a movable sample holder and projection data sets are acquired for different phantom positions. An appropriate combination of these data sets allows a simulation of object motion during the image acquisition. Based on this framework, the influence of different motion parameters such as the position increment and the time point of movement are investigated.

1 Introduction

Breast cancer is the most common malignant tumor in Europe, with up to 70000 new cases per year in Germany [1]. Microcalcifications are one of the most important signs of malignancy and the key finding to detect in situ and early invasive cancers in screening mammography and digital tomosynthesis. Various criteria such as the number of calcifications, the spatial distribution or the morphology of single microcalcifications can help to estimate the probability of malign disease and avoid unnecessary biopsy. Moreover the experience of the reporting physician is an important criteria for a correct diagnosis [2].

Digital tomosynthesis (DT) is an x-ray based limited angle imaging technique that allows a reconstruction of cross-sectional images from a limited number of low dose two-dimensional projections. Compared to conventional mammography, the benefits of DT include a reduction of overlapping structures leading to an improved sensitivity and specificity of breast cancer detection. However, DT requires a longer scan time (up to 25 seconds) and motion artifacts as a result of patient movement during the acquisition process are more likely to occur. These artifacts appear as blurring in the reconstructed images. As a consequence, the reduced visibility of small structures such as microcalcifications may cause misdiagnosis and inappropriate treatment decisions.

In order to define an upper limit of patient motion that can be tolerated during an acquisition without reducing the diagnostic value of the tomographic images, this contribution presents a systematic study on the influence of motion on the image quality and the depiction of microcalcifications. The paper is organized as follows. In section 2 a description of the imaging geometry, the phantom and motion simulation is given. Furthermore, the simultaneous algebraic reconstruction technique (SART) is introduced. Finally, the results are presented in section 3.2 and an outlook is given.

2 Material and Methods

As far as known for patient data, motion parameters and artifact-free ground truth images are not available. In order to investigate the influence of patient motion on the resulting image quality, we therefore used a microcalcification phantom that was placed on a movable sample holder and acquired projection data sets for different phantom positions.

An appropriate combination of these data sets allows a simulation of object motion during the acquisition. This approach offers the important advantage that the exact phantom position is known for each projection image. A detailed description of the used materials and method is given below.

2.1 Imaging Geometry and Phantom

For this study, all data sets were acquired using a digital mammography system (Siemens Mammomat Inspiration). The system uses a half-cone x-ray tube, which is moved along a $50°$-arc above a stationary detector and acquires

25 equidistantly distributed low-dose projection images. A schematic illustration of the tomosynthesis device is given in Fig. 1. Additionally, relevant technical parameters are summarized in Table 1.

To mimic microcalcifications, the used phantom consists of eggshell fragments of different sizes embedded in a turkey breast which appropriately emulates human breast tissue. Because microcalcifications consist of different calcium compounds [3] and eggshells of calcium carbonate, both are depicted in the same way in tomosynthsis images. A cross-sectional image of the phantom is shown in Fig. 2.

Table 1: Tomosynthesis geometry and imaging parameters

System parameter		Value
Acquisation	Angular range	50° -arc
	Projections	25
Distances	Source to iso-center	608 mm
	Iso-center to table	24 mm
	Table to detector	17 mm
Detector	Detector length x-axis	240 mm
	Detector length y-axis	305 mm
	Dexel size	0.085 mm
Reconstruction	Pixel size	0.085 mm
	Image size / pixels	3584x2816
	Slice thickness	1 mm

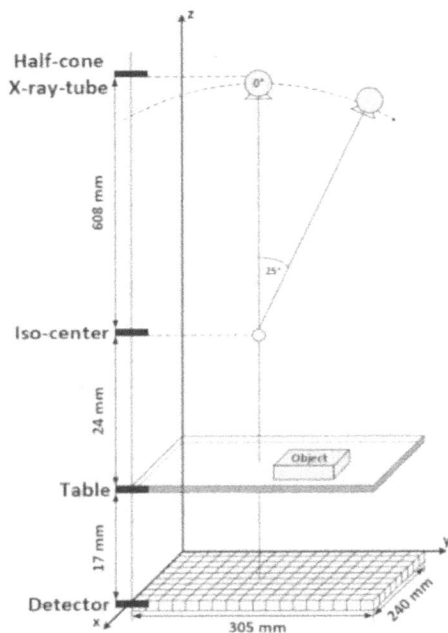

Figure 1: Schematic drawing of the Siemens Mammomat Inspiration geometry

2.2 Motion simulation

In order to mimic motion effects, the phantom is placed on a sample holder, which is depicted schematically in Fig.

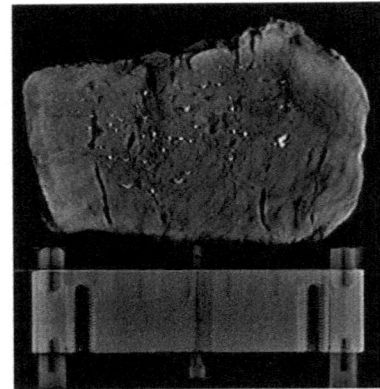

Figure 2: A 121 mm × 113 mm section of a cross-sectional image of the microcalcification phantom.

3. Using an adjustment screw the position of the sample holder can be changed horizontally with a step size of 0.1 mm. Using this setup, a database consisting of projection images acquired for different phantom positions is generated. Besides the ground truth projection data $P_0 = \{p_1^0, p_2^0, \ldots, p_{25}^0\}$, where $p_i^0 \in \mathbb{R}^{3584 \times 2816}$ is the i-th projection image, this database includes the projection data $P_d = \{p_1^d, p_2^d, \ldots, p_{25}^d\}$, where $d = 0.1, 0.2, 0.3, \ldots, 2$ is the translation in mm with respect to the ground truth position.

Data sets with motion can then be constructed from the projection data P_d of different object positions. If the object is translated over a distance d after the acquisition of the k-th projection, the corresponding data set is given by

$$\hat{P}_d^k = \left\{p_i^0\right\}_{i=1,\ldots,k} \cup \left\{p_i^d\right\}_{i=k+1,\ldots,25} . \qquad (1)$$

This approach follows the method mentioned in [4].

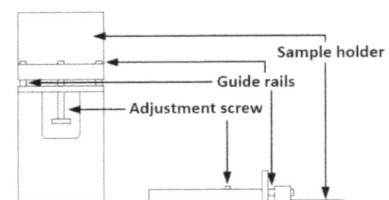

Figure 3: Schematic drawing of the motion device

2.3 Image Reconstruction

Once the projection data sets $\hat{P}_d^k$ are generated, cross-sectional images containing motion artifacts can be calculated. In this work, the simultaneous algebraic reconstruction technique (SART) is used for image reconstruction.

The tomographic reconstruction problem can be formulated as a system of linear equations

$$Af = p \qquad (2)$$

with a discrete representation of an object $f = (f_1, \ldots, f_N)^T \in \mathbb{R}^{N \times 1}$, a system matrix $A \in \mathbb{R}^{M \times N}$

which represents the acquisition process and geometry and a vector $\boldsymbol{p} = (p_1, \ldots, p_M)^T \in \mathbb{R}^{M \times 1}$ that contains the measured projection values. N is the total number of voxels and M is the number of detector elements multiplied with the number of views. The SART algorithm iteratively reduces the error between the measured projection values p_j and the forward projection of the image estimate $\tilde{p}(\boldsymbol{f})$. The iterative scheme converges to a weighted least square solution and is given by

$$f_i^{(n+1)} = f_i^{(n)} + \frac{\lambda}{A_{i,+}} \sum_{j \in J_\Theta} \frac{A_{i,j}}{A_{+,j}} (p_j - \tilde{p}_j(\boldsymbol{f}^{(n)})),$$

$$A_{i,+} = \sum_{j \in J_\Theta} A_{i,j},$$

$$A_{+,j} = \sum_{i=1}^{N} A_{i,j},$$

$$\tilde{p}(\boldsymbol{f}) = \boldsymbol{A}\boldsymbol{f}. \tag{3}$$

In Equation (3), the parameter λ denotes a relaxation parameter. Often the relaxation parameter range is recommended to $0 < \lambda < 2$ [5]. $A_{i,+}$ is a normalization for the numbers of rays intersecting each voxel. $A_{+,j}$ normalizes the path through all involved voxels in the current ray. The updating term is applied considering all rays in one angular view $j \in J_\Theta$. The program uses a distance driven approach to determine the weights $A_{i,j}$. A detailed explanation of the approach is given in [5].

For all reconstructions shown in this contribution the iteration process has been stopped after three iterations. This number has been chosen as a compromise between an acceptable image quality and a reasonable computation time. The relaxation parameter λ is set to 1 and the initial image values are 0. The reconstruction is done with a sequential access order.

3   Results and Discussion

To examine the influence of motion on the depiction of microcalcifications projection data sets $\hat{P}_d^k$ with different values for d and k are computed. Then, an analysis of motion effects is performed by visual inspection of the corresponding reconstructions. Additionally, gray value profiles along a line through an eggshell fragment are analyzed.

3.1   Results

Fig. 4 shows reconstructed images of an eggshell fragment for different translation distances d. The reconstruction shown in Fig. 4(b) is based on the projection data set $\hat{P}_{0.1}^{12}$, which means that the phantom is translated over a distance of 0.1 mm in x-direction after the acquisition of the 12th projection image. Analogously, the images shown in Fig. 4(c) and 4(d) are based on the data sets $\hat{P}_{0.6}^{12}$ and $\hat{P}_{1.0}^{12}$. A visual comparison with the ground truth image which is depicted in Fig. 4(a) shows that the shape of the eggshell fragment changes with increasing d. For small values of

d, the size of the fragment is increased in the direction of motion and at a certain $d \approx 1.0$ mm the fragment is reconstructed as two separate points with a reduced intensity.

As shown in Fig. 5, these findings can be confirmed by an analysis of the gray value profiles along a vertical line through the eggshell fragment. Moreover, this graph reveals an increased maximum gray value for a translation of 0.1 mm. This unexpected behavior could not be clarified yet and will be in the focus of future work.

In order to analyze the influence of the time point, when the movement occurs, data sets $\hat{P}_d^k$ with a constant translation $d = 1$ mm but different values $k = 5, 10, 15, 20$ were generated. According to equation 1, the parameter k is directly related to the time point of movement. The corresponding gray value profiles through an eggshell fragment are illustrated in Fig. 6. It can be seen that all graphs have two local maxima and the position of the global maximum is shifted from left to right with a decreasing value k. Furthermore, it is shown that the difference between the gray values at the local maxima is decreased if the motion occurs in the middle of the acquisition process, while it is increased if the object is translated at the beginning or the end of the measurement process.

In addition, not only motions at a single moment, but also multiple motions during the acquisition process are investigated. For this purpose, the phantom was translated over a constant distance d after the acquisition of the 5th, 10th, 15th and 20th projection image. For $d = 0.2$ and $d = 0.4$ the total translation is given by 0.8 mm and 1.6 mm. The corresponding gray value profiles are shown in Fig.7. It can be seen that the maximum gray value decreases and the width of the gray value profile is increased when the total translation is increased.

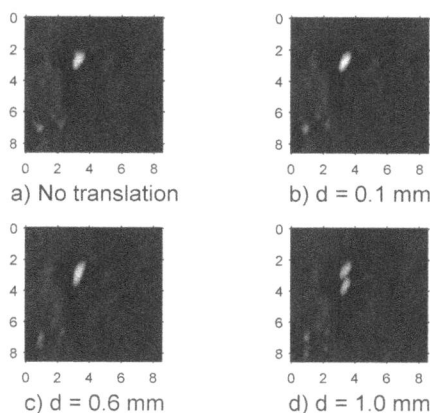

a) No translation

b) d = 0.1 mm

c) d = 0.6 mm

d) d = 1.0 mm

Figure 4: 8.5 mm × 8.5 mm section of a reconstructions of an eggshell fragment with an increasing degree of motion during the image acquisition. The ground truth image without any motion is shown in a). The images b)-d) are based on the projection data sets $\hat{P}_d^{12}$ with $d = 0.1, 0.6, 1.0$ mm, meaning that the object is translated vertically over a distance d after the acquisition of the 12th projection image.

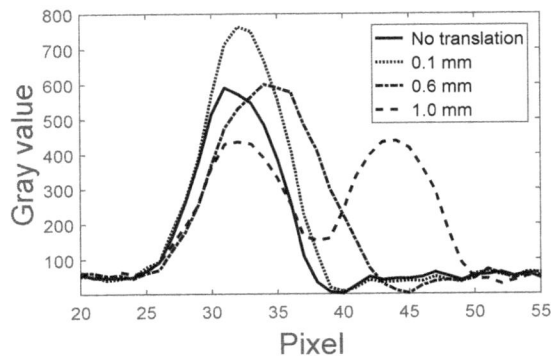

Figure 5: Gray value profile along a vertical line through the eggshell fragment shown in Fig. 4. A 35 pixel long part from the 20-th to the 55-th pixel of the line is illustrated.

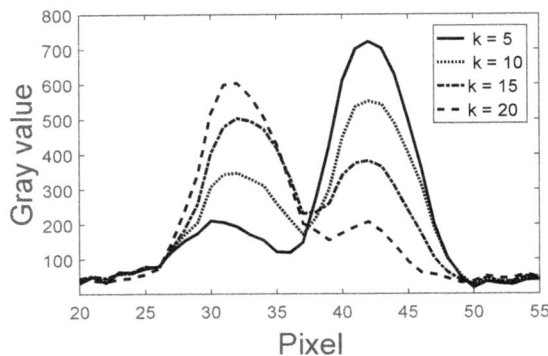

Figure 6: Gray value profile along a line through the eggshell fragment in dependence of the parameter k, which is directly related to the time point when the motion occurs. A 35 pixel long part from the 20-th to the 55-th pixel of the line is illustrated.

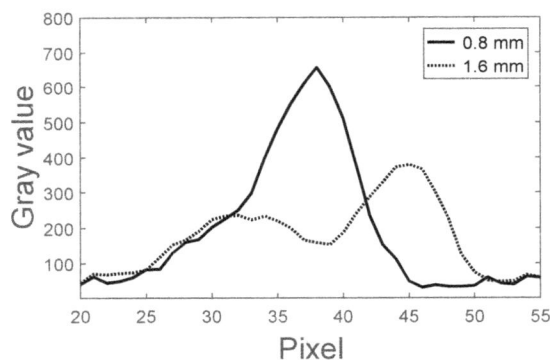

Figure 7: Gray value profile along a line through the eggshell fragment for multiple motions during the acquisition process. The total translations with respect to the ground truth position are 0.8 mm and 1.6 mm.

3.2 Discussion

It could be shown that the shape and the intensity of the eggshell fragment in the reconstructed images is directly in-

fluenced by the extent of object movement during the measurement. With an increasing translation distance d the gray values of an eggshell fragment are more widely spreaded until a distribution with two local maximums is reached. This manifest itself in a duplicate depiction of one fragment in the image. At the same time, the intensity of the eggshell fragment in the reconstructed image is reduced. Furthermore, it could be shown that the extent of image quality degradation by motion artifacts is reduced if the object movement occurs closer to the beginning or the end of the acquisition process.

Based on these results, it can be expected that even slight movements of less than 1 mm can influence the depiction of microcalcifications in a negative way. In the presence of more complex tissue structures of the human breast, that are not taken into account in this study, the reduced visibility may cause misdiagnosis. In order to further analyze the influence of the surrounding tissue, more complex phantoms with microcalcifications of known size should be designed. Additionally, more complex movements should be taken into account for future studies. This requires the construction of a sample holder that does not only allow a translation of the object in a single direction but also allows the simulation of a rotation or non-rigid motion.

Acknowledgement

The work has been carried out at the Klinik für Radiologie und Nuklearmedizin, Universitätsklinikum Schleswig-Holstein Campus Lübeck supervised by the Institute of Medical Engineering, Universität zu Lübeck.

4 References

[1] P. Kaatsch, C. Spix, A. Katalinic et al. *Cancer in Germany*, Robert Koch Institute, Berlin, vol. 10, pp. 74-77, 2016.

[2] E. M. Fallenberg, L. Dimitrijevic, F. Diekmann et al. *Impact of magnification views on the characterization of microcalcifications in digital mammography.* In: Fortschr Röntgenstr 2014, Georg Thieme Verlag, Stuttgart, pp. 274-280, 2014.

[3] L. Corbic, *The significance of the geometric magnification view for the differential diagnosis of microcalcifications*, Medizinischen Fakultät Charite – Universitätsmedizin Berlin, Berlin, 2015.

[4] S. Ens, *Bewegungsdetektion und -korrektur in der Transmissions-Computertomographie*, Springer Vieweg, Wiesbaden, 2015.

[5] Y. Levakhina, *Three-Dimensional Digital Tomosynthesis Iterative reconstruction, artifact reduction and alternative acquisition geometry*, Springer Fachmedien, Wiesbaden, 2014.

Discriminating breast microcalcifications with MARS

K. Müller[1], C. J. Bateman[2], T. Kirkbride[3], A. P. Butler[4], and P. H. Butler[5]

[1] Medizinische Ingenieurwissenschaft, Universität zu Lübeck, k.mueller@student.uni-luebeck.de
[2] University of Otago, New Zealand, christopher.bateman@otago.ac.nz
[3] Ara Institute of Canterbury, New Zealand, tracy.kirkbride@ara.ac.nz
[4] University of Otago, New Zealand, anthony@butler.co.nz
[5] University of Canterbury, New Zealand, phil.butler@canterbury.ac.nz

Abstract

This study deals with the feasibility to spectrally resolve calcium hydroxyapatite from calcium oxalate using MARS spectral CT. For the diagnosis of breast cancer it is important to identify such calcifications. Different microcalcifications are indicators of whether a tumorous growth is benign or malignant. At present, mammography has limited ability to resolve a difference in the composition of microcalcifications. This could result in the need for biopsy to achieve diagnosis. Spectral CT offers the potential to reach the same diagnosis non-invasively. To examine MARS' ability to discriminate these two calcium compounds, several scans were performed with different settings. The measured attenuation coefficients have been analysed to gather spectral information. Preliminary results presented here show that under certain circumstances one can observe spectrally distinct signals (different mass attenuation coefficients) for these two materials. However, further research needs to be performed.

1 Introduction

Microcalcifications (MCs) in breasts can be a sign of early breast cancer. At present only mammography is used for detecting calcifications as it is the most common imaging technique for breast examination [1]. Detecting the tumor in an early stage increases the chances for a cure. Some breast cancers can only be detected by mammography when they show MCs. Analysing calcifications within the breast is first focussed on the morphology [2]. A major problem with detecting MCs with mammography is tissue superposition. This is especially caused by high breast tissue density and the compression during the examination due to taking 2D-images of a 3D-object. Tissue superposition makes it more difficult to spot and analyse MCs in order to make a diagnosis. To gain a confident diagnosis, it is sometimes necessary to perform a biopsy. One way to overcome the limited abilities of mammography could be breast CT where the breast does not undergo compression [1].

For several years it is noted that different types of calcifications may be present in benign and malignant breast tumors. Calcium hydroxyapatite (CaHA) is associated with malignant tumors whereas Calcium oxalate (CaOx) is assigned to benign tumors [3]-[5]. For now, these two types of MCs cannot be distinguished reliably by mammography [5].

MARS is a multi-energy CT system allowing X-ray based spectral molecular imaging. It works with an energy-discriminating photon-counting detector Medipix3RX that has been developed by CERN, originally for modern high-energy physics experiments. Using such detectors has been adapted to other sciences including medical X-ray radiography. MARS enables discrimination of different materials based on their energy dependence of their attenuation coefficients. By sorting the measured X-ray spectrum into energy bins, MARS makes a spectroscopic material discrimination possible. The system offers a high spatial resolution.

2 Material and Methods

To assess the feasibility of distinguishing between CaHA and CaOx the measured attenuation coefficients μ of both CaHA and CaOx for various scans have been analysed. For comparison, the μ of a ROI (region of interest) of 100 voxels for each material have been plotted as a function of energy bins. The ROI was picked in the center of each tube (except for phantom 1 as it is not completely filled). To simplify the spectral analysis, the last energy bin was used as an anchor point. This means that the attenuation of CaOx and CaHA is scaled to be the same in that energy bin. This corresponds to an in- / decrease of the material's density. If the CaHA curves do not overlap the CaOx curves, their differences are spectrally resolved.

The first phantom (see Fig. 1) scanned is a perspex cylinder (acrylic glass) with ducts that are filled with the material of interest. It has been previously used and was now extended by one tube filled with pure CaOx and one with CaHA (patient's sample). The other rods in the phantom contain different concentrations of both CaHA and $MgSO_4$ (however $MgSO_4$ is not relevant for this work).

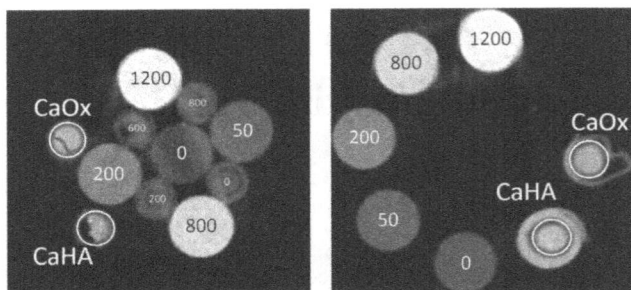

Figure 1: Left: first phantom with five CaHA rods of different concentrations (large diameter, 0-1200 mg/ml) and four with MgSO$_4$ (small diameter, 0-800 mg/ml) filled tubes, a pure CaOx tube and a CaHA tube (patient's sample). Right: second phantom with five different concentrations of CaHA rods (left half, 0-1200 mg/ml), a more homogeneously filled CaOx tube and the CaHA tube.

The first scan on MARS-4 (scanner nr. 4) was conducted with the settings listed in table 1 as well as the parameters of the scan in MARS-10 (scanner nr. 10) and a final scan on MARS-4 after protocol commissioning. To perform a material decomposition (MD), five different concentrations of CaHA rods have been added to the phantom as shown in Fig. 1. The MD was realised by a custom-made MATLAB function. The MD method is based on a PCA (Principal Component Analysis). The function requires the measured mass attenuation coefficients of two different concentrations of the same material.

Table 1: Scan settings for both phantoms on both scanners.

Settings	on MARS-4 (first phantom)
X-ray detector	CdTe-Medipix3RX
Tube voltage	80 kVp
Exposure time	150 ms
Tube current	30 μA
Energy bins (keV)	7-20, 20-30, 30-40, 40-50, 50-80
	on MARS-10 (second phantom)
X-ray detector	CZT-Medipix3RX
Tube voltage	78 kVp
Exposure time	300 ms
Tube current	60 μA
Energy bins (keV)	18-30, 30-45, 45-57, 57-80
	on MARS-4 (second phantom)
X-ray detector	CdTe-Medipix3RX
Tube voltage	80 kVp
Exposure time	200 ms
Tube current	20 μA
Energy bins (keV)	7-18, 18-30, 30-45, 45-57, 57-80

3   Results and Discussion

Previous scans showed that MARS might be able to measure a spectral difference in CaOx and CaHA. The methods as mentioned above, were part of a developing process for

further research into the capability of the MARS CT system of distinguishing two spectrally similar materials.

3.1   Scan with the CdTe detector in MARS-4

The first results are shown in Fig. 2. It shows the 100 voxels plots for two different slices that generally represent all slices regarding their result. The attenuation coefficients for both CaOx and CaHA are mostly the same. Because there is a strong overlap of the graphs for both compounds which makes them look the same material. However, there is a dependency on the regions that have been chosen for comparison. The more grey values visible in the image, the higher the fluctuations in the graphs. The bright voxels correspond to the upper curves in the graphs and the darker voxels to the lower curves. The CaOx values are comparatively low spread and the CaHA values are widely spread with big differences in attenuation.

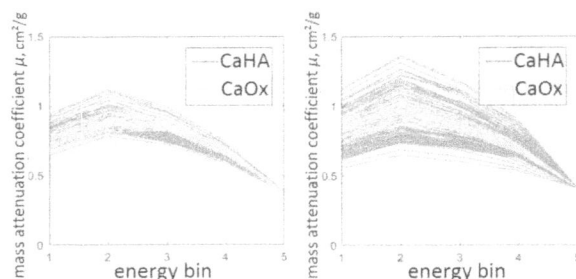

Figure 2: Results of the first scan of the first phantom. These slices generally represent all slices regarding their result. The plots show a high similarity and a strong overlap between CaOx and CaHA. The high fluctuations of CaHA are due to an inhomogenously filled tube.

Due to the material's properties of the CdTe detector layer, a fluorescence peak of fluorescence photons affects the signal. In the first energy bin of 7-20 keV the energy signal is primally fluorescence. Between 20 keV and 30 keV a lot of fluorescence photons contribute to the signal and only a little bit of the actual low energy signal. The counted photons between 30 keV and 80 keV are true photons (and a little pile-up). As this project is interested in energy information from the energy range of 20-30 keV, fluorescence photons are problematic because they obliterate the actual low energy signal and the important information are not obtainable. Using a GaAs detector can solve this problem because GaAs does not produce fluorescence photons. The ratio of fluorescence induced signal to true photons of the CdTe detector is quite high in the second energy bin. This is equivalent to a low attenuation. This component can be removed using a GaAs and allows a better seperation.

3.2   Scans with the CZT detector in MARS-10

MARS-10 holds a camera with a CZT detector. A CZT layer also induces fluorescence photons like the CdTe detector. This means that the problem of a superimposing fluorescence signal still remains. Both detector materials have their fluorescence peak at 23 keV but with a different amplitude. This amplitude depends on the number of Cd atoms. Even with the issue of fluorescence photons the spectral differences between CaOx and CaHA are visible for many slices with the new protocol. Especially the difference in the last two energy bins, without fluorescence photons, is completely resolved which was not possible earlier. The 100 voxels plots for the phantom are shown in Fig. 3. The lines for CaOx and CaHA do overlap for some slices in energy bin 1 and 2. This could be the result of the fluorescence photons. It could also be due to an inhomogenously filled ROI used for comparison. Particularly dark and brighter voxels within the phantom caused the inhomogeneity. Still, several slices show a major or minor overlap. These slices (Fig. 3) generally represent the result of all slices. The best result is shown left whereas one of the worst results is shown on the right side of the image. All other slices are somewhere in between. Compared to the results in Fig. 2 these results look much more promising.

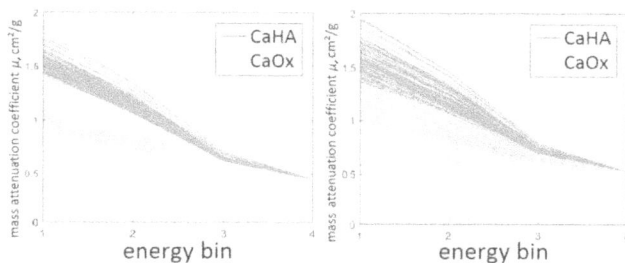

Figure 3: Scan results of the second phantom with MARS-10. These graphs show a major improvement in seperating between both compounds as there is mainly no overlap. However, there is also an overlap for some slices (right).

For following scans the phantom has been extended by a new CaOx and CaHA tube. Fig. 4 shows the improvement of the CaOx content where a much more compressed compound has been achieved. The CaOx signal is very stable throughout the phantom. In several slices (like the right one), the CaOx signal seems to be a completely different one compared to the old CaOx. This one has a lot of low attenuating voxels in the ROI. This is an important outcome for further analysis as this shows that the old CaOx contains a lot of "false" information. If this new CaOx tube would have been used for the first MARS-10 scan it could have affected the results of the 100 voxels plots. Most likely in positive terms.

The subsequent improvement of the CaHA tube on the other hand was not that successful. No satisfying homogeneity could be reached for CaHA. A reason for this could be that the CaHA, although declared to be pure, is a patient's sample. It is not clearly known if there are other materials within the specimen that could have induced this effect. If available, another pure CaHA sample needs to be scanned that does not have lumps or maybe contamination with other materials that may be present.

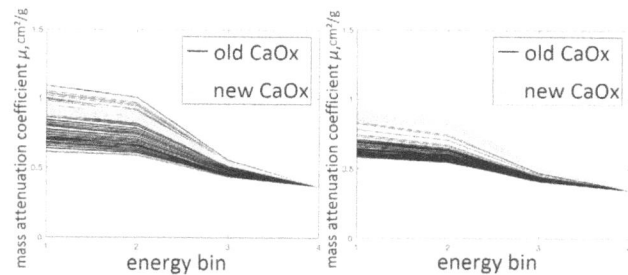

Figure 4: Spectral analysis comparison of the old and new CaOx tube. The CaOx signal is now constant throughout the phantom and shows only little variations for a ROI.

3.3 Scan with the CdTe detector in MARS-4 and a material decomposition

The MD results for the CaHA and CaOx channel are depicted in Fig. 5. These slices represent the important outcomes of the MD. The assignments of the compounds to either CaHA or CaOx are very varying throughout the phantom and show good and worse results. Only within 2-3 slices the MD show a reasonable result where both compounds are assigned correctly. CaOx and CaHA in all other slices of the phantom could not be discriminated accordingly. It is apparent that there is a little by little transition of the MD from good to worse and vice versa. These patches may be a result of inhomogeneously filled tubes. Another conspicuousness is that at first the outer parts of the compounds are assigned differently before the middle is picked up (either wrong or correctly).

To investigate possible reasons for this, the focus was laid on comparing the MD images to the 100 voxels plot for the materials. The results in Fig. 6 show a change in the shape of the spectral analysis plots of the CaHA rods. Up to a certain slice, the curves for HA_{1200}, HA_{800} and also the patient's CaHA sample tend to change in their curvature. The dip of the curvature is around the second energy bin (18-30 keV). Scanning an object in CT is associated with Beam hardening. This means that due to X-ray absorption the low-energy photons are first absorbed and an X-ray beam with higher energy exits the object, being a "harder" beam. As a result, there is more fluorescence relative to the overall photon number, especially in the fluorescence bin of 18-30 keV. This looks like a lot of low attenuation, so the decreased μ induce the dip in the graphs.

Sliding through the phantom, there is a good MD result in the middle of the phantom and a worse result at the beginning and end. The successfulness seems to have a maximum peak somewhere in the phantom. When comparing the MD results with the 100 voxels plots of all CaHAs, no connection has been found yet to explain this. One would assume that the 100 voxels plots behave accordingly like the MD result but this is not the case. The MD images do not show the best discrimination when any of the CaHAs is e.g. a horizontal lane between energy bin 1 and 3. This would lead to the assumption that there is no relation between the dipping and MD result, at least not based on this single scan. There is quite an overlap for both the CaOx and the patient's CaHA in the 100 voxels plots. The result is worse than with MARS-10, but better than with MARS-4 and its new protocol. The graphs also change their curvature at the same behaviour. But like with the results in Fig. 6, no connection to the MD behaviour has been found so far.

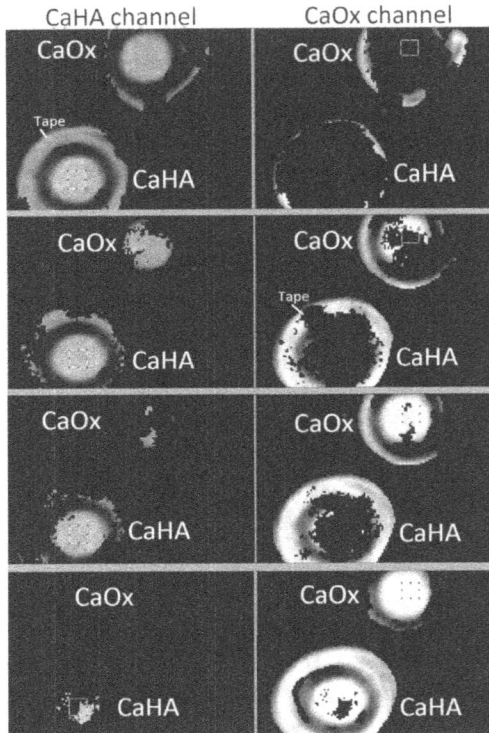

Figure 5: MD results for four different slices in the CaOx and CaHA channel. The best result is in slice 3 (third row). In the CaHA channel mostly only CaHA is picked up and barely no CaOx. And also in the CaOx channel the majority of the CaOx tube is picked up and almost no CaHA.

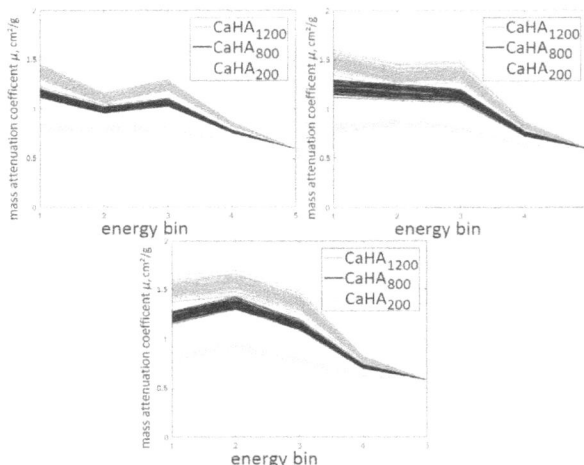

Figure 6: Comparison of attenuation for different CaHA rods for three slices. Somewhere in the phantom, the curves for HA_{1200} and HA_{800} change from a v-shaped to an inverted v-shaped appearance, so the dip disappears.

4 Conclusion

This work is aimed at determining the feasibility of distinguishing between different breast microcalcifications such as Calcium hydroxyapatite and Calcium oxalate using MARS spectral CT. Therefore, a spectral analysis has been done by comparing the measured mass attenuation coefficients of both compounds and using the MARS material decomposition (MD) technique. Preliminary results presented here show that under certain circumstances one can observe spectrally distinct signals (different mass attenuation coefficients) for these two materials. Inhomogenously filled calcium sample tubes led to high signal fluctuations. Also, fluorescence photons due to the detector layer's properties occured. They complicate the analysis by overlapping with the object's actual signal.

Using the MARS MD technique still presents challenges because the compounds have an intrinsic similarity and variation with beam hardening within the phantoms. Creating better compounds with little variation would help to improve the signal. Furthermore, using a GaAs detector layer to avoid fluorescence photons should give better results. Further research still needs to be done to investigate if this discrimination is achievable while scanning breast sized objects.

Acknowledgement

The work has been carried out at the University of Canterbury, Christchurch, New Zealand and supervised by the Institute of Medical Engineering, Universität zu Lübeck.

5 References

[1] C. J. Lai et al., *Visibility of microcalcification in cone beam breast CT: Effects of X-ray tube voltage and radiation dose.* Medical Physics, vol. 34, no. 7, pp. 2995-3004, 2007.

[2] R. Kreienberg, W. Jonat, T. Volm, V. Möbus and D. Alt, *Management des Mammakarzinoms.* Springer, Berlin/Heidelberg, 2006.

[3] L. Frappart et al., *Structure and composition of microcalcifications in benign and malignant lesions of the breast: Study by light microscopy, transmission and scanning electron microscopy, microprobe analysis, and X-ray diffraction.* Human Pathology, vol. 15, no. 9, pp. 880-889, 1984.

[4] C. M. Büsing, U. Keppler and V. Menges, *Differences in microcalcification in breast tumors.* Virchows Archiv A, vol. 393, no. 3, pp. 307-313, 1981.

[5] N. Stone and P. Matousek, *Advanced transmission Raman spectroscopy: A promising tool for breast disease diagnosis.* Cancer Research, vol. 68, no. 11, pp. 4424-4430, 2008.

Infinite Science
Publishing

www.ingramcontent.com/pod-product-compliance
Lightning Source LLC
Chambersburg PA
CBHW080748250326

R18028000001B/R180280PG41598CBX00027B/3